Pipeline Infrastructure Renewal and Asset Management

About the Author

Mohammad Najafi, Ph.D., P.E., F. ASCE is Professor and Director of the Center for Underground Infrastructure Research and Education (CUIRE) and Director of Construction Engineering and Management at the Department of Civil Engineering at The University of Texas at Arlington. He is both founder and chief editor of ASCE's *Journal of Pipeline Systems Engineering and Practice.* He is also author of these McGraw-Hill titles: *Trenchless Technology: Pipeline and Utility Design, Construction, and Renewal; Trenchless Technology Piping: Installation and Inspection;* and *Trenchless Technology: Planning, Equipment, and Methods.*

About NASSCO

NASSCO is the leading organization in the trenchless technology industry. Established in 1976, NASSCO is dedicated to setting industry standards for the assessment and renewal of underground infrastructure through training, education, and networking opportunities. Go to www.nassco.org or call (410) 442-PIPE.

Pipeline Infrastructure Renewal and Asset Management

Mohammad Najafi

McGraw Hill Education

New York Chicago San Francisco
Athens London Madrid
Mexico City Milan New Delhi
Singapore Sydney Toronto

Cataloging-in-Publication Data is on file with the Library of Congress.

McGraw-Hill Education books are available at special quantity discounts to use as premiums and sales promotions, or for use in corporate training programs. To contact a representative please visit the Contact Us page at www.mhprofessional.com.

Pipeline Infrastructure Renewal and Asset Management

1 2 3 4 5 6 7 8 9 LPI 21 20 19 18 17 16

ISBN 978-0-07-182334-0
MHID 0-07-182334-4

The pages within this book were printed on acid-free paper.

Sponsoring Editor
Lauren Poplawski

Editorial Supervisor
Donna M. Martone

Acquisitions Coordinator
Lauren Rogers

Project Manager
Anupriya Tyagi,
Cenveo® Publisher Services

Copy Editor
Cenveo Publisher Services

Proofreader
Cenveo Publisher Services

Indexer
Robert Swanson

Production Supervisor
Lynn M. Messina

Composition
Cenveo Publisher Services

Art Director, Cover
Jeff Weeks

I would like to dedicate this book to my parents, my brothers and sisters, my daughters and their families, my professors and teachers, my friends and colleagues, and my current and former students who have since become my colleagues and leaders in pipeline technology.

Contents

Foreword xix
Preface xxi
Acknowledgments xxiii

1 Introduction and Overview 1
1.1 Introduction 1
1.1.1 Pipeline Defects 2
1.1.2 Modes of Pipeline Failures 4
1.2 Extending Service Life 7
1.3 Trenchless Renewal Methods 10
1.3.1 Cured-in-Place Pipe 11
1.3.2 Sliplining 11
1.3.3 In-line Replacement 13
1.3.4 Modified Sliplining 15
1.3.5 Spray-in-Place Pipe 16
1.3.6 Close-Fit Pipe 20
1.4 Sewer Manhole Renewal 20
1.5 Selection of a Renewal Method 23
1.6 Current Design Concepts for Pipeline Renewals 25
1.6.1 Partially Deteriorated Pipe 25
1.6.2 Fully Deteriorated Pipe 27
1.7 Emerging Design Concepts for Pipeline Renewal Systems 29
1.8 Long-Term Test Methods 29
1.9 Common Terms and Practices in Pipeline Renewal 30
1.9.1 Contract Documents 30
1.9.2 Technical Specifications 30
1.9.3 Design 31
1.9.4 Inspection 31
1.9.5 Quality Assurance/Quality Control 32
1.10 Chapter Summary 33
1.11 Review Questions 34

2 Pipeline Asset Management 35
2.1 Introduction 35
2.2 Benefits of Asset Management 36

2.3 Major Questions in an Asset Management Program ... 37
2.4 Asset Management Framework ... 37
2.4.1 Step One: Identify Objectives ... 37
2.4.2 Step Two: Outline Asset Management Plan Structure and Content ... 39
2.4.3 Step Three: Write the Pipeline Asset Management Plan ... 39
2.4.4 Step Four: Review of the Plan ... 45
2.4.5 Step Five: Update the Plan ... 45
2.5 Government Accounting Standards Board Statement 34 ... 45
2.5.1 Depreciation ... 46
2.5.2 GASB-34-Modified Approach Asset Management Requirements ... 47
2.6 Condition Assessment Tools and Methods ... 51
2.7 Risk Management in Piping Systems ... 52
2.7.1 Likelihood of Failure ... 54
2.7.2 Consequence of Failure ... 55
2.7.3 Managing Asset Risk ... 62
2.8 Asset Management of Oil and Gas-Related Pipelines ... 63
2.8.1 Introduction ... 63
2.9 Chapter Summary ... 90
2.10 Review Questions ... 91

3 Design Considerations for Trenchless Renewal Methods ... 93
3.1 Introduction ... 93
3.2 TRM Selection Process ... 94
3.2.1 Initial Planning ... 94
3.2.2 Pipeline Integrity Assessment ... 95
3.2.3 Develop Renewal Solutions ... 95
3.2.4 Implement and Monitor ... 95
3.3 Renewal Method Depends on the Existing Pipe Conditions ... 95
3.4 Renewal Method Depends on Site and Project Conditions ... 97
3.5 Selection of a Renewal Method (a 10-Step Process) ... 98
3.5.1 Step 1 ... 98
3.5.2 Step 2 ... 100
3.5.3 Step 3 ... 102
3.5.4 Step 4 ... 102
3.5.5 Step 5 ... 105

3.5.6 Step 6 105
3.5.7 Step 7 105
3.5.8 Step 8 105
3.5.9 Step 9 105
3.5.10 Step 10 105
3.6 General Design Considerations 106
3.6.1 Flow Bypassing Considerations 106
3.6.2 Reinstatement of Service Laterals 106
3.6.3 Open-cut 107
3.7 Structural Design of CIPP 107
3.7.1 Major Terms and Design Variables 109
3.7.2 Design Background 110
3.7.3 Partially Deteriorated Gravity Condition 112
3.7.4 Fully Deteriorated Gravity Condition 113
3.7.5 Partially Deteriorated Gravity Flow CIPP Design 114
3.7.6 Fully Deteriorated Gravity Flow CIPP Design 117
3.7.7 Total External Pressure on CIPP 118
3.7.8 Hydrostatic Water and Soil Loads 119
3.7.9 Superimposed or Live Loads 120
3.7.10 Partially Deteriorated Internal Pressure Pipe 125
3.7.11 Fully Deteriorated Internal Pressure Pipe 128
3.7.12 Hydraulic Design of CIPP 131
3.8 Chapter Summary 134
3.9 Review Questions 134

4 Pipeline Prioritization, Inventory and Condition Assessment 137
4.1 Introduction and Background 137
4.2 Pipeline Condition Assessment 139
4.3 Model Input Parameters 140
4.4 Data Collection, Analysis, and Management 140
4.5 Condition Assessment 140
4.5.1 Asset Registry 142
4.5.2 Geographical Considerations 142
4.5.3 Asset Naming and Numbering Conventions 143
4.5.4 Inventory Tools 145

4.5.5 Case Study: Mapping and Inventory of Water and Wastewater Infrastructure at Edgecombe County, NC 146
4.6 Condition Assessment Technologies for Gravity and Pressure Systems 149
4.6.1 Overview 149
4.6.2 Camera Technologies 149
4.6.3 Acoustic Technologies 156
4.6.4 Electrical and Electromagnetic Technologies 159
4.6.5 Other Technologies 162
4.6.6 Case Study: City of Marco Island's Existing Sewers Assessment, and Rehabilitation 164
4.7 Condition Assessment Technologies Specific for Water Distribution System 170
4.7.1 Overview 170
4.7.2 Internal and External Inspection ... 170
4.7.3 Nondestructive Testing and Inspection 175
4.7.4 Electromagnetic Methods 177
4.7.5 Environmental Testing 182
4.7.6 Leak Detection Technologies 182
4.8 Data Collection and Manipulation 187
4.8.1 Overview 187
4.8.2 System Analysis 188
4.8.3 Data Content for Water and Sewer Systems 189
4.8.4 Point-Score Protocols for Water and Sewer Systems 190
4.8.5 Data Sharing and Distribution 191
4.9 System Integration and Operation 194
4.10 Chapter Summary 196
4.11 Review Questions 197

5 Pipe and Pipe Installation Considerations 199
5.1 Introduction 199
5.2 Pipeline Construction Using Open-Cut Method 200
5.3 The Pipe-Soil System 201
5.3.1 Rigid Pipes and Flexible Pipes 201
5.3.2 Soils 204
5.3.3 Pipe-Soil Interaction 205
5.3.4 Behavior of Rigid Pipes 210
5.3.5 Behavior of Flexible Pipes 212

5.4 Common Modes of Pipeline Failures 213
5.4.1 Failure Modes in Rigid Pipes 214
5.4.2 Failure Modes in Flexible Pipe 218
5.5 Pipe Selection Considerations 222
5.6 Cement-Based Pipes 222
5.6.1 Concrete Pipes 223
5.6.2 AC Pipe 231
5.7 Vitrified Clay Pipe 233
5.7.1 Manufacturing 234
5.7.2 Applicable Standards 234
5.7.3 Joint Types 234
5.7.4 Advantages and Limitations 238
5.8 Plastics Pipes 239
5.8.1 Properties of Viscoelastic Construction Materials 240
5.8.2 PVC Pipe 241
5.8.3 PE Pipe 251
5.8.4 Glass-Reinforced Pipe 258
5.9 Metallic Pipes 261
5.9.1 Ductile Iron Pipe 261
5.9.2 Steel Pipe 266
5.10 Corrosion Protection: General 270
5.10.1 Corrosion Protection: Ductile Iron Pipe 270
5.10.2 Corrosion Protection: Steel Pipe 272
5.10.3 Corrosion of Concrete, Composite, and Plastic Pipes 272
5.11 Chapter Summary 273
5.12 Review Questions 273

6 Cured-in-Place Pipe 275
6.1 Introduction 275
6.2 Site Compatibility and Applications 275
6.3 Impacts of Existing Pipeline Defects on CIPP 276
6.4 Main Characteristics 276
6.5 Method Description 285
6.5.1 Initial Setup: Cleaning, Televising, and Resin Impregnation (Wet-out Process) 285
6.5.2 Installation 287
6.5.3 Curing 289
6.6 Inspecting Installation of CIPP 290
6.7 Pipe Plugging or Bypass Pumping 290

6.8 Quality Assurance and Testing ... 293
 6.8.1 CIPP Inspection and Acceptance ... 293
 6.8.2 Workmanship ... 293
 6.8.3 Quality Control Issues ... 294
 6.8.4 Final Steps ... 296
6.9 Major Advantages ... 298
6.10 Major Limitations ... 298
6.11 Scheduling and Costs ... 299
 6.11.1 Scheduling ... 299
 6.11.2 Costs ... 300
6.12 Chapter Summary ... 304
6.13 Review Questions ... 304

7 Sliplining ... 305
7.1 Introduction ... 305
7.2 Site Compatibility and Applications ... 305
7.3 Main Characteristics ... 307
7.4 Sliplining Methods ... 307
 7.4.1 Continuous Sliplining ... 307
 7.4.2 Segmental Sliplining ... 310
7.5 Installation ... 312
 7.5.1 Installation: Continuous Sliplining ... 312
 7.5.2 Installation: Segmental Sliplining ... 314
 7.5.3 Connecting Service Laterals ... 314
 7.5.4 Grouting ... 314
7.6 Design Considerations ... 315
7.7 Sliplining Gaslines ... 316
7.8 Advantages ... 317
7.9 Limitations ... 318
7.10 Case Study ... 318
 7.10.1 Sliplining a 42-in. PCCP Pipeline ... 321
7.11 Chapter Summary ... 327
7.12 Review Questions ... 327

8 In-line Replacement and Pipe Bursting ... 329
8.1 Introduction ... 329
 8.1.1 Method Applicability ... 329
 8.1.2 Range of Applications ... 331
 8.1.3 Unfavorable Conditions ... 331
8.2 Construction Considerations ... 332
 8.2.1 Service Laterals ... 332
 8.2.2 Insertion and Reception Pits ... 333
 8.2.3 Bursting Operation ... 333
 8.2.4 Reconnection of Service ... 333

8.3 Pipe Bursting Methods 334
8.3.1 Pneumatic Pipe Bursting (PB-PN) . . . 335
8.3.2 Static Pipe Bursting (PB-ST) 336
8.3.3 Pipe Splitting (PB-PS) 337
8.3.4 Pipe Insertion (PB-PI) 337
8.4 Pipe Removal Methods 339
8.4.1 Pipe Reaming (PR-PR) 339
8.4.2 Pipe Eating (PR-PE) 340
8.4.3 Pipe Ejection or Extraction (PR-PX) 341
8.5 Pipe Materials 341
8.5.1 Existing (Old) Pipe Material 341
8.5.2 Replacement (New) Pipe Material . . . 342
8.6 Pipe Bursting versus Other Alternatives 343
8.7 Pipe-Bursting Project Classification 344
8.8 Advantages and Limitations 344
8.9 Preliminary Surveying 346
8.10 Preparatory Work 346
8.11 Design Considerations 347
8.11.1 Pre-design Phase 347
8.11.2 Design Phase 348
8.11.3 Utility Survey 349
8.11.4 Investigation of Existing Pipe and Site Conditions 350
8.11.5 Insertion and Pulling Shaft/Pit Requirements 350
8.11.6 Geotechnical Investigation Requirements 350
8.11.7 Maximum Allowable Operating Pressure 352
8.12 Risk Assessment Plan 352
8.12.1 Ground Movement Associated with Pipe Bursting 354
8.13 Plans and Specifications 356
8.13.1 Submittals 357
8.13.2 QC/QA Issues 358
8.13.3 Dispute Resolution Mechanisms . . . 359
8.13.4 Permitting Issues 359
8.13.5 Typical Bidding Form 359
8.14 Construction Considerations 360
8.14.1 Typical Pipe Bursting Operation Layout 360
8.14.2 Shoring the Insertion and Pulling Shafts/Pits 362

8.14.3 Matching System Components to Reduce Risk of Failure 362
8.14.4 Existing Nearby Utilities 363
8.14.5 Bypass Pumping Considerations 363
8.14.6 Dewatering Considerations 364
8.14.7 Ground Movements Monitoring Program 364
8.14.8 Pipe Connection to the Manhole 364
8.14.9 Pipe Connection to Other Pipes ... 366
8.14.10 Pipe-Bursting Water Mains 366
8.14.11 Service Connections 369
8.15 Groves on the Outside Surface of the Pipe ... 371
8.16 As-built Drawings 372
8.17 Contingency Plan 374
8.18 Safety Considerations 374
8.19 Completion Work 374
8.20 Replacing Laterals 375
8.21 Pipe Bursting with Segmental Pipes 375
8.22 Quality Assurance and Documentation 375
8.23 Cost Estimating 376
8.24 Potential Problems 377
8.25 Pipe Design 378
8.26 Case Study: Asbestos Pipe Replacement 378
8.26.1 Two Categories of non-friable ACM 381
8.26.2 Asbestos Concerns 382
8.26.3 Future Handling 383
8.26.4 Case Study Summary 383
8.27 Chapter Summary 384
8.28 Review Questions 384

9 Modified Sliplining 385
9.1 Introduction 385
9.1.1 Background 385
9.2 Primary Characteristics 386
9.2.1 Project Planning 388
9.2.2 Modified Sliplining Applications 388
9.3 Preliminary Inspection 389
9.4 Preparatory Work 389
9.5 Safety Issues 390
9.6 Relining with Steel Pipe 393
9.7 Project Delivery 394
9.8 Advantages 395

9.9 Limitations ... 396
9.10 Chapter Summary ... 396
9.11 Review Question ... 396

10 Spray-in-Place Pipe ... 397
10.1 Introduction ... 397
10.2 Material Type and Purpose ... 397
10.2.1 Cementitious ... 398
10.2.2 Polymers ... 398
10.2.3 Epoxies ... 399
10.2.4 Polyurea and Polyurethane ... 399
10.2.5 Polyesters ... 400
10.3 Installation Process ... 401
10.3.1 Ideal Properties ... 401
10.3.2 Site Investigation and Preparation ... 401
10.3.3 Pipe Inspection ... 402
10.3.4 Water Pipe Cleaning Methods ... 403
10.3.5 Surface Preparation ... 404
10.3.6 Repair Materials ... 408
10.3.7 Equipment and Process ... 410
10.4 Inspection and Testing ... 416
10.4.1 Visual Inspection ... 416
10.4.2 SIPP Thickness ... 416
10.4.3 Spark Testing ... 416
10.4.4 Adhesion ... 417
10.4.5 Vacuum and Exfiltration Testing ... 417
10.5 Chapter Summary ... 417
10.6 Review Questions ... 418

11 Close-Fit Pipe ... 419
11.1 Introduction ... 419
11.1.1 Primary Characteristics ... 419
11.2 CFP Applications ... 420
11.3 Construction Considerations ... 421
11.3.1 General ... 421
11.3.2 Preliminary Evaluation and Inspection ... 422
11.3.3 Preparatory Work ... 423
11.3.4 Installation Phase ... 423
11.3.5 Project Inspection and Delivery ... 424
11.4 Quality Assurance and Documentation ... 425
11.4.1 Advantages ... 425
11.4.2 Limitations ... 426
11.5 Review Questions ... 426

12 Sewer Manhole Renewal **429**
12.1 Introduction 429
12.2 Manhole Problems 433
12.3 Classification of Manhole Problems 435
12.3.1 Type I: Frame and Cover 435
12.3.2 Type II: Structural Defects 442
12.3.3 Type III: Corrosion Problems 442
12.3.4 Type IV Severe Structural Damage 448
12.4 Manhole Renewal Methods 450
12.4.1 The Traditional Method 450
12.4.2 Lining 450
12.4.3 Panels 450
12.4.4 Manhole Inserts 451
12.4.5 Grout Injection 451
12.4.6 Spray lining 451
12.5 Chapter Summary 452
12.6 Review Questions 452

13 Sanitary Sewer Lateral Renewal **453**
13.1 Introduction 453
13.2 Service Lateral Renewal Methods 455
13.2.1 Preparatory Work 456
13.2.2 Cured-in-Place Pipe 457
13.2.3 Pipe Bursting 460
13.2.4 Close-Fit Pipe (CFP) 461
13.2.5 Chemical Grouting 461
13.2.6 Spray-in-Place Pipe 462
13.2.7 Other Renewal Considerations 462
13.3 Chapter Summary 463
13.4 Review Questions 463

14 Localized Repairs **465**
14.1 Introduction 465
14.2 Primary Characteristics 466
14.3 Robotic Repairs 467
14.3.1 Primary Characteristics 467
14.3.2 Method Description 469
14.3.3 Advantages 469
14.3.4 Limitations 470
14.4 Grouting 470
14.4.1 Primary Characteristics 470
14.4.2 Chemical Grouting 470
14.4.3 Resin Injection System 476
14.4.4 Fill-and-Drain Systems 477
14.4.5 Cement Grouting 477

14.5 Internal Seal 479
14.5.1 Primary Characteristics 479
14.5.2 Mechanical Joint Sealing 479
14.5.3 Method Description 482
14.5.4 Advantages 483
14.5.5 Limitations 483
14.6 Point CIPP 483
14.6.1 Primary Characteristics 483
14.6.2 Method Description 484
14.6.3 Advantages 486
14.6.4 Limitations 486
14.7 Chapter Summary 486
14.8 Review Questions 487

A Acronyms 489

B Glossary of Terms 495

C Abbreviations for Organizations Related to Trenchless Technology 519

D Related Documents 523

E Conversion Table 529

F Members of Center for Underground Infrastructure Research and Education (CUIRE) 531

References 535

Index 541

Foreword

Dr. Najafi's invitation to the National Association of Sewer Service Companies (NASSCO) to participate in this latest reference for trenchless technologies offered an opportunity to incorporate our knowledge of the industry into a valuable resource that will foster continued acceptance and growth of trenchless technologies well into the future. This latest book provides the student, engineer, and utility provider with a foundation of knowledge of these technologies, while offering online resources for continued education and advancement.

The book is organized in a user-friendly format to first provide the reader with a comprehensive overview of the importance of assessing and managing our underground assets, followed by the most appropriate actions needed to sustain our infrastructure. Choosing the most suitable rehabilitation technology is important, but is only part of the equation. Comprehensive design and execution of the selected method are critical to the success of each project. This book addresses each of these key issues and decision points to provide the reader with a thorough understanding of what is required for a successful project.

The reference contains significant contributions from Lynn Osborn, P.E., NASSCO's Technical Director, who has over 40 years of practical experience in the technologies presented in this book. There were also numerous contributions from dedicated NASSCO members, including Gerry Muenchmeyer, P.E., Tim Back, P.E., and Jim Harris, P.E., who volunteered considerable time to ensure quality content.

We are honored that Dr. Najafi—whose commitment to setting standards for the asset management and renewal of underground pipeline infrastructure has made significant industry impact—has included NASSCO in the collaboration of this important reference.

Ted DeBoda, P.E.
Executive Director, National Association
of Sewer Service Companies (NASSCO)

Preface

Managing and renewing/replacing thousands of miles of deteriorated underground pipeline assets present many challenges and require a wide range of resources. An Asset management strategy requires knowledge of degradation/defects of the pipeline, analysis of impacts from asset failure, and a set of solutions (i.e., repair or renew/replace methods). The management of underground pipelines and utilities has been traditionally overlooked because of their out-of-sight-out-of-mind nature. Considering failure to be a condition that prevents the designed use of the pipeline system, it is not often the oldest pipe that has failed first. Factors, such as pipe type and material, depth from ground surface, diameter, pipe section length and types of joints, installation method, pipe environment and corrosiveness (soil and ground conditions), type of application and fluid, extent of maintenance, and so on, have a major role in pipeline failures.

The foundation of an effective asset management of pipeline assets is accurate inventory and condition assessment. Buried infrastructure systems, such as sanitary sewers, water, oil, and gas pipes, are particularly challenging because they are both hidden and distributed over large geographic regions. Cost-effective management of these assets requires proper inventory and condition assessment/monitoring to collect required data, analyze the information, and make proper decisions.

Pipeline owners and operators are faced with major expenditures to renew and extend the life of their buried infrastructure. They are faced with the task of allocating limited resources to their systems, with sparse data to back up their decisions. Given a budget, these decisions include the following overall questions:

- What is our worst problem?
- What is the source of the problem and what is its extent?
- How do we correct the problem in a timely and cost-effective manner?

Some problems, such as sanitary overflows (SSOs) or low water pressure or water quality issues can be detected.[1] Others, such as pending sewer collapse or a gas, oil, or water main breaks, are hidden problems that can result in extremely expensive impacts and solutions. It is unlikely that all emergency repairs will be avoided, but by developing an asset management system, the number of emergency repairs can be minimized and budgets can be allocated for proper renewal and replacement work.

Development of an asset management system requires preparation and analysis of necessary data. The immediate process includes determining current status and the goals to be achieved.

This book addresses two major issues: asset management and renewal/replacement of old and deteriorated pipeline systems using trenchless technology methods. After an introduction to asset management and trenchless renewal/replacement methods in Chap. 1, Chap. 2 presents more in-depth information on asset and risk management for water/sewer oil and gas pipelines. Chapter 3 provides an overview of pipeline renewal design with method selection criteria for both gravity and pressure pipelines. Chapter 4 presents methods for inventory, inspection, and condition assessment of pipelines as well as prioritization strategies for renewal and replacement.

The choice of a specific pipe has a major influence on the efficiency and longevity of the pipeline asset and is one of the most important decisions for cost-effective and efficient pipeline asset management and operation and maintenance. Long life-cycle pipes build asset value and provide economical service, minimize operation and maintenance costs and reduce impacts to consumers, environment, and the public. Chapter 5 addresses this important issue and describes different pipe materials with advantages and limitations.

The description of pipeline renewal methods starts with CIPP (Chap. 6), and continues with sliplining (Chap. 7), inline replacement and pipe bursting (Chap. 8), modified sliplining (Chap. 9), SIPP (Chap. 10), close-fit (Chap. 11), manhole renewal (Chap. 12), sanitary sewer lateral renewal (Chap. 13), and localized repairs (Chap. 14). Some of the materials presented in Chap. 8, inline replacement and pipe bursting, is general in nature and can be used for other renewal methods.

A list of acronyms and abbreviations used in the text is presented in App. A, and a glossary of terms in App. B. Appendix C provides a list of organizations related to trenchless technology, App. D contains a list of related documents, and App. E contains a conversion table.

It is hoped that this book can be used as a reference for experienced pipeline professionals as well as educational material for young engineers and university students. Case studies and end-of-the-chapter review questions and problems can help with understanding of concepts presented in each chapter.

---

[1]The recent water quality and corrosivity issues in the City of Flint, Michigan, started with a money-saving strategy 2 years ago by switching Flint's water supply from Lake Huron (which they were paying the city of Detroit for), to the Flint River, a tributary that runs through the city and is known for its quality issues. Because the water from Flint River was not properly treated, lead from aging pipelines to homes began leaching into the Flint water supply and created a major health crisis for residents.

Acknowledgments

This book would not have been possible without the help, support, and patience of Lauren Poplawski, Acquisitions Editor; Donna Martone, Editorial Supervisor; and Lynn Messina, Production Supervisor; McGraw-Hill Education. The author would like to thank McGraw-Hill Education and the editorial staff for their support of this publication. The cooperation and hard work of Anupriya Tyagi, Senior Project Manager, Cenveo Publisher Services and her colleagues are acknowledged.

Many pipeline and trenchless technology professionals and industry leaders contributed and helped in preparation of this book. Specifically, I would like to thank the following assisting authors:

Alan Atalah (Secs. 8.5 through 8.25)
Jonathan Faughtenberry (Sec. 2.8)
Jim Harris (Sec. 2.7)
Lynn Osborn (Sec. 1.9 and questions for Chaps. 9, 11, and 12)
Alhad Panwalkar (Secs. 5.1 through 5.5)
Mohammad Shahidur Rahman (Secs. 4.5 through 4.9)
Shah Rahman (Secs. 5.5 through 5.10)
Amanda Rae Simpson (Sec. 2.4)

Reviewers

Sam Arnaout (Sec. 5.6)
Jeff Boschert (Sec. 5.7)
Steve Cooper (Sec. 5.7)
Wayne Geyer (Sec. 5.9)
Bruce Hollands (Sec. 5.8.2)
Brent Keil (Sec. 5.9)
Larry Kiest (Chap. 13)
Jeff LeBlanc (Sec. 5.8.4)
Mike Leathers (Sec. 5.8.4)
Tom Marti (Sec. 5.8.2)

Don McGriff (Sec. 5.8.3)
Richard Mueller (Sec. 5.6)
Lynn E. Osborn (Chaps. 1, 3, 5, 6, 9, 10 through 12)
Larry Petroff (Chap. 8)
Camille Rubeiz (Chaps. 8 and 11)
Chris Sundberg (Sec. 5.9)
Mike Whitehouse (Sec. 5.8.3)

I would like to thank my former students who have since become my colleagues and leaders in pipeline technologies and contributed to preparation of this book. I also would like to thank my current students, Mohammadreza Malek Mohammadi, Summit Manda, Saleh Mehraban, and Ramtin Serajiantehrani, who assisted with references, definitions, and acronyms. Niloofar Rezaei, CUIRE Program Manager, was helpful in preparation of drawings and illustrations.

I would like to thank my colleagues at the University of Texas at Arlington and Department of Civil Engineering for their support of this book. Guidance and contributions of the industry board members of the Center for Underground Infrastructure Research and Education (CUIRE) (www.cuire.org), whose names appear in App. F, were instrumental in preparation of this book.

Comments, suggestions, and questions are welcome and encouraged, and they should be addressed to the author at najafi@uta.edu or 817-272-0507.

Dr. Mohammad Najafi, P.E., F. ASCE
Professor and Director, Center for Underground
Infrastructure Research and Education (CUIRE)
Department of Civil Engineering
The University of Texas at Arlington
Arlington, Texas
April 2016

CHAPTER 1

Introduction and Overview

1.1 Introduction

Pipelines are complex systems and require constant maintenance. The deterioration process in pipelines is caused by many factors, such as structural, hydraulic, environmental, functional, age of the pipe, quality of initial construction, and type of pipe material. The intensity of structural failures depends on the size of the defect, soil type, interior hydraulic regime, groundwater level and fluctuation, corrosion, method of construction, and loading on the pipe. In gravity pipes, hydraulic failures are caused by infiltration and inflow (I/I) problems. These I/I problems reduce the planned hydraulic capacity of gravity pipes and lead to system backups with increasing the potential for collapse.

Pipelines must be designed and constructed to current standards to ensure planned performance, as well as for protection of public and the environment. Pipe failure can be broadly determined when the pipe can no longer achieve its intended functions for a number of reasons due to burst, cracks and holes, collapse, or rupture. Over time, corrosion and stresses due to outside and/or inside forces can degrade a pipeline performance to the point that a surface collapse, fluid leakage, or gas release might occur. Oil and gas pipelines are comprised of cylindrical metal sections with varying wall thicknesses determined by the expected outside loads and operating pressures. According to the Pipeline and Hazardous Materials Safety Administration,[1] "defect or anomaly is a deviation from the original configuration of the pipeline." This could be a change in wall thickness due to metal loss, a deformation of the pipe wall, or a crack.

[1]Available at http://primis.phmsa.dot.gov/comm/FactSheets/FSPipeDefects.htm?nocache=4854

Pipes are prone to certain types of failures based on the type of material as well.

A set of complex interactions of various mechanisms that occur within and around the pipeline can cause distress and collapse of a pipe. The physical design, age, operation size, complexity, topography, external and internal environments and temperatures, rainfalls, and storms are additional important parameters that impact pipe failure. The failure of a pipeline can cause different types of consequences, depending on the type (sanitary and storm sewer, oil and gas, water, etc.) and size of the pipe (such as transmission and distribution), its internal pressures, location, depth, surface, and subsurface conditions. While it is sometimes difficult to predict when a pipe will fail, with proper inspection and monitoring, it is feasible to estimate whether a pipe has deteriorated sufficiently for failure or rupture to occur.

1.1.1 Pipeline Defects

Pipeline defects can also be classified as built-in or long term. Built-in defects are generated during pipeline construction and represent conditions that affect the performance of pipes after installation. The built-in defects can result in pipe failures and ruptures shortly after installation and service. Long-term defects are caused as a result of the deterioration process and may take many years to show up. Construction-related or built-in defects can be like offsets in alignment, joints loosely fitted or loosened by vibrations, flattened or ovaled pipes, sags due to settlement of poorly compacted soils, stresses caused by dynamic loadings of backfill, removal of trench sheathing and pilings, overburden compaction, and so on. Joints can experience construction defects, such as pinching of rubber gaskets, misalignment of gaskets, and squeezing, due to "overshoving" of one pipe into another. A structural failure can be a crack, break, split, cavitation of the pipe opening, or separation at a joint (Najafi, 2005). Figure 1.1 illustrates a built-in pipe defect that was discovered during inspection and testing immediately after installation.

The main cause of long-term pipeline deterioration, which may occur over an extended time period and is not a result of construction practice, is corrosion resulting in loss of pipe wall thickness. Examples of the causes of long-term pipeline deterioration are surface damage due to pipe gases, excessive hydraulic flows, structural failures, leaks and infiltrations, and erosions.

Surface damage caused by sulfate corrosion from sewer gasses is an example of long-term pipe deterioration, although the actual rate of deterioration can fluctuate widely based on condition in the pipe. Bacteria in the wastewater stream convert the sulfates to hydrogen sulfides which, when released into the pipe air space, become oxidized into sulfuric acid. Sulfuric acid is reactive to some pipe materials,

Figure 1.1 A built-in pipe defect that was discovered during inspection and testing immediately after installation. (*Source*: Reuters.)

such as the cement in concrete, mortar in brick pipes, and metal, making it to corrode. Severe corrosion can jeopardize the structural integrity of a pipe or manhole and lead to collapse. As a safety note, it should be noted that hydrogen sulfide, a common byproduct of sewage, is a rapid-acting systemic poison that causes respiratory paralysis with consequent asphyxia at high concentrations. Inhalation of high concentrations may cause coma after a single breath and may be rapidly fatal. Prolonged exposure may cause pulmonary edema.[2]

The state of the surrounding soil is the most important parameter in the structural performance of a pipe. The main factors that affect the rate of soil loss include pipe defect size, hydraulic conditions (watertable and frequency and magnitude of surcharge), and soil properties (cohesive or noncohesive soil). Severe defects (larger than 4 in. or 100 mm), high watertable (above pipe level), frequent and high magnitude of hydraulic surcharge, and soil types (silts, silty fine sands, and fine sands) can have serious effects on soil loss. Loss of side support will allow the side of the pipe to move outward when loaded vertically, and collapse will likely occur once the pipe deformation

[2]National Institute for Occupational Safety and Health. (1977), Criteria for a Recommended Standard Occupational Exposure to Hydrogen Sulfide. Cincinnati, Ohio: National Institute for Occupational Safety and Health, 1977. DHEW Publication No. NIOSH 77-158.

exceeds 10% in rigid pipes such as concrete, vitrified clay, and so on. Flexible pipes, such as steel, polyvinyl chloride (PVC), polyethylene (PE), and corrugated metal pipes (CMP), can deform to a greater extent while continuing to resist external loading prior to forming cracks or other structural damage. See Chap. 5 for more information on rigid and flexible pipe materials and their properties. Uneven loading of pipes due to joint displacement also accelerates the pipe deterioration process.

1.1.2 Modes of Pipeline Failures

Pipe failure or rupture is likely to occur when the environmental and operational stresses act upon pipes whose structural integrity has been compromised by corrosion, degradation, inadequate installation, or manufacturing defects. Pipeline failures can be sudden and catastrophic, impact lives of hundreds or thousands of people, and cost millions of dollars. A complete failure can be defined as when pipe is completely collapsed or ruptured and/or can no longer carry its functions. When a large-diameter water-transmission pipeline (usually 24 in. and larger) breaks, it can flood neighborhoods and surrounding areas. It may take a week or more before the pipeline owner can fix the pipeline, and during this period there will be a water shortage. The failure of pipe may cause a sinkhole, disrupting traffic, and damage road embankment and pavement. More importantly, water pipeline failures waste millions of gallons of precious water, a resource that is hard to replace. Figures 1.2 to 1.5 present water, gas, and sewer pipe failures.

Pipe breakage types were classified by O'Day et al. (1986) into three categories: (1) circumferential breaks, caused by longitudinal stresses; (2) longitudinal breaks, caused by transverse stresses (hoop stresses); and (3) split bell, caused by transverse stresses on the pipe joint. This classification may be complemented by an additional breakage type, that is, holes due to corrosion or impact. Circumferential breaks due to longitudinal stress are typically the result of one or more of the following occurrences: (1) thermal contraction (due to low temperature of the sewage in the pipe and the pipe surroundings) acting on a restrained pipe; (2) bending stress (beam failure) due to soil differential movement (especially clayey soils) or large voids in the bedding near the pipe (resulting from leaks, I/I, etc.); (3) inadequate trench and bedding and embedment practices; and (4) third-party interference (i.e., accidental breaks, etc.).

Pipeline corrosion and failure is an environmental, safety, economic, and operational issue. These factors are the basis for the trenchless renewal of pipeline and also have significant influence on the decision to renew the pipeline. In selecting an appropriate renewal method, it is important to consider such factors as renewal method durability, reliability, strength to carry loads (hydrostatic and soil/traffic

A

B

FIGURE 1.2 A sinkhole 40 ft in diameter and 16 ft deep, shut down traffic Thursday afternoon, February 7, 2008, on I-25 northbound just south of the 58th Avenue exit (exit 215) in Adams County, Colorado. (*Source*: Mark T. Osler, The Denver Post.)

FIGURE 1.3 A car at the bottom of a sinkhole caused by a broken water line in Toledo, Ohio, on July 3, 2013. (*Source*: AP Photo/Lt. Matthew Hertzfeld, Toledo Fire and Rescue.)

FIGURE 1.4 Flames go up on November 22, 2011, after a gas pipeline explosion near the Athens-Morgan County line in Southeastern Ohio. (*Source*: Cheryl Powers, Associated Press.)

FIGURE 1.5 Service road sanitary sewer collapse on the Michigan State University (MSU) campus. (*Source*: MSU physical plant.)

in addition to internal pressures), social impacts, constructability and bypassing requirements, pipe section length, types of joints, as well as suitability of a specific method, and availability of qualified contractors. The aim of the selection process is to consider all factors to arrive at the most cost-effective, technically viable solution. Table 1.1 presents a summary of factors impacting pipeline renewal and replacement. Many of the factors are interrelated and one condition may impact other conditions of pipelines.

1.2 Extending Service Life

Service life extension can be broadly defined as any technology that can be applied to an existing, aging, and deteriorated infrastructure system to increase its useful life. In the pipeline industry, the threshold for new service design life is commonly set at 50 years, although some methods and new pipe products under certain conditions may provide a service life of up to 100 years.

Since the 1980s, a number of methods have become available to in-situ renew existing and deteriorated underground pipelines. These methods belong to the family of "trenchless technology" methods. Trenchless technology methods include all the methods that can install a new pipeline or renew/replace an existing and deteriorated pipeline with minimum surface or subsurface excavation and disruptions. Trenchless renewal/replacement methods not only can prolong

Construction and installation/ renewal and replacement	Quality of installation Bedding and embedment Costs Open-cut methods Trenchless technology methods Gap analysis of data availability, such as inspection data and GIS Coordination with local and regional stakeholders Funding opportunities from public and private sources (i.e., public private partnerships) New technologies Risk Assessment and management
Operational and maintenance	Pipelines inspection methods Real-time monitoring methods Asset management strategies Repairs and renewal issues Capacity problems (hydraulics, biofilm, etc.)
Structural	Internal hydraulic pressures: • Operating • Peak • Transients • Cyclic (fatigue) • Control devices/techniques Dead loads (earth) Live loads (traffic) Groundwater loads Aboveground pipelines–support spacing Ground movements Seismic loads Thrust restraint Frost loads
Environmental	Corrosion Contaminated ground (permeation) Live, groundwater, and dead (soil) loads Seismic loads Intrusions Temperature differences Soil pH
Material issues	Corrosion Durability (life expectancy) Reliability Stress crack propagation Appurtenances Joints Repairs and maintenance
Sustainability	Life cycle costs Environmental impacts Social impacts Energy requirements (materials, manufacturing, delivery, installation, repairs, renewals, replacements, operation and maintenance, etc.)

TABLE 1.1 Summary of Parameters Impacting Pipeline Renewal and Replacement

life of the existing pipe or install a new pipe, they can reduce operation and maintenance costs. The trenchless methods can lower the life cycle cost of the project, and achieve economic savings in emergency repairs and eventual replacement costs associated with "open-cut," which may involve enormous social and economic costs, often inherent in open-cut replacement and repair methods.

The term "renewal methods" refers to all the methods that can extend the service life of a pipeline beyond its original design life. Renewal methods are used in this book to collectively refer to "rehabilitation," "upgrade," "renovation," and "replace" a deteriorated pipeline with trenchless technology methods. The term "repair" refers to all the methods that can fix localized defects, without adding to the service life of the pipeline (Najafi, 2005). The repair methods are included in Chap. 14.

Figure 1.6 is a conceptual plot of the life expectancy. The decision to renew a pipeline is a complicated issue and requires much technical and managerial information. While age of the pipeline is an important factor, other parameters, such as failure or break rates, hydraulic capacity, quality of service, consequences of failures, environmental issues, location of the pipeline, budgetary and regulatory concerns, and so on, need to be considered. The lack of knowledge of the condition of pipes and remaining service life present difficulty in terms of determining which trenchless renewal method to use and when to use these methods, to ensure pipeline sustainability and make the best asset management option at the lowest cost.

The maintenance of assets needs to be balanced between planned and unplanned maintenance schedules so that the cost of maintenance activities is minimized. Figure 1.6 also illustrates the deterioration and renewal strategies for an asset, which results in the effective management and operation. This figure demonstrates several maintenance activities during the life of the asset with continued deterioration process until the point that a major renewal/rehabilitation/

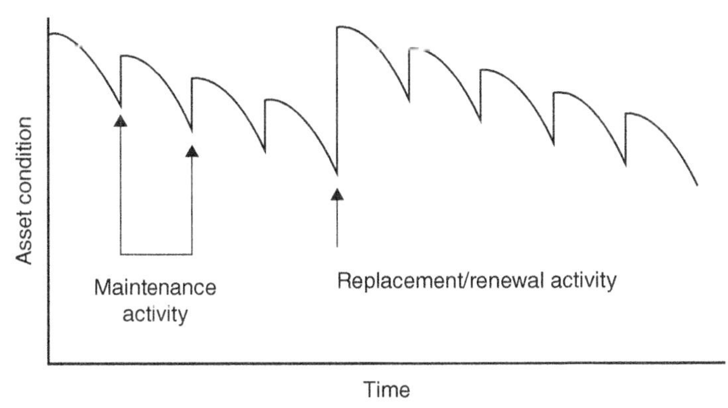

FIGURE 1.6 Asset deterioration and renewal.

replacement is required. After the renewal process, the asset starts a new life with a new cycle of maintenance activities as the result of deterioration process.

1.3 Trenchless Renewal Methods

Many trenchless renewal methods (TRMs) are available. These methods can be used to replace, rehabilitate, upgrade, repair, or renovate (collectively called renewal) existing pipelines. They can also be used to replace and enlarge existing pipes. The basic TRMs can be categorized into the following types (Fig. 1.7):

1. Cured-in-place pipe (CIPP): Chap. 6
2. Sliplining (SL): Chap. 7
3. In-line replacement (ILR) and pipe bursting (PB): Chap. 8
4. Modified sliplining (MSL): Chap. 8
5. Spray-in-place pipe (SIPP): Chap. 10
6. Close-fit pipe (CFP): Chap. 11
7. Sewer manhole renewal (SMR): Chap. 12

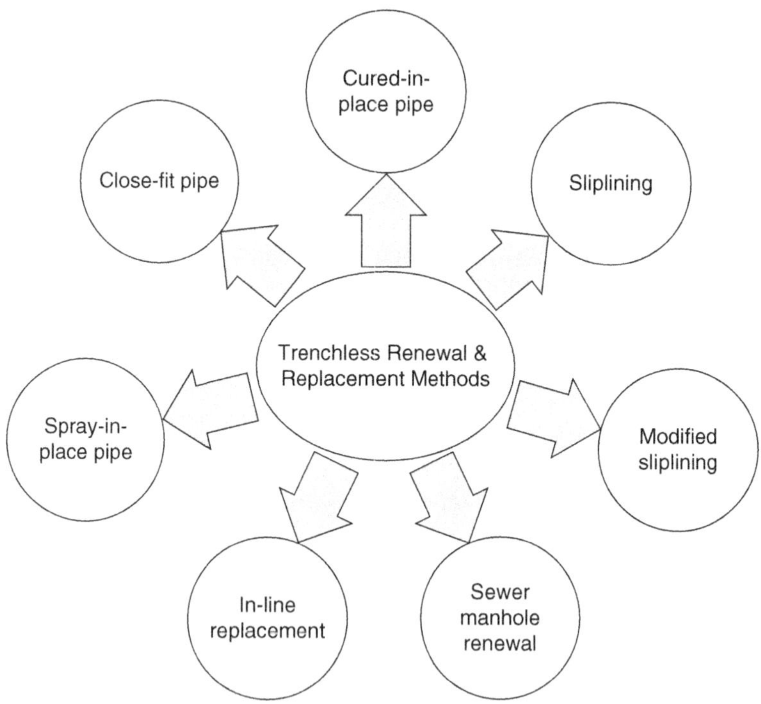

FIGURE 1.7 Basic trenchless renewal methods.

The choice of trenchless pipeline renewal methods depends on the physical conditions of the existing pipeline system, such as pipeline length, type, material, size, type and number of manholes or access points, service connections; and the nature of the problem or problems involved. The problems with the existing pipe may include structural or nonstructural, infiltration or inflow, exfiltration or outflow, pipe breakage, joint settlement, joint or pipe misalignment, capacity problems, and so on. Other features of the renewal systems, such as applicability to a specific project, constructability, cost factors, and life expectancy, should also be considered. Figure 1.7 presents main divisions of pipeline renewal methods.

1.3.1 Cured-in-Place Pipe

The most common renewal method is cured-in-place pipe (CIPP) which involves the insertion of a resin-impregnated fabric tube into an existing pipe by use of water inversion or winching. The fabric is a polyester material, fiberglass reinforced, or similar. Usually water or air pressure is used for the inversion process. Hot water, steam, or ultraviolet (UV) light can be used for curing. The pliable nature of the resin-saturated fabric prior to curing allows installation around curves, filling of cracks, bridging of gaps, and maneuvering through pipe defects. The cured-in-place-pipe method can be applied for structural or nonstructural purposes. Table 1.2 presents the main characteristics of CIPP methods. These characteristics are subject to project and site conditions. Figures 1.8 and 1.9 illustrate the CIPP installation process for inverted-in-place and winch-in-place, respectively.

1.3.2 Sliplining

Sliplining (SL) is mainly used for structural applications when the existing pipe does not have joint settlements or misalignments. In this method, a new pipeline of smaller diameter is inserted into the existing pipe and usually the annulus space between the existing

Method	Diameter Range, in (mm)	Maximum Installation, ft (m)	Liner Material	Applications
Inverted-in-place	3–120 (75–3,000)	3,000 (1,000)	Thermoset resin/fabric composite	Gravity and pressure pipelines
Winched-in-place	4–54 (100–1,500)	1,000 (300)	Thermoset resin/fabric composite	Gravity and pressure pipelines

Table 1.2 Main Characteristics of CIPP Methods

Figure 1.8 CIPP inverted-in-place installation method. (*Source:* Insituform Technologies.)

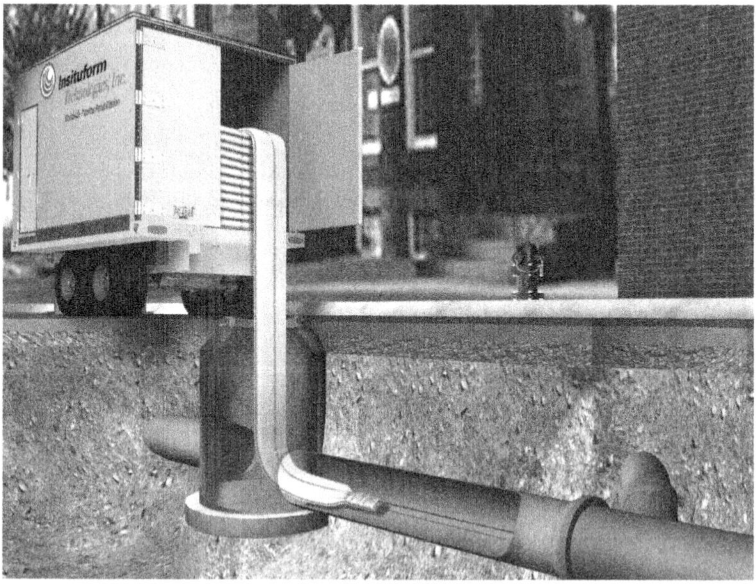

Figure 1.9 CIPP winched-in-place installation method. (*Source:* Insituform Technologies.)

Method	Diameter Range, in (mm)	Maximum Installation, ft (m)	Liner Material	Applications
Segmental	4–158 (100–4,000)	1,000 (300)	HDPE, PVC, and GRP	Gravity and pressure pipelines
Continuous	4–63 (100–1,600)	1,000 (300)	HDPE and PVC	Gravity and pressure pipelines

TABLE 1.3 Main Characteristics of Sliplining Methods

pipe and new pipe is grouted. This installation method has the merit of simplicity and is relatively inexpensive. However, there can be a significant loss of hydraulic capacity. Table 1.3 presents the main characteristics of sliplining methods. Figures 1.10 and 1.11 illustrate the two variations of sliplining methods: segmental sliplining method and continuous sliplining method.

1.3.3 In-line Replacement

When capacity of deteriorated pipelines is found to be inadequate, then in-line replacement (ILR) should be considered. There are two categories representing ILR: pipe bursting and pipe removal. Pipe bursting, as the name implies, uses a pulling head (static pipe bursting) or a vibrating hammer (pneumatic pipe bursting) to break the

FIGURE 1.10 Segmental sliplining method. (*Source*: Hobas Pipe USA.)

FIGURE 1.11 Continuous sliplining method. (*Source*: Insituform Technologies.)

existing pipe and force particles in the earth while a new pipe is pulled and/or pushed in its place simultaneously. Pipe removal on the other hand, can be performed by use of horizontal directional drilling (HDD) (also called pipe reaming), horizontal auger boring (HAB), or microtunneling (MT) (also known as pipe eating) equipment. In this method, the existing pipe is broken into small pieces and taken out of borehole by means of slurry (for HDD or MT methods) or auger (for the HAB method). Table 1.4 presents the main characteristics of the ILR methods. Figure 1.12 illustrates pneumatic pipe bursting and pipe reaming methods, respectively.

When using the HDD equipment to remove the existing pipe (pipe reaming), the HDD rig with minor modification pushes the drill rods through the existing pipe and connects the rods to a special

Method	Diameter Range, in (mm)	Maximum Installation, ft (m)	Liner Material	Applications
Pipe bursting	4–140 (100–3,500)	750 (250)	PE, clay, PVC, GRP, and DI	Gravity and pressure pipelines
Pipe removal	Up to 36 (900)	300 (100)	PE, PVC, GRP, and DI	Gravity and pressure pipelines

TABLE 1.4 Main Characteristics of the ILR Method

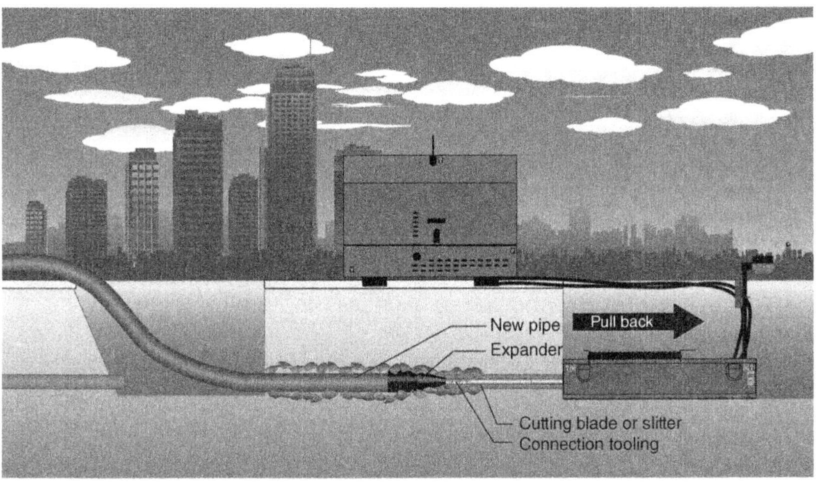

FIGURE 1.12 Pneumatic pipe bursting method. (*Source*: HammerHead.)

reamer (see Fig. 1.13). Then, the new pipe string is attached to the reamer via a swivel and pulling head. As the drill rig rotates the drill rod and simultaneously pulls back the drill rod, the existing pipe is grinded and replaced by the new pipe. Refer to Najafi (2013) for more information on HDD operation.

1.3.4 Modified Sliplining

The modified sliplining MSL includes methods that pipe segments or plastic strips are installed in close-fit with the existing pipe and the annular space is grouted. There are two variations of the MSL method: panel-lining method (PLM) and spiral wound method (SWM).

Panel linings can be used to structurally renew large diameter (more than 48 in. or worker-entry) gravity pipelines. This method can

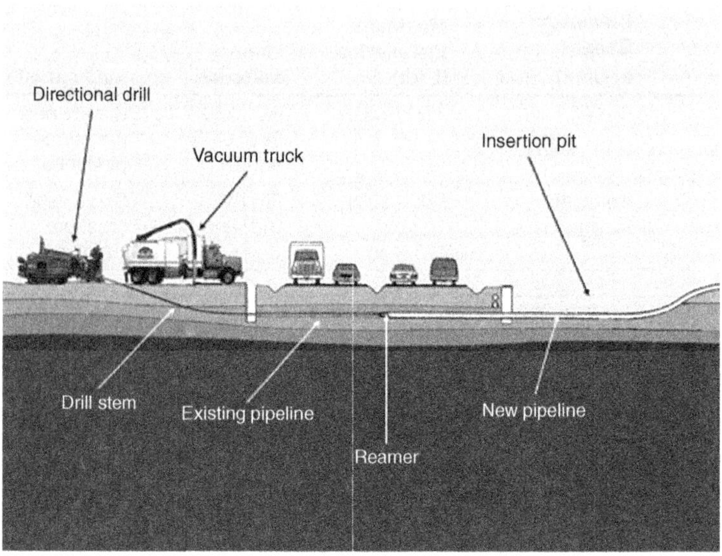

FIGURE 1.13 Pipe reaming method. (*Source*: Nowak Pipe Reaming.)

accommodate different shapes, such as noncircular pipelines. The main type of material for this method is fiberglass composites. Figure 1.14 illustrates a PLM.

The spiral wound method, also for large diameter gravity pipelines, uses a layered composite PVC liner and cementitious grout to renew the existing pipe. The combination of the ribbed profile on the PVC liner and the fluid nature of the grout produce a highly integrated structure with the PVC liner "tied" to the existing pipe through the grout. The structural strength of the renewed pipe is determined by the grout characteristics. Table 1.5 presents the main characteristics of modified sliplining methods. Figure 1.15 presents the spiral wound installation.

1.3.5 Spray-in-Place Pipe

1.3.5.1 Cementitious Materials

Cementitious materials are used in both small and large diameter pressure and gravity pipelines primarily for its most basic characteristic—cost-effectiveness in corrosion protection of metal pipes. In water applications, cement mortar lining (CML) is most common and serves two main functions: (1) the alkalinity of the cement inhibits further corrosion (especially in case of a cast iron pipe) and (2) the relatively smooth internal surface reduces hydraulic roughness and improves flow characteristics. It should be noted that cement mortar lining is also applied to many new iron water pipes to inhibit corrosion. In this application, the CML does not fulfill a structural function, rather it

Figure 1.14 Panel-lining process. (*Source*: Channel Line.)

C

Figure 1.14 Panel-lining process. (*Continued*)

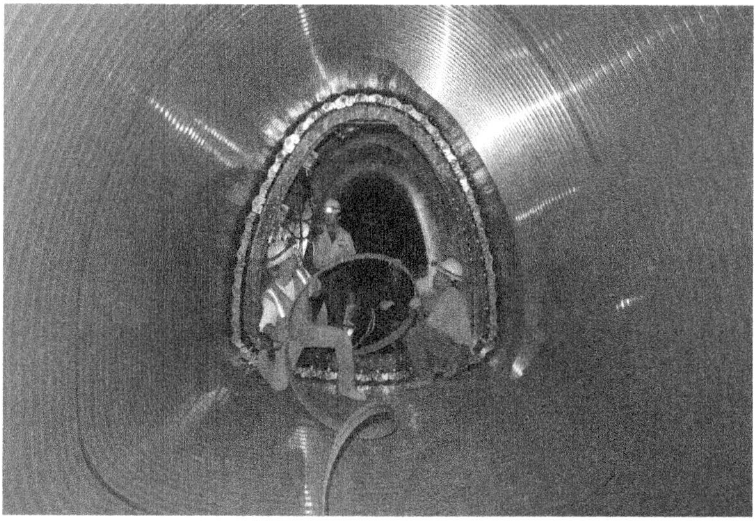

Figure 1.15 Spiral wound installation machine. (*Source*: SPR.)

Method	Diameter Range, in (mm)	Maximum Installation, ft (m)	Liner Material	Applications
Panel lining	>48 in. (1,200)	Varies	GRP	Gravity pipelines
Spiral wound	4–100 (100–2,500)	1,000 (300)	PE and PVC	Gravity pipelines

Table 1.5 Main Characteristics of the Modified Sliplining Method

reduces the rate at which the pipe will deteriorate. This lining is not appropriate for pipes which leak or where corrosion has reduced the wall thickness significantly. It is also important to apply sufficient thickness of CML to create the alkaline environment at the CML and iron interface for water pipes. As with steel reinforcement in concrete structures, inadequate cover to the metal will allow the onset of corrosion, which will cause the CML to crack and spall.

In sanitary sewer applications, cementitious linings comprise various base materials, such as Portland cement and calcium aluminate. These materials are used to line underground structures to prevent infiltration and rebuild structures deteriorated because of age and erosion from I/I. Calcium aluminate cements are frequently used to provide some sulfide resistance in renewing sewer structures where mild corrosion is evident. Cementitious materials are also frequently used as a basecoat material for polymer linings in more corrosive sanitary sewer environments.

1.3.5.2 Polymers (Epoxy, Polyurethane, and Polyurea) Materials

Solvent-free 100% solids epoxy, polyurethane, and polyurea coatings have unique advantages for successful renewal and corrosion protection of pipelines. These thermoset linings are essentially inert plastics when cured and therefore resistant to most corrosive elements found in these environments. They have longer life expectancies when properly applied by professionally trained applicators using specialized spray equipment to achieve optimal performance as defined by the product manufacturers. Because these products contain no solvents, they require no evaporative process to cure and do not emit destructive volatile organic compounds (VOCs). Polymer formulations with 100% solids are safer in confined spaces and can also be formulated for thicker, structure-enhancing applications. Such products now dominate the pipeline renewal methods because they can provide fast renewal solutions while providing semi-structural solutions, such as spanning holes and cracks and protecting existing pipelines from aggressive corrosive environment.

Table 1.6 presents the main characteristics of SIPP. Figure 1.16 illustrates the schematic diagram of spray-in-place lining.

Method	Diameter Range, in (mm)	Max Inst., ft (m)	Liner Material	Applications
SIPP	Dependent on the type of product and manufacturer	N/A	Portland cement, epoxy, polyurethane, and polyurea	Pressure and gravity pipelines

Table 1.6 Main Characteristics of SIPP

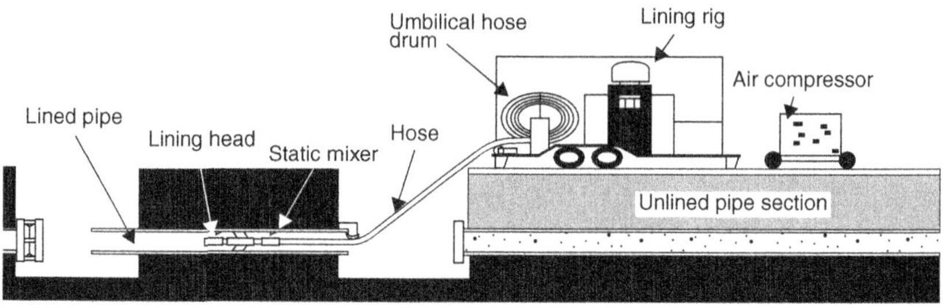

Figure 1.16 Spray-in-place pipe lining. (*Source*: 3M Water Infrastructure.)

Method	Diameter Range, in (mm)	Maximum Installation, ft, (m)	Liner Material	Applications
Fold and formed	4–30 (100–800)	700 (200)	PVC	Gravity and pressure pipelines
Drawdown	3–60 (75–1,500)	1,000 (300)	HDPE and MDPE	Pressure pipelines
Rolldown	3–24 (75–600)	1,000 (300)	HDPE and MDPE	Pressure pipelines

Table 1.7 Main Characteristics of Close-fit Methods

1.3.6 Close-Fit Pipe

Close-fit pipe (CFP) renewal temporarily reduces the cross-sectional area of the new pipe before it is installed, then expands it to its original size and shape after placement to provide a close-fit with the existing pipe. This method is more common for pressure pipeline applications and can be used for both structural and nonstructural purposes. The lining pipe can be reduced on-site or in the manufacturing plant and reformed by heat and/or pressure or naturally. There are three versions of this approach: fold and formed (F&F) using polyvinyl chloride (PVC) pipe, and drawdown (DD), and rolldown (RD) both using polyethylene (PE) pipe. Table 1.7 presents main characteristics of close-fit lining methods. Figure 1.17 illustrates a close-fit lining installation.

1.4 Sewer Manhole Renewal

Sanitary sewer manholes require renewal to prevent or reduce surface water inflow and groundwater infiltration, to repair structural damage, and to protect surface from corrosion. When renewal

FIGURE 1.17 The close-fit lining using PE pipe. (*Source*: United Pipeline Systems.)

methods will not solve the problems cost-effectively, trenchless manhole replacement should be considered. Selection of a particular renewal method should consider the type of problems, physical characteristics of the structure, location, condition, age, and material type of existing manhole. Extent of successful manhole renewal experiences with a specific product should also be considered in the method selection process. The sewer manhole renewal (SMR) can be divided into the following methods: (1) cementitious, (2) cast-in-place, (3) cured-in-place, and (4) profile PVC. Table 1.8 presents main characteristics of SMR methods. Figure 1.18 shows the manhole lining process.

Method	Diameter Range, in (mm)	Maximum Installation, ft, (m)	Liner Material	Applications
Sanitary Sewer Manhole renewal and Replacement	Any	N/A	SIPP, PVC, PE, CIPP, and composites	Structural, semi-structural, and nonstructural

TABLE 1.8 Main Characteristics of SMR Methods

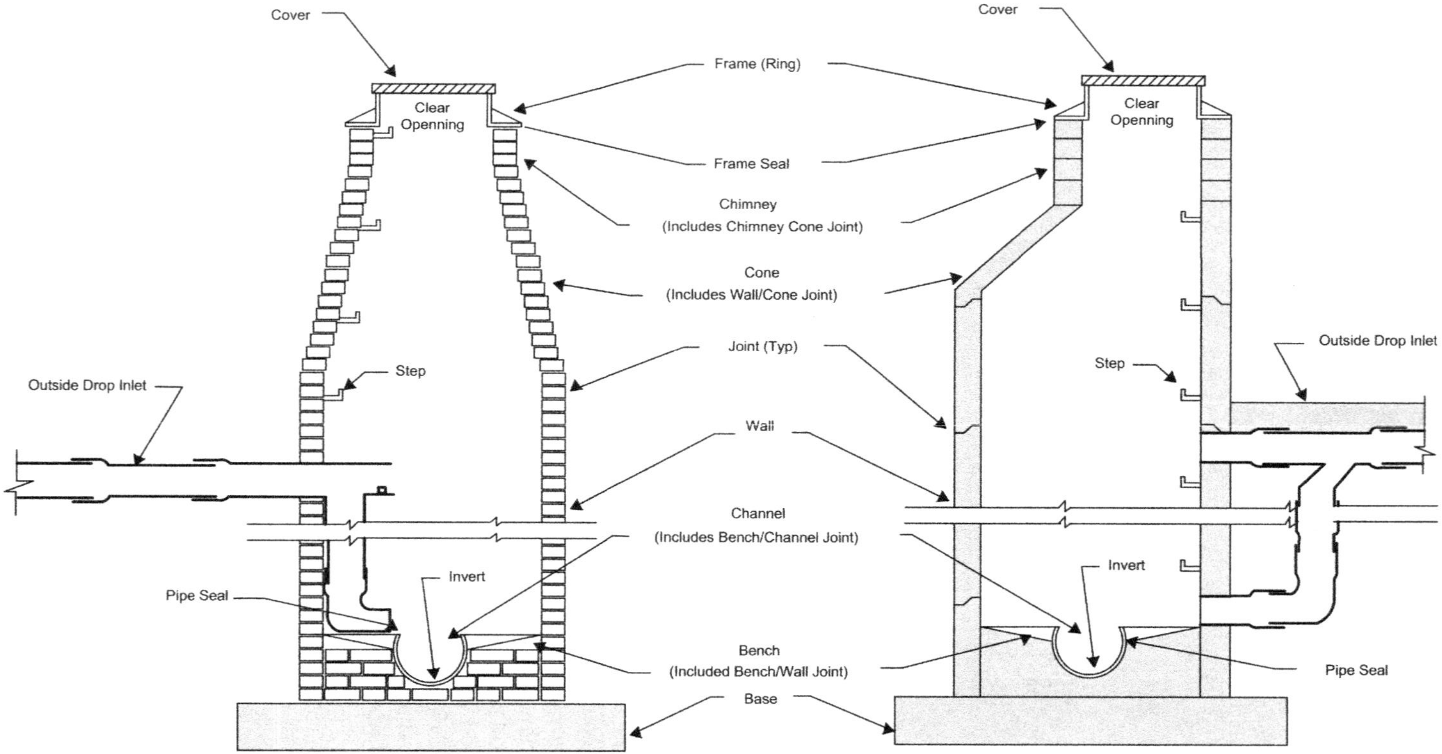

FIGURE 1.18 Manhole lining process.

1.5 Selection of a Renewal Method

There are many factors which need to be identified prior to pipe renewal, such as pipe age and history, cause of problems or failures, type of application and fluid quality, internal and external pressures, extend of cracks, holes, and leakage, existing pipe material and remaining pipe wall thickness, hydraulic capacity, current and future plans for the pipeline, surface and subsurface conditions, and capability of specific renewal method, to address identified problems and defects. Table 1.9 presents the renewal methods and application to service life extension of pipes.

Without a deep and complete understanding of the overall conditions of the existing pipeline, it is very difficult to select appropriate renewal methods. The selection method can be summarized in a two-step process with more information provided in Chap. 3: (1) identification and classification of the deteriorated pipes and definition of the problems and (2) selection of the renewal method. These steps are described below:

- **Step 1:** Identification and classification of the deteriorated pipe is the first step in the selection of a particular renewal method. The availability of various closed-circuit television (CCTV) inspection equipment and other inspection methods as well as computer models for flow analysis and water quality computation, geographic information system (GIS), global positioning system (GPS), and other software can help pipeline owners to collect, maintain, and analyze data on a system-wide basis. For example, NASSCO's Pipeline Assessment Certification Program® (PACP®) provides a standardized protocol for collection and management of data from internal pipeline inspection, and has become the standard of existing sewer pipe inspection and classification. Some of the factors that must be evaluated, which will govern the selection of the renewal method, include age, existing pipe material, history of previous maintenance or repairs, leaks, number and problems with joints, bends, tees, pipe size and pipe section length, and number of service laterals. Other pipeline problems include the following:
    - **Hydraulic Capacity Problems:** Hydraulic problem is a failure originating from a decrease in pipe flow capacity or increase in demand. Infiltration and inflow increases the flow, as well as increased number of service connections. A decrease in capacity can also be result of reduction in diameter due to corrosion, sediments, tree roots, increased number of structural defects grease, and other factors.

Renewal Method	Joint Problems	Corrosion	Cracks/Holes	Inflow, Infiltration, and Exfiltration	Structural Problems	Inadequate Hydraulic Capacity	Extension of Service Life of Existing Pipe
CIPP	Marginal*	Yes	Yes	Yes	Yes	Marginal	Yes
SL	Marginal	Yes	Yes	Yes	Yes	No	Yes
CFP	Marginal	Yes	Yes	Yes	No	Marginal	Yes
SIPP	No	Yes	Marginal	Yes	Marginal	No	Marginal
MSL	Marginal	Yes	Yes	Yes	Yes	No	Yes
ILR	Yes	Yes	Yes	Yes	Yes	Yes	Yes

*Marginal means special preparation (such as a point repair) may be required before installation of the renewal method and/or the renewal solution may be effective dependent on the specific product (manufacturer), quality of application, and project conditions.

TABLE 1.9 Renewal Methods and Application to Service Life Extension of Pipes*

- **Water Quality Problems:** Water, in contact with plastic pipes, surface linings, or other materials, can be affected by migration of components that make the water quality unacceptable with respect to aesthetic effects or health. Volatile organic compounds (VOCs) leach from defective or cracked pipes and linings. These VOCs can add taste and odor to the drinking water and can cause the water to fail the acceptable drinking standard limits.
- **Structural Problems:** Structural defect failure mechanisms include cracks, fractures, and breaks which are caused by pipe material defects during manufacturing, delivery and installation, change in pipe loadings, and excessive or cyclic internal or external pressures. Tree roots, broken or cracked pipes, missing or broken cleanout caps, or undersized gravity pipes, can cause blockages. Infiltration and inflow have great impact on pipe capacity and are caused when groundwater enters the pipe system through pipe defects and leaking joints. Many surface collapse failures are associated with degraded but functioning pipes that fail owing to a large one-time event. Smaller cyclical dynamic loads include load transfer from above-ground activities, such as routine truck, machinery, and bus or train traffic, or in-ground movements, such as those caused by expansive soils or frost heave.

- **Step 2**: Once the pipe to be renewed is selected, the flow diagrams in Figs. 1.19 to 1.21 can be used to suggest appropriate renewal methods.

1.6 Current Design Concepts for Pipeline Renewals

The objective of pipe renewal design involves a set of equations considering the factors such as groundwater, soil, traffic loads, and other operational loading conditions. As for any structural design, there are many variables and parameters used in the design of the pipe wall thickness which provide the required performance requirements to resist the loads. There is no single design equation that can be used for all different conditions that must be taken into account for the proper design of a renewal method, so it is necessary to divide these conditions into different groups. For both gravity flow and internal pressure, design equations have been categorized into partially and fully deteriorated on the basis of ASTM F1216. These piping conditions are defined as follows.

1.6.1 Partially Deteriorated Pipe

A partially deteriorated existing pressure pipe is defined as one where the existing pipe may be cracked or corroded with a diameter

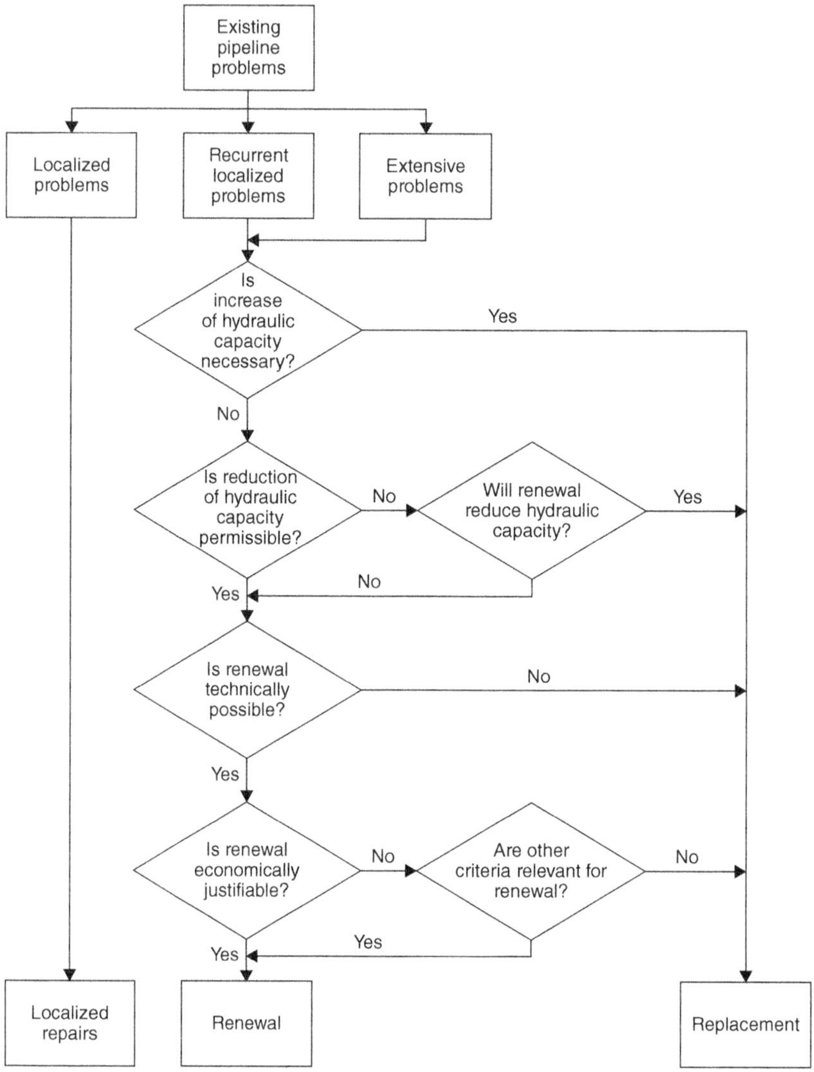

FIGURE 1.19 Decision support system for selection of a trenchless renewal, replacement, or repair method.

the original pipe can support the soil and surcharge loads throughout the design life of the renewed pipe. The soil adjacent to the existing pipe must provide adequate side support. The pipe may have longitudinal cracks and up to 10.0% distortion of the diameter. If the distortion of the diameter is greater than 10.0%, alternative design methods are required.

For the partially deteriorated *gravity* condition, the CIPP is designed to support the hydraulic loads due to groundwater, since the

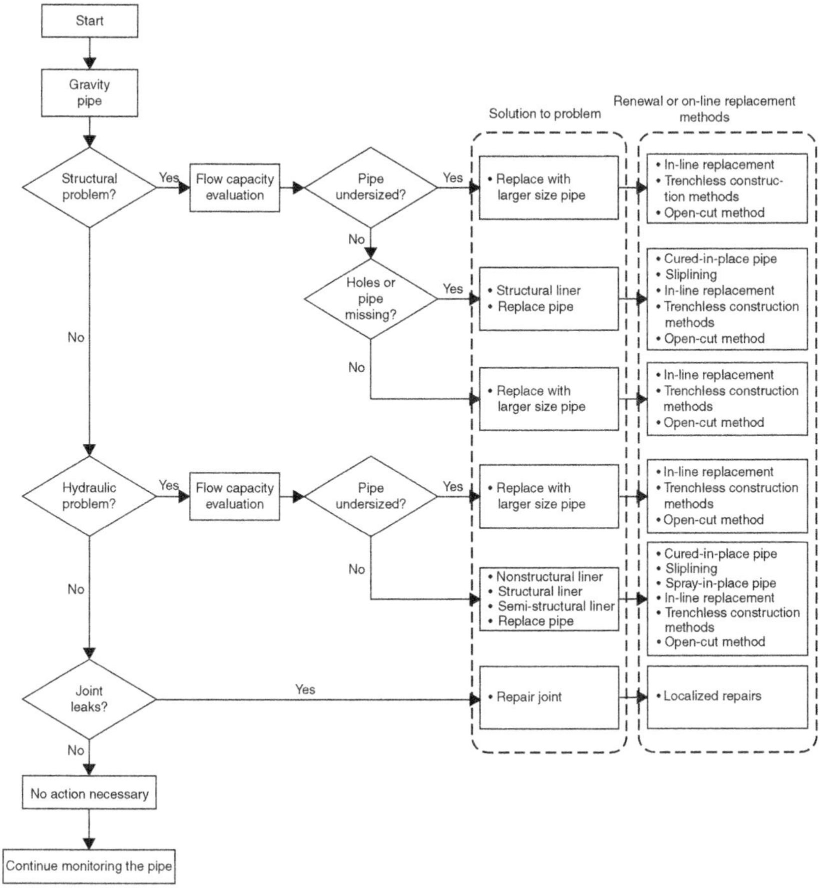

FIGURE 1.20 Decision support system for gravity pipe renewal methods.

soil and surcharge loads can be supported by the existing pipe. The groundwater level should be determined by the pipeline owner or the design or consulting engineer, and the thickness of the CIPP should be sufficient to withstand this hydrostatic pressure without collapsing.

For the partially deteriorated *Pressure* Condition, the CIPP installed in an existing underground pipe is designed to support external hydrostatic loads due to groundwater as well as withstand the internal pressure in spanning across any holes in the existing pipe wall.

Refer to Chap. 3 and Appendix X1, ASTMF1216-09, for design equations.

1.6.2 Fully Deteriorated Pipe

In this condition, the existing pipe is not structurally sound and cannot support soil and live loads or is expected to reach this condition over the design life of the renewed pipe. This condition is evident

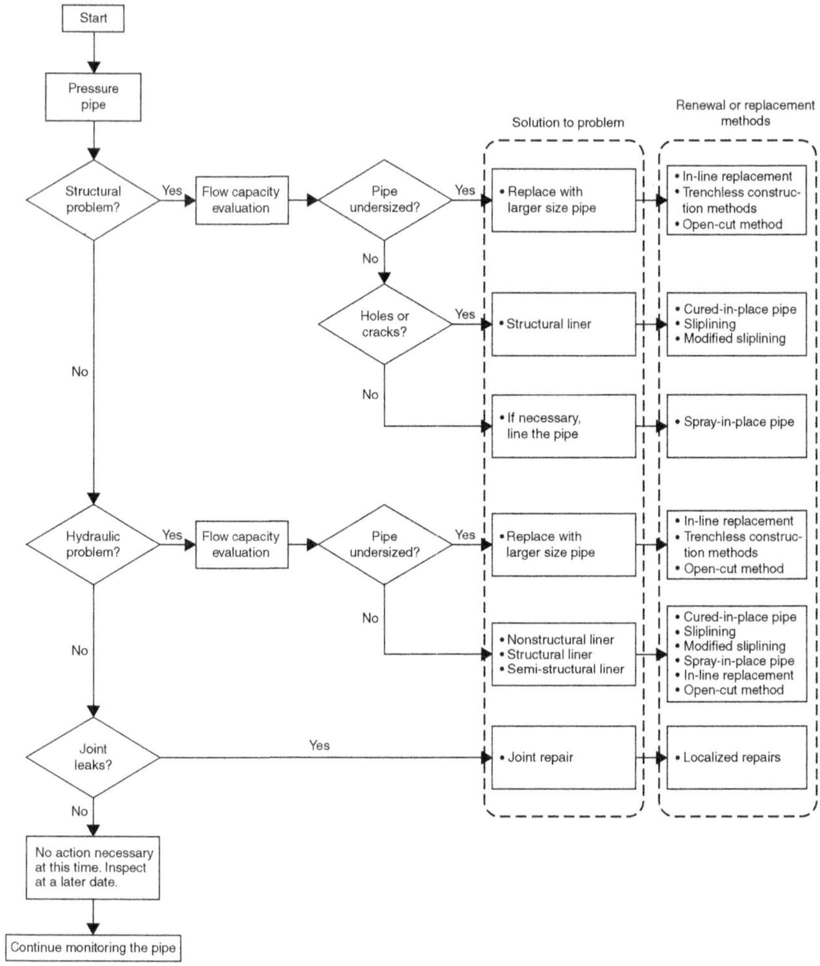

FIGURE 1.21 Decision support system for pressure pipe renewal methods.

when sections of the existing pipe are missing, the pipe has lost its original shape (excessively deformed), or the pipe has severely corroded due to the effects of the fluid, atmosphere, soil, and/or applied loads.

For the fully deteriorated *gravity* pipe condition, the CIPP is designed to support hydrostatic, soil, and live loads. The groundwater level, soil type and depth, and live loads should be determined by the pipeline owner or the design or consulting engineer, and the CIPP thickness should be calculated to withstand above loads without collapsing.

For the fully deteriorated *pressure* pipe condition, the CIPP is designed to withstand all external loads and the full internal pressure.

Refer to Chap. 3 and Appendix X1, ASTMF1216-09, for design equations.

1.7 Emerging Design Concepts for Pipeline Renewal Systems

A conventionally buried pipe is structurally designed to "carry" or "transfer" all service loads, which include the weight of soil backfill placed over it, hydrostatic pressures, vacuum, internal working pressures, and loads applied at the ground surface. As stated in the previous section, current methods for structural design of flexible gravity pipelines by Appendix X1 to ASTM F1216 (2009), generally, consider two external load cases for designing of the liner pipe. The first is sustained hydrostatic pressure due to groundwater acting in the annular space between the liner pipe and the existing pipe. The second load case assumes that earth and traffic loads will, in due course, be transferred from the existing pipe–soil structure to the liner pipe. This excessive conservatism is particularly true in the treatment of the "fully deteriorated pipe condition" by ASTM F1216, for two reasons.

First, in the above concept, the soil load reaching the liner pipe is overestimated by treating the liner as if it had been directly buried in a trench. The more correct liner pipe condition, even in circumstances where the existing pipe may continue to deteriorate after renewal, is a tunnel lining situation (Schrock and Gumbel, 1997). Second, the formula used to describe liner response to transferred soil load, and hence calculate the required wall thickness, has been incorrectly modified from an already conservative theory for direct-bury applications, which entail further irrational safety factors (Gumbel, 1998).

The basic design concept to develop buckling pressure or safe water head charts for each renewal technique incorporates imperfections, but allows variation of the relevant conditions according to the condition of the pipe to be renewed. The most appropriate form of chart may vary according to the type of renewal technique and/or liner material. The reader is referred to ASCE emerging concepts for the design of pipeline renewal systems for more information.[3]

1.8 Long-Term Test Methods

Recently the design life of pipelines has gained more interest due to early deterioration of some pipe materials as well as frequent pipe breaks in the news and unexpected and excessive repair and

[3]Available at http://www.uta.edu/ce/cuire/culvert.php

maintenance costs. At present, there are limited established design methodologies for pressure pipe lining systems. The existing ASTM standard F1216-09 provides some information on design methods and procedures for the renewal of gravity and pressure pipes, using the CIPP lining method. Some short-term tests have been conducted to evaluate the key mechanical properties of the pipe linings and their constituents. The data obtained from these tests can be used for defining the properties of the liner during numerical analyses. For long-term testing and prediction of structural failure of flexible pipes when subjected to an external hydrostatic pressure, various alternatives are available. The most widely used approach involves the modification of the Timoshenko equation covering the buckling of unconfined pipe (Timoshenko and Gere, 1961).

Additionally, ASTM D2990-09, "Standard Test Methods for Tensile, Compressive and Flexural Creep and Creep-Rupture of Plastics," provides a test of simply supported beam samples of plastics where "creep modulus" is measured over 10,000 hours with extrapolating to long-term values. These data are used to define an apparent long-term flexural modulus of elasticity for use in the buckling equations for plastic pipes. This allows for a long-term design that compensates for shape changes as a result of material creep. However, the actual flexural modulus, as shown by retroactive testing (EPA, 2012), demonstrates minor changes over the life of a CIPP in gravity sewer service.

ASTM D1598-15a, ASTM D2992-12, and ASTM D2837-13e1 are used to test pipe samples that are subjected to a varying internal pressure to fail the lining pipes in 10,000 hours. A graph of hoop stress with respect to time is plotted on the log scale and extrapolated to determine long-term hoop stress on the pipe and the lining. These test results are used in the design of pipe lining systems as described in Chap. 3.

1.9 Common Terms and Practices in Pipeline Renewal

1.9.1 Contract Documents

Following the procurement process, the project is awarded to the successful contractor. During this process, the contractor signs a contract with the owner or with the owner's representative to complete the work. Included as a part of this contract are the contract documents. The contract documents spell out in detail all aspects of the work to be completed, such as bonds, regulatory requirements, special conditions, general conditions, the technical specifications, and so on.

1.9.2 Technical Specifications

The technical specifications for the pipeline renewal projects are the instructions for the contractor to furnish a specified product or completed project that meets the contract requirements. The technical specifications also instruct the inspector as to what needs to be observed, inspected, measured, tested, and documented to ensure that the product meets the contract requirements. For the pipeline renewal work, performance specifications are preferred where the contractor defines the means and methods for installation of the work. Performance specifications allow the contractor to use innovative means that are at his or her disposal to deliver the specified product at a defined level of quality and at the lowest cost for the owner. A well-written specification defines:

1. existing conditions
2. the selected renewal product
3. work required by the contractor and the inspector
4. submittal requirements, including design
5. the product quality level required
6. required quality controls, including specific ASTM requirements
7. required quality assurance practices during construction
8. the required measurements through inspection and testing
9. the warranty inspection procedure

1.9.3 Design

For pipeline renewal products such as CIPP, design often means calculating the minimum wall thickness required based on the physical properties of the material selected, the design parameters, and the purpose of the project. Design also includes selecting the proper materials for the project and the appropriate installation approach. Design can be completed by the specifier or, in a performance specification, can be completed by the contractor (or the contractor's engineer) and submitted to the owner or the owner's representative for approval.

1.9.4 Inspection

Proper jobsite inspection is critical to having a quality project. There should be a technology-trained inspector onsite to observe, measure, inspect, document, and test the pipe materials provided and confirm the quality of the delivered product. This includes all the measureable requirements described in the contract documents and technical specifications. The inspector also ensures that the contractor's submitted means and methods are properly implemented. The inspector should prepare a warranty checklist consisting of 10% to 15% of the renewal work. The checklist represents work quality issues observed and documented during installation. Pipelines on the checklist should be reinspected at least one month before the end of the warranty period. If deficiencies are found, the amount of pipes to be re-inspected can be increased, if necessary. Extended warrantees may be applied for portions of the project that have not met the requirements of the contract, are defective, or have been repaired after renewal.

1.9.5 Quality Assurance/Quality Control

A key element of a successful project is to achieve a quality installed product. To help ensure this, a trained and knowledgeable inspector should be on the job site. To define the inspector's role and authority, quality controls and assurances must be outlined in the contract documents, and all submittals, specification requirements, and referenced standards should be available to the inspector. The following are suggested quality checks before, during, and after product installation.

1.9.5.1 Inspections and Quality Checks before and during Installation

- *Materials.* Manufactures' technical information and testing data, including shipping, storage, and handling recommendations are reviewed. Safety data sheets (SDS) are received and reviewed.
- *Amount of resin.* For CIPP, resin quantity, per unit length for each size pipe on the project, is determined through information supplied by the manufacturer. Each wet-out tube is checked for the proper amount of resin by reviewing the tube wet-out sheet and comparing the amount of resin used to the manufacturer's required amount.
- *Condition of the existing pipe.* The existing pipe is inspected, typically through CCTV preferably using the NASSCO's Pipeline Assessment and Certification Program (PACP), just prior to product installation. The inspector should be knowledgeable on different conditions that can occur in a pipeline and should be able to determine if a pipeline can be renewed to specified standards or if it must first be cleaned and/or spot repaired (see Chap. 14).
- *Curing schedule (for CIPP).* Since in CIPP, the cured resin builds the liner pipe, correct curing is very important. The cure schedule is based on the manufacturer's requirements and contractor performance for different field conditions. Cure is monitored and compared to the submitted manufacturer's recommended schedule.
- *Workmanship of lateral connections.* The inspector verifies that the contractor leaves a clean, smooth service lateral connection within tolerance, meeting the contract specified requirements.

1.9.5.2 Inspections and Quality Checks after Installation

- *Visual inspection.* Inspection, typically by CCTV, of the final installed product verifies a defect-free and functionally operating system. Any defects located must be repaired immediately.
- *Physical properties.* The physical properties of the installed product are determined through independent sample testing and compared to the values specified or submitted for wall thickness design. CIPP samples must be taken and kept according to technical specifications. The contract documents should define a remedy if the test results are below-specified requirements.
- *Thickness.* Wall thickness is measured and compared to the approved design for each installed product. The contract documents should define a remedy to be used if the tested thicknesses are below-approved thickness.
- *Chemical resistance.* Resistance to corrosive agents in the waste stream is certified by the manufacturer in accordance with ASTM requirements.
- *Water tightness.* Leakage is tested as specified in the contract documents. Testing options include an exfiltration test, low-pressure air test, and a visual inspection for infiltration.

To ensure a quality product installation, quality control and quality assurance requirements should be specified in the contract documents. The contractor installs the products in accordance with the contract documents and the submitted procedures, and a trained and knowledgeable inspector then observes, inspects, measures, tests, and documents before and during liner installation and then assures final product quality once the product is installed.

1.10 Chapter Summary

With the continued aging of existing pipelines and underground infrastructure, the need for renewal of these pipelines is also increasing. The needs for improving quality of life, reduction of inconvenience to the society, and less impact on environment make trenchless renewal methods (TRMs) a sustainable choice for pipeline owners and public agencies. The main benefits of TRMs are not only being cost-effective, but also these methods provide environmentally friendly and sustainable construction operations. A proper selection of a renewal method reduces the possibility of pipeline failures and massive costs of emergency repairs. Understanding the causes of pipeline failures helps in the selection of an appropriate renewal method and extended service life of the pipe. Once renewal methods have been determined and executed, its assessment and effectiveness need to be performed. This will determine the suitability of that particular method for future applications.

This chapter presented an overview of the pipeline deterioration mechanism and a summary of different TRMs with factors to consider when selecting a specific renewal method. These methods have become method of choice to traditional open-cut pipeline installations and replacements. Trenchless renewal methods stop leaks, resist corrosion and abrasion, install a new pipe in place of the existing and deteriorated pipe, and provide a new design life. An overview of current and emerging design concepts also was presented.

1.11 Review Questions

1. Name five types of pipeline defects.
2. Define pipeline deterioration process.
3. Describe partially deteriorated pipe condition.
4. Describe fully deteriorated pipeline condition.
5. Compare renewal and repair methods.
6. Name and briefly describe capabilities and limitation of seven trenchless pipeline renewal methods.
7. In what conditions trenchless replacement methods are recommended?
8. Name different types of trenchless replacement methods.
9. Describe emerging concepts in pipeline renewal.
10. Describe long-term testing.
11. Define contract documents.
12. Define technical specifications.
13. Define design.
14. Define inspection.
15. Define QA/QC.
16. Name inspections and quality checks before and during renewal installation.
17. Name inspections and quality checks after renewal installation.

CHAPTER 2

Pipeline Asset Management

2.1 Introduction

Asset management is a very broad term that can encompass several meanings. This chapter identifies recognized definitions and methodologies of asset management in the pipeline industry.

The International Infrastructure Management Manual (IIMM, 2006) defines asset management as "the combination of management, financial, economic, engineering, and other practices applied to physical assets with the objective of providing the required level of service in the most cost-efficient manner."

Asset management is a systematic process of operating, maintaining, and upgrading physical assets cost-effectively. It combines engineering and mathematical analyses with sound business practice and economic theory. Asset management systems are goal-driven and, like the traditional planning process, include components for data collection, strategy evaluation, program selection, and feedback. The asset management model explicitly addresses integration of decisions made across all program areas. Its purpose is simple—to maximize benefits of a pipeline system to its customers and users, based on well-defined goals within available resources.

Asset management can be defined as "simply bringing relevant data and analytic tools together with systematic implementation processes to ensure that the defined goals of the system are attained as efficiently as possible." The American Association of State Highway and Transportation Officials (AASHTO) and the Federal Highway Administration (FHWA) define asset management as "a systematic process of maintaining, upgrading, and operating physical assets cost-effectively. It combines engineering principles with sound business practices and economic theory, and it provides tools to facilitate a more organized, logical approach to decision-making. Thus, asset management provides a framework for handling both short- and long-range planning."

Similar to above, the Transportation Research Board (TRB) defines asset management as "a systematic process for maintaining, upgrading and operating the physical assets of a transportation system. Asset management employs engineering principles, economic theory, sound business practices, and information systems to determine short and long term resource allocations."

While there is no universally accepted definition of asset management, all the definitions of asset management have the same key points: to meet a level of service that is satisfactory to the customers and to provide that service in the most cost-effective manner.

2.2 Benefits of Asset Management

Having an asset management (AM) system will allocate funding to the most crucial parts of the pipeline infrastructure and maintain service level in a cost-effective manner.

- It reduces catastrophic failures, causing disruption of service, possibility of injury and death, and damage to environment, properties, and quality of life of citizens.
- It reduces cost of emergency repairs which might be 10 times more than planned repairs and renewal work.
- It defines an orderly and logical program of repair and replacement of capital facilities and equipment.
- It develops a comprehensive inventory of capital assets.
- It creates a valuable financial planning tool intended to smooth financial programming and maintain realistic replacement spending.
- It creates logical justifications to help elected officials understand utility's need to maintain reserves.
- It provides necessary funds for operating personnel to maintain equipment and perform major cyclic repairs without having to compete for scarce public resources.
- It provides lower overall rates for the public.
- It reduces rate shock.
- It creates better equipment managing and results in longer equipment service.
- It creates a better matching of resources and financial needs.

All of these benefits allow the pipeline owner to provide a proactive rather than a reactive response to issues affecting pipeline assets.

Additionally, according to the Water Infrastructure Network,[1] investing in water and wastewater systems will pay substantial dividends to the public health, environment, and the economy. By preventing pollutants from reaching the rivers, lakes, and coastlines, water treatment plants prevent waterborne disease, make America's waters safe for swimming and fishing, and preserve natural treasures such as the US Great Lakes. Having an AM system in place will assure that scarce funding is allocated to the assets that need it the most and will maintain a standard of service that is acceptable to customers.

2.3 Major Questions in an Asset Management Program

Table 2.1 presents a list of questions to start an asset management program. These questions can be used to gather and organize the data that will be needed to create an asset management strategy.

2.4 Asset Management Framework

All pipeline owners practice some form of AM. However, a formal AM system indicates a dedication to a cost-effective, systematic, and sustainable infrastructure. The following are key elements of an asset management program:

1. Taking a lifecycle approach;
2. Developing cost-effective management strategies for long term;
3. Providing a defined level of service and monitoring performance;
4. Managing risks associated with asset failures;
5. Sustainable use of physical resources; and
6. Continuous improvement in asset management practices.

Figure 2.1 presents a five-step approach for preparing an asset management plan (AMP). The following sections further explain these steps.

2.4.1 Step One: Identify Objectives

The first step determines who the target audience of the plan is, what that audience already knows about the organization, what the desired information from the plan is, and how the information is intended to

[1]Available at http://www.win-water.org/reports/

Question	Description
What do you own?	Asset identification and geographic location. Typically, this question involves a complete inventory of all assets, cataloged by type and providing a level of detail about the component to be maintained. It must also include the physical attributes of the asset (size, depth, materials, and so on).
What is it worth?	Asset financial data, including replacement cost.
What is its condition?	Physical description of pipeline data, including all subsystems (such as manholes and access points, pumps, vaults, joints, valves, fittings, services, and so on). Operational performance data and condition monitoring, including maintenance backlog. For each component, historical performance data, maintenance history, failure incidents, materials, installation, cost, age, material type manufacturer, and other information would also be included.
What is the remaining service life?	Estimations of useful physical and economic life of pipeline asset.
What is the maintenance strategy?	Operational procedures, preventive/predictive/condition-based maintenance schedules, criticality of the asset to the system performance in order to meet business objectives, target and/or designed level of service, and so on.
What are other current practices?	Relevant decision support methods in use for repair-versus-replace decisions for assets, for selection of asset quality, for maintenance program for maintenance strategy, for data management, and so on.
What is the replacement strategy?	Consideration of the estimated replacement or renewal cost that is ahead of useful physical or economic life of the pipeline asset, ahead of decline in unacceptable level of service, and before unacceptable maintenance/repair costs occur.
What levels of service do you want to provide?	Minimum performance and/or service standards, such as pressure and flow, response time, system reliability, monitoring strategy and controls, and so on.
What are the existing and future performance demands?	Estimations of projected population growth, business expansion, consumer usage trends, and so on.

Table 2.1 List of Major Questions to Start an Asset Management Program

Figure 2.1 Five-step approach for preparing asset management plans.

be used. This step also identifies the level of detail and extent of information to which the target audience will have access. This step outlines boundaries the asset management program will have.

2.4.2 Step Two: Outline Asset Management Plan Structure and Content

The outline of an AMP structure and content can be very detailed or very general, dependent on the set objectives of the AMP. This step requires the asset management team to define how assets will be grouped for reporting purposes. For example, factors influencing the decision on how to categorize assets include (1) the number and the value of assets if the assets are managed in similar ways, (2) the scale of maintenance and operational costs, where the assets are in their lifecycle, (3) the depth of asset information available, and (4) the structure of management and service delivery contracts.

2.4.3 Step Three: Write the Pipeline Asset Management Plan

There is no ideal structure to an AMP; however, IIMM (2006) outlines the following nine steps that can be included in an AMP or to serve as a guideline for developing one:

1. Executive summary;
2. Introduction;
3. Levels of service;
4. Future demand;
5. Lifecycle management plan;
6. Financial summary;
7. Asset management practices;
8. Plan for improvement and monitoring; and
9. References and appendices.

The above steps are discussed in more detail below.

2.4.3.1 The Executive Summary

Executive summary should be a separate document that gives an overview of the entire AMP and includes general information emphasizing the key issues of asset management.

2.4.3.2 Introduction

The introduction to the AMP should provide the reasons for preparing the pipeline AMP and a justification for owning and operating the assets covered in the AMP. It includes the following:

- *Background:* It covers the purpose of the plan, relationship with other planning documents, infrastructure assets included in the plan, key stakeholders in the plan, and the organization structure.
- *Goals and Objectives of Asset Ownership:* It includes the reasons and justification for pipeline asset ownership and links to organization vision, mission, goals, and objectives.
- *Plan Framework:* It includes the key elements of the plan.
- *Basic and Advanced Asset Management:* It outlines a basic to advanced approach of AM, and defines the sophistication and limitations of the particular AMP.

Having a basic understanding of the AMP and the reasons why it is being implemented is the purpose of the introduction.

2.4.3.3 Level of Service

The level of service section should clearly define the categories that are proposed for the AM plan and should include the following information:

- Customer needs and expectations;
- Strategic and corporate goals;
- Legislative requirements;
- Current level of service; and
- Desired level of service.

Details within each of these categories should be provided so that the desired outcome of the AMP is specific and quantifiable.

2.4.3.4 Future Demand

The future demands section involves predicting the growth needs of the pipeline assets. This will be beneficial for planning purposes and asset allocation. This section addresses those needs with the following categories:

- *Demand Forecast:* Factors influencing demand should be mentioned in this section, along with a detailed projection of growth or decline of demands on services, anticipated changes in customer expectations, and the impact of changes in demand on asset utilization.

- *Changes in Technology:* Use of new technology and its effects on providing future services should be outlined in this section, along with the obsolescence of current practices.
- *Demand Management Plan:* This section should describe non-asset solutions available as alternatives to asset-based solutions (i.e., demand management, insurance, and managed failures) and also summarize new works programs and costs.

Understanding the future demand of an asset system will allow for good planning and resource allocation.

2.4.3.5 Lifecycle Management

This section outlines exactly what is planned to manage and operate the assets at the agreed level of service while optimizing lifecycle costs. The five subsections to the lifecycle management plan are as follows:

- Background data;
- Routine maintenance plan;
- Renewal/replacement plan;
- Creation/acquisition/augmentation plan; and
- Disposal plan.

2.4.3.5.1 Background Data The background data should include the physical parameters of the assets, such as age, size, mix, material, location, and current issues. Also, this section should include a summary of total asset parameters in table or graph formats, an overall plan of asset system or network, and how to obtain part-by-part asset information. The capacity and performance of the asset, condition of the asset, valuations of the asset, and historical data should also be included in the background data of the lifecycle management plan.

2.4.3.5.2 Routine Maintenance Routine maintenance includes the regular day-to-day work that is necessary to keep the pipeline assets operating, including emergency repair for portions of assets. The maintenance plan should include the current trends and issues of the assets (i.e., spending and complaints), current and past levels of service, and the maintenance of decision-making process, both planned and unplanned. Also included in this section should be standards and specifications defining the materials, methods, and service standards that have been set in order to meet the level of service, and the risks associated with alternative standards. A summary of future costs should also be included, such as forecasting planned and unplanned maintenance work and costs, noting deferred maintenance as an associated risk, and outlining how maintenance will be funded.

2.4.3.5.3 Renewal/Replacement Renewal and replacement as defined in this section consists of major work that restores, rehabilitates, replaces, or renews an existing pipeline asset to its original capacity. It does not increase the asset's design capacity. Work that increases design capacity is considered new works expenditure and does not apply in this section. This plan should follow the similar format of the maintenance plan and include a renewal plan that shows how replacements/renewals are identified and to what standards they are placed, the end of life projections, and the renewal decision-making process. The renewal standards need to be set defining the materials, methods, and service standards to meet required levels of service and the risks associated with alternative standards. This section also includes a summary of future costs, including a forecast program of replacements and costs, cash flow forecast of costs, deferred renewals, risk analysis, and identification of how replacements will be funded.

2.4.3.5.4 New Construction The new construction or creation/acquisition/augmentation plan is for new projects, which include any upgrade or improvement, which exceeds the previous asset's capacity. Upgrades and improvements are a result of growth and/or social and environmental needs. This process should begin with selection criteria that outline a formal procedure for ranking pipeline asset creation/acquisition projects, have standards and specifications that define the materials, methods, design standards, and risks associated with alternatives, and have a summary of future costs.

2.4.3.5.5 Disposal Plan It is the final step in an asset's lifecycle and includes any activity related to disposing of a decommissioned pipeline asset, including the sale, demolition, or relocation of the asset. For the disposal plan, future disposal needs should be forecasted and include the time and cost of the disposal, and a cash flow forecast of income/expenditure from asset disposal should also be formulated. Figure 2.2 illustrates the components of the lifecycle management plan that is described earlier.

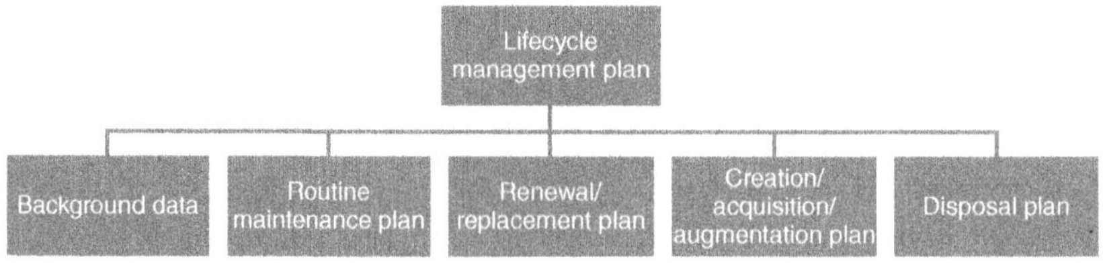

FIGURE 2.2 Lifecycle management plan components.

2.4.3.6 Financial Summary

The financial summary should contain the financial requirements from all the information previously presented. The financial summary should include the following:

- *Financial Statements and Projections:* It should be prepared for a minimum of 10 years and include cash flow forecasts by year, breakdown of expenditure by service groups, breakdown of expenditure into routine maintenance, renewal and new works expenditures, and trends from the previous two to three years.
- *Funding Strategy:* It should provide details of how expenditure will be funded and determine whether any planning is needed to smooth out variations in cash flow.
- *Valuation Forecasts:* It provides forecast of future value of pipeline asset and valuation methodology and forecast of depreciation.
- *Key Assumptions Made in Financial Forecasts:* It is essential for the reader to understand the accuracy of the financial forecasts and how they can be improved.

The financial summary provides detailed information on the current and future value of pipeline assets and outlines how future expenditures will be funded.

2.4.3.7 Asset Management Practices

The asset management practices section will outline the information available on the pipeline assets, the information systems used, and the processes used to make decisions on how these asset will be managed. The accounting and financial systems portion must detail the accounting system and any changes that have been made as a result of the AMP, define the differences between maintenance, renewal, and new works expenditure, and provide details of accounting standards and guidelines that must be complied with. The asset management systems portion must detail the types of data available for AM decision making, state the quality, reliability, and adequacy of the data, define the software (if any) used to store and analyze data, and state where the information is stored and how often data is collected.

Information flow requirements and processes must also be defined, stating the key information flows to and from the AMP, the processes used to make decision on AM for replacements, renewals, and acquisitions. Additionally, these should state the formal project ranking system (if any), define the process to make the best decision, and state if the process takes into account risk cost, lifecycle costs, performance prediction, and optimized decision making. Also, key standards and guidelines influencing AM attributes should be stated in this section.

2.4.3.8 Plan for Improvement Monitoring

A key portion of an AMP is to have a means for monitoring and improving asset. This section details the performance measures, improvement program, and the monitoring and review procedures that will be used to continuously maintain and improve the AMP.

2.4.3.9 Reference and Appendices

The AMP will undoubtedly be created using outside documents and resources. This section of the AMP should include any references used along with any necessary appendices, such as asset data, capital expenditure programs, and so on, which would be important references for others who are being introduced to the AMP.

Figure 2.3 shows the nine sections discussed earlier that outline a written AMP.

These nine sections outlined the requirements of an AMP. The AMP is probably the most time- and resource-consuming step in the asset management process, but it entails the majority of the "work"

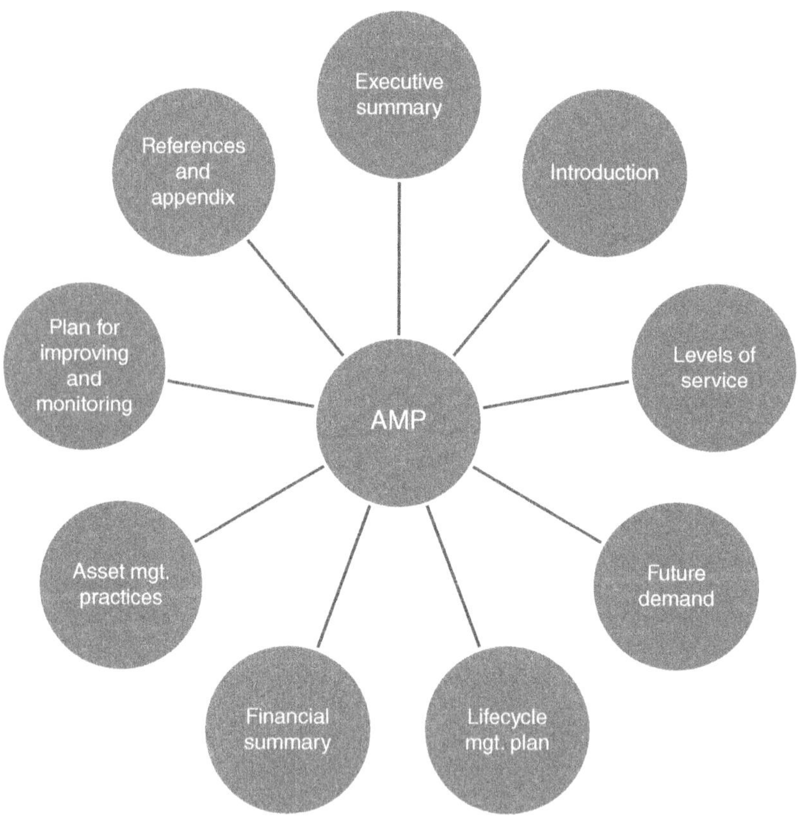

FIGURE 2.3 Components of an AMP.

that needs to be put forth to implement an asset management system. The AMP should be a working document within the asset management process and is not expected to be complete at the beginning. The following two steps are the last in the AMP process.

2.4.4 Step Four: Review of the Plan

Once the plan has been compiled, a person with expertise in asset management should look over the plan and identify its strengths and weaknesses as well as its ability to meet any disclosure or other needed criteria.

2.4.5 Step Five: Update the Plan

The AMP should be a working document that is constantly updated as customer expectations change, AM systems improve, assets are added to the network, and if any other changes occur that would affect the AMP.

2.5 Government Accounting Standards Board Statement 34

The previous sections identified strategies for implementing an AMP. This section addresses the requirements and importance of the Government Accounting Standards Board Statement 34 (GASB-34) as it relates to asset management.

The GASB is an independent agency responsible for developing standards for state and local governmental accounting and financial reporting that result in useful information for users of financial reports and guide and educate the public.

The GASB-34 was issued in June 1999. While there are many reporting standards required within the GASB-34, this chapter presents the specific aspect which defines the requirements for reporting major capitol infrastructure assets in government financial statements. According to the GASB-34, pipeline assets must be reported in the agency's financial statements. The GASB-34 does not have the power of law; however, the GASB defines the generally accepted accounting principles (GAAP). Financial statements must comply with GAAP to receive an unqualified opinion from auditors, which is extremely important to agencies, especially if bonds have been or are intended to be issued. Before the GASB-34, most organizations used cash accounting to account for their assets, in which physical assets appear on the books only during the year in which they are constructed. The GASB-34 requires that major capital assets to be accounted by using accrual accounting, meaning that their costs must be charged or depreciated over the life of the asset. The GASB-34 allows the accrual accounting to be calculated one of two ways, either by straight-line depreciation or by the modified approach depreciation.

2.5.1 Depreciation

An asset is a resource with economic value that can be converted to cash. When the asset is new, it has a dollar value equal to purchase price or cost of building the asset. Overtime, depreciation will apply to dollar value of asset as a non-cash expense to the owner. Depreciation can be considered in different terms, such as decline in value of an asset due to deterioration and aging, decline in value of an asset to the owner because of performance and/or maintenance costs, or decline in market value.

Depreciation can be determined by systematic allocation of the cost of a pipeline asset divided by its depreciable life. It can be calculated by different methods for different objectives, such as taxes, financial reports, and cost accounting. To calculate depreciation, the present worth or cost of asset (P), its useful life (usually number of years, N), and its salvage value at disposal (S) must be determined. One of the most important benefits of depreciation calculation for pipeline owners and agencies is valuation and determination of remaining life of their asset at any given time. By calculating depreciation, the pipeline owners and government agencies/utilities become aware of cost of their asset and its remaining life, thereby develop an AMP including a financial plan for asset disposal and replacement.

2.5.1.1 Straight-line Depreciation

The simplest and most common method of depreciation in the GASB-34 is the straight-line depreciation, where

$$\text{Depreciation rate, } Rn = \frac{1}{N}$$

where N is number of years (useful life of asset)

Annual depreciation amount at the end of new year,

$$D_n = \text{Value at the end of the previous year} \times \frac{1}{N}$$

This method depreciates the asset over its useful life and is reported in the statement of activities portion of the financial report. In determining the asset's service life, the pipeline owner or agency should consider its present condition and how long the asset is expected to meet service demands, along with the general guidelines from professional or industrial organizations, information on comparable assets, internal information, and manufacturer's recommendations.

2.5.1.2 Modified Approach

The second method for the GASB-34 is the "modified approach," which incorporates the benefits, or values, of maintenance activities into the asset value reporting process. This method requires that the agency to have an asset management system that maintains an up-to-date inventory of all eligible assets, performs condition assessments, summarizes the results using a measurement scale, and estimates each year the annual amount required to maintain and preserve the eligible assets at the condition level to be self-set by the pipeline owner or the governing agency. Additionally, the owner or the agency must document that the eligible pipeline system is being preserved at or above the condition level which was self-established. This method allows a more realistic asset value to be recorded, because pipeline assets may not depreciate as quickly as calculated in the straight-line method, since they are maintained and preserved to meet or exceed service standards. This is due to the fact that as parts of the asset may be renewed, it can be considered for the asset to maintain its value.

The choice of method used is at the discretion of the pipeline owner. Because some local governments and utilities may not have an ongoing comprehensive asset management system in place, they may choose the straight-line depreciation method since it is easier to implement. However, the GASB-34 is flexible and allows an agency to switch from one reporting method to another. Since the modified approach provides more detailed and accurate information about the condition and funding requirements for the assets, it is more beneficial if pipeline owners choose this method. The modified approach implies that the agency has a working asset management system in place. Table 2.2 presents the pros and cons of both reporting methods.

2.5.2 GASB-34-Modified Approach Asset Management Requirements

As said in the previous section, the GASB-34-modified approach requires the agency and the pipeline owner to manage infrastructure assets using an asset management system that has the following characteristics:

1. Provide an up-to-date inventory of eligible pipeline assets.
2. Perform condition assessments of the eligible pipeline assets and summarize the results using a measurement scale.
3. Estimate each year the annual amount to maintain and preserve the eligible pipeline assets at the condition level established and disclosed by the agency or pipeline owner.

The agency and pipeline owner must also document that the eligible pipeline assets are being preserved in accordance with the following requirements:

1. Complete condition assessments of eligible infrastructure assets are performed in a consistent manner at least every three years.

Straight-line Depreciation		Modified Approach (Asset Preservation)	
Pros	**Cons**	**Pros**	**Cons**
Easier to implement, requires less data and deliberation	Does not address actual asset condition	Provides information on asset condition (pro if condition is good. Some managers/boards may consider this as con if condition is bad.)	Initial investment setting up the process is more costly, especially if no databases exist from other sources
Finance personnel and auditors are more familiar and comfortable with this approach	Gives perception that assets are being allowed to deteriorate	Sets quantified targets for asset condition (pro if met. Some managers/boards may consider this as con if targets are not met.)	More of an intellectual challenge
Relatively simple financial reporting	Does not account for investments in maintenance and rehabilitation	Provides documented explicit costs of preserving assets	If condition level falls to below targets, must revert to depreciation method
Lower risk of negative audit report	Does not represent real value of the asset. Represents unreasonable low cost of service in older systems	Real value of asset is documented	If condition levels fall below targets, it may be seen as failure of management
	Depreciation factors bear little relationship to actual changes in asset condition	Quantified performance measures allow trends to be checked and projections to be made	Provides information on asset condition (pro if condition is good. Some managers/boards may consider this as con if condition is poor.)
	Does not address “overhang” liabilities arising from deferred maintenance and rehabilitation	Variable future funding needs can be easily translated into rates	Sets quantified targets for asset condition (pro if met. Some managers/boards may see this as a con if targets are not met.)

Choosing a Reporting Method			
Straight-line Depreciation		Modified Approach (Asset Preservation)	
Pros	Cons	Pros	Cons
	Meaningless to customers and public. Viewed as unreliable or incomplete way to report costs	Involves policy makers and all utility managers, not just finance personnel	Failure to maintain stated condition level could result in the need to secure funding for reinstating the level, political issues, or where infrastructure failure occurs, and liability situations
	Not a measure of management effectiveness	Provides meaningful information for customers and public	
		Satisfies asset management expectations of CMOM[1]	
		Probably, it will be required under any new federal financing program	
		Eventually, it will be expected by lending institutions and bonding monitors	

[1]http://www3.epa.gov/npdes/pubs/cmom_guide_for_collection_systems.pdf

TABLE 2.2 Choosing a Reporting Method

2. The results of the three most recent condition assessments provide reasonable assurance that the eligible pipeline assets are being preserved approximately at or above the condition level established and disclosed by the agency or pipeline owner.

Table 2.3 compares the requirements for meeting the depreciation method of reporting versus the modified approach of reporting under the GASB-34.

If eligible pipeline assets meet the requirements of GASB-34 paragraphs 23 and 24 and are not depreciated, all expenditures made for those assets (except for additions and improvements during renewal work) should be expensed in the period incurred. Additions and improvements to eligible pipeline assets should be capitalized as they increase the capacity, efficiency, or extend the life of pipeline assets rather than preserve the useful life of the assets.

In selecting methodologies for pipeline material and management and maintenance strategies, distribution systems and transmission mains are evaluated as separate subsets due to the extent of the consequences when one fails over the other. The failure of a distribution system is often less eventful than the failure of a transmission main due to the size of failure, lack of service to numbers of people, and time to repair. Separating the piping into two subsets is due to the cost of corrosion protection, evaluation of risk, and cost and consequence of failure (CoF). Failure due to corrosion can be controlled and predicted to some degree of certainty, and this ability to document the method of controlling corrosion on pipe with metallic components is critical to the use of the modified approach.

Due to the smaller size of distribution systems, a qualified corrosion professional can predict the frequency of failures in a system to some degree of accuracy, and with this information, a maintenance repair and replacement schedule can be implemented that will keep the subset at a condition established as acceptable. This would be considered as maintenance costs for GASB-34, and meets the requirements of the modified approach.

Transmission mains are larger and not a part of a periodic replacement program, and because their failure would yield much larger consequences than a distribution system, a corrosion protection system would be a beneficial investment. The lifecycle cost-benefit of a corrosion protection system ranges from 7 to 42 times benefit-to-cost ratio, making it a good investment for large diameter pipes. The corrosion protection system can be installed either during initial construction or at the first sign of unacceptable corrosion, and a maintenance schedule can be implemented to maintain the pipeline at an acceptable conditions. This maintenance and periodic replacement of the protection system can be identified, scheduled, and quantified, making the GASB-34-modified approach a feasible reporting method when this strategy is used.

<table>
<tr><th>Straight-line Depreciation</th><th>Modified Approach</th></tr>
<tr><td>Summary
Infrastructure assets will be reported in the statement of net assets based on historical costs and will be depreciated over their useful lives</td><td>Summary
a. Infrastructure assets will be reported in the statement of net assets based on historical costs
b. Report all preservation costs as expense and no report of depreciation
c. Demonstrate that assets are being preserved and an asset management system is in place
d. Require condition information and required versus actual maintenance/preservation</td></tr>
<tr><td>Requirements
Historical cost includes
a. Estimated historical cost
b. Replacement or current cost, which is then deflated back to the acquisition date, is an acceptable method for estimating original cost</td><td>Requirements
An asset management system must
a. have an up-to-date inventory of eligible infrastructure assets
b. perform condition assessments of the eligible infrastructure assets and summarize the results using a measurement scale
c. each year estimate the annual amount to maintain and preserve the eligible infrastructure assets at the condition level established and disclosed by the organization</td></tr>
<tr><td>Procedures
Depreciation expense should be reported in the statement of activities. It should be measured by allocating the net cost of depreciable assets (historical cost less estimated salvage value) over their estimated useful lives in a systematic and rational manner. It may be calculated for
a. a class of assets;
b. a network of assets;
c. a subsystem of a network; and
d. individual assets.</td><td>Procedures
The organization must document that assets are being preserved through normal expenditures at an acceptable level established by that government entity.
a. Complete condition assessments of eligible infrastructure assets are performed in a consistent manner at least every three years.
b. The results of the three most recent condition assessments provide reasonable assurance that the eligible infrastructure assets are being preserved approximately at or above the condition level established and disclosed by the organization.</td></tr>
</table>

TABLE 2.3 Comparing Reporting Requirements for Straight-line Depreciation and the Modified Approach

2.6 Condition Assessment Tools and Methods

One of the key aspects of asset management is to know what conditions the pipeline assets are in, and to know when they will need to be renewed or replaced. Condition assessment techniques for all types of pipe materials and applications have been developed. Condition assessment is a three-step process.

1. Plan and prioritize through risk assessment—identify areas of low, medium, and high risks.
2. Conduct field inspections.
3. Provide engineering analysis.

In consideration of risk, criticality (consequence of segment failure) and vulnerability (probability of occurring) must be considered. Criticality can be such parameters as repair costs, loss of service, health and safety impacts, and environmental impacts. Figure 2.4*a* and *b* presents examples of low and high criticalities. Examples of vulnerability include sanitary sewer overflows (SSOs), inflow and infiltration (I/I), history of failure, age, pipe material, soil, bedding and embedment conditions, traffic loads, and soil corrosivity. Pipe materials fail differently and require customized assessment techniques. As described in Chap. 5, pipe materials include prestressed concrete cylinder pipe, reinforced concrete pipe, steel pipe, ductile iron pipe, cast iron pipe, polyvinyl chloride, high-density polyethylene (HDPE), and asbestos cement pipe.

2.7 Risk Management in Piping Systems

Federal and state regulations, such as GASB-34, have increased the need for utilities to competently and expediently maintain assets at an acceptable level of service (LOS). To accomplish this, utilities and pipeline owners need facts (data) regarding their pipeline assets. These data are included in maps, databases, and reports, which must relate to each other and form a foundation for planning. As a standard tool in this process, methodologies, such as National Association of Sewer Service Companies (NASSCO) Pipeline Assessment and Certification Program (PACP®) survey data and condition ratings, enable utilities to more efficiently prioritize, budget, and schedule various operational, maintenance, replacement, and renewal tasks.

National Association of Sewer Service Companies Pipeline Assessment and Certification Program (NASSCO PACP®) was established in 2002 to provide standardization and consistency to the way the buried wastewater asset industry evaluates underground infrastructure. At that time, there was no standardized protocol in the United States for the collection and management of data collected from internal pipeline inspection. Collection system owners either created their own individual systems or simply allowed each operator to collect data using no standard at all.

This lack of standards severely limited the value of observations collected by tractor-mounted closed circuit television cameras which

FIGURE 2.4 Examples of low criticality (a) and high criticality (b). (*Source*: Pure Technologies.)

have been in use since the early 1960s. Inconsistent observations made it impossible to compare the condition of one pipe to another even within the same network. Standardization not only made it more practical to compare the condition of multiple segments within a pipe network but also allowed the industry to benchmark these conditions to provide a better understanding of the deterioration mechanisms affecting underground infrastructure. The PACP® was developed by NASSCO to be a flexible, customizable program to provide its members with a standardized method to code defects, assess asset condition, and plan for the rehabilitation or replacement of sewers. It has now been adapted for use with manholes and laterals and other piping systems, such as storm water systems, and dam and levee piping systems.

As part of the PACP®, the NASSCO has developed a rating system where every observation is assigned a 1 to 5 numerical rating for which the larger rating indicates a more significant or defective observation. These ratings are separately assigned to observations that are structural (cracks, breaks, etc.) and observations that are related to operations and maintenance (deposits, rots, I/I, etc.). These ratings are also categorized by type, such as sanitary pipes, manholes, storm systems, and dams/levee pipes.

This process drives a risk-based methodology that is compatible with the size and resources of individual utilities. It demonstrates how the PACP data, along with general knowledge of the utility, can be used to develop considerable portions of an AMP. The following sections demonstrate how a rating system, such as the one developed for the NASSCO PACP® data, can be used to assess the risk of pipe failure. While these methods describe risk management as related to wastewater pipelines, they can be modified for use for any category of asset.

As stated earlier, risk relates to the probability of a pipe failure happening, or the likelihood of a negative impact occurring, and how this can be avoided. Failure can be defined as a pipeline system not accomplishing its agreed LOS. Sometimes risk is used to describe the severity of consequence(s) related to a potential failure. Risk assessment relates the combination of both likelihood of failure (LoF) and consequence of failure (CoF). Risk takes into account the asset's physical conditions as well as the impact that its failure would have on system performance and stakeholders. Mathematically, the risk from an occurrence may be expressed as the product of likelihood and CoF as quantified in the following equation:

$$\text{Risk} = \text{LoF} \times \text{CoF}$$

2.7.1 Likelihood of Failure

Likelihood of failure (LoF) refers to a calculated numerical representation (rating) that denotes the probability of failure based upon an asset's physical condition. There are several ways to establish a rating for a buried pipe asset. Each has deficiencies. Condition assessment methodologies typically result in 1 to 5 rating for each observation that rating increases as the significance of the defective observation increases. This increased significance is proportional to pipeline's LoF.

A basic method of rating a pipe asset is to sum these individual observation ratings. However, this can allow for a rating to be overly influenced by a large quantity of minimally rated observations such as fine holes or small cracks. A better method may be to index this overall total rating by dividing it by the total number of particular observations that contributed to the total. This analysis still can result in an indexed rating that is not representative of any of the contributing ratings if an asset were to have a few significant ratings and a large quantity of less significant ones.

Another way of rating a buried pipe is to summarize in a "quick rating" the quantities of observations associated with the most significant ratings. For example, considering a quick rating of 4735, this figure would indicate seven grade-4 defects and five grade-3 defects, with no grade-5 defects. There may be less significant grade 1 or 2 observations. If there are more than nine defects of a particular grade, then alpha characters are used.

A "weakest link in the chain argument" leads to the conclusion that the most effective method of using condition assessment to predict LoF is to establish, using a quick rating, a number based upon the quantity of the most significant defective observations related to each particular pipe asset. This accounts for the logic that a pipe with the most significant defective observations is more likely to fail than one with less significant observations and more so with additional observations.

For example, using the previously considered quick rating of 4735, the LoF rating would be 4.7. If the number of grade-4 defect observations exceeded nine, then the LoF rating would be 5.0. In this system, the maximum rating would be 6 and would mean that there were at least 10 grade-5 defective observations.

2.7.2 Consequence of Failure

The failure of an asset to meet its agreed LOS—in this case, buried pipes not conveying wastewater to the wastewater treatment plant—results in various consequences. CoF is the combination of direct and indirect impact on the vicinity and community due to a potential pipe failure.

In recent decades, this type of impact has typically been expressed in "Triple Bottom Line" (TBL) terms—also known as people, planet, and profit. This concept emerged in the early 1980s as a criterion for measuring economic, ecologic, and societal success. The TBL expands the traditional organization's financial bottom line to view other nontraditional social and environmental factors. Consideration of these factors encourages an organization to also focus on social and environmental aspects rather than solely upon the (more visible) economic aspects of management. This also helps an organization to consider its stakeholders.

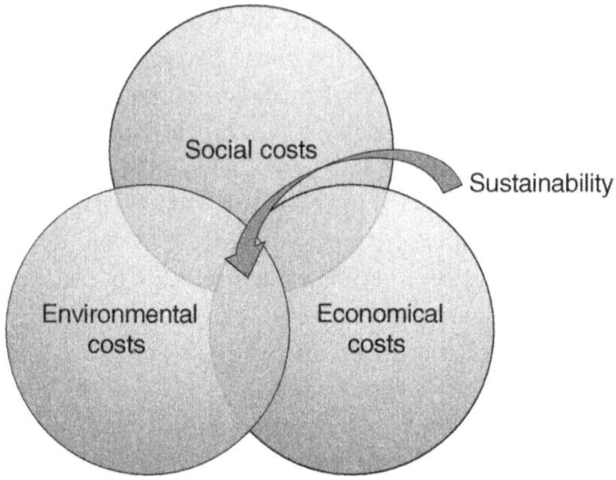

FIGURE 2.5 Consideration of sustainability in social, environmental, and economical costs.

As shown in Fig. 2.5, the TBL concept focuses not only upon direct economical costs but also on social costs and environmental costs. The goal is sustainability of all the pipeline assets in a balanced manner.

2.7.2.1 Economical Costs

The economical CoF encompasses the impacts of direct and indirect economic losses to the affected organization and third parties due to asset failure. Economic factors are typically expressed in dollars and include property damage, repair costs, and production losses among others. Regarding utilities, economical cost considerations may include those related to pipe material, diameter, depth, and length of the pipe segments; direct or indirect interaction with other infrastructure elements, such as roads and bridges, in the close distance for repair; effects of local topography on required access; and possible access restriction created by such barriers as walls and fences. Direct costs of failures include such costs as those for asset repairs, legal fees, and fines. Indirect consequences include environmental cleanup costs and loss of business revenue to the community. In addition, consideration must be given to other external costs that are not directly measureable but result from the failure, such as loss in property value, utility credibility, and increased insurance rates. Both direct and indirect costs are considered when measuring the economical CoFs.

2.7.2.2 Social Costs

The social consequences of failure represent the impact on society due to asset failure. These considerations may include the number and types of affected properties, such as hospitals, schools, parks, businesses, traffic disruptions, or critical services as defined by the owner; the duration of the failure; and the public image. In addition, there must be consideration for health and safety issues that may directly or indirectly create possibility for public exposure to health-threatening problems, injuries, or even fatalities. Public image and the utility's credibility are extremely important due to the magnitude of failure on the negative public exposure, criticism, and legal actions the pipeline owner may face due to the failure.

2.7.2.3 Environmental Costs

The environmental CoF considers the impact to ecological conditions occurring as a result of pipe failure. Examples are contamination of soil, groundwater, and surface water. Environmental cost considerations may include proximity to wetlands and waterways, proximity to Federal Emergency Management Agency flood zones, possible contamination of potable water sources, and the sensitivity of nearby soils. The main consequence is the SSO resulting from a failure. Also, fines from consent decrees and federal and state agencies may result in economic and social costs.

2.7.2.4 Rating Methodology

Consequence of failure is determined by considering the location and demographics of an asset. For example, a 56-in. combined trunk sewer 100 ft downstream of a combined sewer overflow that crosses a body of water obviously has a higher CoF than an 8-in. sanitary sewer at the upstream end of the system that only serves one resident. Likewise, a different 56-in. sewer would have a lower CoF than another 56-in. combined trunk sewer that is installed deeper, has a higher dry weather flow, and crosses under a high-traffic roadway.

To present this concept in a useful format, these considerations should be broken into separate scoring parameters using the same scoring system as LoF so that the two components equally contribute to risk. This can be done by several methods; however, the method selected must be a result of sufficient consultation with the utility and pipeline owner. This should include a consideration of how each of the three cost structures should be weighted and to what extend the various components of these costs are present and important to the community.

When assigning weighting factors, the asset management team should consider how much the parameter contributes to the economic, social, and environmental impacts in the event of a failure. Some parameters may impact just one of the three TBL categories, while some may have varying degrees of impacts. An example of this would be a sewer line that crosses a waterway. This clearly can impact the environmental aspect of the TBL, considering the likelihood for contamination of the stream. There may also be some social impacts, with respect to an interruption in recreational use of the waterway, and economic impacts that result from penalties and fines. All of this must be thoroughly discussed with the system owner.

The following procedures should be considered as a guide regarding discussions with a utility and pipeline owner and as a possible and reasonable scenario of a CoF rating procedure. The utility and pipeline owner may desire additional considerations or may not consider some of these issues pertinent.

The considerations that result in a CoF rating should be conducted prior to condition assessment and should apply to all subsequent condition assessments intended to be applied to asset management. Much, if not all, of these considerations may be determined through the use of a geographic information system mapping process.

2.7.2.5 Relative Network Position of Pipe

Consequence of Failure for a pipe due to its relative network position could be a simple function based on just the diameter of a pipe or the complex upstream trace function based on a number of pipes connected (discharging) to the pipe. In general, large diameter pipes are located at the downstream portion of the system, while small diameter pipes are located at the upstream portion. CoF will be higher when a blockage or failure occurs on a large diameter pipe compared to a small diameter pipe.

Consequence of failure may be assigned to a scale from 1 to 6, with 6 being the highest consequence and 1 being the least. We can establish six groups based on the diameter of pipes within the system. As an example, Table 2.4 may be a reasonable distribution. All of the examples that follow can be adapted to any particular assessment plan. Table 2.4 presents an example of CoF for pipe diameter.

This consideration should affect both economic and social costs. Larger diameter pipe generally costs more to repair, replace, or renew and impacts more customers if a failure were to occur.

The depth of pipe can also reflect an economic cost since deeper pipe is generally more expensive to repair, resulting in a distribution such as shown in Table 2.5.

Diameter, in.	CoF Factor
<8	1
≥8–<10	2
≥10–<15	3
≥15–<21	4
≥21–<30	5
≥30	6

Table 2.4 An Example of CoF for Pipe Diameter

Depth, ft	CoF Factor
<6	1
≥6–<10	2
≥10–<14	3
≥14–<18	4
≥18–<24	5
≥24	6

Table 2.5 An Example of CoF for Pipe Depth

Instead of diameter, relative network position of a pipe can also be used to assign a CoF factor. This approach generally requires more sophisticated/complex maps of the system. Relative network position can be calculated as sum of relative position of all pipes contributing to a downstream pipe. This is a social cost consideration since the larger number of segments should relate to more customers upstream. The six groups based on relative network position may be rated as shown in Table 2.6.

A similar approach can also be used to define the CoF factor based on the number of customers serviced by the pipe. Cumulative sum of services (laterals) connected to a pipe can be calculated by adding cumulative sum of services connected to all pipes discharging at upstream structure of a pipe plus number of services directly connected to the pipe.

2.7.2.6 Location of Pipe

Failure of a pipe, regardless of relative network position, may have a higher consequence in terms of disruption to the community and the cost of managing emergency repairs of a pipe. Failure of a pipe on a major road causes more disruption to traffic compared to failure of a pipe on a local road. Challenging accessibility can also have a significant impact on the cost of emergency repairs. Geographic information

Relative Network Position of Pipe	CoF Factor
≤10	1
11–30	2
31–70	3
71–120	4
121–150	5
>150	6

Table 2.6 An Example of CoF for Relative Network Position of Pipe

Location of Pipe (Class of Road)	CoF Factor
Unpaved	1
Minor local	2
Major local	3
Collector	4
Arterial/building/pool	5
Highway/waterway	6

Table 2.7 An Example of CoF for Pipe Location

system-based mapping, with layers, including street network, buildings, waterways, and other datasets, can be used to assign CoF based on, for example, the class of road the pipe is located. This is both an economic cost consideration in which pipe under more major roads generally costs more to repair and a social consideration due to the increased disruption of traffic. Table 2.7 presents a reasonable consideration.

2.7.2.7 Proximity to Environmentally Sensitive Features

Pipes are not always installed within road right-of-way. There are various reasons to install a pipe at a close distance to environmentally sensitive features like rivers, creeks, drinking water source area, and lakes. This is, of course, an environmental cost consideration. A higher CoF factor is set based on the distance between a pipe and an environmentally sensitive feature as shown in Table 2.8. This will, of course, be affected by the nature of the sensitive environment. Geographic information system algorithms/models can be set up to assign CoF based on calculated distance of a pipe from predefined set of environmentally sensitive features.

2.7.2.8 Service to Customer of Significant Importance

Every utility will have a set of customers who are very significant for the well-being of the community. These customers may include hospitals,

Distance Between Pipe and Waterway (ft)	CoF Factor
150 or more	1
100–150	2
75–100	3
50–75	4
25–50	5
<25	6

Table 2.8 An Example of CoF for Pipe Distance to Waterway

Distance Between Downstream Pipe to a Service Lateral for Customer With High Importance (ft)	CoF Factor
20,000 or more	1
15,000– 20,000	2
10,000–15,000	3
5,000–10,000	4
1,000–5,000	5
<1,000	6

Table 2.9 An Example of CoF for Pipe Distance to Downstream Pipe for High Importance Customers

schools, manufacturing facilities, emergency services, and so on as determined by the utility. Providing uninterrupted service to these facilities may be a priority for the utility. Geographic information system based upon a downstream trace from a service lateral of a customer can be performed to find pipes servicing this customer. Based on pipe's proximity to these customers, a higher social CoF can be assigned as shown in Table 2.9.

2.7.2.9 Accessibility for Maintenance and Inspection

Access to manholes and pipes is very important for inspection and repairs. Large construction equipment is sometimes required to repair the failure of a pipe. Response time for a service crew may be significantly higher, if access to the pipe is difficult. The failure of such a pipe may cause significant damage to the environment as well as private properties due to delays in response created by difficulties in accessing the failure point. A higher CoF should be assigned to these pipes. This will affect the economic costs directly due to the difficulty and perhaps social costs if the property needs to be disrupted to gain access as shown in Table 2.10. Spatial pipe and utility corridor layers can be used to assign CoF for accessibility of pipe.

Accessibility of Pipe	CoF Factor
On right-of-way—no traffic control	1
On right-of-way—with traffic control	2
On public lands with vehicle access	3
On public lands without vehicle access	4
On private lands without vehicle access	5
Behind built structures and no vehicle access	6

Table 2.10 An Example of CoF for Pipe Accessibility

The above examples should not be considered either complete or absolute. These types of factors can only be determined with considerable input from utilities, pipeline owners, and their consultants. However, these examples can serve as guidelines for various types of considerations.

2.7.2.10 Calculation of Overall CoF

Each of the above factors may have a different weight depending on the goals and priorities of a utility. In the following section, an overall CoF score is calculated as a weighted average of all individual CoFs.

2.7.3 Managing Asset Risk

The overall risk associated with a failure event is a function of the likelihood of a failure event and its consequence. Not all highly probable events need the same attention since they may not have equally high consequence (impact) to the community. Any mitigation and/or response plan shall be designed based on overall risk due to a failure event. Upon establishing standardized definitions and a scoring system for LoF and CoF, a risk matrix can be developed and plotted for review and analysis. Figure 2.6 is a schematic of a risk matrix and suggested LoF and CoF categories that are consistent with prior discussions. Increased LoF should result in more aggressive maintenance

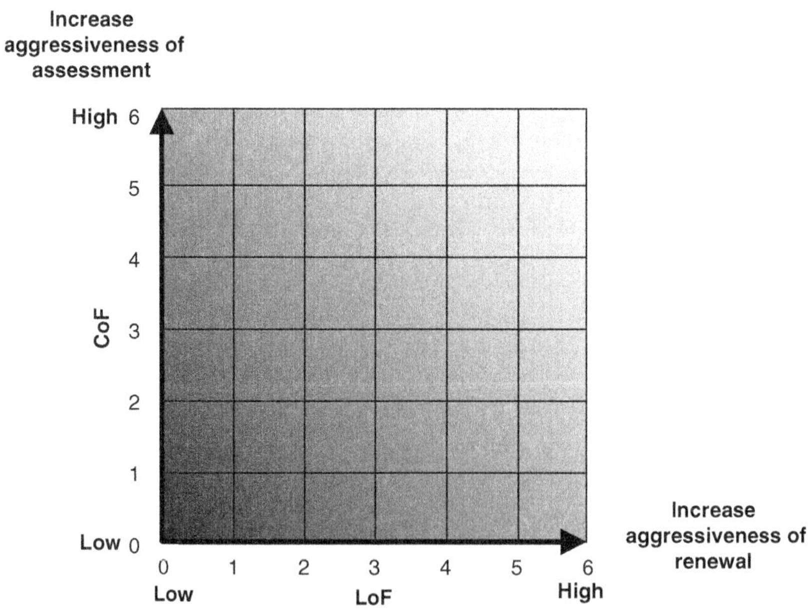

FIGURE 2.6 A risk matrix and suggested LoF and CoF categories.

and repair. Increased CoF should result in increased assessment. The resulting matrix can provide the basis for an economically efficient, balanced, and proactive assessment, maintenance, replacement, and renewal program.

Figure 2.6 illustrates risk management strategy for a pipe asset based on LoF and its CoF, each with a range of 0 to 6. As with most of these concepts, this should be considered a guide for a reasonable approach. There are many variations that can be made to address unique needs of an individual pipeline. The specific approach used for a pipeline must be developed with sufficient input from utility staff and must be implemented consistently throughout the pipeline owner's asset management system. Since asset failure can result in more expensive, socially difficult, and environmentally destructive consequences, increased CoF should lead to more aggressive assessment. Within the graph, those segments with a CoF in the darker green range may need to be cleaned and inspected less frequently. Those in the lighter green/red range should be inspected more frequently.

Pipe renewal priority and methodology are directly related to the condition of the pipe, which is the determining LoF factor. If a pipe exhibits more severe defects, then it generally should be renewed more aggressively. However, this renewal strategy must be based upon the overall strategy and goals of the pipeline owner and in concert with CoF principles. For example, those pipes with a low CoF (say in a residential cul-de-sac) may only be scheduled for renewal as total failure occurs. In contrast, more crucial pipe (large diameter in protected wet lands) may, for example, be scheduled for renewal when the LoF exceeds three. The AMP should clearly explain the details of the processes associated with procedure.

2.8 Asset Management of Oil and Gas-Related Pipelines

2.8.1 Introduction

The main focus of this section is to understand the current oil and gas-related pipeline asset management issues and practices. Although oil and gas pipelines are predominantly privately owned, the failures to properly manage these assets can affect the public in both cost and safety.

Proper asset management of oil and natural gas-related pipelines begins with determining if the pipeline is a regulated pipeline or non-regulated pipeline. The Pipeline and Hazardous Materials Safety Administration (PHMSA) is responsible for regulating and ensuring the safe and secure movement of hazardous materials to industry and consumers by all modes of transportation, including pipelines (PHMSA, 2013). The PHMSA has authored a set of regulation for the oil and gas pipeline industry, and they are known as the DOT 192 and 195 pipeline codes. DOT 195 is the code requirement for transportation of hazardous and/or combustible liquids by way of pipeline, and Department of Transportation 192 (DOT 192) is the code requirement for the transportation of combustible and/or hazardous gases by way of pipeline (PHMSA, 2013). The DOT codes are minimum requirements for proper asset management of oil and gas pipelines and also the standard for certain operational practices to be implemented for sufficient asset management operations.

Majority of oil and gas pipeline companies use an HDPE pipe as the primary material for conveyance of salt and freshwater. They primarily use the Plastic Pipe Institute's (PPI) manual on polyethylene pipe design and construction (PPI, 2008) for this application. The American Water Works Association (AWWA) also provides a polyethylene design and construction manual (AWWA M55) that is widely used. While the PPI and AWWA M55 guidelines for polyethylene pipe provide criteria for design and construction of fresh and saltwater pipelines, the minimum standards for operations and maintenance of the polyethylene pipe must be established by the pipeline company/operator. The oil and gas pipeline asset management can be broken down into the following categories:

1. Economics [Life cycle cost analysis (LCCA)].
2. Engineering.
3. Finance.
4. Management.
5. QA/QC for construction and procurement operations.
6. Operations and maintenance.

2.8.1.1 Economics

Economic analysis compares revenue and subsidies of a project with its financial cost. Economic analysis of public works projects compares societal values in terms of economic benefits and economic costs, and these maybe different from their corresponding financial values (Khankarli, 2013). This analysis compares project values in terms of benefits and costs. In case study practices/surveys of two different oil and gas pipeline companies, economics were a key factor in the planning phase of a pipeline project. Below are three case study examples of how economics were incorporated into the planning phase of oil and gas pipeline projects.

2.8.1.1.1 Case Study Internal corrosion is one of the main threats to a gas pipeline due to the compounds that are mixed in with the natural gas. Internal corrosion can attack the inner walls of a pipeline if proper asset management techniques are not implemented (see Figs. 2.7 and 2.8). In this case, the inner wall of a 24-in. natural gas pipeline was corroded because pigging was not used to prevent bacteria and water buildup in the pipeline.

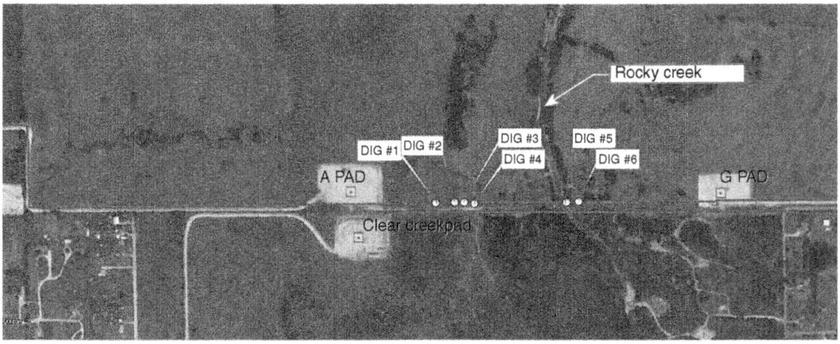

FIGURE 2.7 Corroded pipeline (24 in.).

An economic model was prepared for renewal of this pipeline (see Table 2.11). Since there was 62% wall loss, the pipe had to be renewed, but two options were explored to determine the most economical solution to renewal. The first option was to immediately renew the pipe, and the second option was to keep the pipe as is, increase maintenance requirements, and renew the pipe five years from now. To replace the pipe now, the capital cost was estimated at $1 million, which includes all engineering, labor, project management, and construction costs. If it was decided to wait five years, considering inflation, the complete cost to repair would increase to $1.25 million. The revenue stream is based upon a well that is estimated to provide

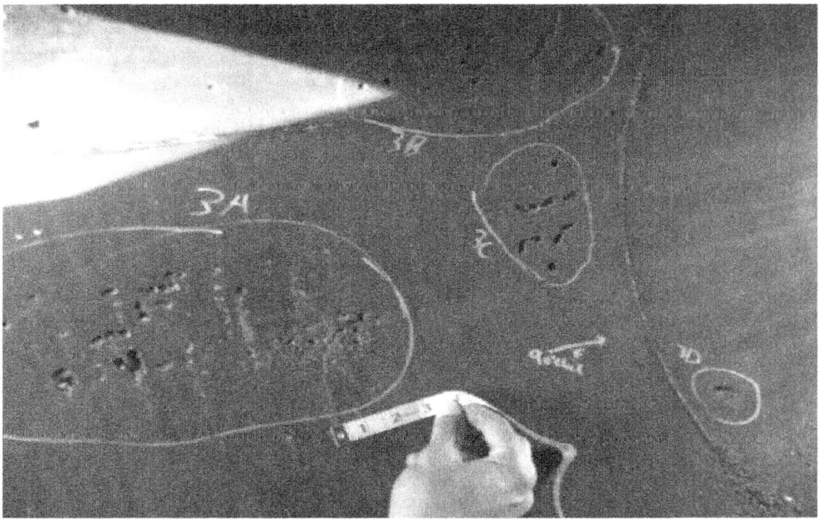

FIGURE 2.8 Interior corrosion of the 24-in. pipeline.

Economics—Case A	
Repair	
Revenue (MCF)*	$1.00
OPEX (MCF)**	$0.25
CAPEX****	$1,000,000
Flow rate (MCF/Day)***	3,000

	1	5	10	15	20	25	30
Total MCF	1,095,000	1,095,000	1,095,000	1,095,000	1,095,000	1,095,000	1,095,000
Revenue	$1,095,000	$1,095,000	$1,095,000	$1,095,000	$1,095,000	$1,095,000	$1,095,000
OPEX	$(273,750)	$(273,750)	$(273,750)	$(273,750)	$(273,750)	$(273,750)	$(273,750)
Capital	$(1,000,000)						
Net	$(178,750)	$821,250	$821,250	$821,250	$821,250	$821,250	$821,250
Payback (years)	1.22						
NPV (10%)	$6,832,762.60						
Rate of return	459%						

*Based upon contractual agreement of $1.00/MCF.
**Based upon need for chemical injection, pigging operations, and electrical needs for facilities.
***Based upon average production of wells over lifetime.
****CAPEX basis cost to replace pipe. Pipe should last the life of the well.
MCF: thousand cubic feet of natural gas
CAPEX: Capital expenditure
OPEX: Operating expense
NPV: Net present value

Table 2.11 Sample Pipe Renewal Economics

Keep As Is	
Revenue (MCF)*	$1.00
OPEX (MCF)**	$0.50/MCF
CAPEX****	$1,250,000
Flow rate (MCF/Day)***	3,000

	1	5	10	15	20	25	30
Total MCF	1,095,000	1,095,000	1,095,000	1,095,000	1,095,000	1,095,000	1,095,000
Revenue	$1,095,000	$1,095,000	$1,095,000	$1,095,000	$1,095,000	$1,095,000	$1,095,000
OPEX	$(547,500)	$(273,750)	$(273,750)	$(273,750)	$(273,750)	$(273,750)	$(273,750)
Capital		$(1,250,000)					
Net	$547,500	$(428,750)	$821,250	$821,250	$821,250	$821,250	$821,250
Payback (years)	6.00						
NPV (10%)	$6,097,951.19						
Rate of return	192%						

*Based upon contractual agreement of $1.00/MCF.
**Based upon need for chemical injection, pigging operations, and electrical needs for facilities.
***Based upon average production of wells over lifetime
****CAPEX basis cost to replace pipe after five years with infiltration. Pipe should last the life of the well.

Table 2.11 Sample Pipe Renewal Economics (*Continued*)

three million cubic feet (MCF) per day of natural gas revenue of $1/1,000 MCF. The operating expense is based upon assumed operational cost at $0.25/1,000 MCF of natural gas. This $0.25 cost includes labor, electrical, chemical injection, and any other operating expenses associated with the pipeline. The LCCA is based upon the life of the well, which is assumed to be 30 years.

Based upon the above assumptions, the option to immediately repair gives a net present value of $6.83 million at an IRR of 459% with a one-year payback and benefit-to-cost ratio of 6.8:1. The option to delay renewal in five years gives a net present value of $6.09 million at an IRR of 192% with a six-year payback and benefit-to-cost ratio of 6:1. The reason for the difference in cost is due to the extra maintenance cost required to keep the pipe in acceptable condition during the five years it is delayed from repair. The extra maintenance is required since the pipe has 62% wall loss. The operating costs were estimated and assumed to be double of what the immediate repair solution would be for the first five years. Based upon the pipe wall loss and the economic analysis, it was recommended that the option to immediately renew a mile of 24-in. pipe should be pursued. Table 2.11 provides the economic analysis of the two options.

2.8.1.2 Engineering

Engineering is a key consideration for asset management of oil and gas pipelines, because it not only determines the system design and life but it also impacts and determines the economics, maintenance and asset management, and safety of the project.

2.8.1.2.1 Design Standards A major consideration in asset management of oil and gas-related pipeline is what standard to utilize because that is the basis on optimizing the assets efficiency and life expectancy. Most companies are using the DOT 192 and 195 federal code regulations integrated with ASME B31.8, AWWA M55, and the Plastic Pipe Institute Handbook. For example, a pipeline company has over 150 miles of steel gas pipelines. Since the commodity of this product is natural gas, and is classified as a regulated pipeline, companies are required by code to adhere to the DOT 192 design requirements. However, ASME B 31.8 is a stricter and detailed guideline to DOT 192; therefore, this company has incorporated ASME B31.8 as their design standard for their pipeline assets. In addition, this company has created their own library of specifications and design standards. By doing this, this company was able to develop a national company-wide standard for design of their pipeline assets, which ensures the company assets are built to a performance-level goal. The life of a gas well is typically 25 to 30 years; therefore, the company design standards are set so that these assets' design life is sufficient to meet a 30-year design life.

2.8.1.2.2 Planning of Maintenance Facilities As said earlier, oil and gas pipelines are required by DOT codes 192 and 195 to follow a regular maintenance schedule. Although the code is not specific on how they should maintain the asset, it does state that there must be maintenance performed and recorded on the asset. The following is an instruction from the code of federal regulations: "Each operator shall prepare and follow for each pipeline, a manual of written procedures for conducting operations and maintenance procedures, including emergency response (DOT 192.605 (a))." The most widely adopted maintenance program for oil and gas-related pipelines is a regular pigging and in-line inspection process; therefore, these facilities must be planned for in the design phase of the pipeline project.

A pigging facility is an above-ground facility that utilizes the natural pressure in the pipe to launch either a foam brush pig or a more durable plastic device. The planning of these pigging and in-line inspection facilities starts at the engineering and design phase of the project. Figure 2.9 shows a simple design and layout of a removable pig facility.

A pipeline company designed pig facilities to be built on a skid and had the flexibility to be removed and relocated for other applications within the shale play gathering area if the pipeline extends out and prevents multiple pig facilities. By incorporating good design practices with the use of this pig in the engineering and design phase, overall labor cost can be reduced; and well shut time is alleviated because the system incorporates a bypass, and it also reduces capital cost in evaluating the life cycle costs in asset management planning.

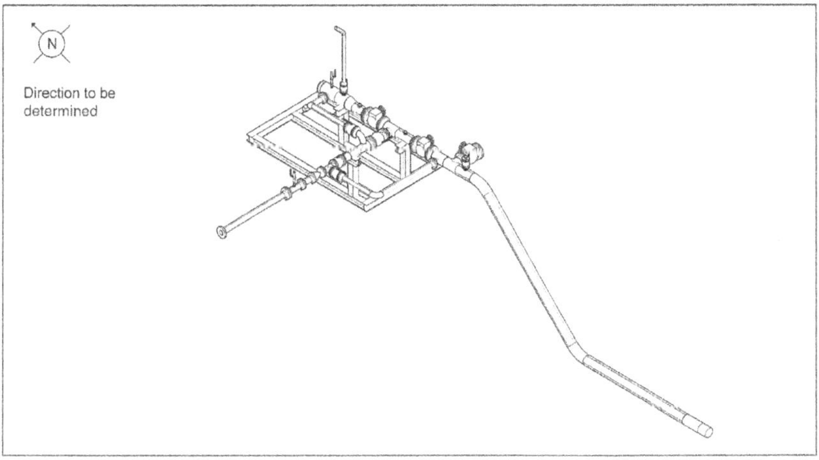

Figure 2.9 Pig launcher.

Trying to quantify the cost and labor time savings of incorporating this pig facility into the design phase was difficult to do because most shale drilling areas are developing, pipelines are constantly expanding, and pipeline lengths and sizes vary. However, the cost for a 4-in. to 24-in. pig facility can be quantified, thus showing the capital expenditure costs saved by having a pigging device that can be removed and utilized on other projects.

2.8.1.2.3 Pipeline Design Improper pipeline design can lead to excessive pipe renewal, which increases asset management and maintenance costs. A common improper pipe design is fittings for pigging that are designed with a bend that is less than a 1.5 diameter of the pipe. A company had to replace several locations of 10- to 16-in. steel pipeline because several bends in the alignment were not designed properly. The company had to dig up several locations since the pig devices could not negotiate the pipe bends. Figure 2.10 shows an actual bend that was dug up, removed, and replaced.

Other engineering considerations are that wyes should be designed and built between 30° and 60°. Wyes at less than 30° create a large bore space opening that prevents the pig/tool from being able to move through the wye, and at angles greater than 60°, it cannot make the turn through the bore because angle is too sharp. Figure 2.11 provides a picture of wye designed with an angle less than 30° at an intersection that had to be replaced because an in-line tool could not make the pass through. Wyes with angles less than 30° increase the length over which the bore section merges (Fig. 2.12), and that would increase the probability of a pig coming to rest in the wye

Figure 2.10 Less than 1.5 diameter for a 16-in. pipe bend.

FIGURE 2.11 Less than 30° wye for a 16-in. pipe.

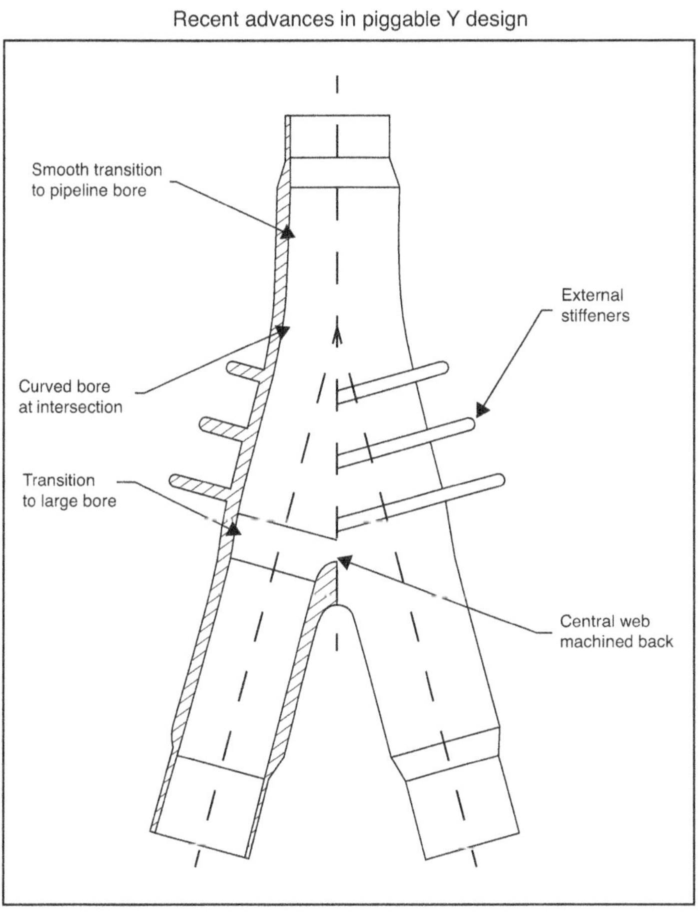

FIGURE 2.12 A piggable wye.

with flow passing around it. Larger 60° angles mean that the pigs have to turn more sharply into the outlet, with correspondingly larger impact forces and accelerations. Model tests indicate that 30° is the optimum angle (Tiratsoo, 1999). In addition, a finite element analysis should be performed for the structural integrity of the wye under operational pressures.

2.8.1.2.4 Hydraulic Planning A case study involving the engineering design and its impact to asset management involves a company selling freshwater for fracking. This company has built a freshwater pipeline distribution network which provides production fracking water to well sites. Figure 2.13 provides a rough sketch of a particular area with improper engineering planning, due to a lack of resources which resulted in an underdesigned freshwater distribution network.

For example, this system consists of 60 miles of 10-in. HDPE DR 21 pipe which has a 101 psig pressure rating with demands at each well site ranging from a flow of 30 barrels per day (bpd) to 20,000 bpd. During the planning phase of this project, more thought was given to the cost of purchasing 60 miles of 10-in. HDPE DR 21 pipe versus 10-in. HDPE DR 11 pipe without considering the engineering/hydraulic or economic impacts. The 10-in. DR 21 pipe was purchased at $7.50/ft versus the 10-in. DR 11 pipe which could be purchased at

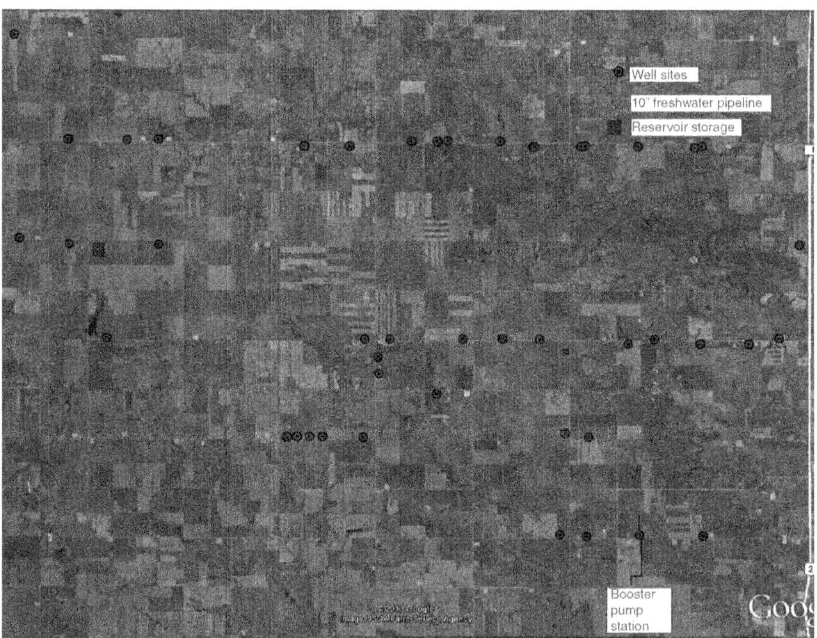

FIGURE 2.13 The improper freshwater system.

$10/ft. The cost difference between DR 21 and DR 11 for 60 miles of pipeline was $792,000. Based on cost savings of $792,000, the 10-in. HDPE DR 21 pipe was selected, constructed, and laid in trench with other salt, crude, and oil pipelines. A year later, there was need for more freshwater, and facilities such as a booster pump station as shown in Fig. 2.13 were needed. During the sizing and selection of the booster pump station through hydraulic modeling software, it was determined that due to the limits of the DR 21 pipe pressure class, a flow of 20,000 bpd could not be delivered without additional pump stations and reservoir facilities to help boost the water without over pressuring the pipe network. Figure 2.14 is a map representing the allowable flows with the one planned booster pump station and DR 21 pipe.

From the engineering modeling of the system, a flow rate of 20,000 bpd could not be achieved throughout the majority of the system, and in some cases, there was no ability to provide flow to the system. Due to this, two additional pump stations and reservoir facilities had to be constructed to maintain the 20,000 bpd flow rate. This resulted in an additional capital expenditure of $4,000,000. This is $3,208,000 above the $792,000 in savings as originally thought. The lesson learned from this case study is the importance of having sound hydraulic and engineering planning studies when developing pipeline systems, and to not compare capital cost of the assets alone.

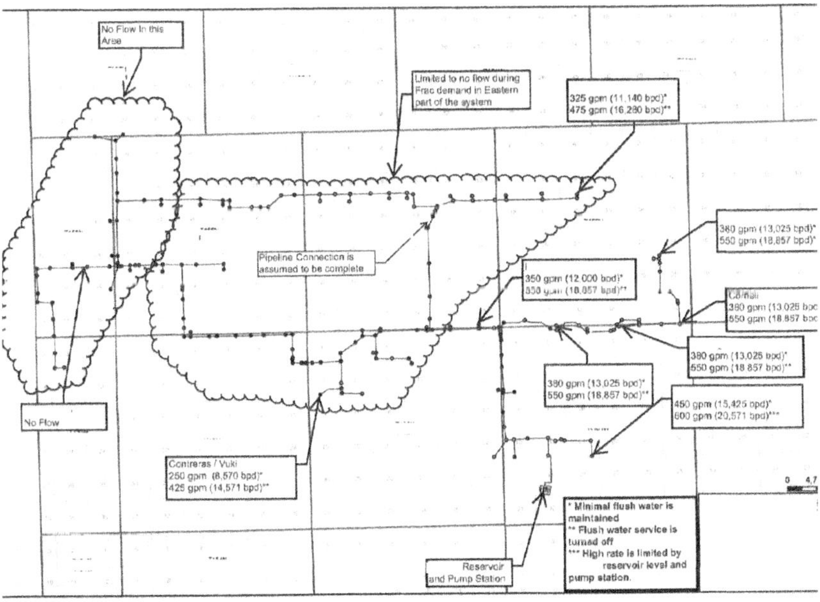

FIGURE 2.14 Flow capacities of 10-in. HDPE DR 21 of a freshwater system.

The additional pump station and reservoirs also resulted in more maintenance and operating expenses, such as pump and equipment maintenance and electrical costs.

2.8.1.2.5 External Load Design External load design is very important in the engineering phase of asset management planning of oil and gas-related pipelines, because if the pipe is improperly designed for future external loads due to new land development or building new facilities such as highways or railroads, it could require the operator to repair and replace the pipe at the operator's expense. For example, if a pipeline is to transverse a future highway, and the operator obtains the approval necessary to cross the highway and does not account for the loads imposed on the pipe, the pipe could obtain structural damage due to collapse or crushing, causing the pipe to require replacement. Since the operator was aware of the future highway crossing and obtained the necessary approval from the jurisdiction having authority over the highway, the operator would have to replace the pipe at the operator's expense. Not only would the cost of the pipe be an added expense but also the cost associated with damage to the highway pavement would have to be accounted and paid for. Most oil and gas pipeline operators use the API 1104 guidelines for design of highway and rail crossings.

2.8.1.2.6 Cathodic Protection Cathodic protection plays a key role in the engineering and planning phase of asset management, because a sound cathodic protection system is one of the most reliable ways to alleviate pipe corrosion and pipe renewal. Additionally, DOT 192.112 subsection (f) states the following:

1. The pipe must be protected against external corrosion by a non-shielding coating.
2. Coating on pipe used for trenchless installation must be non-shielding and resist abrasions and other damage possible during installation.
3. A quality assurance inspection and testing program for the coating must cover the surface quality of the bare pipe, surface cleanliness and chlorides, blast cleaning, application temperature control, adhesion, cathodic disbondment, moisture permeation, bending, coating thickness, holiday detection, and repair.

Subsection .461 of DOT 192 also provides a more detailed breakdown as to the requirements of cathodic protection for steel pipeline in natural gas conveyance. Subsection .500 of DOT 195 requires cathodic protection and specifies the type of coatings to use as well. Cathodic protection is considered a must when it comes to owning and operating a steel pipeline. The most popular method of cathodic protection for steel oil and gas pipelines is an impressed current system, with a 20-mil fusion-bonded epoxy external coating. For bore pipe installation, a 40-mil abrasion-resistant coating is the most widely specified coating. The most widely used inspection method for coating application is a jeep, pull, or holiday testing to verify the bonding of the external coating to the steel pipe.

2.8.1.3 Finance

Financing is typically set up by the way the company is structured. An oil and gas pipeline company will either be a private or public corporation. A private company is not publicly traded on the stock market and is owned by individuals. A privately owned company receives its finance mainly by loans, and revenues generated by the company. A publicly owned company is typically owned by the stock investors and controlled by board of directors. The main form of financing for publicly owned oil and gas pipeline company is through investors who purchase shares of stocks or bonds through the company. The more value a company is worth, the higher the company's shares of stock can sell for. This generates more cash revenues, which help to finance the company. The value of the company is determined by the "EBITDA" system, which stands for "earnings before interest, taxes, depreciation, and amortization." EBITDA measures the financial performance impacted by company operations. It evaluates this without having to consider other factors, such as financing costs (interest), accounting practices (depreciation and amortization), or tax tables. EBITDA is what helps establishes the company worth (Investing Answers, 2013). Once a public oil and gas pipeline company has determined its worth, the value of the stock is then determined by the number of stocks the company will initially offer when it goes to initial public offering (Investopedia, 2009). For example, a company worth $10 million who offers a million shares will value its stock at $10/share. The shares are then purchased, and as the company grows and more shares are in demand for the company, the company can offer more shares of the company for investors to purchase, which then generates the revenue for the company.

2.8.1.4 Management

The management portion of asset management is the execution of the economics, financials, operations, and maintenance, and implementation of the day-to-day plans set forth in the asset management plan. For oil and gas-related pipelines, this includes but is not limited to operations and maintenance procedures, record keeping, process and documentation, construction oversight, procurement oversight, system performance and tracking review, and inventory.

2.8.1.4.1 Operations and Maintenance Procedures Good pipeline asset management involves knowing the latest technologies for maintenance and inspection. Current methods for oil and gas pipeline maintenance are as follows:

- **Pigging procedures**
  - Pigging is a method where a brush/foam/heavy-duty plastic tool is pushed down the pipe by using backpressure from the flowing gas.
- **Cathodic protection**
  - The two main methods for gas pipeline cathodic protection are impressed current and sacrificial anodes.
- **Good O&M process, procedures, and documentation**
  - This is the best way to ensure proper operational asset management of a pipeline system.
- **Chemical treatment**
  - Two main methods for gas pipeline chemical injection are methanol injection, which prevents hydration buildup, and anti-bacteria agents for preventing bacteria buildup in the pipeline.

Current methods for pipeline inspection are as follows:

- **Caliper pig**
  - It is the pigging tool that determines wall thickness and pipe ovality.
- **Gyro pig**
  - It determines pipe profile and bends through the use of a gyroscope and helps to verify horizontal directional drilling (HDD) bend radiuses and the exact profile of the pipe.
- **Gauge plate**
  - It determines pipe ovality and damage to pipe wall.
- **Magnetic flux leakage**
  - It determines changes in wall thickness by inducing magnetic flux in pipe to detect changes in amplitude due to wall thickness change, and it is an inferred measurement.
- **Ultrasonic**
  - It determines wall thickness by use of a transducer to send and receive sound waves. The difference in time of flight determines wall thickness.

2.8.1.4.2 Inventory Inventory is the first step in any asset management program (EPA, 2009). If asset locations are not known, then management is not possible through ArcGIS software, which allows all the pipeline information to be displayed in one central software program. The ArcGIS software maps show pipeline location based upon as-built data. Through the use of the software system, information, such as material, wall thickness, operations and maintenance procedures, maximum allowable operating pressures, lengths, coordinate locations, valves, sizes, capacities, terrain, joint methods, installation dates, and any other data the operator needs can be included.

In addition, during the construction of the pipeline, an on-site surveyor will survey the location of the pipe as it is constructed. This survey information is then incorporated onto an alignment sheet that is stored in the company's databases. According to DOT 192 and 195 codes, all pipelines are to be mapped according to DOT 192 subsections .491 and .600 and DOT 195 subsections .500 and .600. The geographic information system mapping system is also reported to the state's one-call system so that records of utilities can be kept for the purposes of identifying and locating.

2.8.1.4.3 Construction Oversight Construction oversight is a crucial step needed on the management side of asset management in oil and gas pipelines. There are a lot of moving parts during the construction phase of an oil and gas pipeline installation, and therefore, pipeline companies utilize resources for ensuring proper QA/QC during construction. DOT 192.300 subpart G, 192.600, and 195.266 require a job book to be kept and maintained at all times. These records include but are not limited to non-destructive testing records, welders' qualifications, as-built alignment sheets, as-built drawings, safety documentation, personnel qualifications, and other information relative to the safe operation of the pipeline. Usually, a full-time company representative is on site. Company's representative can be one chief inspector, one or two senior inspectors, and multiple utility inspectors. In addition, a safety and environmental inspector is on site for every project. The inspectors are responsible for the day-to-day observation of the contractor to ensure that proper installation, testing, payment, safety, and record keeping were met. The inspectors also ensured materials being delivered to the site were as specified and in a good condition. The inspectors are responsible for transmitting the day-to-day project construction information to the company representative who is responsible for compiling the inspectors' information into one job book.

2.8.1.4.4 Procurement Oversight To ensure the proper procurement of materials delivered to a pipeline construction project, every oil and gas pipeline company has a procurement representative who is responsible for managing the proper procurement of materials. The procurement representative ensures all factory material test information is received and acceptable, and a pre-shipment site inspection of the materials prior to delivery to the jobsite is performed. This information is documented and included in the construction job book.

Type	Natural Gas Pipelines	Oil Pipelines
Workers qualifications	192.507	195.507
Integrity management	192.947	195.589 and 195.452
Threats and risk assessment	192.947	195.452
Assessment condition	192.947	195.947
Operations and maintenance Procedures and schedules	192.603	195.402
Corrosion control methods	192.605 and 192.476	195.589
Tests	192.517	195.310
Construction records	192.605 and 192.476	195.266

TABLE 2.12 Partial List of 192 and 195 Records and Code References

2.8.1.4.5 Record Keeping, Process, and Documentation Under the DOT 192 and 195 codes, certain records are required to be kept. Table 2.12 is a partial list of 192 and 195 records and code references to the documents that are required. More detail to the exact documentation needed could be founded within the code section.

2.8.1.4.6 System Performance and Tracking Review The methods for system performance and tracking review may vary with different pipeline companies. However, the basic concept of system performance and tracking review does not vary from one company to another, which are monitoring capital and operating costs, and evaluating the pipeline performance (i.e., volume and revenue production).

2.8.1.4.7 Case Study An oil and gas pipeline company primarily focuses on saltwater and freshwater production sales as its primary revenue resource. The company will charge a producer a cost per barrel of saltwater and freshwater to be piped to a location for either disposal or injection. To determine if the asset is economically feasible to build, the company will perform an economic analysis as shown in Table 2.13 to determine if the asset will have positive economics generally over a 5- to 10-year lifespan. In the economic analysis, a monthly water volume projection is provided, and a price per barrel of commodity conveyed is set between the oil and gas pipeline company and producer. The only unknown in the analysis is the operating expenses that the pipeline company will have to pay each month on per barrel of commodity conveyed. However, for the analysis, an estimate can be made as to what the operating expense may become, and is input into the analysis. If the economics are feasible, the company will most likely pursue development of the asset. Once the asset is built, the company will then track volumes on a monthly

Midstream Services-Revenue

One Month Revenue

SWD-Pipeline Asset	SWD-PL Volumes (bbl)					Total Revenue				
			Injection Breakdown			$1.00	$2.00	$3.00	$4.00	
	Pipeline	Injection	Production	Flowback	Other	Pipeline	Production	Feedback	Other	Total
Basin A	189,708	273,701	256,968	14,410	2,323	$189,708	$513,935	$43,230	$9,294	**$756,167**
Basin B	80,081	105,422	97,843	7,579	—	$80,081	$195,686	$22,736	$—	**$298,504**
SWD Pipeline Totals	**269,789**	**379,123**	354,811	21,989	2,323	$269,789	$709,622	$65,966	$9,294	**$ 1,054,671**
FW-Pipeline Asset	**Production**		**Frac**		**Other**					
	Volume	**Fee**	**Volume**	**Fee**	**Volume**	**Fee**	**Production**	**Frac**	**Other**	**Total**
Basin A	27,344	$1.00	53,462	$2.00	381	$3.00	$27,344	$106,924	$1,143	**$135,411**
Basin B	—	$1.00	—	$2.00	—	$3.00	$—	$—	$—	**$—**
FW Pipeline Totals	27,344		53,462		381		$27,344	$106,924	$1,143	**$135,411**
Total OMS Revenue	**$1,190,082**									

Table 2.13 Saltwater and Freshwater Revenue Example Worksheet

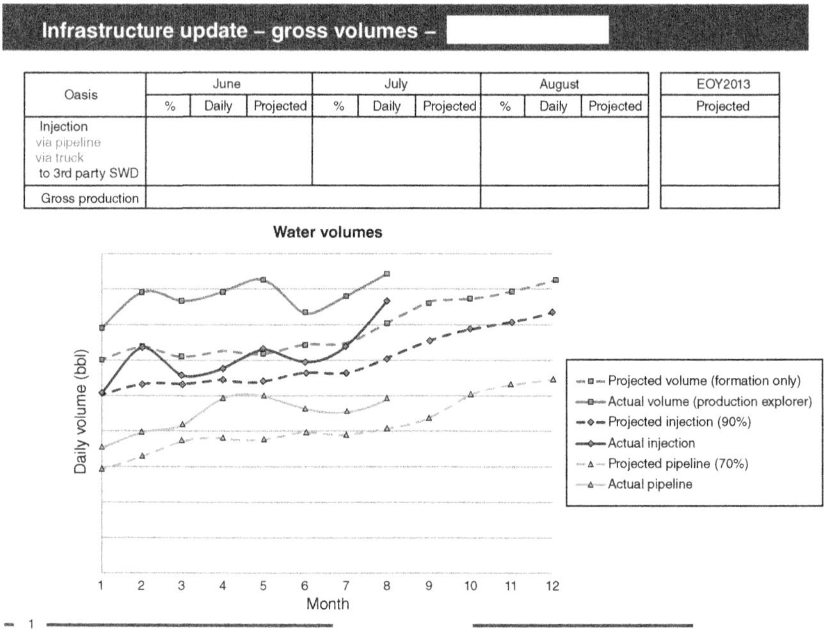

FIGURE 2.15 Saltwater volume projections example worksheet.

basis, determine its revenue, and along with that track capital and operating expenses. At the end of each month, a monthly economic analysis on revenue and expenses is performed and compared with the original 5- to 10-year model to track the system performance. Figure 2.15 provides an example of tracking revenue and volumes for system performance.

This company's method of system tracking and performance review allows it to compare projected values to existing values. From the comparison graphs shown in Fig. 2.15, the "actual performance" is outperforming the "projected performance." This ensures the company that not only is the system performing well, but it is also producing positive economics for the company. If "actuals" were less than "projected," then it either needs to be renewed or additional capital needs to be spent to allow the system to perform at a positive level.

2.8.1.5 Risk Management

DOT 192 and 195 codes require natural gas and oil pipeline operators to identify the risk and threats associated with an oil and/or gas pipeline. Additionally, for unregulated pipelines (generally class 1, salt water or freshwater pipelines), it is a prudent action for an operator to identify the risk and threats that can attack these pipelines as well. The challenges of pipeline asset management present themselves in several different forms, but can be broken down into the following

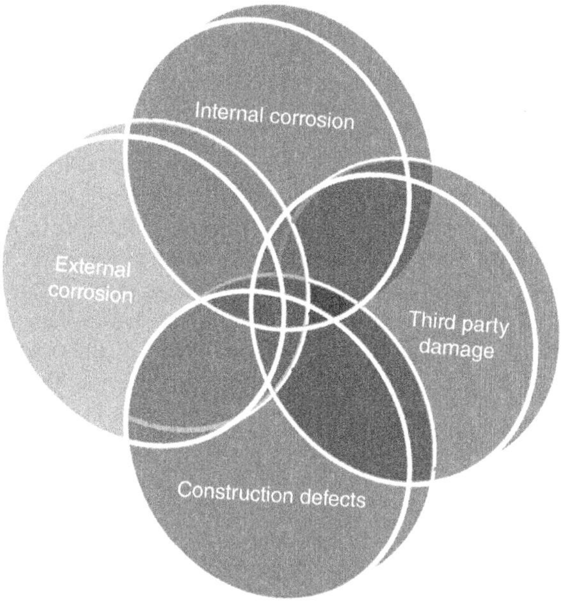

Figure 2.16 Four main threats to oil and gas pipelines.

four main threats: external corrosion, internal corrosion, third-party damage, and construction defects. Figure 2.16 provides a structure of how these threats interact with oil and gas-related pipelines.

This section will discuss the threats, how and why they occur, and what are the current methods for detecting and preventing these threats through case studies.

2.8.1.5.1 Case Study: Internal Corrosion In Section 2.8.1.1.1, a case study was provided as an economic example in asset management planning. However, not only is this a good example of how economics play into pipe rehabilitation and maintenance cost, but it is also a main threat in oil and gas pipeline risk and threat assessment. Internal corrosion is one of the main threats to a gas pipeline due to the compounds that are found mixed in with the natural gas. These compounds mainly consist of bacteria from the well flowback, saltwater from the ground formation as a result of the wellbore drilling, hydrocarbons, and oxygen. These four components can attack the inner walls of a pipeline if proper asset management techniques are not implemented. This case study is an example of what can happen to the inner walls of a gas pipeline if the proper facilities such as pigging are not in place to prevent bacteria and water buildup in the pipeline.

After running an in-line inspection tool (ILI) on this 24-in. pipeline, the results indicated 62% wall loss in over a hundred different locations throughout a one-mile length of pipe. The initial investigation by the ILI tool indicated that the pipeline was installed without any pigging capabilities for 24 months; therefore, the pipe set idle/deadleg and allowed bacteria scaling and water to build up in the pipe and corrode the inner walls of the pipe. This form of pipe design and installation is known as a deadleg because it does not have any pigging facilities to flush out water and inject chemicals to kill bacteria.

Another case study provides an economic example in asset management planning for internal corrosion that occurs when the proper asset management of facilities is in place, but not utilized. In this case, a 12-in. pipeline was discovered to have six leaks after a pipeline patrol indicated a gas leak along the pipe right-of-way. Once the leak was detected, an ILI tool was run through the pipe and confirmed that the pipe had six leaks with defects up to 1 in. in width (Fig. 2.17). Since the pipeline had sufficient pigging facilities in place, it was determined that proper operation and maintenance activities were not being practiced for this stretch of pipe by lack of operation and maintenance records. The combination of low flows, non-pigging operation, and lack of chemical injection allowed the buildup of water and bacteria which ultimately corroded the pipe.

An economic model was prepared for this renewal project. There were two options to repairing this pipeline, with first being the immediate repair or wait for five years. The second option was to keep the pipe as is, increase maintenance requirements, and renew the pipe five years from now. Although the economics favored the immediate repair alternative, it was recommended that the option to immediately repair the 12-in. pipe be pursued, because of the safety issues around a leaking gas pipeline. The conclusion to this case study reiterates the importance of having sound operations and maintenance procedures and well-documented operation and maintenance as part of the overall asset management.

2.8.1.5.2 Case Study: External Corrosion External corrosion is another top threat to natural gas pipelines. This case study presents external corrosion of a 12-in. pipeline that was determined to have 50% wall loss after running an ILI tool through the pipe. After exposing the pipe at the location of the wall loss, it was determined that the wall loss was due to improper application of joint weld coating. As shown in Fig. 2.18, the joint coating was not properly and adequately applied to the joint, and/or the joints were coated too quickly, causing hydrogen gas to build up during the drying and not allowing a sufficient coating. Therefore, the defective coating exposed the iron pipe, allowing a direct corrosion point. The proper asset management technique that could have avoided this situation is to have good field inspection and proper crew training to apply field joint coatings. Due to the safety issues that required this pipe to be repaired, an economic analysis was not prepared, because regardless of the outcome,

A

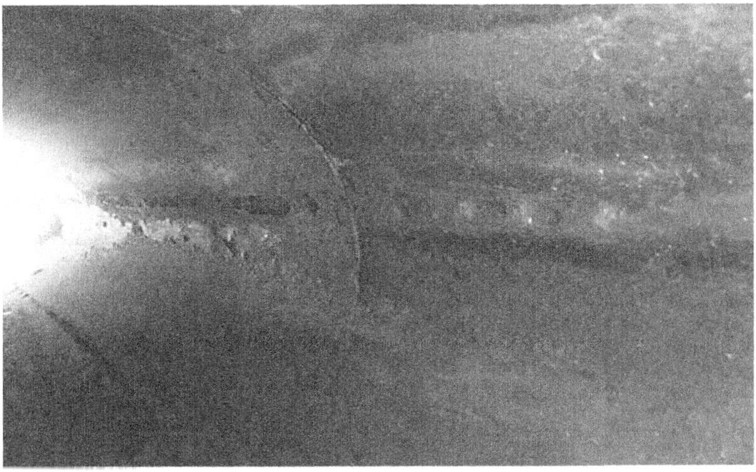

B

FIGURE 2.17 Interior of 12-in. corroded pipeline.

the pipe had to be immediately repaired. The repair method for this corroded joint was a clockspring composite wrap (Fig. 2.19), which was the least expensive option. Other option explored was to cut a joint of pipe, but this required more labor for cutting and welding, and there would be loss of gas in the pipeline when the pipe is required to be blown down for cut in repair. The conclusion to this case study reiterates the importance of incorporating sound field and construction inspection into an oil and gas pipeline asset management plan.

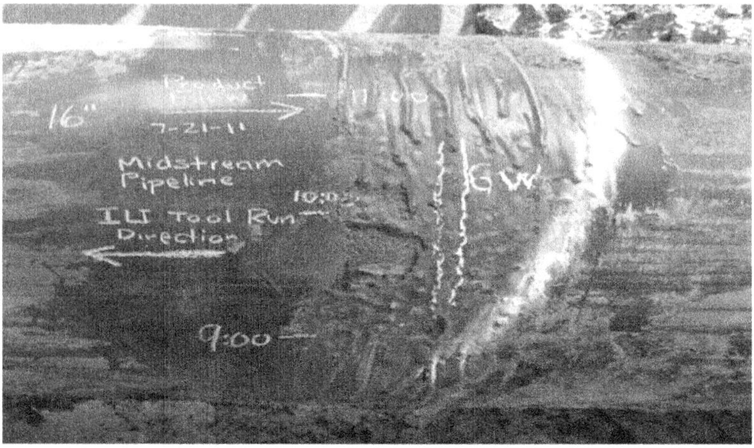

Figure 2.18 Exterior of corroded joint.

2.8.1.5.3 Case Study: Material Defects The third major threat to a natural gas pipeline is material defects. These material defects are due mainly to poor mill quality/production and/or poor handling once received at the construction site. Material defects such as cracked pipe or insufficient pipe coating can cause major problems to the pipe such as internal and external corrosion. Figures 2.20 and 2.21 show an example of poor pipeline coating during jeep testing that led

Figure 2.19 Clockspring wrap for joint repair.

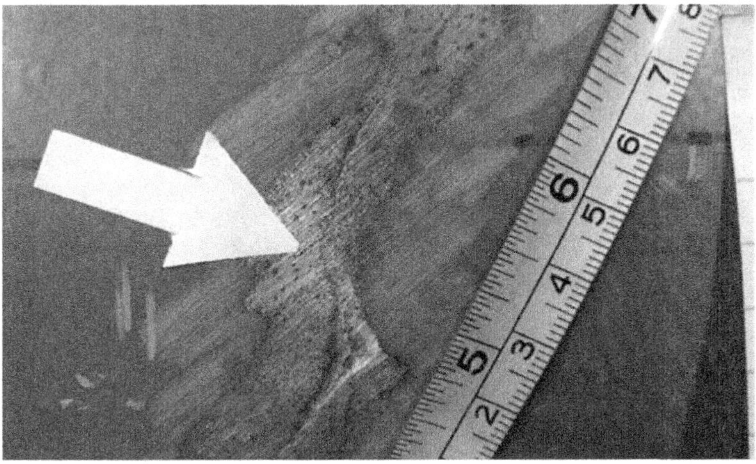

FIGURE 2.20 Cracked pipe discovered during jeep testing.

to the discovery of a cracked pipe. It was concluded that there was insufficient mill quality oversight and testing due to a lack of inspection at the mill. The mitigation method for this pipe was to quarantine the pipe and send it back to the mill. The conclusion to this case study reiterates the importance of incorporating sound field construction oversight and mill inspection into an oil and gas pipeline asset management plan.

2.8.1.5.4 Case Study: Construction Defects The fourth main threat to a natural gas pipeline system is construction defects caused the

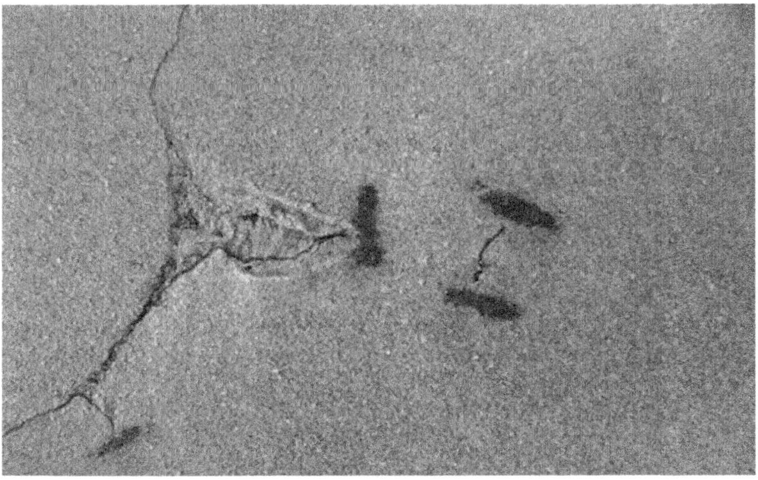

FIGURE 2.21 Sandblasted cracked pipe during jeep testing.

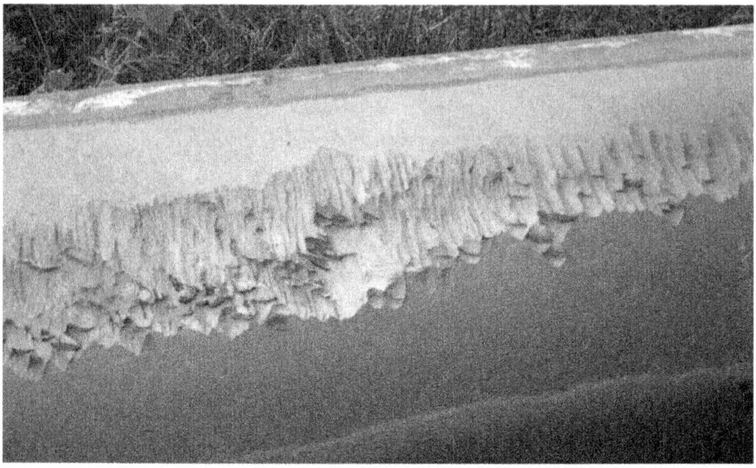

FIGURE 2.22 Damaged pipe due to HDD reamer.

mechanical damage. Figure 2.22 shows a pipeline that has been damaged during reaming operation of an HDD operation to install a parallel pipeline during the reaming stage installation of a 12-in. pipe. This could have been prevented by the use of proper field line locates and as-built records. The repair method for this damage was to replace 47.5 ft of the 12-in. pipe.

2.8.1.5.5 Case Study: Hydrate Formation Another threat to natural gas pipelines is the formation of hydrates when hydrocarbon gases occur in the gas composition usually greater than 1,180 British thermal unit (BTU). When hydrocarbon gases greater than 1,180 BTU occur in the gas composition, hydrates can form when pressure increases, temperature decreases, and low flows occur. Under these circumstances, hydrates can cause ice-type crystals to form in the pipeline (Fig. 2.23). These hydrates clog the pipe and result in an insufficient product flow through the pipeline system. Most hydrates are discovered when there is a pressure drop in the pipeline, low flows occur, or an ILI or pig run discovers the hydrates. Proper asset management to prevent these types of issues is to maintain system pressures by adding adequate compression where needed, methanol injection, and proper pigging, and to maintain temperature and pressure under the dewpoint requirement for the gas. An economic and operational analysis should be performed to alleviate hydrate formation as a basis for selecting a suitable method. The conclusion to this case study reiterates the importance of sound operation and maintenance procedures and incorporating engineering into an oil and gas pipeline asset management plan.

Figure 2.23 Hydrate build.

2.8.1.5.6 In-line Inspection Tools The in-line inspection is a key component to asset condition and level of service because it provides the information needed to determine pipe current condition. The different types of recommended ILIs are as follows:

- **Caliper pig**
  - It is the pigging tool that determines wall thickness and pipe ovality.
- **Gyro pig**
  - It determines pipe profile and bends through the use of a gyroscope and helps to verify HDD bend radiuses and exact profile of the pipe.
- **Gauge plate**
  - It determines pipe ovality and damage to pipe wall.
- **Magnetic flux leakage**
  - It determines changes in wall thickness by inducing magnetic flux in pipe to detect changes in amplitude due to wall thickness change, and it is an inferred measurement.
- **Ultrasonic**
  - It determines wall thickness by use of a transducer to send and receive sound waves. The difference in time of flight determines wall thickness.

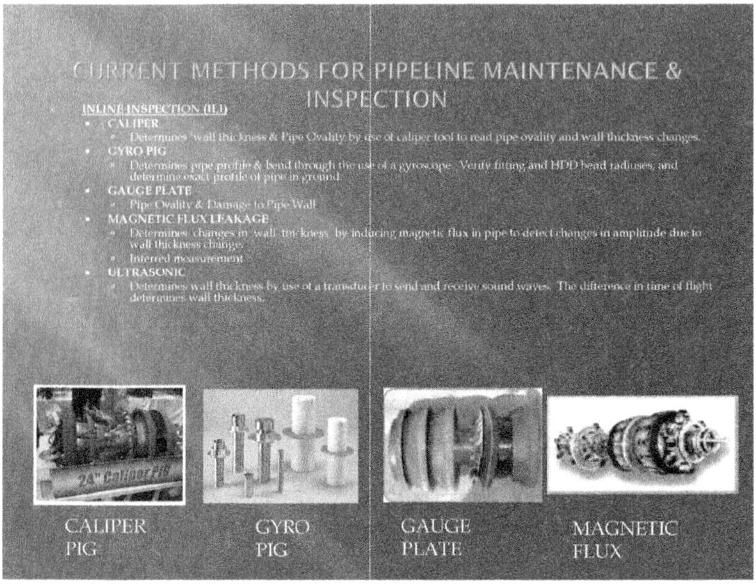

FIGURE 2.24 Different in-line inspection and pigging tools.

Figure 2.24 provides a picture of the different types of ILIs.

By utilizing an ILI on oil and gas pipelines, the United States Environmental Protection Agency (USEPA)'s question of what is its condition, what is its useful life, and what is its value can be answered. In addition, an ILI will ensure that the EPA's best practices of developing a condition assessment, assessing the remaining useful life, and determining the asset values and replacement costs are followed.

2.8.1.6 Life Cycle Cost Analysis (LCCA)

The LCCA can be used to select the most cost-effective option among different competing alternatives to build a project when each method is equally appropriate to be implemented on technical consideration. Not only LCCA reviews the capital cost to implement a project, but it looks at all the cost impacts from the projects inception to its disposal by looking at capital cost, revenues, financing, and operation and maintenance costs. The LCCA is a key piece to asset management planning, because it considers the life of the asset, which helps a pipeline operator understand what are the associated risks with the project and how best to manage the risk. If the risks are known, and the project must move forward, the operator can then build an asset management plan for the project using an LCCA model as a basis for tracking and managing O&M, funding, capital expenditures, and revenue streams. For sound LCCA models to be completed, there is specific information that must be known and accounted for, which are operations and maintenance, plans to renew the asset, and plans to expand the project. According to the USEPA' s Small System Asset Management Plan Best Practices Guide, a well-thought-out asset management plan incorporating life cycle cost should ask the following three questions:

1. What alternative strategies exist for managing O&M, personnel, and capital budget accounts?
2. What strategies are the most feasible for my organization?
3. What are the costs of rehabilitation, repair, and replacement for critical assets?

In addition, the USEPA Guide states that best practices for the LCCA should include the following:

- Moving from reactive maintenance to predictive maintenance.
- Knowing the costs and benefits of rehabilitation versus replacement.
- Looking at lifecycle costs, especially for critical assets.
- Deploying resources based on asset conditions.
- Analyzing the causes of asset failure to develop specific response plans.

2.8.1.6.1 Case Study: Managerial Policies A large pipeline company that operates over 5,000 miles of regulated pipelines throughout the United States identified a problem that most of the pipeline projects they constructed were either improperly installed, installed with delay, or needing millions of dollars in pipeline renewal after they were constructed. What the company found regarding the root cause of these issues was that there were not any standard company policies, procedures, and processes on the lifespan of the pipeline projects. The company did not have a standard capital execution plan for implementing projects; therefore, information about the projects was not being communicated to all the team members, resulting in improperly installed pipe or capital projects being delayed sometimes for an average of 730 days (2 years).

In addition, as there were no company-adopted engineering standards, pipe was being designed in many different and incorrect ways. The absence of a standard process for operations and maintenance resulted in both insufficient operations and maintenance procedures or none at all that required millions of pipe renewal repairs. This company took the initiative to put together a team dedicated to developing standard managerial processes, policies, and procedures across the company to facilitate proper development and procedures for pipeline asset management. As a result, the company reduced its capital execution plan timeline for capital projects from 730 days to

365 days, resulting in reduction of its operating costs and pipe repairs by half. The company found that employee morale was increased because policies and processes were put in place to help them get their jobs done efficiently and correctly.

2.9 Chapter Summary

This chapter presented an overview of asset management. Developing an asset management program first comes with the idea that a utility needs to implement a more business-like discipline in how it operates. Once this conclusion has been reached and is accepted among all levels of the organization, it can begin to take root. Among the first steps of an asset management program for buried infrastructure is inventory assessment, which in itself can be a challenge for some utilities that date back more than a century due to recordkeeping that is not diligently maintained. Condition assessment is the next step so that data collected in this stage can be used for decision regarding repair, replacement, and operations and maintenance strategies. For an asset management program to be successful, key service levels must be defined. By defining these service levels, a utility establishes the framework from which decisions can be made. Another component of an asset management program that has been traditionally lacking is the use of "TBL" accounting, a process by which environmental and social ramifications are factored into the decision-making criteria in addition to the economic considerations. Risk management techniques are also an integral part of a successful asset management plan, i.e., identifying the critical infrastructure and ensuring that it functions as designed. The key points of chapter can be summarized as follows:

- Good infrastructure asset management is vital because infrastructure, land, and building assets represent a vast investment worldwide and provide the platform for economic and social development.
- Asset management provides the technique for making management decisions to ensure that the required levels of service are provided in the most cost-effective manner.
- Improved asset management practices have been driven largely by the regulators and managers requiring sustainable financial management, but also by recent network failures highlighting inadequacies in asset planning.
- Asset management can provide process needs to the link from the highest level strategic planning through to day-to-day operational activities (total asset management).
- Asset management can provide benefits relating to accountability, service management, and financial efficiency.

- Asset management plans are key tactical plans which demonstrate how to cost effectively achieve the organizational strategic goals. Asset management techniques can be applied at different levels of the complexity (basic/advanced) depending on the need of the organization and perceived economic benefits from asset management.
- Lifecycle asset management recognizes that asset decisions should be made with consideration of the cost from planning through to disposal.

2.10 Review Questions

1. Define asset management.
2. Name five benefits of asset management.
3. List nine major questions to start an asset management program.
4. Name the six key elements of an asset management program.
5. Name nine sections that according to IIMM (2006) can be included in an AMP or to serve as a guideline for developing one.
6. What are the key elements of an asset management framework?
7. What are the five steps for preparing an asset management plan?
8. Name the five categories of level of service.
9. Define "routine maintenance."
10. Define "depreciation."
11. What are benefits of "depreciation?"
12. Compare "straight-line" with "modified approach."
13. What is "CMOM?"
14. What is "likelihood of failure?"
15. What is "consequence of failure?"
16. Explain "social consequences of failure."
17. Name the components of the financial summary.
18. Describe the components of an "economic model."
19. What is a "regulated pipeline?"
20. What are the consequences of "improper pipeline design?"
21. Why "external load design" is important?
22. Name current methods for oil and gas pipeline maintenance.
23. What are "in-line inspection tools?"
24. Describe "life cycle cost analysis."

CHAPTER 3

Design Considerations for Trenchless Renewal Methods

3.1 Introduction

The most important step in the designing of a trenchless renewal technique is the selection of the most appropriate, cost-effective, and reliable method. Obviously, selection of a solution is based on the recognition of the problem or problems with the existing pipeline system. As there is no universal problem, there is no universal solution and the method selected must be suited for the individual project conditions. Sometimes the best method is found in the application of multiple systems to achieve the most cost-effective solution. The trenchless renewal solution can be applied for addressing problems such as structural, infiltration/inflow, capacity, corrosion, and so on. The problem area can be found through a complete evaluation of as-built drawings and other records, and inspection and monitoring of the pipeline systems. The design of a pipeline renewal system includes (1) identification of pipe conditions and problem recognition and classification, (2) prioritization of problem considering strategies and long-term plans, (3) selection of an appropriate pipeline renewal method, (4) designing renewal methods based on project specific conditions, and (5) implementation and monitoring.

This chapter provides information on method selection and design basics. As mentioned in Chap. 1, throughout this book, the term "renewal" refers to all aspects of pipeline renewal, upgrade, and renovation activities where the design life of the pipeline is extended. The term "repair" refers to a problem that is temporary or permanently fixed without adding to the design life of the pipeline system.

3.2 TRM Selection Process

The trenchless renewal method (TRM) selection criteria involve the following two main steps:

- Assessment of current pipeline-specific conditions, including state of deterioration and identification of problems as well as future conditions and long-term plans for the pipeline.
- Selection of suitable renewal technologies based on current conditions and future plans.

The selection of a renewal method requires knowledge of the pipeline interior conditions (corrosion, deposits, cracks, misalignments, settlements, joint problems, etc.) as well as the underground conditions around the exterior of the pipe and current and future loading conditions. Pipe-specific conditions, such as gravity or pressure, type and composition of fluid, workspace area and bypassing requirements, number and degree of bends, slope, depth, length, diameter, flow rate, internal deposit buildup, and the like, must be evaluated. Accurate classification of defects is essential in selecting a suitable renewal solution. It is also important to estimate rates of deterioration and to predict pipeline failure. When the pipe and the surrounding soil are in equilibrium, this condition might remain stable for a long time with a minimum likelihood of catastrophic failure or collapse. Using an assessment technology that can accurately determine deterioration rates is an important component of the decision-making process. As technologies for pipeline assessment continue to expand, it is important to emphasize that appropriate technologies must be used to obtain a thorough understanding of the condition and the dynamics of the deterioration process.

In general, a trenchless pipeline renewal program can be summarized in four phases as follows:

- Initial planning.
- Pipeline integrity assessment.
- Develop renewal solutions.
- Implement and monitor.

3.2.1 Initial Planning

During this stage, to the extent possible, the pipeline owner, operator or engineer should have information available on

- pipeline inspection results through closed-circuit television (CCTV) or other inspection methods;
- fluid (water, wastewater, oil, and gas) characteristic data (pH, temperature, chemical composition, etc.);

- pipeline-specific conditions (pipe material, depth and diameter, ovality, settlement and deformation, etc.); and
- jobsite conditions, degree of urbanization, ground conditions, and availability of workspace.

3.2.2 Pipeline Integrity Assessment

The focus in this task will be on the following:

- Pipeline inspection and analysis.
- Pipeline hydraulic capacity.
- Pipeline defects (leaks, corrosion, structural, capacity, infiltration or inflow, etc.).
- Pipeline environmental conditions (surface and subsurface surrounding the pipe).

3.2.3 Develop Renewal Solutions

Using the process described in this chapter as a guide, a suitable trenchless renewal/replacement method can be selected. The requirements for candidate methods must be evaluated with pipeline-specific conditions and other project needs, and a final selection is to be made based on all project conditions. The guidelines in this chapter have been developed to allow the preliminary decisions. The final decision requires other considerations, such as available budget, organizational policy and objectives, short-term and long-term plans for the pipe, operation and maintenance requirements, history of previous applications, compatibility with existing systems, availability of experienced and qualified contractors, and level of workability, ease, and confidence with a specific method.

3.2.4 Implement and Monitor

The pipeline owner or operator should assign a project engineer or a manager to carry out the pipeline renewal task. A qualified contractor must execute the pipeline renewal work under supervision of the assigned project engineer. Post renewal monitoring activities should follow organizational operation, reliability, and maintenance (ORM) guidelines.

3.3 Renewal Method Depends on the Existing Pipe Conditions

Among other factors mentioned previously, selection of the appropriate renewal technique depends on the type and application of the existing pipe being renewed and the type of defects that are being remedied. For pipes that are in good structural condition but are judged inadequate with regard to hydraulic capacity, renewal methods can comprise in-line replacement (ILR, Chap. 8) and lining technologies [such as cured-in-place pipe (CIPP, Chap. 6) and close-fit pipe (CFP, Chap. 11)]; and in some cases, open-cut (OC) method may be considered.

For pipes that are in poor structural condition, but otherwise in sufficient hydraulic capacity, renewal methods can range from ILR, sliplining (SL, Chap. 7), modified sliplining (MSL, Chap. 9), CFP, and CIPP.

For pipes that are in good shape, but otherwise have localized defects, any of the localized point repairs [point source repairs, PSRs)], described in Chap. 14 can be selected. For service laterals, methods described in Chap. 13 can be used.

For pipes that require corrosion and abrasion protection and stabilization, any of the spray-in-place pipe (SIPP) methods (Chap. 10) can be selected.

For deteriorated manholes, any of the methods described in Chap. 12, "Manhole Renewal Methods," can be selected.

According to the American Society for Testing of Materials (ASTM) F-1216-09 "Standard Practice for Rehabilitation of Existing Pipelines and Conduits by the Inversion and Curing of a Resin-Impregnated Tube," in general, pipeline renewal encompasses the following two general degrees of methods:

- Those in which the existing pipe continues to bear the loads (partially deteriorated), but has corrosion and infiltration problems.
- Those in which new pipe or other reinforcement is added to help bear the loads (fully deteriorated). Recent studies have shown that the term "fully deteriorated" is fundamentally flawed because the pipe, even in its fully deteriorated state, is holding the soil load and in reality is "not" fully collapsed. The research in this area is ongoing and more information can be found in the American Society of Civil Engineers (ASCE) Pipeline Division committee report on design guidelines for pipeline renewal systems (2005)[1]. New design criteria would enable engineers to provide more cost-effective and reliable pipeline renewal systems. This chapter follows the current design guidelines based on the ASTM F-1216-09.

The key factor in current design guidelines is whether the existing pipe is structurally sound enough to continue to carry the earth, live, and hydrostatic loads imposed on it. If the existing pipe is structurally sound enough to continue to carry the earth and live loads,

---

[1]Available at: http://www.uta.edu/ce/cuire/Emerging%20Concepts%20in%20Pipeline_Renewal_Design%20Final%20Version.pdf

then several internal lining techniques might be applicable, including SL, cured-in-place pipe, internal sealing techniques, and close-fit pipe. The thickness and strength of the lining in these cases are required to be sufficient to resist only the hydrostatic loads imposed by the groundwater. It is assumed that the existing pipe carries the other loads. Although these methods are designed to increase the structural load carrying capacity of a pipe, they cannot prevent continued structural deterioration. Consequently, they are sometimes referred to as "semi-structural" renewal methods.

If a pipeline is deteriorated to the point from where it cannot continue to carry the earth loads and live loads imposed on it, the renewal method is designed to serve as a stand-alone pipe. Sliplining, CIPP, MSL, or CFP might be appropriate, but the wall thickness of the new pipe will be thicker than in the case described above, and the installer must assure that the existing pipe has not deformed to the point at which the new pipe cannot be installed. If this cannot be assured, the new pipe can, under certain conditions, be replaced by ILR technique. If because of the existing pipe conditions, none of the TRMs cannot be employed (e.g., because of shape, internal operating pressure, soil voids and localized collapse, alignments, size, joint settlements, and capacity problems), the pipe might have to be replaced using a parallel line installed by the trenchless construction methods. In extreme cases, a combined number of different TRMs or even a combination of OC and trenchless methods might be considered. The latter approach would be considered when the surface and subsurface disruptions associated with conventional OC technique could be tolerated.

3.4 Renewal Method Depends on Site and Project Conditions

Similar to any other construction project, a renewal project is site and project specific, and many factors for selection of a specific renewal method need to be considered. For example, local laws and regulations, such as restrictions on personnel entry into manholes and pipelines, are site dependent. This is a major issue for selecting a renewal method and should be resolved before initiating a pipeline renewal program. Existing pipeline life expectancy (see Chap. 5) and asset management requirements are important for the selection of a renewal method. Not all methods are proven to provide a design life of, say, 50 years or more. Each pipeline owner or operator needs to make appropriate decisions based on its ORM strategy. Short-term and long-term plans for the use of pipeline, availability of local service providers and experienced contractors, budgetary requirements, new pipe material, and type of fluid and its acidity and corrosivity are examples of other project factors to consider.

3.5 Selection of a Renewal Method (a 10-Step Process)

This section presents a 10-step process for developing solutions for pipeline renewal. Table 3.1 presents a summary of these steps.

The 10-step solution process presented in Table 3.1 will guide the owner and operator to rapidly make quality and cost-effective decisions. Using this process, the user or engineer can select methods to solve specific problems without having to develop a comprehensive background and knowledge of all renewal methods.

3.5.1 Step 1

The proper selection of renewal methods and materials depends on a comprehensive understanding of the specific problems to be corrected, including external as well as internal pipeline conditions. In that regard, it is important to conduct a pipeline integrity assessment

Step	Objective	Task
1	Define the problem.	Conduct complete and accurate pipeline condition assessment.
2	Identify candidate methods.	Review the applicability and technical compatibility parameters, and select methods likely to succeed.
3		Review the limitations of the selected candidate methods and confirm that the selected methods are compatible.
4		Review the associated installation cost range.
5		Review the installation experience factors.
6	Make a final selection.	Select a candidate method.
7	Develop a better understanding of the selected method.	Locate the selected candidate method in Table 3.4 and develop an understanding of how it fits within the complete family of methods. Review the brief description of the method. For more detailed information, refer to the appropriate chapter in this book.
8	Identify technology providers (contractors).	Review the vendors of the selected renewal method.
9		Contact vendors if more information is needed.
10	Provide solution.	Develop the design and contract documents for the final renewal method selected.

TABLE 3.1 How to Select a Pipeline Renewal Method

program to minimize the likelihood of an unexpected collapse, blockage, leakage, spillage, and so on. This step includes the following:

- Initial planning.
- Pipeline integrity assessment.

As mentioned previously, during the initial planning, the operator should collect the available information (to the extent practical) on pipeline, fluid characteristic data, inspection results, and jobsite conditions.

The initial planning information and data are shown in Table 3.2.

Category	Parameters
General	Pipeline maps, including detailed plan and profile view drawings
	Existing pipe shape (circular/noncircular)
	Existing pipe material
	Existing pipeline diameter
	Soil type
	Ground cover depth
	Ground conditions
	Groundwater table
	Major obstacles (roots, joint settlements, straightness, bends, etc.)
	Existing liner (if any) material and thickness
	Joint types
Fluid characteristics data	pH levels
	Solid characterization
	Temperature
	High-pressure or high-velocity flow
Inspection results	Internal pipeline conditions
	Soil condition outside the pipeline
	Voids located around pipeline pipe
Jobsite constraints	Accessibility
	Flow interruption
	Personnel entry
	Excavation
	Operator service interruption
	Safety issues

Table 3.2 Pipeline Data and Information Gathered During Initial Planning

A contractor usually does the pipeline integrity assessment. At the conclusion of this effort, the operator should be able to infer whether the pipeline problems are associated with the following:

- Cracks, fractures, and holes.
- Corrosion.
- Infiltration, inflow, and exfiltration.
- Deformation.
- Line deviation.
- Capacity (hydraulic).
- Joints.
- Debris and obstructions.
- Construction fracture (connections).

Table 3.3 lists some typical findings and conclusions from pipeline assessment effort.

3.5.2 Step 2

After the pipeline assessment has been completed and the problem(s) have been defined, candidate renewal methods need to be identified. This step requires the user to select one or more performance parameters that best represent the pipeline condition. Once the parameters are selected, use Table 3.4 to select preliminary candidate methods

Extent of the damage (localized or continuous)
In-line offsets, valves, other obstructions, and so on (number or sizes)
Circumferential breaks, if yes, maximum gap width (in.)
Longitudinal breaks, if yes, maximum gap width (in.)
Lining condition (structurally sound or breaking up or corroded)
Leaking joints
Separated joints, if yes, maximum gap width (in.)
External and internal pipe corrosion (maximum size hole in pipe now and allowed at the end of design life)
Possible sag in the pipeline pipe
Need of pipeline cleaning prior to renewal
Need of enlargement of the discharge capacity
Is reduction of flow capacity permissible?
Constructability issues involved with a specific renewal method (bypass requirements, access pit requirements, etc.)
Preferred new pipeline materials
Number of service laterals or connections

TABLE 3.3 Information and Conclusions from Pipeline Integrity Assessment

Description of the problematic pipeline condition* (See Sec. 3.5.11 for detailed descriptions)							
Renewal Methods*	Infiltration/ Inflow/Exfiltration	Joint Separation	Corrosion	Cracks/ Holes	Joint Problems	Structural Problems	Inadequate Hydraulic Capacity
ILR-PB	Yes	Yes	Yes	Yes	Marginal	Yes	Yes
ILR-PR	Yes	Yes	Yes	Yes	Yes	Yes	Yes
SL	Yes	Yes	Yes	Yes	Marginal	Yes	No
CFP	Yes	Yes	Yes	Yes	Marginal	Yes	No
MSL	Yes	Yes	Yes	Yes	Marginal	Yes	No
CIPP	Yes	Yes	Yes	Yes	Marginal	Yes	No
SIPP-CM	Marginal	Yes	Yes	Yes	Marginal	No	No
SIPP-PL	Marginal	Yes	Yes	Yes	Marginal	No	No
SIPP-SH	Marginal	Yes	Marginal	Yes	Marginal	Yes	No
SIPP-GU	Marginal	Yes	Marginal	Yes	Marginal	Yes	No
LOR-RR	Yes	Yes	No	Marginal	Marginal	Marginal	No
LOR-GR	Yes	Yes	No	Marginal	Marginal	No	No
LOR-IS	Yes	Yes	Yes	Yes	Yes	Yes	Yes
LOR-CIPP	Yes	Yes	Yes	Yes	Marginal	Yes	No

*For a list of abbreviations (see *Acronyms*) at the end of this book.

Table 3.4 Applicability of Various Renewal and Replacement Methods

that are generally applicable and suitable for the performance factors selected. Table 3.4 illustrates applicability of different renewal methods for specific pipeline conditions. It should be noted that localized repair (LOR) or PSR solutions (also called spot repairs) should always be considered first because they can be an important cost-saving alternative to renewing an entire section of a pipeline. Based on actual renewal costs, some decision makers have developed guidelines for determining when to use pipe section solutions and when to use LOR solutions. One suggestion is that if three or more LOR solutions would be required in a pipeline section, then it would be better to use pipe section renewal instead of LOR. This guideline is intended to be used only when the actual cost and effectiveness data are not easily available. Of course, if better data are available, they should always be used.

Main Pipeline Problems

Corrosion. Pipeline corrosions due to corrosive fluid and/or soil.

Cracks or Holes. Cracks owing to pipeline defects or lack of proper bedding or installation or operational overstress (built-in and long-term defects).

Infiltration/Inflow/Exfiltration. Water entering or exiting a pipeline system.

Inadequate Hydraulic Capacity. Hydraulic capacity insufficient to carry more flow.

Joint Problems. Joint settlement, misalignment, sags, and offsets owing to lack of proper bedding or installation or material.

Joint Separation. Joints with infiltration or inflow problems that have caused a void in underlying soil.

Structural Problems. Inadequate structural capacity to carry soil and/or traffic loads.

See Acronyms and Glossary of Terms at the end of this book.

3.5.3 Step 3

After selecting the potentially applicable methods (Step 2), use Table 3.5 to learn about the limitations of the selected candidate methods.

3.5.4 Step 4

Use Table 3.6 to check the candidate methods for cost feasibility and receive proposals.

Note: Many factors influence project cost, which is dependent on specific project conditions. The cost information in this table is the best judgment from municipal projects and includes the overall cost, including cleaning, inspection, bypassing, and sludge disposal.

Renewal Methods*	Must Grout Annular Space	Must Seal Liner Ends and/or Laterals	OC Needed to Reconnect Laterals	Need Entrance Pit	Need Bypass	More than 10% Loss of Diameter	Many Joints in New Pipe	Difficult to Install if Existing Pipe Is Misaligned	Might Damage Nearby Utilities	Not Suitable for all Pipe Materials	Range of Diameters (in)	Personnel Entry into Pipeline Is Needed	No Structural Support
ILR-PB		X	X	X	X			X	X	X	4–42		
ILR-PR		X	X	X	X				X	X	12–36		
SL	X	X	X	X	X[a]	X	X	X			4–100		
CFP		X	X[b]	X[c]	X	X		X			4–63		
SL	X	X		X[d]	X		X	X			More than 36	X	
CIPP		X			X						3–120		
SIPP-CM		X[e]		X	X			X			3–24		X
SIPP-PL				X	X			X			3–24		X
SIPP-SH		X		X	X	X					42–144	X	X[f]
SIPP-GU		X		X	X						42–108	X	X[g]
LOR-RR		X	X[h]		X						48–252		X
LOR-GR					X						3–180		X
LOR-IS		X			X						3–180		X[i]
LOR-CIPP		X			X						6–54		

*For a list of abbreviations, see *Acronyms and Abbreviations* at the end of this book.
Note: See also note to Table 3.4.
[a]Depends on the project. Many SL projects are completed without bypass.
[b]OC not needed for gravity fold and formed.
[c]No pit required for most fold and formed.
[d]Pit may or may not be required depending on size of access opening. In sanitary sewers, may have to remove manhole cone.
[e]No sealing required with cement mortar lining.
[f]Depends on the design. Shotcrete and gunite can be structural depending on reinforcing fibers/bars used.
[g]Depends on the design. Shotcrete and gunite can be structural depending on reinforcing fibers/bars used.
[h]Depends on the repair.
[i]Depends on the repair used.

Table 3.5 Limitations of Various Pipeline Renewal and Replacement Methods

Renewal Method	Renewal Method	Reference Chapter	Experience Indicator	Cost Indicator
PB	Pipe Bursting	8	Medium	Medium
PR	Pipe Reaming	8	Low	High
SL	Sliplining	7	High	Medium
CFP	Close-fit Pipe	11	High	Low
MSL	Modified Sliplining	9	Medium	Medium
CIPP (dia. <18 in.)	Cured-in-Place Pipe	6	High	Low
CIPP (dia. >18 in.)	Cured-in-Place Pipe	6	High	Medium
SIPP-CM	Cured-in-Place Pipe—Cement Mortar	10	Medium	Medium
SIPP-PL	Cured-in-Place Pipe—Panel Lining	10	Medium	Medium
SIPP-SH	Cured-in-Place Pipe—Shotcrete	10	High	Low
SIPP-GU	Cured-in-Place Pipe—Gunite	10	High	Low
LOR-RR	Localized Repair—Robotic Repair	14	High	High
LOR-GR	Localized Repair—Grouting	14	High	Low
LOR-IS	Localized Repair—Internal Seal	14	Medium	Medium
LOR-CIPP	Localized Repair—Cured-in-Place Pipe	14	High	High

Table 3.6 Cost Experience Indicators for Various Renewal Methods

This cost information should be used with caution and for comparison purposes only. Typically, the cost of an industrial pipeline renewal project is two to five times the cost of a municipal project. Most of the experience has come from the municipal sector.

Cost and experience indicators:

- High cost more than \$10/ft per in. diameter
- High experience more than 20 years
- Medium cost \$5 to \$10/ft/in. diameter
- Medium experience 10 to 20 years
- Low cost up to \$5/ft/in. diameter
- Low experience less than 10 years

3.5.5 Step 5

Use Table 3.6 to check the candidate methods for overall and project-specific installation experience.

3.5.6 Step 6

Select a final candidate method or, if there are still several candidates, make a list of the final candidates. To get to this point, the user needs to initially know about the candidate methods but should have an understanding of the overall pipeline conditions.

3.5.7 Step 7

Develop a further basic understanding of the installation process, performance characteristics, and limitations of the selected method. Table 3.4 allows the user to view the classification system of renewal methods used in this book and identify the section where each candidate method is discussed. At this point, it is possible that the user is not satisfied with any of the methods resulting from Step 6. The user must go back to Step 1 and confirm that the pipeline condition assessment is thorough and accurate. Then the user can repeat Step 2. It might be necessary to evaluate possible renewal candidate methods that are rated as marginally suited (Table 3.4) to solve specific problems. When special design considerations are met, these marginal solutions might prove to be acceptable. After new methods are selected, the user should repeat Steps 3 to 7.

3.5.8 Step 8

Use a trenchless technology directory to find local vendors or contractors who provide the selected method.

3.5.9 Step 9

Contact vendors of the selected method to obtain more detailed information and confirmation that the method is indeed appropriate for the intended application.

3.5.10 Step 10

Develop the design and contract documents to procure the applicable method to solve the pipeline problem.

Note: It is likely that in addition to the tables in this book, other sources of information will also be available to the decision makers. The decision process on the suitability of a specific method is a complex matter that can depend on many factors including the information provided in this book as well as the cost-benefit analysis, lifecycle costs analysis, and long-term goals for a specific renewal method. When selecting a renewal method, all available data should be considered—not just the tables in this book. The final recommendation that is submitted for approval should include proper justification such as a summary of the pipeline condition survey as well as conclusions based on the tables in this book and other data used in the evaluation.

3.6 General Design Considerations

Selection of the appropriate renewal technique depends, among other factors, on the type of pipe material being repaired and the type of defect that is being remedied. In cases involving installation of a slightly smaller liner within an existing pipe, the question of hydraulic capacity of the newly renewed pipe needs to be considered. Fortunately, most of the in-situ lining techniques produce a pipe with a lower friction coefficient than the original existing pipe, partially or totally compensating for the reduction in pipeline cross-sectional area. The issue of hydraulic capacity of the renewed pipe is not discussed in this book except to point out that the pipeline hydraulics need to be checked for those renewal techniques that rely on internal liners.

3.6.1 Flow Bypassing Considerations

Most renewal technologies require that the flow in the line be temporarily rerouted during the time the repair work is being conducted. Depending on the circumstances, this can be a significant cost consideration, amounting to up to one-third of the total cost for a pipeline renewal job. In the municipal sector, there are a few renewal technologies, which do not, as a rule, require flow bypassing. These include the segmental SL method, mechanical seals in pipelines, and internally applied grouting in manholes. The key consideration for industrial pipelines is simply whether the safety limitations would preclude personnel from entering the line while it is in service. For lines that carry mainly storm runoff, this may be feasible in an industrial setting, although even here the safety considerations and reviews would be of paramount importance. Other than this exception, the general assumption for industrial pipeline renewal is that the flow in the pipeline will have to be bypassed and rerouted during the time the line is being repaired.

3.6.2 Reinstatement of Service Laterals

For pipelines other than interceptors, the reinstatement of smaller diameter service laterals is another task for most renewal techniques. For some renewal techniques, this is done by cutting from the inside of the main pipe using a CCTV-guided robotic device, and sealing the joint from within the main line using another set of robotic tools. The complete set of tools to effect reinstatement and sealing of a lateral may all be mounted on a single robotic tractor or platform, requiring only one trip through the line to complete the reinstatement of laterals.

For techniques which reestablish the laterals from inside the main, the laterals must first be located. Renewal techniques using flexible liners are amenable to visual location of the laterals via CCTV by looking for a "dimple" on the renewed pipe wall. For liners that produce a stiffer wall, location of the laterals is done either by careful inspection using CCTV before lining or by using more sophisticated sensing techniques after lining renewal. Some techniques involve placing an electronic or radioactive source in the lateral that can be detected by an instrument inside the newly lined pipe.

When property ownership issues do not restrict access to the upstream end of the laterals, the laterals can sometimes be drilled from the lateral into the renewed pipe rather than from inside out. This option is probably more feasible for industrial pipelines than for municipal pipelines as the property ownership issue would generally not be a problem within an industrial operator.

Regardless of how the laterals are located or opened, sealing the connections to them is generally more difficult than sealing the line. If hydraulic integrity is the primary concern, this factor needs to be considered. In general, reference to the allowable leakage rates when conducting a hydrostatic or air pressure proof test on a new or renewed line is based on the assumption that the testing is done before the laterals are reattached.

3.6.3 Open-cut

Although not a trenchless method, open-cut (OC) always needs to be considered as an option during a renewal project. However, compared to a trenchless method, OC is often more costly and takes more time to complete. Most of the extra cost of OC over trenchless methods (about 70 percent of the OC cost) is because of the restoration of the site. Also, the possibility of pipeline damage because of shoring removal, backfill, and compaction (built-in defects) during the OC should be considered. OC involves disturbing the ground (both surface and subsurface) and requires potentially costly environmental disposal of any contaminated soil (e.g., from previous leaks) that might be discovered. Conventional dig-and-replace method can also disrupt surface traffic and other surface operations (refer to Najafi and Gokhale (2005) for more information on social costs of the OC method).

3.7 Structural Design of CIPP

The CIPP method is the most widely used of all the pipeline renewal methods (Fig. 3.1). For this reason, this section presents an overview of CIPP design procedure. This presentation also serves as an example for the design of other renewal methods in which same principles are usually applicable. The CIPP should be designed in accordance with ASTM

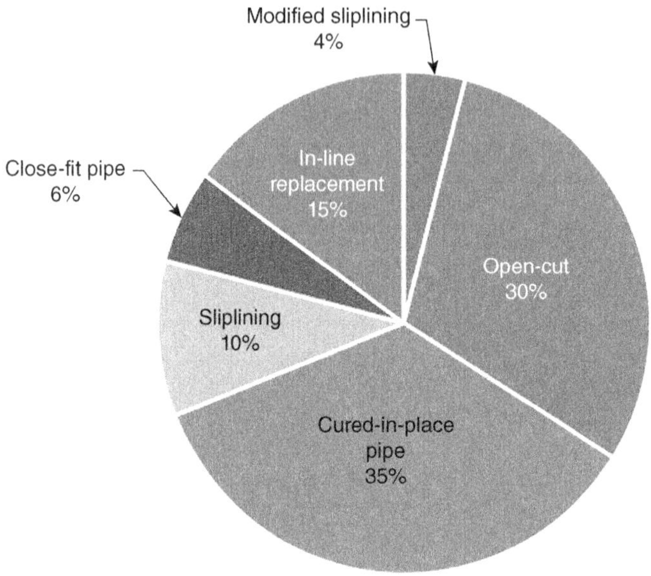

Figure 3.1 Usage statistics of pipeline renewal methods.

F-1216 entitled "Standard Practice for Renewal of Existing Pipelines and Conduits by the Inversion and Curing of a Resin-Impregnated Tube." This standard has a nonmandatory design appendix, which addresses the method used for calculating wall thickness. Virtually all CIPP suppliers and contractors recognize this standard. A copy of this standard can be purchased from the ASTM. The standard for the CIPP is all-inclusive covering material requirements, construction methods, and design parameters. The equations and definitions used in this section are taken from the ASTM F-1216.

As for any other renewal method, the first step in designing a CIPP project is identifying the condition of the existing pipe. The ASTM F-1216-09 divides existing pipe conditions into two classes: partially deteriorated condition and fully deteriorated condition. The assignment of a partially or a fully deteriorated design procedure depends upon the existing condition of the existing pipe or its expected structural contribution over the CIPP design period. An engineer with experience in designing CIPP is most qualified in determination of a design procedure.

Alternative designs for CIPP have merits and potentially offer more accurate predictions of performance. However, it is not within the scope of this book to provide alternative design equations until they have been accepted and adopted by specific organizations, such as ASTM, AWWA, and ASCE. In order to provide a basis for these design models, a review of the development of design theory used in ASTM standards will be discussed.

3.7.1 Major Terms and Design Variables

As for any structural design, there are many variables and parameters used in the design of the pipe wall thickness, which provides the required performance requirements (strength, stiffness, etc.) to resist the loads. For a proper design, it is important to use quantities that are representatives of the site conditions for each pipe section designed. If precise values are not known, good engineering judgment should be used to estimate the magnitudes of some of these variables. Design engineers with familiarity and experience with CIPP product and its design can offer required engineering judgments. Some of these parameters are specified below (see Table 3.7 for minimum recommended design properties for CIPP):

3.7.1.1 Terms

Tube. The fabric material product used to carry and hold the thermoset resin (dry product prior to wet out) in place.

Liner. A fabric tube that has been saturated with a liquid resin system.

CIPP. The final installed product once the liner is cured into a pipe.

3.7.1.2 Design Variables

Flexural Modulus of Elasticity, E. The correct resin or tube selection for the effluent and exposure will determine the associated physicals used for that system. Fillers or reinforcements should be evaluated and used appropriately as needed. All of this must be considered prior to assigning a modulus value. Typical values for E are 250,000 pound per square inch (psi) (minimum accepted value) to about 500,000 psi.

Long-term Flexural Modulus of Elasticity, E_L. Independent analysis should be used to determine the 50-year retention modulus value used in the design calculations. Typically, a reduction of the short-term modulus of 50 percent is used for long-term design values.

Soil modulus, E_S'. The type of soil, depth of burial, and surface use will determine the range for soil modulus. If little is known about

Test Property	Epoxy Resin	Vinyl Ester	Isophthalic Polyester	Filled Isophthalic Polyester
Flexural modulus, ksi	300	250–300	250–350	400–500
Flexural strength, ksi	5.0	4–5	4–5	4.0
Tensile strength, ksi	4.0	2.5–3.5	2.5–3	2–2.5

Table 3.7 Minimum Recommended Design Properties for CIPP

the conditions, a 700-psi value should be used. For example, if the pipe runs under a roadway at a depth of 10 ft or more, a minimum value of 1,000 psi should be used.

Ovality, q. The existing pipe out of roundness significantly contributes to the wall thickness calculation of CIPP. Estimates of the percentage of ovality should be made using the best pipe dimension information available to ensure sufficient wall thickness. In existing pipes, requiring renewal that has not been closely evaluated, 2% ovality is commonly used to estimate the required wall thickness. To quantify q, in the large-diameter existing pipes where personnel entry is possible, actual in-pipe measurements should be used. Laser profiling can also be used in most pipe sizes.

Factor of Safety, N. Usually, the factor of safety for pipeline TRMs is 2. However, in large-diameter, personnel-entry pipes where more accurate data and thorough existing pipe investigations data, such as ovality, groundwater pressure, and so on, can be quantified, a value of 1.5 is acceptable.

3.7.2 Design Background

The objective of structural buried pipe design evolves around the ability to develop a set of equations that can model resisting loads, such as groundwater, soil, traffic, and other loadings. Through practical experience, it has been determined that buried cylindrical linings usually fail by buckling when exposed to an external load. Timoshenko et al. (1988) have published some of the earliest proven buckling theories in the early 1900s. Equations provided by them were subsequently modified to take into account long tubes having practical thickness consistent with the pipe and structures available at the time. One of the first practical applications of this work was the successful development of the first submarines. The unrestrained buckling equation that was developed for long, thin tubes is given in Eq. (3.1) as follows:

$$P_W = \frac{2Et^3}{(1-v^2)D_m{}^3} \tag{3.1}$$

where P_W = hydrostatic water pressure, psi
E = modulus of elasticity, psi
t = pipe wall thickness, in.
v = Poisson's ratio, typically = 0.3
$D_m = (D_o - t)$ = mean pipe diameter, ft
D_o = mean outer CIPP diameter, ft

In the 1940s, Spangler published work that was conducted on flexible piping systems. This work was the basis on which stiffness of flexible pipes was derived. The measurement of pipe stiffness has been standardized with ASTM D2412 and is determined at a pipe deflection of 5 percent. This is a relatively simple test and is performed on a freestanding, unsupported pipe placed between two parallel plates that are pressed toward each other at a controlled rate. Spangler also developed a model for the deflection of buried flexible pipe that took into account factors such as dead load forces, pipe bedding, and soil modulus.

Work by these early pioneering engineers was extremely important in laying the foundational basis of design equations used for CIPP today. However, it is important to understand that there is a very little similarity between the loading experienced by installed CIPP and that of buried rigid or flexible pipe. CIPP is installed into existing pipe that has typically been buried for many years. As such, the soil has long since consolidated and the soil pipe system is typically stable. Therefore, installed CIPP is supported by the soil pipe system, and the subsequent pipe deflections can be expected to be minimal. Also, when CIPP is installed into an existing pipe, the surrounding pipe provides constrained ring support to the CIPP under the influence of uniform hydrostatic water pressure.

When CIPP is exposed to this type of loading, it is under compression. If the load increases to a critical level, the CIPP will eventually deform and fail by buckling. Hydrostatic buckling experiments, carried out by Aggerwal and Cooper (1984), Lo and Zhang (1994), and Kleweno (1997), have demonstrated practical ranges of the enhancement that can be obtained by the support provided by the existing pipe. These studies demonstrate that supported CIPP could buckle at pressures that are seven to fifteen times greater than that of unrestrained CIPP. In order to account for this support in the development of buckling equations used in the design of CIPP, this phenomenon was characterized as an enhancement factor and assigned the variable K. The enhancement factor is the ratio between the restrained buckling pressure and the unrestrained buckling pressure. By applying a statistical treatment of the data generated by Aggerwal and Cooper (1984), K was assigned a value of seven. In other words, there is a high statistical confidence that the restrained buckling pressure will be at least seven times greater than the unrestrained buckling pressure.

Even with the application of enhancement factor, CIPP failure may still occur as a result of plastic creep deformation. Materials such as thermoset and thermoplastic resins will undergo slight deformation over time when exposed to a constant load, such as hydrostatic pressure. Given enough pressure and long enough period of time, the CIPP can deform to the extent that it will produce catastrophic failure by buckling. In order to take the long-term effects of creep into account, the modulus of elasticity in buckling equation attributed to Timoshenko was modified to a long-term modulus. In addition, a safety factor and correction for pipe ovality were also added to obtain the restrained buckling equation. By substituting the dimension ratio

(DR) for the mean diameter and rearranging, the equation reduces as given in Eq. (3.2).

$$P_W = \frac{2\,KE_L}{(1-v^2)}\frac{1}{(DR-1)^3}\frac{C}{N} \tag{3.2}$$

where E_L = long-term modulus of elasticity of pipe material, psi
K = enhancement factor, typically $K = 7$
$DR = D_o/t$, dimensionless
D_o = mean outside diameter of the CIPP, ft
N = safety factor
C = ovality correction factor (Table 3.8)

$$C = \left[\frac{D_{min}}{D_{max}{}^2}\right]^3 = \left[\left(1-\frac{q}{100}\right)\Big/\left(1+\frac{q}{100}\right)^2\right] \tag{3.3}$$

$$q = 100\times\frac{(D-D_{min})}{D} \quad \text{or} \quad 100\times\frac{(D_{max}-D)}{D} \tag{3.4}$$

where C = ovality reduction factor (Table 3.8)
q = percentage of ovality of the original pipe
D = inside diameter of the original pipe, in.
D_{min} = minimum inside diameter of the existing pipe, in.
D_{max} = maximum inside diameter of the existing pipe, in.

There is no single design equation that can be used for all different conditions that must be taken into account for the proper design of CIPP, it is necessary to divide these conditions into different groups. For both gravity flow and internal pressure, design equations have been divided into categories of "partially deteriorated" and "fully deteriorated" conditions of the existing pipe to be renewed. These piping conditions are defined in the following sections.

3.7.3 Partially Deteriorated Gravity Condition

Once again, a partially deteriorated gravity flow pipe is one in which the existing pipe may have displaced joints, cracks, or corrosion, but is structurally able to support all soil and surface loads. In this case, the existing pipe is intended to provide structural support over the full circumference of the CIPP. Assuming that a pipe is partially deteriorated, the CIPP will be designed to withstand uniform hydrostatic

Percent ovality	1	2	3	4	5	6	7	8	9	10
Reduction factor, C	0.91	0.84	0.76	0.70	0.64	0.59	0.54	0.49	0.45	0.41

Table 3.8 Ovality Reduction Factor, C

pressure over the full circumference of the CIPP. In addition, as a conservative approach, this design does not assume that the CIPP is attached to the existing pipe in any way. The hydrostatic pressure on the CIPP is calculated as shown in Eq. (3.5):

$$P_W = \frac{H_W \text{ft} \times 62.4 \text{ pcf (unit weight of water)}}{144 \text{ in}^2/\text{ft}^2} \tag{3.5}$$

where H_W = groundwater height above the invert of the pipe and the other variables are defined earlier.[2]

A partially deteriorated pressure pipe is one in which the existing pipe may also have minor corrosion, leaking joints, and/or small holes, and should be free of any longitudinal cracks. In this case, the existing pipe is assumed to be able to withstand the specified internal design pressure over the expected lifetime of the pipe. When assuming that a pressure pipe is partially deteriorated, it is assumed that the CIPP will conform tightly against the existing pipe everywhere (i.e., in the bends or diameter changes, etc.) and uses the strength of the existing pipe to support the stresses. The thickness of the CIPP can be considered to span small holes or leaking joints but will not be of sufficient thickness to withstand design pressures. In addition, if the partially deteriorated pressure pipe is assumed to be leaking, the designer must also be aware of the external hydrostatic pressure to ensure that the minimum CIPP thickness is sufficient to withstand these forces. In most pressure pipe, the internal pressure is greater than external pressure due to groundwater, so resistance to external pressure is only considered when the pressure pipe is out of service or depressurized.

3.7.4 Fully Deteriorated Gravity Condition

A fully deteriorated gravity flow pipe is one in which the existing pipe has insufficient strength to support all soil and surface loads. A fully deteriorated pipe is characterized by severe corrosion, missing pipe, longitudinal cracks, and severely deformed pipe. Assuming a pipe to be fully deteriorated, the CIPP is designed as a pipe that will be able to withstand all hydrostatic, soil, and live loads that may exist in the CIPP-soil system with adequate soil support.

An alternative strategy for fully deteriorated gravity flow pipes is available to the designer in areas where there are isolated sections of missing or severely offset pipe that would otherwise cause it to be classified as fully deteriorated. In these areas, it may be possible to carry out point repairs and renew the pipe as a partially deteriorated

[2]For the partially deteriorated equation, groundwater height is taken from the invert of the pipe. It can be measured from the top of pipe, but then the diameter of the pipe must be added to H_W for the calculation. For fully deteriorated (modified AWWA), height of water is taken from the top of the pipe.

classification. However, each situation must be considered by a knowledgeable professional engineer.

A fully deteriorated pressure pipe is one in which the existing pipe has failed or has insufficient strength to operate at specified design pressures, or both. A pipe may also be classified as fully deteriorated if it is determined that it will not be able to withstand design pressures at some point during the expected lifetime. A fully deteriorated pressure pipe is characterized by significant loss of wall thickness owing to severe corrosion, large holes, missing sections of pipe, and leaking longitudinal cracks. Assuming a pipe to be fully deteriorated, the CIPP is designed as a stand-alone pipe that will be able to withstand all internal pressure. In addition, the designer must also be aware that fully deteriorated CIPP pressure pipe must be capable of withstanding external hydrostatic pressure.

3.7.5 Partially Deteriorated Gravity Flow CIPP Design

While renewing existing pipe that has been classified as partially deteriorated for gravity flow condition, the restrained buckling condition applies. In this case, the classical buckling equation that has been described previously is rearranged in Eq. (3.6) to solve for CIPP thickness.

$$t = \frac{D_o}{\left(\dfrac{2KE_L C}{P_W N(1-v^2)}\right)^{1/3} + 1} \qquad (3.6)$$

where D_o = mean outer CIPP diameter, in.
K = enhancement factor, typically $K = 7$
E_L = long-term modulus of elasticity of the pipe material, psi
C = ovality reduction factor (Table 3.8)
P_W = external water pressure measured above the pipe invert, psi
N = safety factor, typically $N = 1.5$ to 2
v = Poisson's ratio, typically $v = 0.3$, dimensionless

For partially deteriorated design conditions where the groundwater is below the invert of the pipe, the hydrostatic pressure is equal to zero, and the restrained buckling equation cannot be used to calculate the CIPP thickness. For this special design case, the calculated thickness of the CIPP must be equal to or greater than that which will produce a maximum DR = 100. This ASTM requirement will require a CIPP pipe with stiffness in the range of 1 to 2 psi. When this special design condition exists, the CIPP thickness is determined by Eq. (3.7).

$$t = D_o/100 \qquad (3.7)$$

While designing for circular partially deteriorated pipe, the CIPP is under constant compressive hoop stresses. If the existing pipe is out of round or has localized ovalization, bending moment forces may predominate on the CIPP. For this special case, the CIPP must be checked to ensure that the bending forces do not exceed the long-term flexural strength of the CIPP. To determine this, the bending stresses on the CIPP are determined by the Eq. 3.8.

$$\frac{S}{P_W N} = [1.5q/100(1+q/100)DR^2] - [0.5(1+q/100)DR] \qquad (3.8)$$

where Eq. (3.4) defines q, and the other parameters have been defined previously.

Table 3.9 presents sample design thicknesses for different values of H_w (or P_w) using particular values for E_L, C, N, ν, and K.

For example, determine the minimum wall thickness required for the following existing gravity piping condition (Fig. 3.2):

- Existing pipe classification = partially deteriorated
- Mean outer CIPP diameter (D_o) = 24 in.
- Minimum pipe diameter (D_{min}) = 23.1 in.
- External water above invert (H_W) = 8 ft
- Minimum CIPP flexural modulus (E) = 350,000 psi
- Minimum CIPP flexural strength (S) = 5,500 psi
- Long-term modulus (E_L) = 175,000 psi
- Long-term strength (S_L) = 2,750 psi

Step A. Determine hydrostatic pressure acting on CIPP [Eq. (3.5)].

$$P_w = \frac{8 \text{ ft} \times 62.4 \text{ pcf (unit weight of water)}}{144 \text{ in}^2/\text{ft}^2} = 3.46 \text{ psi}$$

Step B. Calculate the pipe ovality.
Determine q using Eq. (3.4).

$$q = \frac{100(24 \text{ in.} - 23.1 \text{ in.})}{24 \text{ in.}} = 3.75\%$$

Determine ovality reduction factor using Eq. (3.3).

$$C = \left[\frac{(1-3.75/100)}{(1+3.75/100)^2}\right]^3 = 0.715$$

Step C. Determine minimum CIPP design thickness using buckling Eq. (3.6).

$$t = \frac{24 \text{ in.}}{\left(\frac{2(7)\times 175,000 \text{ psi } (0.715)}{2(3.46 \text{ psi})(1-0.3^2)}\right)^{1/3} + 1} = 0.36 \text{ in.}$$

		E_L		175,000 psi								K				7			
		C		0.84								$1 - \nu^2$				0.91			
		N		2															
Variables		CIPP thickness (in.) for each pipe diameter (in.)																	
Water depth (H_W), ft	Water pressure (P_W), psi	6	8	10	12	15	18	21	24	27	30	33	36	42	48	54	60	66	72
1	0.4380	0.04	0.06	0.07	0.09	0.11	0.13	0.15	0.17	0.20	0.22	0.24	0.26	0.30	0.35	0.39	0.43	0.48	0.52
2	0.8700	0.06	0.07	0.09	0.11	0.14	0.17	0.19	0.22	0.24	0.27	0.30	0.33	0.38	0.44	0.49	0.54	0.60	0.65
3	1.3000	0.06	0.08	0.10	0.13	0.15	0.19	0.22	0.25	0.28	0.31	0.34	0.37	0.44	0.50	0.56	0.62	0.69	0.75
4	1.7300	0.07	0.09	0.11	0.14	0.17	0.20	0.24	0.27	0.31	0.34	0.37	0.41	0.48	0.55	0.61	0.69	0.75	0.82
5	2.1700	0.07	0.10	0.12	0.15	0.17	0.22	0.26	0.30	0.33	0.37	0.41	0.44	0.52	0.59	0.66	0.74	0.81	0.88
6	2.6000	0.08	0.10	0.13	0.16	0.20	0.24	0.27	0.31	0.35	0.39	0.43	0.47	0.55	0.63	0.70	0.78	0.86	0.94
7	3.0300	0.08	0.11	0.14	0.17	0.20	0.25	0.29	0.33	0.37	0.41	0.45	0.49	0.57	0.66	0.74	0.82	0.91	0.99
8	3.4600	0.09	0.11	0.14	0.17	0.22	0.26	0.30	0.34	0.39	0.43	0.47	0.52	0.60	0.69	0.77	0.86	0.94	1.03
9	3.9000	0.09	0.12	0.15	0.18	0.22	0.27	0.31	0.36	0.40	0.44	0.49	0.54	0.63	0.71	0.80	0.89	0.98	1.07
10	4.3300	0.09	0.12	0.15	0.19	0.23	0.28	0.32	0.37	0.42	0.46	0.51	0.56	0.65	0.74	0.83	0.93	1.02	1.11
11	4.7600	0.09	0.13	0.16	0.19	0.24	0.29	0.33	0.38	0.43	0.48	0.52	0.57	0.67	0.76	0.86	0.95	1.05	1.15
12	5.2000	0.10	0.13	0.17	0.20	0.24	0.30	0.34	0.39	0.44	0.49	0.54	0.59	0.69	0.78	0.88	0.98	1.08	1.18
13	5.6300	0.10	0.13	0.17	0.20	0.25	0.30	0.35	0.40	0.45	0.50	0.56	0.61	0.70	0.81	0.91	1.01	1.11	1.21
14	6.0600	0.10	0.14	0.17	0.20	0.26	0.31	0.36	0.41	0.46	0.52	0.57	0.62	0.72	0.83	0.93	1.03	1.13	1.24
15	6.5000	0.11	0.14	0.18	0.21	0.26	0.31	0.37	0.42	0.48	0.53	0.58	0.63	0.74	0.85	0.95	1.06	1.16	1.27
16	6.9300	0.11	0.15	0.18	0.22	0.27	0.32	0.38	0.43	0.48	0.54	0.59	0.65	0.76	0.86	0.97	1.08	1.19	1.30
17	7.7360	0.11	0.15	0.19	0.22	0.28	0.33	0.39	0.44	0.50	0.55	0.61	0.66	0.77	0.88	0.99	1.10	1.21	1.32
18	7.7900	0.11	0.15	0.19	0.22	0.28	0.33	0.39	0.45	0.50	0.56	0.62	0.67	0.78	0.90	1.01	1.12	1.23	1.34
19	8.2300	0.11	0.15	0.19	0.23	0.28	0.34	0.40	0.46	0.51	0.57	0.63	0.69	0.80	0.91	1.03	1.14	1.26	1.37
20	8.6600	0.11	0.15	0.19	0.23	0.29	0.35	0.41	0.46	0.52	0.58	0.64	0.70	0.81	0.93	1.04	1.16	1.28	1.39

TABLE 3.9 Partially Deteriorated Gravity Condition

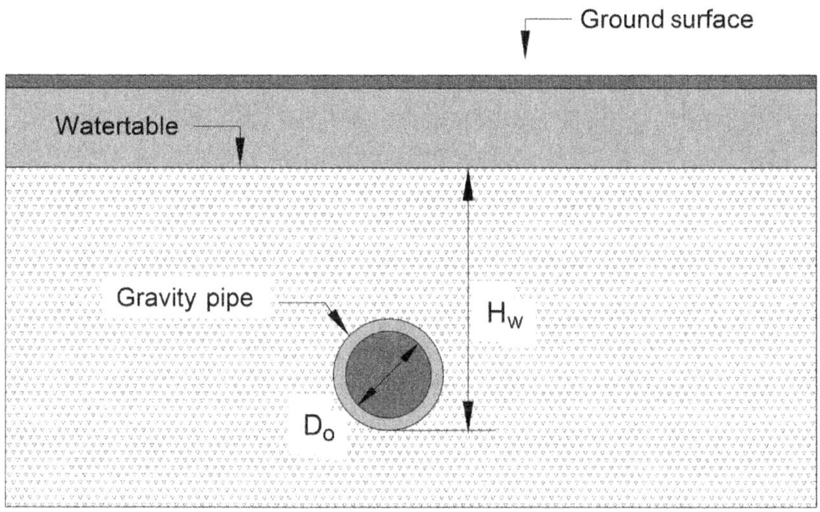

Figure 3.2 Partially deteriorated gravity design example.

Step D. Because the pipe is out of round, bending stresses must be calculated to ensure they do not exceed the long-term flexural strength of the CIPP.

Determine S using Eq. (3.8).

$$DR = D_o/t = 24/0.36 = 66.67$$

$$S/(3.46 \text{ psi})2 = [1.5(3.75/100)(1+3.75/100)66.67^2] - [0.5(1+3.75/100)66.67]$$

$$S = 6.92[259.4 - 34.55]$$

$$S = 1,556 \text{ psi}$$

Step E. The minimum CIPP design thickness is 0.36 in. because the bending stresses are less than the long-term CIPP flexural strength. However, if the calculated bending stress is greater than the long-term flexural strength, Equation 3.8 would control. In this case, solving for the proper thickness can be accomplished by trial and error. Starting by choosing DRs that are smaller than those previously used until the bending stress is less than the long-term flexural strength of the CIPP.

3.7.6 Fully Deteriorated Gravity Flow CIPP Design

While renewing existing pipe that has been classified as fully deteriorated in a gravity flow condition, ASTM F-1216 and ASTM F-1743 specify the use of a design equation from American Water Works

Association (AWWA) C-950 that has been modified by adding the ovality reduction factor and the consideration of long-term effects due to creep. In this case, the modified AWWA C-950 equation from ASTM F-1216 has been rearranged to solve for CIPP thickness as given in Eq. (3.9).

$$t = 0.721 \times D \times \left(\frac{(NP_t)^2}{E_L \times C \times R_W \times B' \times E'_S} \right)^{1/3} \tag{3.9}$$

where P_t = total pressure due to water, soil, and live load acting on pipe, psi

R_W = buoyancy factor, dimensionless

B' = empirical coefficient of elastic support, dimensionless

E' = modulus of elasticity of adjacent soils or modulus of soil reaction, psi

The CIPP designed by the modified AWWA C-950 formula is required to have a minimum stiffness EI/D_o^3, which is 50 percent of the AWWA minimum stiffness. The AWWA C-950 specifications call for EI/D_o^3 to be equal to 0.186, and 50 percent of this value is 0.093. In the following equation, this means that pipe designed with a flexural modulus of elasticity E = 350,000 psi would have a DR equal to 68 for fully deteriorated pipe. If the CIPP stiffness is too low, the wall thickness must be increased accordingly to ensure that the design condition in Eq. (3.10) is met.

$$\frac{EI}{D_o^3} = \frac{E}{12(DR)^3} \geq 0.093 \tag{3.10}$$

where E = flexural modulus of elasticity of the CIPP, psi

I = moment of inertia, in^4, $I = t^3/12$

While designing fully deteriorated CIPP where the existing pipe is out of round or the CIPP may have localized ovalization, bending moment forces may predominate on the CIPP. For this special case of the fully deteriorated design, the CIPP must be checked to ensure that the bending forces do not exceed the long-term flexural strength of the CIPP. To determine this, the bending stresses on the CIPP are determined by modifying Eq. (3.8) and substituting total pressure (P_t) to produce Eq. (3.11).

$$\frac{S_L}{P_t N} = [1.5q/100(1+q/100)DR^2] - [0.5(1+q/100)DR] \tag{3.11}$$

where Eq. (3.4) defines q, and the other parameters have been defined previously.

3.7.7 Total External Pressure on CIPP

Several new parameters are introduced for the design of fully deteriorated gravity flow pipe. While determining fully deteriorated designs, all loads acting on the CIPP must be estimated to determine the total pressure (P_t). Estimating the contribution of each individual load and adding them together accomplish this. The total load is typically made up of hydrostatic water pressure (P_W'), buoyancy corrected soil load (P_S), superimposed or live loads (P_L), and other loads such as a vacuum load (P_V). Loading due to vacuum is a special case and will not be discussed in this book. The total pressure on the pipe can be represented as shown in Eq. (3.12).

$$P = P_W' + P_S + P_L + P_V \tag{3.12}$$

3.7.8 Hydrostatic Water and Soil Loads

Initially, groundwater and soil heights must be determined or estimated to begin the design process. For the fully deteriorated design condition, the groundwater and soil heights are determined from the top of the pipe and not the invert. The hydrostatic pressure is determined with Eq. (3.5) (see Sec. 3.7.3) rewritten below.

$$P_W' = H_W\left(0.433\frac{\text{psi}}{\text{height of water (ft)}}\right)$$

The contribution related to soil loading involves many different parameters. The soil prism loading pressure is determined by Eq. (3.13).

$$P_S = \frac{\gamma \cdot H_S \cdot R_W}{144\left(\dfrac{\text{in}^2}{\text{ft}^2}\right)} \tag{3.13}$$

where γ = soil unit weight, lb/ft³ (see Table 3.10 for sample soil types and unit weights)

H_S = soil height above top of pipe, ft

R_W = water buoyancy factor, dimensionless

$$R_W - 1 - 0.33(Hw/H_S) > 0.67 \tag{3.14}$$

Other related design parameters are the modulus of soil reaction or elastic support (E') and the coefficient of elastic support (B'). The modulus of soil reaction values used for CIPP design should typically represent stable undisturbed soils that would have E' values in the

Soil Type	Unit Weight, γ(lb/ft³)
Sand and gravel	110
Saturated topsoil	115
Ordinary clay	120
Saturated clay	130

TABLE 3.10 Soil Types and Unit Weights

range of 700 to 3,000 psi. Most typically, a value of 700 psi is recommended for unknown soil conditions. Where the pipe is buried deep and the soil conditions stable, values of 1,000 to 1,500 psi may be applicable. In areas known to have weak and unstable native soils, a value of 300 psi may be more appropriate. The coefficient of elastic support (B') is determined with Eq. (3.15). Table 3.10 presents elastic support coefficients at various soil depths.

$$B' = 1/(1 + 4e^{-0.065H_s}) \tag{3.15}$$

3.7.9 Superimposed or Live Loads

For the fully deteriorated design condition, dynamic live load pressures (P_L) occur frequently and are a standard design condition for the parameter P_L. Live loads may be classified as either concentrated or distributed, depending on the soil pipe conditions and the depth the pipe is buried. In some cases, the live load may be characterized by impact factors. Impact loading is generally only applicable for pipes that are relatively shallow (i.e., 2 to 5 ft). A number of maximum live load conditions have been studied and recommended for pipe buried beneath highways, railways, and airport runways. The generally accepted guidelines for determining these live loading conditions are provided by the American Association of State Highway and Transportation Officials (AASHTO) Standard Specifications for Highway Bridges, HS-20-44 highway loading, American Railway Engineers Association (AREA), and the Federal Aviation Agency (FAA) respectively. The most frequently encountered design condition is that for pipes buried under active roads or highways. For highway HS-20 loading, the live load becomes insignificant beyond 7 ft of soil height cover over the top of the pipe. Live load pressures (P_L) associated with the aforementioned piping conditions are presented in Table 3.11. Buried pipelines under airport runways and taxiways must be able to support the large aircraft live loads as well as the dead load from the soil and runway pavement. Additional wheels on the aircraft landing gear will spread the aircraft weight over proportionally more area to limit stress in the pavement, but the load on buried structures may increase because of the overlap of individual wheels or groups of wheel loads. The design of buried pipeline is dependent on the pavement material and thickness, distributed live load magnitude, shape and size of the pipe and depth of fill over the top of the pipeline.

For example, determine the minimum wall thickness required of the following fully deteriorated gravity piping conditions Fig. 3.3:

- Existing pipe classification = Fully deteriorated
- Mean outer CIPP diameter (D_o) = 48 in.
- Minimum pipe diameter (D_{min}) = 47.04 in.

Soil Height, H_s	Highway, HS-20-44		Railway, E-80	
	Pressure, psi	Impact Factor	Pressure, psi	Impact Factor
0–1	>15.1	0.3	NA	0.4
1	15.1	0.3	NA	0.4
2	10.9	0.2	26.4	0.36
3	5.3	0	23.6	0.28
4	2.2	0	18.4	0.24
5	1.7	0	16.7	0.20
6	1.3	0	15.6	0.16
7	1.1	0	12.2	0.12
8	1.0	0	11.1	0.08
9	*	0	9.4	0.04
10	*	0	7.6	0
12	*	0	5.6	0
15	*	0	4.2	0

Table 3.11 Live Load Pressure and Impact Factors

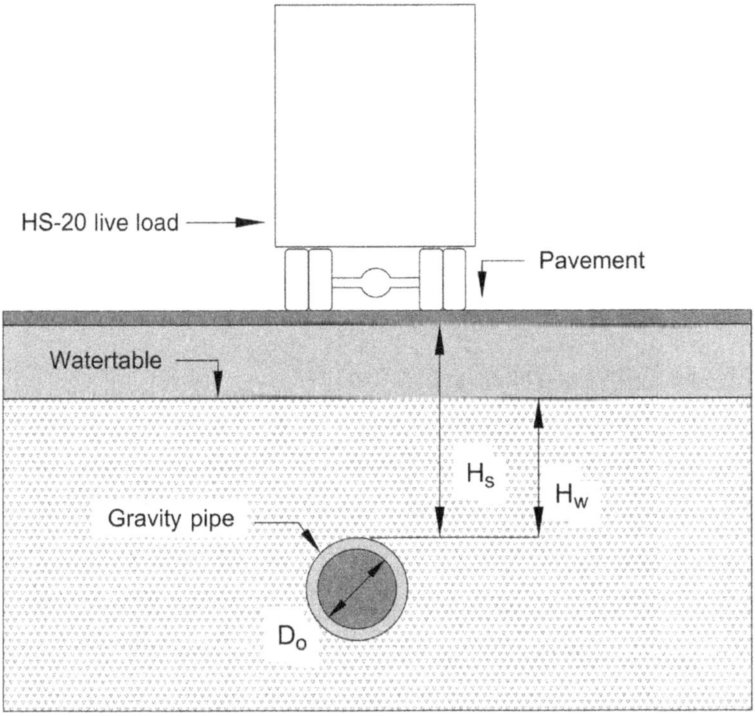

Figure 3.3 Fully deteriorated gravity design example.

- External water above pipe (H_W) = 8 ft
- Depth of soil cover above pipe (H_S) = 15 ft
- Type of soil = Ordinary clay (120 lb/ft^3)
- Soil modulus (E_S') = 700 psi
- Live load = HS-20
- Minimum CIPP flexural modulus (E) = 350,000 psi
- Minimum CIPP flexural strength (S) = 5,500 psi
- Long-term modulus (E_L) = 175,000 psi
- Long-term strength (S_L) = 2,750 psi

Step A. Determine the total load [Eq. (3.12)]

$$P_t = P'_W + P_S + P_L$$

Hydrostatic water pressure [Eq. (3.5)]

$$P'_W = 0.433(H_W) = (0.433 \text{ psi/ft})(8 \text{ ft}) = 3.46 \text{ psi}$$

Soil load

$$P_S = wH_S R_W / 144 \text{ in}^2/\text{ft}^2$$

$$R_W = 1 - 0.33(H_W/H_S) = 1 - 0.33(8\text{ft}/15\text{ft}) = 0.824 \text{ which is} > 0.67$$

Ordinary clay soil unit weight = 120 lb/ft^3 (Table 3.10)

$$P_S = 120(\text{lb/ft}^3) \times 15(\text{ft}) \times 0.824/144(\text{in}^2/\text{ft}^2) = 10.3 \text{ psi}$$

The soil pressure can also be determined by multiplying the soil prism pressure provided in Table 3.12 and the buoyancy correction factor (R_W) presented in Table 3.14.

$$P_S = 0.824(12.5) = 10.3 \text{ psi}$$

Live load (P_L) from Table 3.11 is ~ 0.0 psi.

Total load

$$P_t = 3.46(\text{psi}) + 10.3(\text{psi}) = 13.76 \text{ psi}$$

Step B. Calculate coefficient of elastic support (see Table 3.13)

$$B' = 1/(1 + 4e^{-0.065 H_S})$$

$$B' = 1/(1 + 4\ e^{-0.065(15\text{ft})}) = 0.40$$

Step C. Calculate pipe ovality using Eq. (3.3)

$$q = \frac{100(48 - 47.04)}{48} = 2.0\%$$

Height of Water, H'_w or Soil, H_s, ft	Hydrostatic Pressure P'_w, psi	Soil Prism Pressure, psi Soil Unit Weight, γ, lb/ft³				
		100	110	115	120	130
1	0.43	0.7	0.8	0.8	0.8	0.9
2	0.87	1.4	1.5	1.6	1.7	1.8
3	1.30	2.1	2.3	2.4	2.5	2.7
4	1.73	2.8	3.1	3.2	3.3	3.6
5	2.17	3.5	3.8	4.0	4.2	4.5
6	2.60	4.1	4.6	4.8	5.0	5.4
7	3.03	4.9	5.3	5.4	5.8	6.3
8	3.46	5.6	6.1	6.4	6.7	7.2
9	3.90	6.3	6.9	7.2	7.5	8.1
10	4.33	6.9	7.6	8.0	8.3	9.0
11	4.76	7.6	8.4	8.8	9.2	9.9
12	5.20	8.4	9.2	9.6	10.0	10.8
13	5.63	9.2	9.9	10.4	10.8	11.7
14	6.06	9.9	10.7	11.2	11.7	12.6
15	6.50	10.7	11.5	12.0	12.5	13.5
16	6.93	11.5	12.2	12.7	13.3	14.4
17	7.37	12.2	13.0	13.6	14.2	15.3
18	7.79	13.0	13.8	14.4	15.0	16.3
19	8.23	13.8	14.5	15.1	15.8	17.2
20	8.66	14.5	15.3	16.0	16.7	18.1
22	9.53	15.3	16.8	17.5	18.3	19.9
24	10.4	16.8	18.3	19.2	20.0	21.7
26	11.3	18.3	19.9	20.8	21.7	23.5
28	12.1	19.4	21.4	22.3	23.3	25.3
30	13.0	20.8	22.9	24.0	25.0	27.1

TABLE 3.12 Soil Prism Pressure as a Function of Water or Soil Height and Soil Density

Step D. Determine the minimum CIPP thickness for buckling:

$$t = 0.721 \times D \times \left(\frac{(NP_t)^2}{E_L \times C \times R_W \times B' \times E'_S} \right)^{1/3}$$

$$= 0.721(48)\left(\frac{(2 \times 13.76)^2}{175{,}000(0.824)(0.84)(0.40)(700)} \right)^{1/3} = 0.97 \text{ in.}$$

$$t = 0.97 \text{ in.}$$

$B' = 1/(1 + 4e^{-0.065H_s})$	
Soil Height (H_s, ft)	**Elastic Support (B')**
1	0.20
2	0.21
3	0.23
4	0.24
5	026
6	0.27
7	0.28
8	0.30
9	0.31
10	0.32
11	0.34
12	0.35
13	0.37
14	0.38
15	0.40
16	0.41
1	0.43
18	0.45
19	0.46
20	0.48
22	0.51
24	0.54
26	0.58
28	0.61
30	0.64

Table 3.13 Coefficient of Elastic Support, B'

Step E. Check for minimum pipe stiffness, using Eq. 3.10

$$DR = 48/1 = 48$$

$$350,000/12(48^3) = 0.26 > 0.093$$

Check for pressure due to bending stresses [Eq. (3.11)]

$$S_L = 13.76(2)\{[(1.5(2)/100)(1+2/100)48^2] - [0.5(1+2/100)48]\}$$

$$S_L = 1{,}266 \text{ psi}$$

Step F. Therefore, the design thickness for the CIPP is

$$t = 0.97 \text{ in.}$$

The calculated bending stress (1,266 psi) is less than the estimated long-term bending strength of the resin (i.e., 2,700 psi), so bending stresses does not control the design thickness.

3.7.10 Partially Deteriorated Internal Pressure Pipe

While designing for internal pressure, it is critical to obtain a proper evaluation of the condition of the pipe being renewed. Other important parameters include the operating pressure(s), test pressures, surge pressures, and/or water hammer that may significantly exceed operating or test pressures. In addition, the project engineer and contractor must be aware that a pipe requiring heavy cleaning may change from partially to fully deteriorated condition after such cleaning. After cleaning, it is recommended that the pipe classified as partially deteriorated be tested for operating or test pressure to verify the condition of the pipe prior to lining. If the pipe is able to maintain the specified pressure, then it can be classified as partially deteriorated without question. However, this may not always be possible because of the presence of small holes in the pipe that do not allow it to maintain pressure. When the condition of the pipe, or the operating parameter, or both are not well defined, it is recommended that the pipe be classified as fully deteriorated. Pressure pipe presents a higher risk of application of CIPP; therefore, it is recommended that the contractor has experience in this area of technology to ensure success.

The partially deteriorated design equation for internal pressure pipe given in ASTM F-1216 was derived with the assumption that the CIPP acts like a uniformly pressurized round flat plate with fixed edges covering an existing hole in the pipe. CIPP is designed with the assumption that the aforementioned condition prevails and that bending stresses at and around the hole (if one exists) control the design thickness. This design assumption is more conservative than that of a square or rectangular plate. The equation in ASTM F-1216 is rearranged to solve for CIPP thickness in Eq. (3.16).

$$t = \frac{D_O}{[5.33/P_i(D_o/D_h)^2(S_L/N)]^{0.5} + 1} \tag{3.16}$$

where D_o = mean outer CIPP diameter, in.
P_i = internal pipe pressure, psi
D_h = hole diameter in the existing pipe, in.
S_L = long-term flexural bending strength for the CIPP, psi
N = safety factor, $N = 2$ minimum

In order for the circular flat plate design condition to be valid, the criteria defined in Eq. (3.17) must be satisfied. If this condition is not met, then the CIPP cannot be considered a circular flat plate, and ring tension or hoop stress will dominate, and the CIPP will be designed as a fully deteriorated internal pressure pipe.

$$\frac{D_h}{D_o} \leq 1.83(t/D_o)^{0.5} \tag{3.17}$$

Once the CIPP thickness has been calculated, this value must be compared with the thickness calculated from Eq. (3.6) to confirm that the external hydrostatic water pressure does not dominate the design condition. The larger thickness is then selected for the design.

3.7.10.1 Partially Deteriorated Pressure Flow CIPP Design Example

Determine the CIPP thickness for the following piping conditions (Fig. 3.4):

- Existing pipe classification = partially deteriorated
- Existing pipe inner diameter, $D = 15$ in.
- Existing pipe maximum diameter, $D_{max} = 15.2$ in.
- Internal pressure, $P_i = 80$ psi
- External water, H_w = 5 ft above pipe invert for partially deteriorated pipe
- Maximum pipe hole diameter, $D_h = 1$ in.
- Minimum CIPP flexural modulus, $E = 350{,}000$ psi
- Minimum CIPP flexural strength, $S = 5{,}500$ psi
- Minimum long-term modulus, $E_L = 175{,}000$ psi

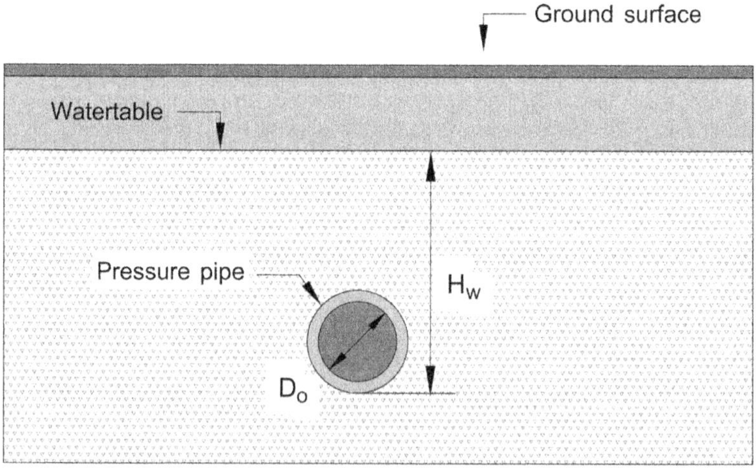

Figure 3.4 Partially deteriorated internal pressure CIPP design example.

- Minimum long-term strength, $S_L = 2{,}750$ psi
- Minimum long-term tensile strength, $S_{tL} = 1{,}750$ psi

Step A. Determine the pressure pipe thickness using Eq. (3.16).

$$D_O = \frac{D + D_{\max}}{2} = \frac{15.0 \text{ in.} + 15.2 \text{ in.}}{2} = 15.1 \text{ in.}$$

$$t = \frac{15.1}{\left[\left(\frac{5.33}{80}\right)\left(\frac{15.1}{1.0}\right)^2\left(\frac{2750}{2.0}\right)\right]^{0.5} + 1}$$

$$t = 0.1 \text{ in.}$$

Step B. Check thickness with Eq. (3.17).

$$\frac{1}{15.1} \leq 1.83\left(\frac{0.1}{15.1}\right)^{0.5}$$

$$0.066 \leq 0.149$$

The condition of Eq. (3.17) is met.
Step C. Check thickness for external pressure using Eq. (3.6).

$$q = 100\frac{(15.2 - 15.0)}{15.0} = 1.33\%$$

$$C = \left[\left(1 - \frac{1.33}{100}\right) \Big/ \left(1 + \frac{1.33}{100}\right)^2\right]^3 = 0.89$$

$$P_W = 5(4.33) = 2.2$$

$$t = \frac{15.1}{\left[\frac{(2)(7)(175000)(0.89)}{(2.2)(2)(1 - 0.3^2)}\right]^{1/3} + 1} = 0.18$$

The thickness for external pressure is greater than that required for internal pressure and the $DR = (15.1/0.18) = 84$ is less than 100 as required in Eq. (3.7).

Step D. As the pipe is slightly out of round, the bending stresses must be checked using Eq. (3.8).

Note: This the case when there is no internal pressure

$$S_L = 2.2(2)[(1.5(0.0133)(1 + 0.0133)(84)^2) - (0.5(1 + 0.0133)84)] = 440.35$$

$S_L = 440$ psi, which is less than the long-term flexural strength of the CIPP.

Therefore, the final CIPP thickness is $t = 0.18$ in.

3.7.11 Fully Deteriorated Internal Pressure Pipe

As discussed previously, it is critical to understand the physical conditions of the pipe and the operating parameters of the system when designing for fully deteriorated pressure pipe. This pipe classification assumes that the existing pipe has no capability to hold any of the pressure, and the CIPP must be designed to hold all internal and external pressures as a stand-alone pipe.

The design equation for fully deteriorated pressure pipe given in ASTM F-1216 assumes pressure pipe is a thick-walled cylinder as given below. *Thick-wall cylinder* and the internal pressure is calculated in Eq. (3.18).

$$P_i = \frac{2S_{tL}}{(DR-2)N} \tag{3.18}$$

The equation for thin-walled cylinders is given as follows:

$$P_i = \frac{2S_{tL}}{(DR-1)N} \tag{3.19}$$

When rearranged to solve for thickness, Eq. (3.19) becomes

$$t = \frac{D_O}{(2S_{tL}/P_iN)+1} \tag{3.20}$$

where all the design parameters for the pipe have previously been defined.

Although the differences between a thin-wall and a thick-wall cylinder are relatively small, the solution for a thin-walled cylinder is a more conservative approach than what ASTM F-1216 recommends. The CIPP thickness for external pressure should be checked against Eqs. (3.8) and (3.9) for fully deteriorated gravity flow pipe. The greatest thickness is chosen for the pressure pipe.

3.7.11.1 Fully Deteriorated Pressure Flow CIPP Design Example

Determine the CIPP thickness for the following pipe design conditions (Fig. 3.5):

- Existing pipe classification = fully deteriorated
- Existing pipe inner diameter, $D = 15$ in.
- Existing pipe maximum diameter, $D_{max} = 15.2$ in.
- Internal pressure, $P_i = 80$ psi
- External water, $H_W = 5$ ft above top of the pipe
- Soil height, $H_S = 10$ ft above top of the pipe
- Soil type = saturated topsoil
- Soil Modulus, $E' = 1{,}000$ psi

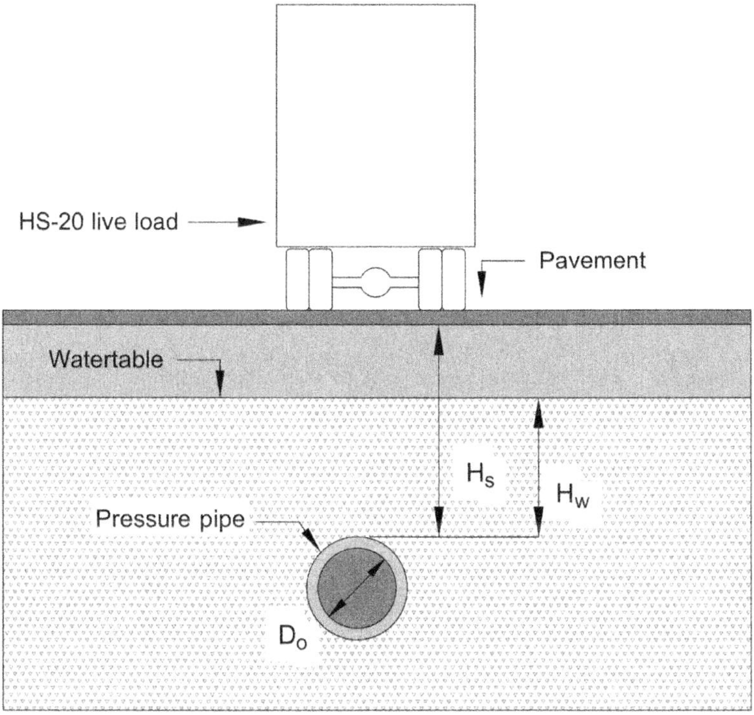

Figure 3.5 Fully deteriorated internal pressure CIPP design example.

- Minimum CIPP flexural modulus, $E = 350{,}000$ psi
- Minimum CIPP flexural strength, $S = 5{,}500$ psi
- Minimum long-term modulus, $E_L = 175{,}000$ psi
- Minimum long term strength, $S_L - 2{,}750$ psi
- Minimum long-term tensile strength, $S_{tL} = 1{,}750$ psi changed to reinforced pipe for 5,000 psi and thickness goes down

Step A. Determine the CIPP thickness using Eq. (3.20).

$$D_o = 15.1 \text{ (from previous example)}$$

$$t = \frac{15.1}{[2(1750)/80(2)+1]} = 0.66 \text{ in.}$$

Step B. Check the CIPP thickness against the fully deteriorated gravity flow design condition [Eq. (3.9)].

Determine the total pressure $P_t = P_W' + P_S + P_L$

$$P_W' = 5(0.433) = 2.2 \text{ psi} \tag{3.21}$$

$$P_S = \frac{115(10)(0.84)}{144} = 6.7\text{ psi} \tag{3.22}$$

$$P_L = 0 \quad \text{(Table 3.11)}$$

$$P_t = 2.2 + 6.7 = 8.9$$

Determine pipe ovality (see Eq. 5.4)

$$q = 1.33\% \ \ [\text{Eq. (3.5)}]$$

$$C = 0.89$$

Determine CIPP thickness [Eq. (3.9)]

$$R_w = 0.84 \ \ \text{(Table 3.14)}$$

$$B^1 = 0.32 \ \text{(Table 3.13)}$$

$$t = 0.721 \times D \times \left(\frac{(NP_t)^2}{E_L \times C \times R_W \times B' \times E'_S} \right)^{1/3}$$

$$= 0.721(15.1)\left(\frac{(2 \times 8.9)^2}{175{,}000(0.89)(0.84)(0.32)(1{,}000)} \right)^{1/3} = 0.21\text{ in.}$$

As 0.21 in. <0.66 in., internal pressure dominates the design.

Step C. Check the CIPP thickness for minimum pipe stiffness [Eq. (3.10)].

$$DR = 15.1/0.66 = 22.9$$

$$\frac{350{,}000}{12(22.9)^3} = 2.43 \geq 0.093$$

Step D. Check CIPP for bending stresses [Eq. (3.11)]

$$S_L = 8.9(2.0)[(1.5 \times (0.0133)(1+0.0133)(22.9)^2) - (0.5(1+0.0133)22.9)]$$

$$S_L = -17.8\text{ psi}$$

As S_L is less than the long-term flexural strength of the CIPP, the bending stresses do not control the design.

Step E. All the checks indicate that internal pressure dominates the design of this fully deteriorated pressure pipe and the specified thickness is $t = 0.66$ in.

$R_W = 1 - 0.33\ (H_W/H_S) \geq 0.67$	
Ratio (H_W/H_S)	Factor (R_W)
0.00	1.00
0.05	0.98
0.10	0.97
0.15	0.95
0.20	0.93
0.25	0.90
0.30	0.90
0.35	0.88
0.40	0.87
0.45	0.85
0.50	0.84
0.55	0.82
0.60	0.82
0.65	0.79
0.70	0.77
0.75	0.75
0.80	0.74
0.85	0.72
0.90	0.70
0.95	0.69
1.00	0.67

TABLE 3.14 Water Buoyancy Factor, R_W

3.7.12 Hydraulic Design of CIPP

3.7.12.1 Gravity Flow

The installation of CIPP typically improves flow characteristics of the pipe being renewed. Flow is improved because the inner surface of CIPP is extremely smooth and continuous, without any joints or discontinuities that create friction to flow. Typically, the Manning's equation is used to predict flow in gravity or open channel piping as follows:

$$Q = VA = \frac{1.486 A R^{2/3} S^{1/2}}{n} \tag{3.23}$$

where Q = flow rate, ft^3/sec (cfs)
V = velocity, ft/sec
A = flow area, ft^2

n = Manning's coefficient of roughness, dimensionless (see Table 3.15)
$R = A/P$ = hydraulic radius, ft
P = wetted perimeter of flow, ft
S = slope of grade line, ratio of vertical drop (h) to horizontal distance over which the drop occurs (L) usually expressed as a percentage (dimensionless).

When the pipe is circular and the flow is full as in a surcharged situation, the Manning's equation may be modified to Eq. (3.24):

$$q = \frac{0.463D^{8/3}S^{1/2}}{n} \tag{3.24}$$

where D = pipe internal diameter, ft, and other parameters are defined previously.

For circular pipe flowing full, the Manning's equation can be abbreviated to produce an easy comparison of flow capacity between CIPP and different piping materials as given in Eq. (3.25).

$$\text{\% Increase in Flow Capacity} = \frac{Q_{\text{CIPP}} \times 100}{Q_{\text{exist}}} = \frac{n_{\text{exist}}}{n_{\text{CIPP}}}\left(\frac{D_{\text{CIPP}}}{D_{\text{exist}}}\right)^{8/3} \times 100 \tag{3.25}$$

The Manning's coefficients are dependent on many factors such as the type and condition of the existing pipe and the quality of CIPP installed. A conservative average Manning's coefficient for CIPP in relatively smooth concrete, clay, or steel pipe has an "n" of 0.010, as shown in Table 3.15. However, this coefficient may change over time as deposits may build up in an uncleaned pipe over time.

3.7.12.1.1 Gravity Flow Design Example

Determine the change in flow capacity when a circular 24-in. concrete pipe is flowing full and is lined with a 0.47-in. (12-mm) thick CIPP.

Step A. Select Manning's coefficients for the piping materials (Table 3.15).

Pipe Material	Range of Manning's Coefficient, *n*	Recommended Manning's Coefficient, *n*
CIPP	0.009–0.012	0.010
Vitrified clay	0.013–0.017	0.014
Concrete	0.013–0.017	0.015
Corrugated metal	0.015–0.037	0.020
Brick	0.015–0.017	0.016

TABLE 3.15 Manning's Coefficients for Typical Piping Materials

(a) n for CIPP = 0.010

(b) n for concrete = 0.015

Step B. Determine inside pipe diameters.

(a) Existing concrete pipe $D = 24$ in.

(b) New CIPP $D = 24 - 2(12/25.4) = 23.1$ in.

Step C. Determine increased flow capacity with Eq. (3.25).

$$\%\text{ Flow capacity} = \frac{0.015}{0.01}\left(\frac{23.1}{24.0}\right)^{8/3} \times 100 = 135\%$$

Therefore, it is estimated that the CIPP increases the flow of concrete pipe by approximately 35 percent compared to that of the existing concrete pipe. This increase in flow was realized, although the inside diameter of the CIPP was slightly smaller than the existing concrete pipe.

3.7.12.2 Pressure Flow

For determining the flow rate of pressure pipelines, the Hazen-Williams equation is commonly used. The CIPP increases the flow capacity of a pipe because of the inherent smoothness of the inner surface. The Hazen-Williams equation is given as follows:

$$Q = 1.318C \times R^{0.63} \times S^{0.54} \times A \tag{3.26}$$

where Q = flow rate, cfs
C = Hazen-Williams coefficient (see Table 3.16)
R = hydraulic radius, ft = A/P
A = flow area, ft^2
P = wetted perimeter of the flow, ft
S = slope of the grade line, previously defined in Eq. 3.23

Pipe Material/Condition	Recommended Hazen-Williams Coefficient, *C*
CIPP	140
New steel or ductile iron (less than 1 year old)	120
Cement lined new steel or ductile iron	140
Steel (2 year old)	120
Steel (10 year old)	100
Cast iron (5 year old)	120
Cast iron (18 year old)	100
Tuberculated steel or cast iron	80

Table 3.16 Hazen-Williams Coefficients for Typical Piping Materials

As shown previously, the Hazen-Williams equation can be simplified to provide a comparison of flow capacity between CIPP and the existing pipe as shown in Eq. (3.27) below:

$$\% \text{ Increase in Flow Capacity} = \frac{Q_{\text{CIPP}}}{Q_{\text{Exist}}} \times 100 = \frac{C_{\text{CIPP}}}{C_{\text{Exist}}}\left(\frac{D_{\text{CIPP}}}{D_{\text{Exist}}}\right)^{2.63} \times 100 \qquad (3.27)$$

Determination of flow capacities of CIPP relative to other existing piping materials is calculated in the same manner as given in gravity flow design example.

3.8 Chapter Summary

This chapter presented a simple methodology for selection of the most appropriate renewal method. Selection of a renewal solution is based on the recognition of the problem or problems with the existing pipeline system. The selected method must be suited for the individual project conditions. Sometimes the best method is found in the application of multiple renewal alternatives to achieve the most cost-effective solution.

The design of a pipeline renewal system includes (1) identification of pipe conditions and problem recognition and classification, (2) prioritization of problem considering long-term plans, (3) selection of an appropriate pipeline renewal method, (4) designing renewal methods based on project specific conditions, and (5) implementation and monitoring. Additionally, this chapter provided the design basics for partially deteriorated and fully deteriorated gravity and pressure pipelines. The ASTM 1216 design guidelines were followed to present sample design problems.

3.9 Review Questions[3]

1. Describe the different stages of TRM selection process.
2. Why pipe defect identification and inspection prior to renewal method selection is so important? Explain.
3. Define "partially deteriorated" and "fully deteriorated" design principles.
4. A 12-in. diameter (D_o) sanitary sewer pipe with some cracking and offset joints is leaking and is in need of renewal (see Fig. 3.2). Through a CCTV inspection, it appears that in at least one area, the pipe is no longer round and has deflected approximately 5 percent. The water table, through investigation, is found to be normally 3 ft below the surface. The pipe is buried at a depth of 15 ft (H_S) to the top of the pipe. Assume

[3]For missing information, make and indicate reasonable assumptions.

a modulus of elasticity for the resin of 300,000 psi (long-term modulus, E_L = 0.5(300,000) = 150,000 psi) and a long-term flexural strength of 2,500 psi. Determine the wall thickness required for a CIPP liner.

5. The 12.-in diameter (D_o) sanitary sewer pipe (see Fig. 3.3) is cracked and leaking. The severity of the crack has progressed to the point that some pieces of the pipe wall have fallen into the pipeline and voids have begun to form around it. The pipe is shallow and located under a busy city street subject to *HS*-20 loading. Assume that the pipe is buried under a clay soil with a unit weight of 120 pcf. As in the previous example, the pipe is no longer round and has deflected approximately 5 percent. The pipe is buried under the ground to a depth of 5 ft (H_S). The water table is normally located 2 ft (H_w) above the top of the pipe. Assume a modulus of elasticity for the resin of 300,000 psi (long-term modulus, *EL* = 0.5 (300,000) = 150,000 psi), a long-term flexural strength of 2,500 psi, and a soil modulus of 1,000 psi. Determine the wall thickness required for a CIPP liner.

6. A 15-in. inner diameter (*D*) internal pressure pipe has external water 5 ft (H_W) above the top of the pipe (see Fig. 3.4). After cleaning and testing, it is found that the pipe is able to maintain the specified pressure it was originally designed for. However, owing to the presence of cracks and out of round cross-sections at certain point, it is recommended that the pipe be renewed. The internal pressure of the pipe is 85 psi. Assume the CIPP flexural modulus as 350,000 psi and the CIPP flexural strength as 5,500 psi. Minimum long-term modulus may be assumed to be 175,000 psi and the minimum long-term strength as 2,750 psi. Also assume the minimum long-term tensile strength as 1,750 psi. Determine the CIPP wall thickness required for the above conditions.

7. An 18-in. inner diameter (*D*) internal pressure pipe has external water 10 ft (H_W) above the top of the pipe (see Fig. 3.5). After cleaning and testing, it is found that there are small holes in the pipe that will not allow it to maintain its design pressure. Moreover, the actual extent of damage of the pipe and some of the operating parameters are not very well defined. Therefore, it is recommended that the pipe be renewed. The pipe is buried 15 ft (H_S) below the ground level. The internal pressure of the pipe is 70 psi. Assume the CIPP flexural modulus as 350,000 psi and the CIPP flexural strength as 5,500 psi. Minimum long-term modulus may be assumed to be 175,000 psi and the minimum long-term strength as 2,750 psi. Also, the minimum long-term tensile strength is 1,750 psi. Determine the CIPP thickness required for the above conditions.

CHAPTER 4

Pipeline Prioritization, Inventory and Condition Assessment

4.1 Introduction and Background

Factors such as population growth, tighter health and environmental requirements, poor quality control leading to inferior installations, inadequate inspection and maintenance, and lack of consistency and uniformity in design, construction and operation practices have adversely impacted pipeline infrastructure. The aging pipeline system is further impacted by corrosion damage, moving soils, changing temperatures, rainfall/snowfall, in-service stresses and the continuous process of structural deterioration. At the same time, an increased burden on pipeline infrastructure due to significant growth in some sectors tends to quicken the ageing process while increasing the social and monetary cost of service disruptions due to pipeline failures. These environmental and operating stresses inevitably lead to a number of pipeline failures throughout the year. Increasing concerns over health, safety and the environment have contributed significantly to raising the visibility of pipeline risk management. The rapidly deteriorating existing pipes and the expansion of present pipeline network due to increasing demands require the pipeline owners to prioritize the renewal, replacement, and new installations. However, inspecting, predicting, and monitoring the condition of pipelines are difficult tasks.

Traditionally, pipeline owners have addressed the maintenance and operation of their systems with a crisis-based approach. This practice results in the inefficient use of limited funds, causing more frequent failures which end in difficult and costly renewal. The emergency cost of pipeline inspection, repair and renewal/replacement may include service disruptions, adverse publicity, and health and safety problems, and is significantly higher than the cost of a planned strategy. The major reason for reactive approaches to pipeline asset management is that these systems are underground and their conditions are not visible. Thus, the actual problems caused by deterioration are not evident until major failures occur. Another issue in pipeline management is the fact that the condition of these underground assets is generally not fully and properly documented. While condition assessment is very important in developing a systematic procedure in effective pipeline management, most owners do not have complete condition data in their management information systems. The lack of data on the past and current condition of pipelines prevents the system-wide assessment of existing pipelines, the development of prediction models, and the evaluation of the effects of renewal on pipeline condition.

Without a predictive approach to pipeline renewal/replacement needs, condition assessment activities will be unfocussed and may overlook high-risk areas. The predictive approach involves routine and systematic pipeline system's structural and hydraulic condition assessments, establishment of a standard condition rating system, and developing and updating prediction models. Predictive modeling permits effective budgeting of inspection and renewal projects.

Traditionally, pipe age and type of pipe material have been the only factors used in prioritizing inspections. This traditional approach is not sufficient and does not take into account a deeper understanding of the variables leading to failure, or of the impact of failures on the community and the environment. To be fully effective, the pipeline asset management system must have performance models that combine the rate of deterioration and change in the actual pipeline condition influenced by factors such as pipe age and type, pipe external and internal pressures, fluid properties, quality of initial installations, pipe manufacturing and type of joints, corrosion damage, pipe wall thickness and existence of holes and cracks, depth of cover, soil corrosivity, soil and groundwater conditions, pipe slope and alignment, surface conditions, temperature changes, etc. Two main aspects of information on pipeline performance that can be used in asset management decision-making process are (1) information on current condition, which is obtained through field inspection, and (2) information on future performance, obtained using deterioration models and forecasting tools. As discussed in this chapter, there are various types of inspection and condition assessment tools available to evaluate the present state of pipelines. A predictive model can prioritize pipeline segments for further investigations and to implement proactive renewal measures. There are physical, statistical (deterministic and probabilistic) as well as computer (expert systems, artificial intelligence, neural networks, etc.) models available that can be used to predict future conditions and performance of a pipeline.

4.2 Pipeline Condition Assessment

The primary components of any asset management program include the identification, location, and condition of assets. Condition assessment provides the critical information needed to assess the physical condition and functionality of a pipeline system, and to estimate remaining service life and asset value. After the field inspection, pipe defects are classified using a standard coding system and pipe condition is assessed using a systematic method to produce consistent and useful information. The US Environmental Protection Agency (USEPA, 2009a; USEPA, 2009b and USEPA, 2010), Water Environment Research Foundation (WERF, 2007), National Association of Sewer Service Companies (NASSCO), Water Environment Research Foundation (WRF), American Water Works Association (AWWA), and other institutions as mentioned in this book, provide conditions assessment and inspection guidelines and pipe defect codes. This chapter provide excerpt information from these sources.

As said earlier, condition assessment information is used to model a pipe's remaining useful life and its long-term performance, and to make decisions about pipe renewal, replacement, and/or further inspections. This will move the pipeline owner from crisis management to proactive decision making through the administration of data. Once data are administered, asset maintenance, renewal, or replacement determinations are planned based upon criticality, condition and maintenance costs versus economic value.

The assessment of the pipeline physical conditions and decision making on pipe renewal include the following elements:

- Physical modeling of the pipe in the soil.
- Understanding of pipe failure modes, including observable or measurable signs (or distress indicators) that point to these modes, as well as inferential indicators that point to potential existence of deterioration mechanisms.
- Inspecting the pipe to determine distress indicators.
- Interpreting distress indicators to determine pipe condition.
- Empirical/statistical modeling of historical failures.
- Assessing failure consequences (direct, indirect, and social/environmental costs).
- Scheduling pipe renewal/replacement to minimize lifecycle costs while meeting or exceeding service levels.

A variety of methods are available for performing condition assessments, ranging from simple to complex. They generally follow a similar progression of steps: setting objectives for the condition assessment, identification of assets and available data, asset inspection, data analysis, and decision making.

Decision making for condition assessment of a pipeline system includes understanding the possible risks and determining at what point a pipeline owner should intervene to avoid a failed condition with an unacceptable cost and/or consequence. It is important to note that condition assessment alone does not provide risk reduction. The follow-up decision-making process, which leads to prioritization and ranking followed by action to address problems will reduce risks.

4.3 Model Input Parameters

As stated in Chap. 2, to identify the condition of a pipeline, the type of defect needs to be classified and severity level assigned. Each distress condition will be assigned severity levels based on the degree and the combination of structural defects found in the pipe segments. Typically this information needs to be documented by the pipeline owners while performing condition assessment and will serve as the input parameters for the prediction modeling. Table 4.1 summarizes the parameters that have substantial impact on pipe deterioration.

4.4 Data Collection, Analysis, and Management

A successful condition assessment requires that the data collected are organized, analyzed, and maintained in a database system. This important step allows a pipeline owner to develop an understanding of deterioration trends.

4.5 Condition Assessment

A comprehensive condition assessment can be divided into a four-stage process.

- *Stage 1.* Identify physical characteristics of a pipeline using historical, environmental, and operational data to prioritize assets for condition inspection based on the consequences of failure and determine what information is required from an inspection program.
- *Stage 2.* Evaluate the possible methods of inspection for their appropriateness and cost effectiveness to provide the required type and level of information. Undertake an inspection using an appropriate technology.

Input	Parameter
Pipe data	Pipe Material
	Pipe Age
	Pipe Diameter
	Pipe Section Length
	Joint Type and Material
	Wall Thickness
Zoning	Residential/Commercial/Industrial
Depth of cover	Typical depth – 8 to 15 ft
	Shallow – less than 8 ft + high live load
	Deep – more than 15 ft
Gradient	Slope of Pipe
Bedding conditions	
Backfill type	Cohesive Soil
	Non-Cohesive (Granular) Soil
Soil characteristics	Soil Type
	Corrosivity
	Resistivity
	pH Content
	Sulfide Content
	Moisture Level
Groundwater condition	High Groundwater (GWT) (pipe crown is below GWT)
	Low Groundwater (pipe crown is above GWT)
Ground movement	Expansive soils, vibrations, etc.
Internal service loads	Operating pressures, surcharges, etc.
External loads	Soil Load
	Traffic Loading
	Other Surface Loads
	Frost Load Factor
Flow characteristics	
Maintenance type and frequency	
Repair history	

Table 4.1 Input Parameters for Pipe Deterioration Modeling

- *Stage 3.* Carry out a final condition assessment based on the information from stage 1 to provide an assessment of the likelihood and consequences of failure. Note that inspection provides data and not assessment which requires interpretation.

Interpretation may include developing life expectancy curves and predictions of time to failure.

- *Stage 4.* Measure the success of the condition assessment program.

4.5.1 Asset Registry

An accurate and complete system inventory is essential for an asset management program. Often, much of this information is available from existing databases or maps with different levels of accuracy. For example, a typical water or wastewater infrastructure system will include structures such as intake structures, storage tanks, pumps, transmission mains, distribution pipes, service lines, meters, valves, blow-off valves, backflow prevention, wells and springs, galleries and tunnels, chlorination and other treatment equipment, hydrants, electrical systems, monitoring equipment, and so on. A typical wastewater collection system may include gravity and/or force-main lines, manholes, service laterals, and lift stations.

4.5.2 Geographical Considerations

The geographical information system (GIS) is a tool for making intelligent decisions. The GIS organizes geographic data so that it is possible to select data necessary for a specific project or task. A thematic map has a table of contents that allows the reader to add layers of information to a base map. A good GIS program is able to process geographic data from a variety of sources and integrate into the GIS database.

Two types of data stored for each item in the database are as follows:

1. Attribute data (material, diameter, length, etc.).
2. Spatial data (location, usually coordinate based; either vector or raster). Vector data provide discrete features (using points, lines, polygons, or areas and aster data provide a continuous surface).

The GIS maps are interactive. On the computer screen, the GIS users can scan the information in any direction, zoom in or out, and change the nature of the information contained in the map. They can choose whether to see the roads, how many roads to see, and how roads should be depicted. Other items can be selected alongside these roads, such as different pipelines and utilities. The GIS also presents the unique capability to measure and compare dimensions and elevations. This digital elevation can be particularly useful when evaluating the hydraulic grades and operating elevations within a pressure system. By incorporating topography into the model elements, valleys and peaks corresponding to high and low pressures can be more readily identified and incorporated into the planning process to mitigate network deficiencies (ESRI, 2009).

4.5.3 Asset Naming and Numbering Conventions

The asset management team will need to develop a plan for assigning identification codes to the pipelines regardless of the type of data management tool is selected. Codes assigned to assets should convey as much information as possible in a concise and efficient manner.

Once the process of asset identification is complete, the process of naming and numbering can begin. According to IIMM (2006), the identification of length of a pipe is established on the basis of "from" node point and a "to" node point. This means that the nodes rather than the pipe sections are numbered. Tangible node points, such as valves, manholes and access points are easy to identify. Sometimes it will be desirable to have additional node points to identify a change in pipe diameter, age, and so on. If this is the case, an "imanginery" node point can be established, which will constitute a "from" and "to" node, but the node point itself may have no physical attributes. Several naming and numbering conventions can exist. Following are four possible asset identification systems for typical water and wastewater asset groups:

4.5.3.1 Option 1: Street Sequential Numbering

This system is suitable where the assets can be located in relation to a street or road. A typical format can be as follows:

STREET ABBREVIATION/(Street)
ASSET TYPE/(WV)
SEQUENTIAL NODE NUMBER/(999)
Example: EMLY WV 025 (Emily St, Water Valve, Node 25).

The first four alpha characters are the street abbreviation. The street abbreviation gives a quick reference to the likely area in which to search for the asset within any plan system. In some instances, street names will have similar abbreviations. In such cases, a town area identifier may be used to precede the street abbreviation. The remaining part of the numbering system should be numeric and form a sequential numbering system for each node within that street. In incrementing node numbers, an increment other than one should be used to accommodate additional nodes in the future.

4.5.3.2 Option 2: Service and Property Identification

This option uses street numbers instead of sequential numbers. This system is suited more to circumstances where the assets can be located in relation to individual properties. This system is probably the most suitable for numbering properties and assets specific to properties, such as sewer laterals and water service pipes. It may be desirable to use this system for laterals and service pipes and another for the rest of the asset register. The recommended format is as follows:

STREET ADDRESS/(NAME999A)
ASSET TYPE/(-WX)
CONNECTION NUMBER/(9)
Example: EMLY 001 A–WT2 (Emily St, St Address 001, Apt or Unit A, Water Toby No. 2).

There are some potential problems with this option. Consideration needs to be given to properties with multiple addresses, services in road carriageways not associated with properties, and services in reserves and properties without street addresses. Consideration also needs to be given to what will apply in the event of subdivision, when street names and numbers can change.

4.5.3.3 Option 3: Grid Reference or Plan Numbering

This system is suited more to circumstances where the organization has record sheets for field identification. The recommended format is as follows:

AREA CODE/(A)
GRID REFERENCE/(Y18)
ASSET TYPE/(WV)
SEQUENTIAL NUMBER/(025)
Example: A Y18 WV 025 (Township A, Grid Reference Y18, Water Valve, No. 25).

It is important to note that this approach does not provide topological information but rather an indication of location.

4.5.3.4 Option 4: Catchment Numbering

This system is suited more to circumstances where asset networks are defined by catchment systems such as water and wastewater systems. Generally, this will be determined by the geography of the locality. The recommended format is as follows:

CATCHMENT ID/(X)
BRANCH ID/(999)
ASSET TYPE/(XX)
SEQUENTIAL NODE NUMBER/999
Example: A 002 SM 025 (Catchment A, Branch 002, Sewer Manhole, Sequence No. 25).

Catchment numbering generally starts at the lowest outlet and follows the direction of flow back up the catchment. A decision is necessary as to how the numbering should be treated. Branches leading off the main trunk normally have the next sequential number. Node points are sequentially numbered (possibly with gaps for future nodes) up each branch. This system is relatively simple. However, there is no property-based interrelationship, and for field identification, there may be the need for two numbering systems, one for nodes and the second for service connections.

The asset ID numbers in the above methods not only identify the assets but also easily locate them within the subject area. An asset identification that codes as much information as possible can make the difference between an inventory that is useful and one that is not. This is a very important component of the asset identification and should be considered carefully. It should be noted that a pipeline owner can develop and modify these arrangements to meet its own needs, but once put into place, it should be strictly followed to avoid any type of discrepancy.

4.5.4 Inventory Tools

The GIS, global positioning system (GPS), computer-aided design (CAD), and relational database management system (RDBMS) provide an accurate picture of the extent of an asset management portfolio. These systems are directly related to each other in the way they communicate and process data. Usually, a GPS device works interactively with a GIS software. Once the data are received, a CAD software then identifies these data using a survey mapping tool. These data can also be utilized using an RDBMS for inquiry and manipulation. All of these systems can be embedded within modeling software (e.g., Multimedia Integrated Modeling System or MIMS, for more information, visit https://www.epa.gov/fera/mims-trim) for greater analyzing capability.

The unique method of storing and manipulating data in GIS differentiates GIS from other drafting or mapping software. The essential elements of a GIS are (1) the ability to perform spatial operations on the database and (2) map creations and editing functions. The users will log GPS positions into GIS servers in real time and then linking them with a database system. Figure 4.1 illustrates such operation. The GIS can provide the spatial display of the system inventory, a host of boundary, and limiting conditions (e.g., pressure zones, incorporated limits, land uses, and maintenance zones) and graphically present the modeling output. In conjunction with the modeling software, the GIS allow for the visual presentation of the pipes and nodes of the network and can trim and reduce these attributes to simplify the modeled network.

The data required to compile a GIS system inventory for pipeline infrastructure are both quantitative and qualitative in nature. This process is dictated by the type of service, location, and asset types. Using sanitary sewers, for example, three primary assets: pipelines,

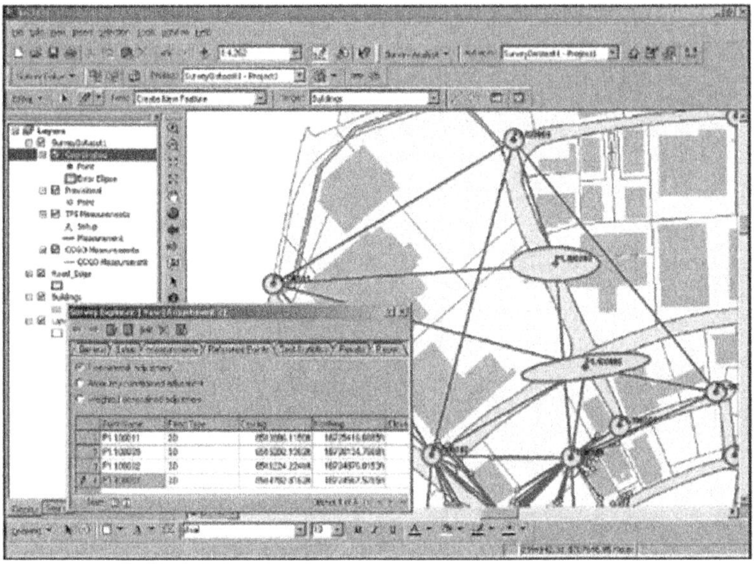

Figure 4.1 GPS observations and total station survey data integrated with local mapping and managed in GIS. (*Source:* ESRI, 2009.)

manholes, and service laterals are defined. Since pipeline segments are defined by upstream and downstream manhole locations, GPS readings are required for each manhole–and although pipe sections may be calculated based on GPS coordinates of manholes, the pipe material and diameter are also required to complete the system inventory–and this can only be determined through manhole and pipe inspections.

4.5.5 Case Study: Mapping and Inventory of Water and Wastewater Infrastructure at Edgecombe County, NC[1]

This project required field mapping more than two dozen feature types and collecting up to 15 attributes per feature along 400 miles of water and sewer transmission mains. Mapping crews had to locate and collect attribute information for every pressurized line, gravity line, manhole, control valve, storage tank, fire hydrant, and pump station.

4.5.5.1 Use of Software

A local engineering firm obtained a bundled hardware-software package, which included a Trimble GeoExplorer and GeoXT[2] equipped with ESRI's ArcPad mobile GIS[3] software and the Trimble

[1]Adapted from: http://www.esri.com/news/arcnews/summer05articles/edgecombe-county.html.

[2]Available at: http://www.trimble.com/Survey/trimbleGeoXRgnss.aspx

[3]Available at: http://www.esri.com/software/arcgis/arcpad

GPScorrect[4] post-processing extension. In addition, the firm acquired Trimble's GPS Analyst, a new ArcGIS extension that creates a seamless workflow between GPS data collected in the field and the enterprise geodatabase in the office.

Edgecombe County's only existing water/sewer system schematic was in CAD format and was not drawn in such a way that it could be readily referenced into the geodatabase. The engineering firm printed the diagrams as field workbooks to assist in locating system features in the field. With guidance from the county, the firm's GIS specialists generated data models, or classification structures, of the attributes that would be collected in the field for each type of infrastructure asset and created a geodatabase based on the data model.

4.5.5.2 Fieldwork

At the beginning of each day's fieldwork, which mostly took place on the shoulders of roads and highways running next to water pipe rights-of-way, the technicians used the GPS Analyst program with ArcGIS on a computer to "check out" the datasets to be mapped. This means the geodatabase files, relating to those infrastructure features that would be mapped or updated, were selected and uploaded cable link from the laptop into the handheld GeoXT (Fig. 4.2).

[4]Available at: http://www.trimble.com/mappingGIS/gpscorrect.aspx

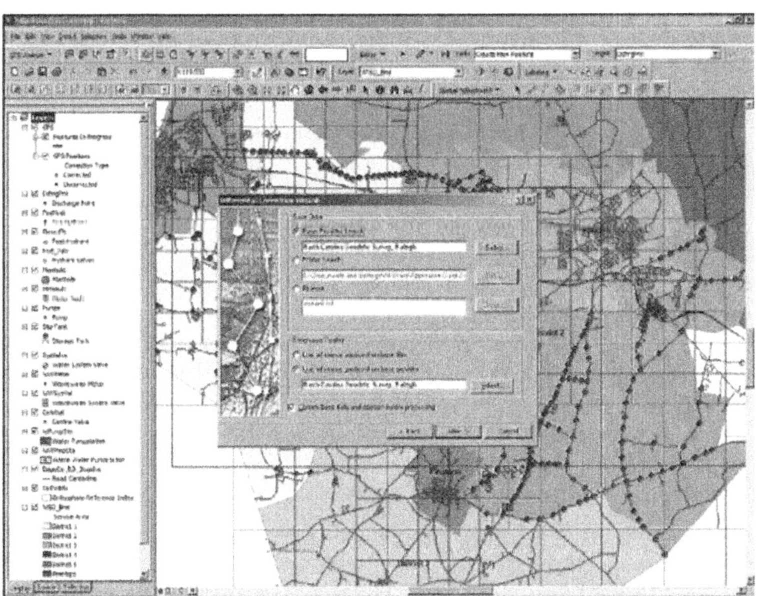

FIGURE 4.2 Checking in raw GPS data using GPS Analyst toolbar. (*Source:* ArcNews Magazine, 2005.)

In the field, ArcPad displayed a schematic of the water system assets that would be mapped. As the field crew found each feature, an on-screen point-and-click menu generated from the data file assisted the user in collecting the desired attribute information. As the crew members entered the descriptive data into ArcPad, the GPS receiver recorded 20 location points for each feature in less than 1 minute. These location data were stored with the other attribute data in the mobile geodatabase inside the GeoXT. Completely transparent to the field technicians, the Trimble GPScorrect extension was also running with ArcPad, collecting additional data from the GPS satellites that would be required later to post-process the location data recorded by the receiver. Differential post-processing allows the crew to enhance the accuracy of the location data to less than 3 ft.

4.5.5.3 Data Processing and Queries

Once the fieldwork was completed for the day, the crew reconnected the GPS receiver with the computer to download field data into the ArcGIS geodatabase. Working in ArcGIS, the GIS specialist then accessed GPS Analyst to differentially correct and edit the GPS points collected in the field. Differential correction of GPS data previously involved exporting the GPS data to an external software package first, but now technicians perform the task of post-processing data within the geodatabase. From the differential correction tool the crew selected from a list of local reference station sources. The differential correction wizard within GPS Analyst assisted in determining which source provided the best coverage for that day's points and then connected with the base data provider to access the appropriate correction data. The software automatically completed the post-processing with ArcGIS within minutes compared with the hour or more it took to export data and correct it externally.

The GPS Analyst extension also allowed the field crews to validate GPS points by visualizing each of the 20 location values collected for every fire hydrant and manhole shown on screen. By looking at the data cluster, they could determine which points are not consistent with the others. These errant points can be deleted to improve the average value of the remaining points and thus the accuracy of the overall feature map.

Once the field data have been collected, the new layer of water and wastewater features is overlaid on a digital orthophotograph of Edgecombe County (Fig. 4.3). Many of the larger assets, such as pump stations and water storage tanks are readily visible on the orthophoto. The GIS specialists compared the location of the features mapped in the field with those visible on the orthophoto. The determined accuracy of the data collection was within 3 ft as mandated by Edgecombe County.

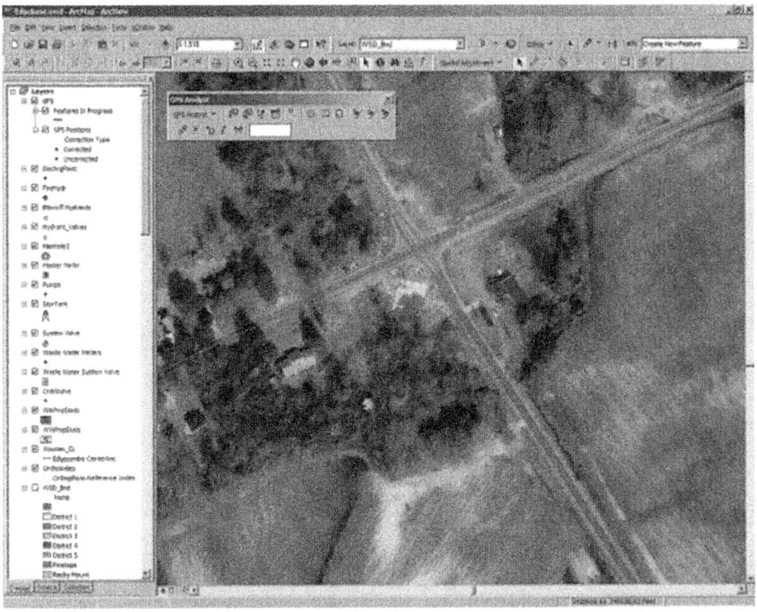

FIGURE 4.3 Overlaying corrected data on sample orthophoto. (*Source:* ArcNews Magazine, 2005.)

4.6 Condition Assessment Technologies for Gravity and Pressure Systems

4.6.1 Overview

Condition assessment primarily involves pipe inspections. Inspection technologies use various methods for detecting pipe defects, structural, operational, and environmental conditions that could potentially impact pipe performance. These technologies have varying abilities for detecting and quantifying specific types of pipe defects and may have limited applications depending on pipe material and/or pipe diameter.

A condition assessment method includes a variety of inspection technologies, based on the specific characteristics of a pipeline network. The wastewater technologies are summarized in Table 4.2.

4.6.2 Camera Technologies

4.6.2.1 Closed-Circuit Television

The closed-circuit television (CCTV) inspection records a close-up observation of pipe surface. It comprises a CCTV camera and lighting apparatus, mounted on a carrier. A winch and pulley system moves

Program Objective	Camera				Acoustic			Electrical and Electromagnetic			Laser	Innovative Technologies						
	CCTV	Zoom Camera	Digital Scanning	Push Camera	In-line Leak Detectors	Acoustic Monitoring Systems	Sonar	Electrical Leak Location	Remote Field Eddy Current	Magnetic Flux Leakage	Laser Profiling	Gamma-Gamma Logging	Ground-Penetrating Radar	Infrared Thermography	Micro-Deflection	Impact Echo/ SASW	Ultrasonic Pulse Velocity	Guided Wave Ultrasonic
Application																		
Pipe type	G	G	G	S	G, F	F	G, F	G, F, S	G, F, S	G, F, S	G, F	G, F, S	G, F, S	G, F, S	G	G	G	F
Pipe material	Any	Any	Any	Any	Any	PCCP	Any	NF	F	F,PCCP	Any	C	Any	Any	B	B,C	C	F
Pipe diameter (in.)	>6	>6	6–60	1–12	≥4	≥18	≥12	3–60	≥2	2–56	≥4	TBD	18–30	TBD	N/A	>6	TBD	>2
Defects Detected																		
Sediment, debris, roots	Y	Y	Y	Y	–	–	Y	–	–	–	Y	–	–	–	–	–	–	–
Pipe sags and deflections	Y	Y	Partial	Y	–	–	Y	–	–	–	Y	–	–	–	–	–	–	–
External pits and voids	–	–	–	–	–	–	–	–	–	–	–	–	–	–	Partial	Y	–	–
Corrosion and metal loss	–	–	Partial	–	–	–	Y	–	Y	Y	Y	–	–	–	–	–	–	Y

Offset joints	Y	Y	Partia	Y	–	–	–	–	–	–	–	–	–	–	–	–	–	–
Pipe cracks	Y	Y	Y	Y	–	–	Y	Y	Y	Y	–	–	–	–	–	Y	Y	Y
Leaks	Y	Y	Y	Y	Y	–	–	Y	Y	–	–	–	Y	Y	–	–	–	–
Broken prestressed	–	–	–	–	–	Y	–	–	Y	–	–	–	–	–	–	–	–	–
Wall thickness	–	–	–	–	–	–	–	–	Y	–	–	–	–	–	–	Y	Y	Y
Service connections	Y	Y	–	–	–	–	–	–	–	–	–	Y	–	–	–	–	–	–
Bedding condition	–	–	–	–	–	–	–	–	–	–	–	Y	Y	–	–	–	–	–
Bedding voids	–	–	–	–	–	–	–	–	–	–	–	Y	Y	Y	Partial	–	–	–
Deteriorated insulation	–	–	–	–	–	–	–	–	–	–	–	–	–	Y	–	–	–	–
Overall condition	–	–	–	–	–	–	–	–	–	–	–	–	–	–	–	–	–	–

Source: USEPA (2010).
Note: Pipe type: G = gravity line, F = force main, S = service lateral.
Pipe material: NF = nonferrous, F =ferrous, B = brick, C = concrete, PCCP = prestressed concrete cylinder pipe.
Others: TBD = to be decided, N/A = not applicable, Y = yes, SASW = spectral analysis of surface waves, CCTV = closed-circuit television.

TABLE 4.2 Typical Applications of Condition Assessment Technologies

the CCTV module through the pipe. Larger modules can use an umbilical cord system, which can provide power and communication from and to the ground station, as well as serve to retrieve the device. The basic steps of the CCTV inspection are as follows:

1. Introduce the carrier with the CCTV camera into the pipe via access points.
2. Let carrier travel along the pipe and use the camera to capture and transmit the images to a ground station (inspection truck).
3. Analyze images in the field or office.

A good-quality CCTV camera provides a record of pipe condition and allows the assignment of defect codes. For example, Fig. 4.4*a* shows a concrete pipe in excellent condition, with no apparent cracks or corrosion (Pipeline Assessment and Certification Program, PACP Grade 1). Figure 4.4*b* shows substantial corrosion and is in poor shape (PACP Grade 4).

The inspection by CCTV involves conveying the camera through the pipeline using various technologies, such as pushrod cameras (pushcams)[5] and remote-controlled robot crawlers. The level of optical control on the camera varies. Its ability to pan, tilt, and zoom enables the operator to gain a full circumferential view of the pipe, and this is why CCTV has become the industry standard for sewer inspection.

4.6.2.2 Zoom Camera

Like the traditional CCTV inspection, the zoom camera inspection technology involves the generation of still imagery and/or recorded video imagery of a pipe as illustrated in Fig. 4.5. The key difference in the zoom camera inspection is that the camera mount is stationary. This technology does not require the camera equipment to pass through the entire length of the pipe segments being inspected. Instead, the camera is mounted either on a truck, crane, pole, or a tripod. The equipment is then lowered into a manhole or access point to perform the inspection, and the camera "zooms" down any pipe entering or exiting the manhole.

The zoom camera inspection is not designed to replace the conventional CCTV inspection, but rather to screen and prioritize pipes for further conventional CCTV inspection or cleaning. An inspection crew can move quickly through a service area and highlight segments requiring more detailed inspection.

---

[5]For example, visit http://www.spx.com/en/pearpoint.

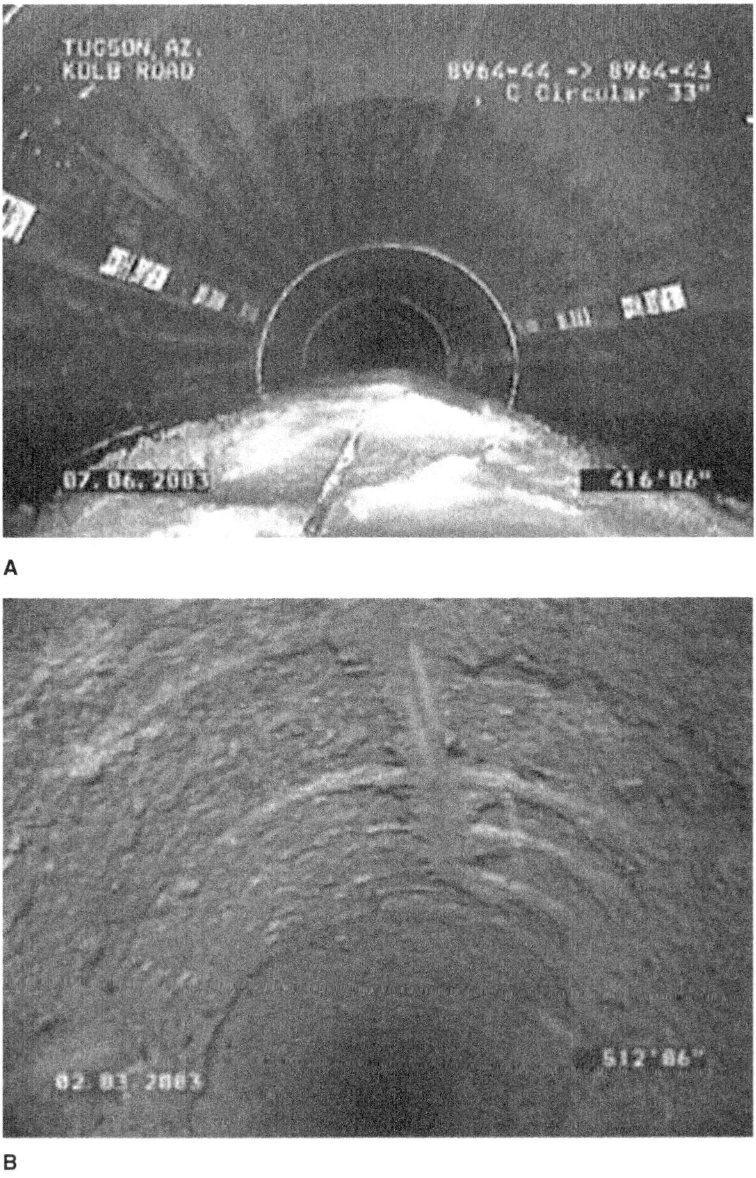

A

B

FIGURE 4.4 CCTV inspection of a concrete pipe. (*Source:* USEPA, 2010.)

4.6.2.3 Digital Scanning

Digital scanning is a state-of-the-art technology within the camera inspection industry. Like conventional CCTV, digital cameras are transported through sewer lines using self-propelled crawlers. Unlike conventional CCTV systems, digital scanning uses one or

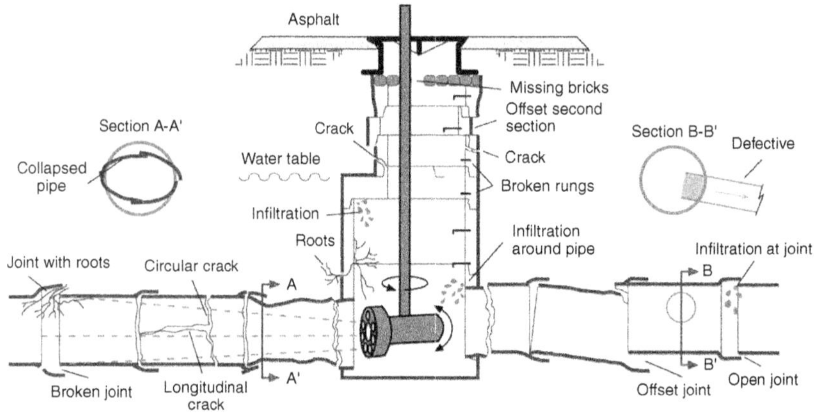

Figure 4.5 Zoom camera images of pipes showing structural defects. (*Source:* USEPA, 2010.)

two high-resolution digital cameras with wide-angle lenses in the front, or front and rear section of the housing (see Fig. 4.6). This configuration allows the generation of two types of images: "unfolded" views of the sides of the pipes and "circular" views of the pipe (similar to CCTV). During pipe inspections, parallel mounted lights are triggered at the same position in the pipe. The scanned hemispherical pictures are put together to form 360° spherical images for later analysis.

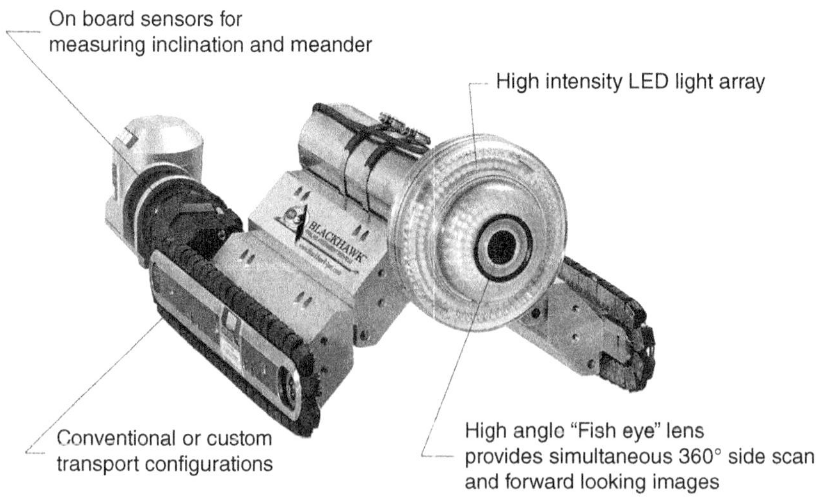

Figure 4.6 Digital scanning camera. (*Source:* Dr. Tom Iseley.)

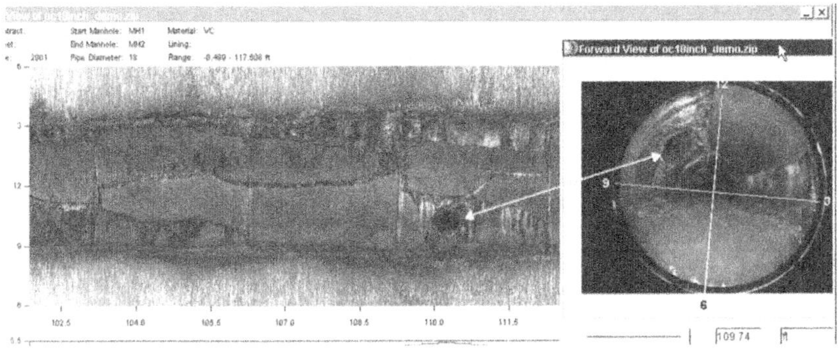

FIGURE 4.7 Example of side scanning results. (*Source:* Dr. Tom Iseley.)

Digital scanning provides advantages over conventional CCTV. Its rate of inspection or production rate is typically two to three times greater than that of CCTV. Because it combines a large number of still digital images, it produces a sharper image than a video. Also, the unfolded view of the inner pipe surface provides an excellent view of pipe conditions as illustrated in Fig. 4.7. The primary advantage, however, is the ability to access and assess the inspection data at a later time using the Sewer Scanning and Evaluation Technology (SSET).

Digital scanning does not rely on the operator panning and tilting to examine defects in the field, because the entire pipe surface is imaged during the inspection and the data are stored. Inspection progresses quickly and the defect coding is done later in the office. During the scanning process, data are transmitted to a computer for where it can be viewed in real time and recorded for later evaluation. Digital scanning develops a full digital image of the pipe segment. This allows the inspector reviewing the images to control the direction of the pan/tilt/zoom features and to stop the image at any point to capture video clips. The "unfolded" inner pipe surface permits computer-aided measurement of defects and objects.

4.6.2.4 Push Camera

Push camera or pushcam technology involves the inspection of pipeline via a small-diameter camera mounted to a pushrod and reel setup, which produces a video of the pipeline. Conventional pushcam systems comprised a camera/probe, cable/reel, and computer/recorder/controller. The probe is used to advance the camera and usually is a semi-rigid rod, constructed of fiberglass. Pushcams are typically used in an environment where it is otherwise impossible to get photographs or video footage such as a small-diameter water, sewer, or drain pipes (6 in.) for service laterals that are typically introduced into the sewer through a cleanout.

The PANORAMO® developed by Rapidview-IBAK, USA, is a good example of a pushcam technology. It uses two integrated

FIGURE 4.8 PANORAMO system. (*Source:* Rapidview-IBAK, USA.)

scanning units, one at the front end and one at the rear end as shown in Fig. 4.8. Each scanning unit consists of a 186° fish-eye lens and a high-resolution digital camera. The two units take hemispherical images and create 360° spherical images.

An unfolded two-dimensional view of the entire section and a three-dimensional view of the pipe allow the viewer to pan the angle of view in all directions. This pan and tilt scanning of details can be done in the office without actually operating the camera during inspection. The operator can pan and rotate a virtual camera like a real one. Another advantage of the PANORAMO system is that it can operate at a relatively high speed of 69 ft/min.

Table 4.3 lists a summary of the camera inspection technologies described earlier.

4.6.3 Acoustic Technologies

Acoustic technology in general terms uses measuring devices to detect vibrations and/or sound waves. In pipeline assessment, acoustic sensors are used to detect signals emitted by defects and are utilized by a variety of commercially available products. Acoustic technologies are used extensively for inspection of water mains; therefore, this category of inspection technology can also be used for force-main inspection. Three distinct classifications of acoustic technologies are as follows:

1. In-line leak detectors, which are used to detect the acoustic signals emitted by pipeline leaks.
2. Acoustic monitoring systems, which are used to evaluate the condition of prestressed concrete cylinder pipe (PCCP).
3. Sonar/ultrasonic systems, which emit high-frequency sound waves and measure their reflection off the pipe wall in order to detect a variety of pipe defects.

4.6.3.1 In-line Leak Detectors

Leak detectors are devices used to detect the sound or vibration produced by leaks in pressurized waterlines or in sewers. These

System	Typical Use	Advantages	Limitations
CCTV	• Gravity sewers only	• Creates a permanent video record of sewer pipe conditions • Able to identify almost every problem inside a pipe	• Cannot image the portions of the pipe that are underwater • Does not provide structural data on pipe wall integrity or a view of the soil envelope supporting the pipe
Zoom camera	• Gravity sewers only	• High production rate • Effective/efficient at prioritizing segments requiring more detailed inspection/ maintenance	• Cannot inspect manhole to manhole for average diameter lines • Potential to miss significant defects
Digital scanning	• Gravity sewers • Limited applicability for force mains and service laterals	• Increased Quality Assurance/Quality Control (QA/QC) • Project personnel can review/control data imagery • Able to make digital measurements of defects • Can compare data directly from one inspection to the next	• More costly than CCTV • Lower production rate compared to traditional CCTV • Only works above waterline
Push camera	• Gravity sewers • Limited applicability for force mains and service laterals	• Used in applications where crawlers/ robotic camera vehicles are unable to function due to their larger size	• Image quality • Lighting • Inability to move over obstructions

TABLE 4.3 Camera Inspection Summary

include handheld listening devices, such as listening rods, underwater microphones (also known as aqua phones, sonoscopes, water phones, or hydrophones), and geophones (ground microphones); leak noise correlators; and in-line devices, which remotely collect information on leaks. Listening devices and leak noise correlators are widely available and have been used for leak detection for decades.

4.6.3.2 Acoustic Monitoring Systems

Acoustic monitoring systems are installed along a PCCP line to provide continuous monitoring of the general condition of the pipe. The PCCP has been used historically for large-diameter raw water transmission and force mains and are subjected to failure due to internal or external corrosion. The acoustic monitoring systems work by detecting the acoustic signal produced by breaking prestressed wires within the pipe wall. While these systems do not identify individual defects, they are useful as screening techniques to determine if further condition assessment should be performed.

4.6.3.3 Sonar

Sonar, an acronym for Sound Navigation and Ranging, was developed in 1906, and is widely used for maritime use. The first use of sonar for inspection of pipelines was carried out by Water Research Centre, United Kingdom (WRc) in 1987. Sonar testing involves very high-frequency ultrasonic sound waves that reflect off the material being inspected, allowing for the detection of defects. Sonar inspection results in a detailed profile (cross-sectional or longitudinal) of the pipe wall below the water surface, which can be analyzed by a variety of methods. Figure 4.9 illustrates an example of sonar output in a cross-sectional view.

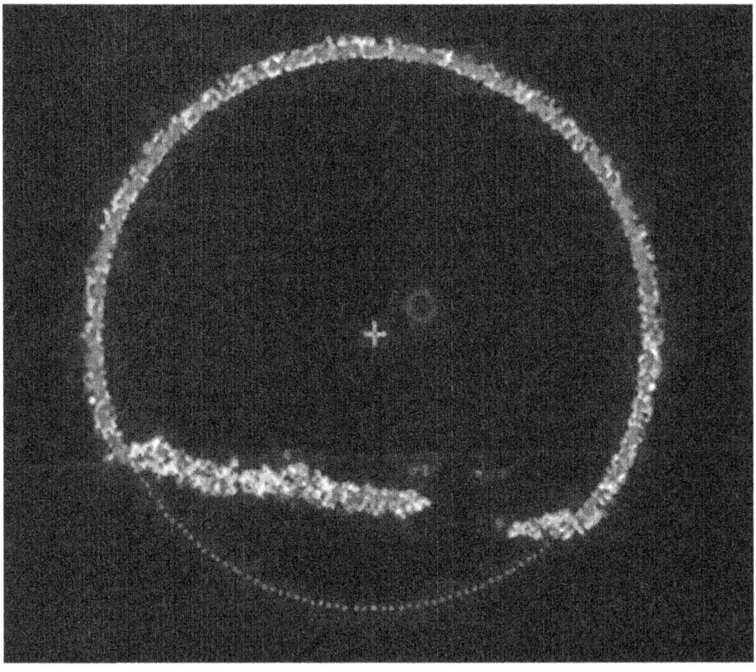

Figure 4.9 Sonar results showing silt deposit in a pipeline. (*Source:* USEPA, 2010.)

System	Typical Use	Advantages	Limitations
• In-line leak detectors	• Pressure and gravity pipelines	• Can detect very small leaks	• Requires minimum flow to be carried through pipe
• Acoustic monitoring systems	• Pressure pipeline only	• Useful as a screening technique prior to more detailed inspection	• Only detects general distress
• Sonar/ ultrasonic systems	• Pressure and gravity pipeline	• Suitable for pipes of any material and a wide range of diameters	• Only inspects pipe below the waterline

Table 4.4 Acoustic Inspection Summary

The sonar method can detect pipe wall deflections, corrosion, pits, voids, and cracks. It can also detect and quantify debris, grease, and silt, and can distinguish between hard and soft debris. However, defects in the pipe wall can sometimes be obscured by grease and debris. Defects greater than 1/8 in. (3 mm) can easily be detected by using the sonar technology.

Table 4.4 lists a summary of the acoustic inspection technologies described earlier.

4.6.4 Electrical and Electromagnetic Technologies

Several sewer evaluation techniques utilize electrical or electromagnetic currents. The Electrical Leak Location (ELL) method is used to detect leaks in surcharged nonferrous pipes. Field eddy current (FEC) testing and remote field eddy current (RFEC) technology identifies defects in ferrous pipes. Magnetic flux leakage (MFL) inspection is widely used in the oil and gas industry to measure metal loss and detect cracks in ferrous pipelines. Laser is an advanced electrical technology that has recently found its use in wastewater pipe inspection.

4.6.4.1 Electrical Leak Location

Electrical leak location (ELL) is one of the most widely used techniques for detecting leaks in geomembrane liners. This technique involves placing an electrode on either side of the material being tested and connecting voltage to each electrode. Because the material being tested is an electrical insulator, voltage only flows through holes in the material. The areas of defects in the material has high current density, which can be detected by measuring electrical potential in the survey area. Although primarily used for geomembrane liner inspection, the technique is also applicable to pipe inspection.

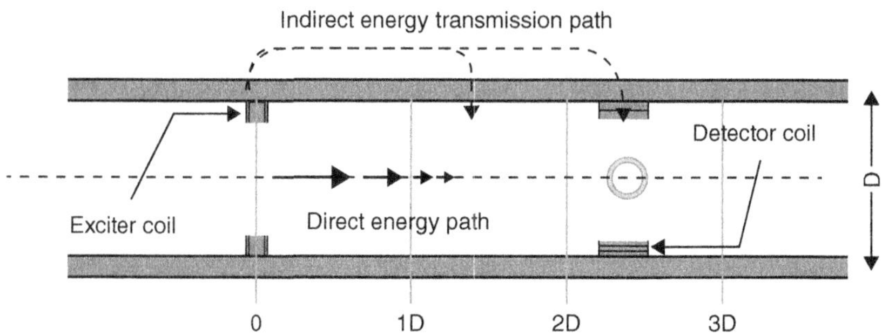

FIGURE 4.10 FEC testing. (*Source:* USEPA, 2010.)

4.6.4.2 Remote Field Eddy Current

Remote field eddy current (RFEC) technology involves the generation of electric currents and magnetic fields to investigate the condition of ferrous material. The FEC pipe inspection involves the use of an alternating current magnetic coil to induce an electric current in conductive pipes. In turn, the electric current generates small magnetic fields or eddy currents in opposition to the coil's magnetic field, which results in a change in the impedance of the coil. As the magnetic coil transverses the pipeline, the change in impedance is measured, allowing for the identification of defects as illustrated in Fig. 4.10.

The RFEC method was developed to overcome the limitations of standard eddy current testing. This method can detect both internal and external defects in pipelines. Remote field eddy current involves the deployment of a probe consisting of multiple magnetic coils, an exciter coil, and one or more detector coils, through the pipeline. As in standard eddy current testing, the currents are induced in the pipe wall. These direct currents quickly attenuate as they flow along the pipe wall towards the detector coil that is typically located approximately two pipe diameters apart from the exciter. A second magnetic field passes from the exciter to the outside of the pipe, flows along the outer pipe wall, and then back into the interior of the pipe to reach the detector. This remote field attenuates very slowly along the outer pipe wall, and is therefore much stronger then the direct field when it reaches the detector. Pipeline defects and pipe wall thickness affect the propagation of the magnetic fields along the pipe walls, thereby altering the signal received by the detector. This allows for the identification of pipeline defects.

4.6.4.3 Magnetic Flux Leakage

The magnetic flux leakage (MFL) detection involves the placement of one or more magnets on the pipe wall, leading to the inducement of a direct current magnetic field in the pipe wall. The strength and direction of magnetic fields are represented by flux lines as illustrated in Fig. 4.11.

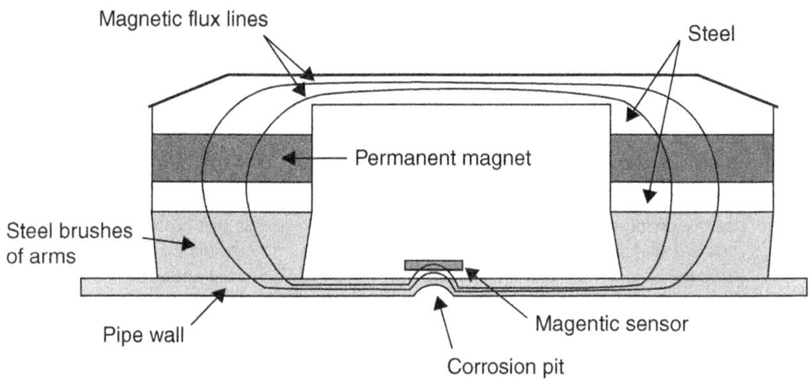

FIGURE 4.11 Magnetic flux inspection. (*Source:* USEPA, 2010.)

When a magnet is near a conductive pipe wall, the majority of flux lines pass through the pipe wall. However, in areas of metal loss, less flux can be carried than the intact, full wall sections of pipe. This leads to leakage of flux from pipe areas, which have undergone metal loss, as well as a change in the shape of the induced magnetic field. Flux leakage is detected by sensors; computer software is then used to determine the type and size of anomalies detected by the sensor.

4.6.4.4 Laser Profiling

Laser profiling generates a profile of a pipe's interior wall. The technique involves using a laser to create a line of light around the pipe wall. For this reason, it is also called the "lightline" method. The laser light highlights the shape of the sewer, allowing for the detection of changes to the pipe's shape, which may be caused by deformation, corrosion, or siltation. Figure 4.12 illustrates such a ring of light with clear deviation from circularity.

Laser inspection can only be used to inspect and measure dry portions of a pipe, which is above the waterline. To assess the entire

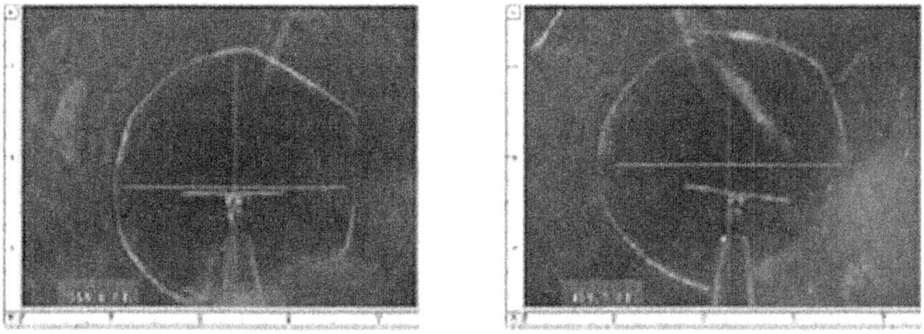

FIGURE 4.12 Example of laser profiling in a pipeline. (*Source:* USEPA, 2010.)

System	Typical Use	Advantages	Limitations
ELL	• Pressure and gravity pipelines, and service laterals	• Suitable for service laterals	• The main application is for electrical detection of leaks in geomembranes
RFEC	• Pressure and gravity pipelines, and service laterals	• Can be used on pipes of most diameters • Can locate a variety of defects	• Limited to ferrous pipes, typically requires post-processing of data by a vendor
MFL	• Pressure and gravity pipelines, and service laterals	• Extensive experience with methods in the oil and gas sector	• Has not been extensively used for assessment of sewer pipes
Laser	• Pressure and gravity pipelines	• Provides better data quality and can be used to create three-dimensional models of pipelines	• Can only detect defects above the water line

TABLE 4.5 Electrical and Electromagnetic Inspection Summary

internal surface of a pipeline, the pipe must be taken out of service. Lasers are often used in combination with other inspection methods, most commonly CCTV and/or sonar. It provides better data quality than CCTV alone, and can be used to create three-dimensional models of pipelines.

Table 4.5 lists a summary of the electrical and electromagnetic inspection technologies described above.

4.6.5 Other Technologies

In addition to the range of commercially available technologies stated above for evaluation of pressure and gravity pipelines, several other innovative technologies are currently under development. While the following technologies may not currently commercially available for condition assessment of pipeline systems, they may be feasible methods of pipeline assessment in the future.

4.6.5.1 Gamma-Gamma Logging

Gamma-Gamma logging is a technique used primarily to evaluate cast-in-place concrete pilings and for vertical borehole investigation in the mining and oil and gas industries. This technique involves the use of a gamma-gamma probe, which consists of a source of gamma radiation such as Cesium-137 and one or more gamma detectors. The detectors are shielded from direct radiation by a heavy metal such as lead. The gamma-gamma probe emits photons which react to surrounding material based on density. The photons are backscattered by the surrounding material, and data are recorded as a density log. An inspection using this technique is accomplished by raising and lowering the probe within a polyvinyl chloride (PVC) inspection tube that is inserted into the concrete piling or borehole. Results of the inspection yield information on the average bulk density of the concrete. Properly constructed structures will have a consistent density. This technique can also be used to locate voids.

4.6.5.2 Ground-Penetrating Radar

The US military originally developed ground-penetrating radar (GPR) to locate underground tunnels and mines. The GPR operates on the same principle as radar. A transmitting antenna emits high-frequency radio waves into the ground. The waves travel through the ground until they reach a material that has a different conductivity and dielectric constant than the earth. The signal is reflected and recorded by a separate receiving antenna. The amount of time it takes for the electromagnetic radio waves to be reflected by subsurface materials can be analyzed to determine the position and depth of features below ground surface such as pipelines and possible voids in pipe embodiments. This method works better in low conductivity soils such as dry sands and gravels.

4.6.5.3 Infrared Thermography

Infrared thermography involves the use of an infrared camera to measure the temperature differential across the surface of an object. Software can then be used to create an image displaying different colors at different temperatures. This allows for the detection of the surface expression of thermal conditions beneath the surface. In this regard, it is a potential method of detecting sewer defects, such as leaks and voids, both of which can result in surface temperature variations when a sufficient internal/external temperature difference exists.

4.6.5.4 Micro-Deflection

Micro-deflection is a nondestructive technology used to evaluate brick, concrete, and clay structures. The method involves the use of a load to create slight deformation, termed a micro-deflection, in the test material. The change in position of the structure is measured, and a graph of load versus deflection is created. Structurally sound test materials would be expected to have a consistent load versus deflection graph, while deteriorated sections of the material would have a different value on the graph.

4.6.5.5 Impact Echo and Spectral Analysis of Surface Waves

Impact echo and spectral analysis of surface waves (SASW) are two related techniques for evaluation of concrete and masonry pipes.

Both methods work by subjecting a pipe to an elastic impact, produced by a device such as a pneumatic hammer, which then propagates through the pipe. The waves are reflected by internal flaws, as well as the surface of the material, and the reflected waves are detected by an acoustic transducer such as a geophone located on the exterior of the pipe. The technique can locate and measure cracks, delamination, voids, and honeycombing.

4.6.5.6 Ultrasonic Pulse Velocity and Guided Wave Ultrasonic

The ultrasonic pulse velocity method and the guided wave ultrasonic testing (GWUT) are potentially suitable for the condition assessment of wastewater collection systems. The ultrasonic pulse velocity method is based on the speed with which an ultrasonic pulse or a compression wave passes through the test material. The GWUT on the other hand induces "plate-type" waves or "guided" waves in the material, permitting detection of cracks and measurement of pipe wall thickness.

The ultrasonic testing assesses the surface as well as the internal features (e.g., thickness and material properties) of the object being tested. For wastewater sewers, ultrasonic testing can measure pipe thickness, detect corrosion, and detect and measure cracks. In its simplest form, ultrasonic inspection uses the pulse-echo method to measure the thickness of materials. The short burst or pulse is induced in the material using a transducer. The pulse passes through the material until it reaches the other side, where its echo is reflected to the surface. The duration and velocity of the pulse are measured and used to determine the distance from the surface to the outer side. The pulse can also be directed into the material at an angle using an angle beam transducer. The angle allows for more accurate detection of cracks or flaws in the material. The conventional ultrasonic testing method typically involves point-by-point measurements along the surface of the material being tested. To ensure sufficient transmission of sound waves, a couplant, either water or a gel, must be applied between the sensor and the test material.

4.6.6 Case Study: City of Marco Island's Existing Sewers Assessment, and Rehabilitation[6]

This study discusses a sanitary sewer renewal program in Marco Island, Florida, as a case study of proactive wastewater collection system management. The existing system was evaluated using the CCTV inspection of pipelines and visual inspection of manholes. Accordingly, a renewal program was developed, which was based on using trenchless technologies to maximum extent.

---

[6]Adapted from Sever and Bradshaw, 2009, available at http://uimonline.com/index/webapp-stories-action?id=310.

4.6.6.1 Scope of Work

In coordination with the consulting engineer, the city started an assessment/rehabilitation program for its sanitary sewers. The goals of this program were to assess the structural integrity of the pipelines, identify sources of infiltration and inflow into the system, and develop a capital improvement plan to prioritize and correct defects that were discovered. Methods used for inspecting the existing system were cleaning the pipelines using a water jet, surveying pipe condition using CCTV, and smoke testing of lateral sewers and select manholes. Defects noted during the CCTV inspection were identified using the PACP coding system. The video records were reviewed by the consulting engineer to verify and evaluate the reported defects.

4.6.6.2 Existing System Inventory

Sanitary sewers in Marco Island consisted of PVC and vitrified clay pipes (VCP), with clay pipes being older than the PVC pipes. The PVC pipes did not demonstrate any notable damage except a few discrete holes and cracks due to rock or miscellaneous object penetration. The major part of the inspected VCP was in sound or moderate condition requiring spot repairs or no repair at all in the foreseeable future. Approximately 16 percent of the existing clay pipes demonstrated extensive damage, such as cracks, fractures, and loose/offset joints. These defects required complete renewal using a structural liner such as cured-in-place pipe.

The cleaning and televising program included the existing sewers, which are distributed over the Island on the northeast, northwest, and southwest. These areas were divided into phases to prioritize and coordinate the project. The overall project included approximately 190,000 ft of sanitary sewer pipes ranging from 8 in. (200 mm) to 12 in. (300 mm) in diameter and 833 manholes, which served 34 percent of the island. Approximately 6 percent of the existing sewers were excluded from the renewal program due to pipes/manholes being new or recent renewal.

4.6.6.3 Sanitary Sewer Pipes

Based on the review of the CCTV records, the consulting engineering company evaluated condition of the pipelines, which was indicated using color-coded plans (Fig. 4.13). Three levels of deterioration, that is, sound, moderate, and poor, were identified by green, yellow, and red, as defined below.

Green. Pipe segments that have few or no observed damage. No renewal was recommended for green pipes.

Yellow. Pipe segments that had moderate amount of defects, such as fractures and cracks. While the yellow pipes, in general, appeared to be sound, immediate renewal of some of them might have been advisable due to localized severe damages (e.g., a hole or a broken pipe).

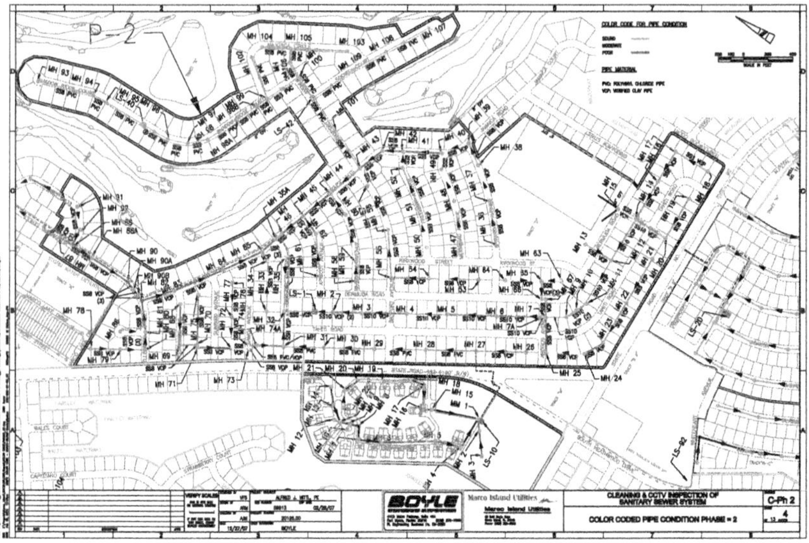

Figure 4.13 Sample color-coded map indicating condition of inspected pipes in a lift station area. (*Source:* Sever and Bradshaw, 2009.)

Red. Pipe segments with frequent cracks and fractures were coded red. Red pipes were not structurally sound; they were prone to failure and they were sources of infiltration. Figure 4.14 shows a breakdown of inspected sanitary sewer pipes based on their condition.

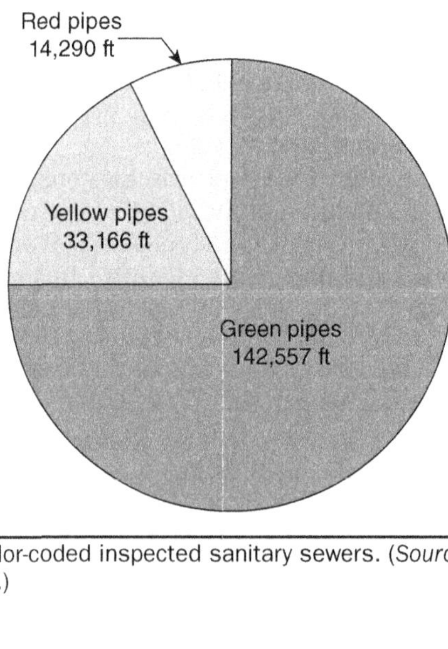

Figure 4.14 Color-coded inspected sanitary sewers. (*Source:* Sever and Bradshaw, 2009.)

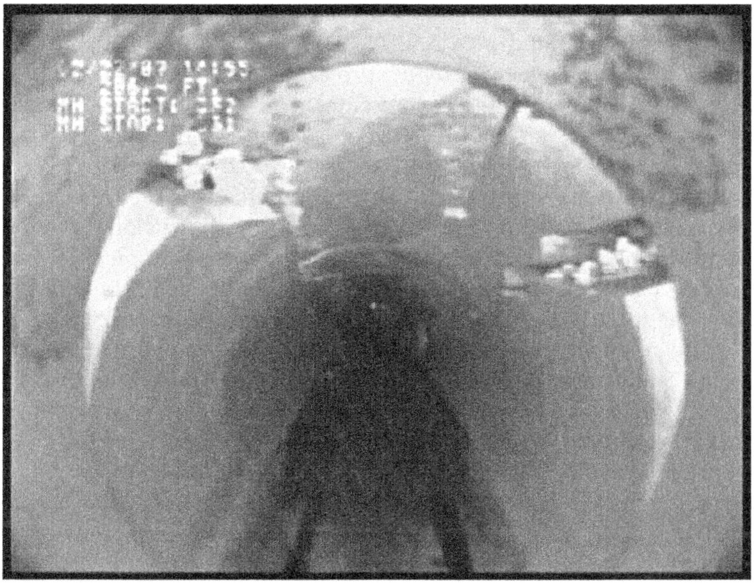

Figure 4.15 View of a fractured clay pipe. (*Source:* Sever and Bradshaw, 2009.)

The condition of the pipelines was highly dependent on the location and material. Accordingly, minor or no damage was noted to the PVC sewers; whereas, the magnitude of damage to the VCP sewers ranged from no damage to extensive (Fig. 4.15). The most common types of defect were cracking, fracture, and root intrusion (at joints) in VCP and sagging in PVC pipes. Pipe segments that had frequent fractures and cracks allowed infiltration or exfiltration depending on the watertable and the location of the pipe. In addition, structurally unsound pipe segments were prone to a collapse, which might have caused sanitary sewer backups and damage to the road above the pipeline.

4.6.6.4 Manholes

The CCTV inspection survey records did not provide a clear view of the manholes; therefore, manholes were visually inspected by the project team. Forty-six percent of the manholes included in the assessment program were included in the inspection (386 of 833). This selection was based on age and location as manholes in low-lying areas (i.e., swales and areas at low elevations). Wet weather flow rates to the Marco Island Wastewater Treatment Plant were significantly higher than dry weather; whereas, based on the CCTV record review, it was understood that infiltration into the collection system was not substantial. Hence, manhole inspection and assessment were aimed at reducing inflow through manholes and enhancing structural integrity of the deteriorated ones. A visual inspection was conducted using a standard form developed by the project team.

Majority of the manholes had defects on the upper section, that is, cover, frame, and chimney; whereas some of them underwent corrosion due to hydrogen sulfide release, which damaged cone and wall. Damage to the bench or trough was observed in only a few manholes, and this was mainly due to root intrusion. Like the sanitary sewer pipes, condition of each manhole was depicted using a color code. Color coding identified a manhole's condition at three levels as sound (green), moderate (yellow), and poor (red).

Green manholes were the ones with minimal or no apparent corrosion on the interior or top (frame and lid). Green manholes did not require any renewal or frame/lid replacement. However, some of them required lid sealing to prevent inflow.

Yellow manholes were the ones with moderate damage. They were structurally sound, that is, minor or no corrosion to the bench, wall, and cone. Yellow manholes typically required replacement of the frame, sealing of the lid, and partial lining (chimney and/or joints) or grouting.

Red manholes were the ones with extensive corrosion throughout the interior and significant inflow (and infiltration for some). Lining the interior of red manholes would have prevented further corrosion and thus a failure or extensive inflow and infiltration could be prevented. Red manholes comprised only 12 percent of the inspected manholes. Figure 4.16 shows a percentage breakdown of inspected manholes based on their condition.

Conditions of the sanitary sewer pipes and manholes were predicted over the long-term if a renewal program was not implemented.

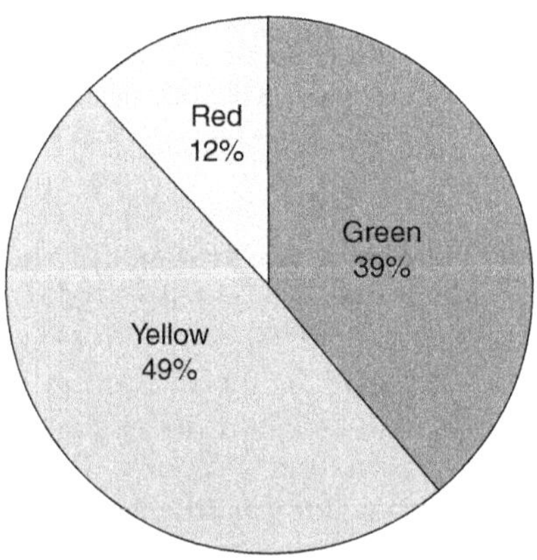

FIGURE 4.16 Percentages of green, yellow, red manholes. (*Source:* Sever and Bradshaw, 2009.)

This prediction was based on a linear correlation between degradation observed in 40 years to date and the degradation that would occur without renewal in 70 years. Condition of the pipelines after 40 years (average age of the inspected pipes and manholes, which varied from 35 to 45 years) was broken down in accordance with the material (clay or PVC). Then it was assumed that the pipe degraded from green to yellow or red over this period in constant increments (i.e., linear degradation), which was a much greater value for clay pipes.

The rate at which clay pipes turned yellow from green was 40 percent; whereas, this rate was 3 percent for PVC pipes. It was more than likely that degradation of a sanitary sewer pipe or manhole was not linear over time as deterioration would have been expected to occur at an accelerated pace when the pipes and manholes reached their design life. However, there was not sufficient data to create a model for the deterioration of the pipes and manholes in long-term in the case of Marco Island, as the pipes were inspected once after 40 years. The nonlinearity in the condition of pipelines vs. time plot is a result of different rate of degradation between PVC and clay pipes. All of the clay pipes were expected to sustain, at minimum, minor damage, and degrade to yellow in 40 years. In other words, there would be no green clay pipes in the system after 40 years without renewal. This have resulted in a decline in yellow pipe quantity as yellow clay pipes proceeded to degrade and turned red; whereas, available green PVC pipes turned yellow then transited to red at a lower rate than clay pipes.

Anticipated benefits of renewal in monetary terms were analyzed over the long-term (70 years). The value of a green pipe segment was taken equal to a new sanitary sewer pipe, which was based on new construction going on in the other parts of the island. The average unit price ($/foot for sanitary sewer pipes) for new construction was amplified by 25 percent to account for relatively low quantity and additional cost of replacing the existing pipes (which could include flow bypassing and extra labor). On the other hand, there was a social cost associated with traffic interruptions and nuisance to the public for new construction or open-trench replacement, which was very complex to define in mathematical terms and excluded from the value of a green pipe. A yellow or red pipe segments was the cost of a green pipe segment (per foot) less the cost of renewal required to bring the pipe condition to green. The cost difference between green and red pipes was much greater for the red pipes that required replacement.

This study assumed that red pipes and manholes required replacement in 70 years, and therefore, they will be of negligible value at the end of this time period without renewal. It was anticipated that with the foregoing assumptions, the total value of 190,000 ft of sanitary sewer pipes and 386 manholes that were evaluated would have decreased from $37.0 million to $32.9 million (2009 dollars) in 70 years without renewal.

4.7 Condition Assessment Technologies Specific for Water Distribution System

4.7.1 Overview

Condition assessment methods for water distribution system can be classified as direct and/or indirect methods. Direct assessment usually requires access to the inside or the outside of the pipe, including visual inspection, destructive testing, and nondestructive testing (NDT). Indirect methods include the analysis of failure history, leakage level, hydraulics, and soil properties. The WERF/WaterRF study "Condition Assessment Strategies and Protocols for Water and Wastewater Utility Assets" lists 85 individual condition assessment tools and techniques (WERF, 2007). Table 4.6 illustrates the basic technologies and their applications together with observations on their limitations and potential application to ferrous water mains.

In common with many other "internal" technologies, the main barrier in undertaking such inspections is the need to gain access to the line, which may require temporary bypassing, service shut down, dewatering, cleaning, and removal of pipe sections or fittings to allow a camera to be put into the line. The costs associated with these support requirements will be many times the cost of the CCTV survey. At the time a line has to be closed down and emptied for maintenance or repairs, then the opportunity to carry out a CCTV survey can be taken (USEPA, 2009a). On the other hand, NDT techniques (also commonly referred to as NDE) include the direct visual observation of defects, such as cracks, corrosion pits, or holes, as well as techniques that provide signals or signatures that are interpreted into distress indicators.

Summaries of the current inspection methods and applications described in Table 4.6 require either external or internal inspection of a pipeline. It should be noted that for external technologies, if the pipe is not exposed, it will be necessary to excavate inspection pits.

4.7.2 Internal and External Inspection

4.7.2.1 Condition Inspection Using CCTV

The new generation of CCTV systems has features that enable an operator to obtain useful information on the internal condition of a pressure main. Key components include the following:

- Pan and tilt head.
- Zoom lens.
- Sonde/probe.
- Crawlers/tractors.

Application	External Integrity Inspection					Internal Integrity Inspection				Leak Detection				
	Pit Depth Measurement	Ultrasonic	Guided Wave Ultrasonic	MFL (ECAT)	BEM	In-line Ultrasonic	In-line MFL	RFT "See Snake"	BEM In-line Pig	Leak Correlators (External)	Listening Sticks (External)	Continuous Acoustic Monitoring (External)	Sahara® (Internal)	SmartBall® (Internal)
Gray cast iron	Yes	Yes	No	Yes	Yes	No	No	Yes	Yes	Yes	Yes	Yes	Yes	Yes
Ductile iron	Yes	Yes	No	Yes	Yes	No	No	Yes	Yes	Yes	Yes	Yes	Yes	Yes
Steel	Yes	Yes	Yes	Yes	Yes	Yes	Yes	Yes	Yes	Yes	Yes	Yes	Yes	Yes
Diameters	Any	Any	2–48 in.	>6 in.	>2 in.	Most	Most	14 in.	>6 in.	Most	Most	<16 in.	>4 in.	>8 in.
Typical length scanned	3–6 ft	3–6 ft	300 ft	3–12 ft	3–12 ft	Miles	Miles	10,000 ft	3,000 ft	300 ft	3 ft	300–500 ft	6,000 ft	Miles
Line in operation	Yes	Yes	Yes	Yes	Yes	Yes (a)	Yes (a)	Possibly	No	Yes	Yes	Yes	Yes	Yes
Scan through coatings/ wrappings/ linings	No	No	Yes	Yes	Yes	No	No	Possibly	Yes	–	–	–	–	–
Loss of metal	No	Yes	Yes	Yes	Yes	Yes	Yes	Yes	Yes	–	–	–	–	–
Pit depth	Yes	No	No	No	No	No	No	No	No	–	–	–	–	–
Graphitization	No	No	N/A	Yes	Yes	Yes	No	Yes	Yes	–	–	–	–	–
Cracks	No	No	Yes	Yes	Yes	Yes	No	Yes	No	–	–	–	–	–
Joint leaks	–	–	–	–	–	–	–	–	–	Yes	Yes	Yes	Yes	Yes

Table 4.6 Summaries of Tools and Technologies Available for Water Pipe Inspection

Application	External Integrity Inspection					Internal Integrity Inspection				Leak Detection				
	Pit Depth Measurement	Ultrasonic	Guided Wave Ultrasonic	MFL (ECAT)	BEM	In-line Ultrasonic	In-line MFL	RFT "See Snake"	BEM In-line Pig	Leak Correlators (External)	Listening Sticks (External)	Continuous Acoustic Monitoring (External)	Sahara® (Internal)	SmartBall® (Internal)
Wall perforation leaks	–	–	–	–	–	–	–	–	–	Yes	Yes	Yes	Yes	Yes
Accuracy for locating small leaks	–	–	–	–	–	–	–	–	–	Good	Fair*	Fair*	Excellent	Excellent
Insertion into line	–	–	–	–	–	–	–	–	–	N/A	N/A	N/A	Via 2 in. tapping	Via valve or tapping
Mobilization costs	Minimal	Low	Medium	Medium	Low	Very high	Very high	Medium	Medium	Low	Low	Low	Medium	Low
Scanning/ processing cost	Low	Medium	High	Medium	Low/ medium	High	High	Medium	Medium/ high	Low	N/A	Medium	Low/ medium	Low/ medium
Suitable for water main investigations	Yes	Yes	No	Yes	Yes	No	No	Yes	Yes	Yes	Possible	Yes	Yes	Yes

*Can be supplemented with a correlator.

Note: BEM = Broadband Electromagnetic; ECAT = External Condition Assessment Tool; RFT = Remote Field Technology.

TABLE 4.6 Summaries of Tools and Technologies Available for Water Pipe Inspection (*Continued*)

The main purpose of such a survey would be to provide some visual indicators to possible defect locations. Where a pipe has an internal lining, it will provide some qualitative information on the condition of the lining. It may also be possible to identify leaks from inward groundwater infiltration when the pipe is empty.

4.7.2.2 Pit Depth Measurements

Pit depth can be measured with a pointed micrometer or needle-point depth gauge. Other methods include a grid with ultrasonic spot measurement, automated ultrasonic scanner, and laser range measurement. The pit depth measurement can be carried out in the field on exposed sections of the pipe (for external corrosion) or in a laboratory on pipe samples (for external and internal corrosion). Before pit measurement, pipe samples are sand/grit blasted to remove corrosion products. Pitting depth measurement can be applied methodically, within a general survey, or opportunistically when a pipe is exposed (e.g., upon breakage repair).

4.7.2.3 Laser-Based Pipe Surface Profiling

Pipe inner and outer profile can be acquired using a spinning apparatus that is needed to control the laser beam. Such a laser range measurement does not require any special illumination and can be carried out in complete darkness. The speed of spinning, sampling rate, and carrier moving velocity determine the accuracy and resolution of the scanning. The inspection is affected by the roughness as well as the color of the pipe surface. The laser ring and camera are typically mounted on a carrier or robotic platform. Gyroscopic position data (i.e., pitch, yaw, and roll) of this platform are needed to achieve the required precision. Currently available laser profiling systems are only used in dewatered pipes. To date, there is no known report on underwater laser profiling for in-service water mains (USEPA, 2012).

Another method makes use of a ring of laser light projected onto the pipe inner. The ring must be strong enough to be "seen" by a camera. The camera is used to capture the images of this projected ring. The laser device moves with the camera through the pipe. The analysis software extracts the laser ring from captured images and reconstructs a digital pipe profile. This profile can be easily unfolded or manipulated for review and analysis. The setup requires the laser ring to fall in the field of view of the camera. The accuracy depends on the fineness of the laser ring and the resolution of the camera (USEPA, 2012).

Table 4.7 lists a summary of the internal and external inspection technologies described above.

System	Typical Use	Advantages	Limitations
Condition inspection using CCTV	• Internal condition inspection of a pipe	• Simple, relatively inexpensive, and suitable for small and large pipes • New systems with multi-camera and/or fish-eye technology can record a full view of a pipe and allow relatively high scanning speed as well as full off-line inspection • Digital recording is convenient for data storage as well as future developments in automatic data interpretation	• Provides information only on defects that are manifested on the pipe inner surface • Inspection results are qualitative and need interpretation • Quantitative rating requires trained inspectors • Not available for in-service water main inspection • Requires a special launching and retrieval chamber in water mains
Pit depth measurements	• External condition inspection of a pipe	• Pipe does not need to be taken out of service during the time of inspection • Direct measurement, no need for interpretation • Provides good indication of sample condition • Does not require special skills, easy to train personnel	• Can be practically applied only to samples, therefore requires some sophisticated statistical analysis to infer general condition of the entire pipe (or pipe segment) • Need to expose the pipe or to cut coupon (destructive testing) • Existing coating needs to be removed. • Original pipe wall thickness must be available for corrosion rate estimation
Laser-based pipe surface profiling	• Internal and external condition inspection of a pipe	• Potential to show the early signs of pipe degradation by corrosion • Provides exact geometric dimensions for rehabilitation options • Enables inspection with minimum lighting requirements • Measures cross-sectional area • Can be applied in a wide range of pipe sizes	• Pipeline needs to be dewatered • Data analysis combines measurements with software and automated processes • No documentation is available on capability to detect cracks

TABLE 4.7 Internal and External Inspection Summary

4.7.3 Nondestructive Testing and Inspection

The following are some basic nondestructive technologies followed by a summary of the purpose, status, source of information, advantages, limitations, performance, breadth of use, and other available information.

4.7.3.1 Sonic/Seismic Technologies

Sonic and seismic technologies generate mechanical waves in a medium and measure the time taken for a reflected wave to reach a transceiver. This time can be correlated to material thickness if the elastic response characteristics of the medium are known. Ultrasonic, on the other hand, measures the propagation time of high-frequency, short-wavelength mechanical waves through a ferrous pipe wall, and correlates this with the nominal thickness of the material. The detection of flaws is based on the reflection of the wave from the interface between materials of different properties, for instance graphite or a cement mortar lining. The resolution is such that small areas of wall loss can be identified, allowing the creation of a map of the wall thickness of a pipe. Ultrasonic waves are at frequencies greater than 100 kHz, but accurate thickness measurements use frequencies in the order of 10 MHz (USEPA, 2009b).

4.7.3.2 External Applications of Ultrasonic

Ultrasonic measurements are among some of the best-established methods for simple external testing of points along a ferrous pipeline wall. The equipment is in the form of handheld instruments where a probe is positioned on the metal as shown in Fig. 4.17. The tool is

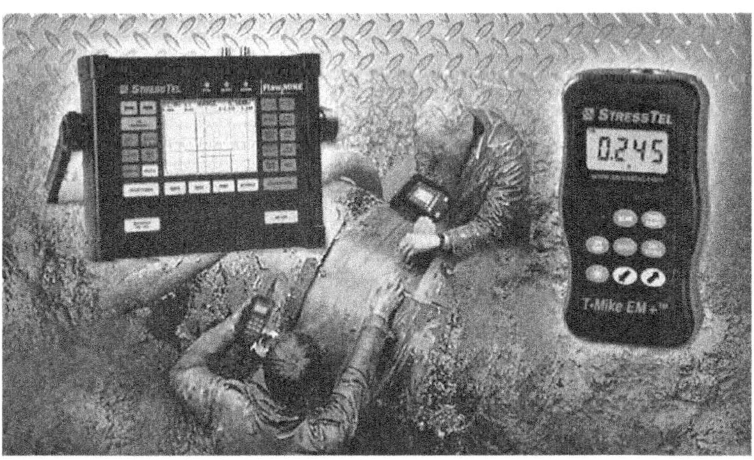

Figure 4.17 Handheld ultrasonic testing tool. (*Source:* USEPA, 2009b.)

calibrated to provide a direct thickness reading, which can then be compared with the original wall thickness. These measurements are point readings of the depth, and there can be substantial variations that are not detected between the chosen points.

Ultrasonic performs best on steel and ductile iron, and less well on thicker grey iron. Most ultrasonic devices require direct contact with clean metal, which involves removal of coatings, linings, and corrosion products. Because ultrasonic tools are inexpensive, they have become popular for in-house inspection tools with some utilities. However, operators still need to be properly trained and understand how the tools work, including their limitations, to avoid generating misleading data (USEPA, 2009b).

4.7.3.3 In-line Investigation Ultrasonic Pigs

The ultrasonic principle is used in intelligent pigs that have been developed for in-line inspection (ILI) of oil and gas pipelines. These tools directly measure the wall thickness as they travel through the pipe. They are equipped with transducers that emit ultrasonic signals perpendicular to the surface of the pipe. These transducers are located in a carrier that covers the internal circumference. Typically, the spacing will be every 8 mm circumferentially and 3 mm longitudinally. An echo is received from the internal and external pipe surfaces. By timing these return signals and comparing them to the speed of ultrasound in a steel pipe, the wall thickness can be calculated. Using ultrasonic in-line requires good contact with the pipe wall to obtain an accurate response (USEPA, 2009b). Typically, the contact medium is a liquid. This contact, or coupling, is usually sufficient for liquid-carrying pipelines. Figure 4.18 illustrates such operation.

This method may be impractical in tuberculated mains and potentially compromised in-lined pipes. It should be noted that the tools used in the oil and gas industry are not suited technically or economically to the inspection of water mains. Various laboratory developments have been reported, and prototypes have been built. However, a known commercial-proven tool is not currently available.

4.7.3.4 Guided Wave Ultrasonic Testing

The GWUT has the potential to inspect portions of pipes with continuity from accessible locations. The GWUT uses the pipe wall as a conduit for the ultrasonic energy. The guided wave pipe inspection technology was developed for external remote detection. The concept is to generate ultrasonic lamb waves and send them along a pipe using one of several transducer types. Discontinuities, such as corrosion, give a differing reflection back to the transducer, which is able to measure and determine the size of the defect. A guided wave propagation and reception collar is mounted on the outer surface of the pipe. This transmits and receives ultrasonic pulses around the full

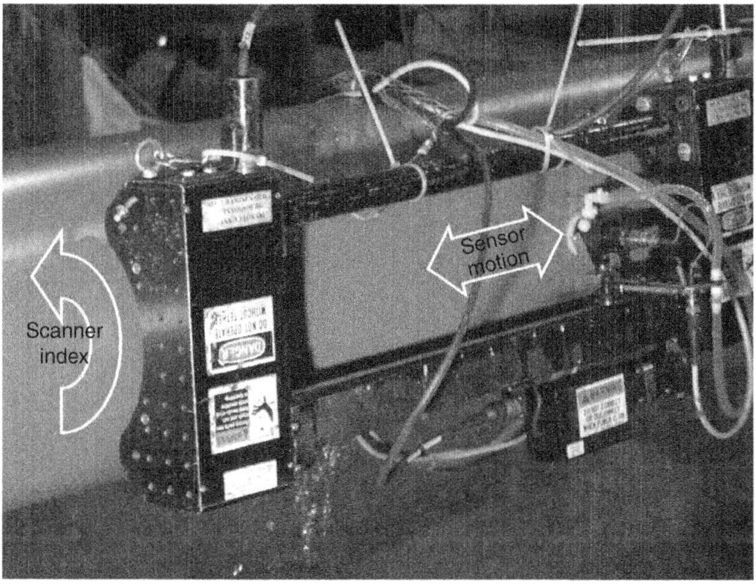

FIGURE 4.18 Automated ultrasonic scanner. (*Source:* USEPA, 2009b.)

circumference of the pipe. Losses of wall thickness can be identified and mapped (USEPA, 2009b). "Teletest" is an example of the GWUT tool as illustrated in Figure 4.19.

To date, commercial tools have been used in the oil and gas industries on pipes with welded steel joints that provide the continuity. The diameter ranges from 2 to 48 in. Typically, the length of pipe that can be scanned depends on pipe condition, coating, and type of soil with maximum distance in the order of 100 ft. The GWUT works for pipes that are insulated or coated. It is particularly useful for buried steel pipelines crossing roads.

4.7.4 Electromagnetic Methods

Several technologies are based on electromagnetic principles, but all are variations on MFL and FEC previously described. They are applicable to ferrous pipes or pipes with a ferrous component.

4.7.4.1 Magnetic Flux Leakage

When a magnet is placed next to a pipe wall, most of the flux lines pass through as the pipe wall is a preferred path for the flux. While most of the flux lines concentrate in the pipe wall, a few pass through the surrounding media. Flux leakage at a metal-loss region is caused by a local decrease in the thickness of the pipe wall. At a metal-loss region, the flux carried by the thin section is less than that carried in the full wall. A sensor positioned on the inside (magnet side) of the

FIGURE 4.19 "Teletest" remote detection tool showing transducer. (Source: USEPA, 2009b.)

pipe is typically used to measure the magnetic field adjacent to the pipe wall. At a metal-loss region, a sensor records a higher flux density or magnetic field, which indicates the presence of an anomaly. An MFL technology is shown in Fig. 4.20.

In this manner, an MFL tool detects an anomaly that causes flux to leak. The measured leakage field depends on the radial depth, axial length, circumferential width, and shape of the anomaly, as well as the magnetic properties of the nearby material. To characterize the anomaly, the measured leakage field must be analyzed.

4.7.4.2 Field Eddy Currents

One eddy current technique, the RFEC, is capable of inspecting the entire wall thickness without the need to use ultra-low frequency. An exciter that is sized to nearly the same diameter as the inside diameter of the pipe is driven with a low-frequency sinusoidal current. A small, magnetic field sensor is positioned some distance away. One portion of the magnetic field, generated by the exciter, travels down the inside of the pipe, with the field directly coupled to the sensor. A second portion of the alternating magnetic field propagates through the material of the pipe, inducing eddy currents as it goes. Once the magnetic field penetrates the outside wall of the pipe, it spreads along the surface of the pipe and re-enters the pipe, again inducing eddy currents to flow in the pipe material. This second path is referred

A

B

Figure 4.20 AES-519 and AES-ECAT pipe wall inspection tools using MFL technology. (*Source:* USEPA, 2009b.)

to as the remote path. The total magnetic field and eddy current flow at any point is the combination of directly coupled and remotely coupled fields. The current main range of tools is based around the name "See Snake" which operates in diameters from 2 to 14 in., although the manufacturer intends building larger versions. The "See Snake" tool modules, illustrated in Fig. 4.21, allow for a large degree of flexibility with a capability of negotiating 90° bends.

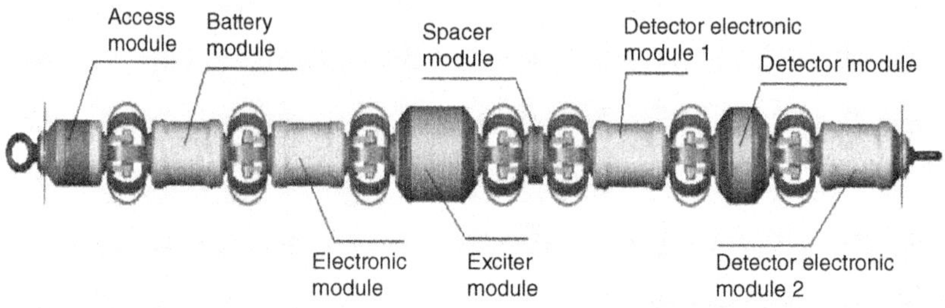

FIGURE 4.21 "See Snake" modules and spacers using the RFEC technology. (*Source:* USEPA, 2009b.)

The spacers center the tool, and direct contact with the pipe wall is not required. The tool can be free swimming or secured tethered on a wire line. Lengths up to 3,000 ft can be inspected from one launch point when wire line joined. The free-swimming version can inspect lengths up to 15,000 ft from the launch point. A "T" or "Y" piece needs to be retrofitted into the line to provide a launch trap. The tool is propelled through the line by the fluid flow or compressed air while the line is in operation (USEPA, 2009b). Reported "See Snake" specifications are:

- Run time: minimum 5 h
- Speed: approximately 10 ft/min
- Pipe thickness: up to 0.5 in. steel; 1 in. for cast iron
- Pipe diameters: 2 to 14 in.
- Clearance is 0.25 to 1 in. around tool
- Tool length is approximately 20 times diameter
- Can negotiate some bends.

4.7.4.3 Broadband Electromagnetic

Unlike the conventional eddy current technique, which uses a single frequency for testing, the broadband electromagnetic (BEM) technique transmits a signal that covers a broad frequency spectrum. A transient input signal generates multiple frequencies, typically ranging from 50 Hz to 50 kHz. The recorded signal from a broadband transmission contains more information and allows detection and quantification of various wall thicknesses as well as the effective conductivity of the complex through wall components of the pipe. Changes in conductivity reflect changes in material properties (USEPA, 2009b).

A transmitter coil passes an alternating current to the pipe surface, which generates an alternating magnetic field. Flux lines from

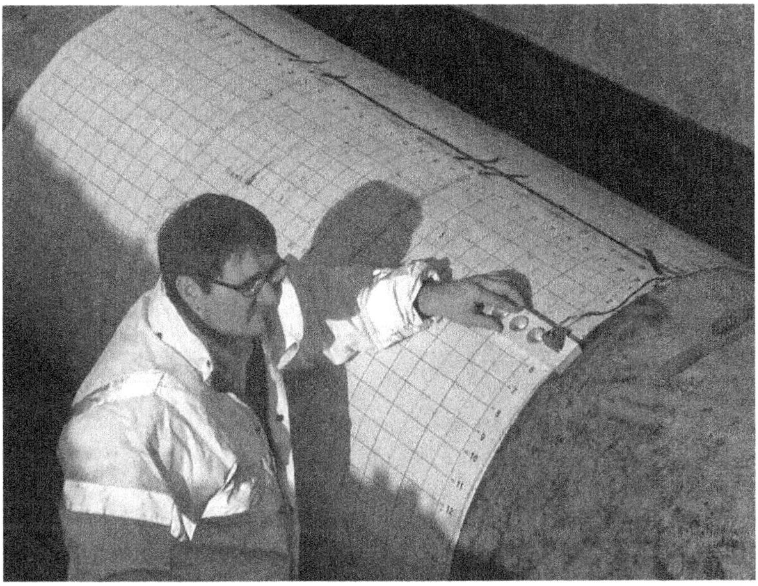

Figure 4.22 BEM handheld tool being used to scan a gray iron pipe. (*Source:* USEPA, 2009b.)

this magnetic field pass through the metallic pipe wall, generating a voltage across it. This voltage produces eddy currents in the pipe wall, which induce a secondary magnetic field. Wall thickness is indirectly estimated by measuring signal attenuation and phase delay of the secondary magnetic field. Figure 4.22 illustrates the BEM technology in use.

External scanning requires excavation of buried pipes. Internal scanning can be carried out with an in-line inspection (ILI) pig, which is driven by hydraulics or by push/pull rod devices. However, the pipeline needs to be out of service, emptied, and cleaned of loose deposits in order to run the pig. In cast iron, BEM can identify and locate metal loss and cracks. It does not require contact with bare metal to detect pits and other metal loss.

4.7.4.4 Radiographic Testing

Radiography is a nondestructive test method that can be used on ferrous pipes. Radiography shows changes in thickness and density that are associated with corrosion. It has been used widely in petrochemical processing plants and outside the United States in water mains. It has limitation that pipes of 15-in. inner diameter and greater must be emptied. It also has considerable health and safety issues. The radiographic testing is expensive and requires specialist operators (USEPA, 2009b).

4.7.5 Environmental Testing

The soil, watertable, and pollutants can create conditions leading to corrosion of unprotected ferrous pipes. Soil environments can be corrosive to ferrous pipes, and even in moderately corrosive soils, the rate of corrosion may be such that the pipe life expectancy will be reduced. Additionally, all types of ferrous pipes can suffer potential damage from contaminated soils. Due to the lack of better indicators of corrosion, the soil environment evaluation is currently the best approach for evaluating external corrosion for unprotected pipes.

Table 4.8 provides a summary of indicators and criticality ratings for soil environment data.

4.7.6 Leak Detection Technologies

Leak detection uses a variety of methods, including acoustic, acoustic with correlation, infrared thermography, chemical (using a tracer gas), and mechanical. Acoustic leak detection is the most widely used method. Since the beginning of the 21st century, a number of new leak detection technologies have been developed, including GPR, combined acoustic logger and leak noise correlator, digital correlator, and radio-frequency interferometer. One of the recent advances is in-line leak detection systems that use the Sahara® and SmartBall® acoustic systems. Some select leak detection technologies are described in the following sections.

4.7.6.1 External Leak Detection

External technologies fall into two groups: acoustic and correlators. The former works on the basis of directly locating the noise of a leak; in its most effective form, and it entails the use of a monitor brought in contact with the main to listen for leakage noise. Other monitors, called hydrophones, do not come into direct contact with the pipe, valves, or hydrants, but are placed over the location of the pipe. There have been numerous acoustic and electronic advances to improve the capability of "stick sounding."

Correlators are based on the velocity of sound made by a leak as it travels along the pipe wall between two hydrophones or similar sensors. The sensors are situated at convenient locations on the pipe some distance away from, and on either side of, the leak point. The difference in time taken for the sound to travel to each sensor allows the difference in path length to be calculated, from which the leak position can be identified. It has long been recognized that correlators are not as effective on large-diameter transmission mains (USEPA, 2009b).

A recent advance in external technology is the deployment of monitoring units designed to monitor acoustic noise over extended time periods including permanent installations. There are several

Indicator	Criticality	Comments
Soil resistivity	High	A function of soil moisture, temperature, and concentrations of ionic soluble salts.
Soil moisture content	Low	Soils with moisture content greater than 20 percent considered more corrosive.
Soil pH	High	Likelihood of corrosion increases in acidic soils with pH values of 4.5 and less.
Chloride ion content	Medium	Use of de-icing salts on paved surfaces can lead to high concentrations of chloride ions in trench. Where concentrations are in the range of 10–1,000 ppm, there is an increasing potential for corrosion.
Sulfate and sulfide	Medium	Presence of sulfide in waterlogged soils indicates that bacteria have been promoting reduction of sulfate ions and indicates Microbiologically Influenced Corrrosion (MIC) activity.
Redox potential	High	Redox potential is an indicator of degree of aeration in soil. A high level is an indicator of aerobic soil and increased potential for corrosion. A low negative redox potential is an indicator of anaerobic conditions and potential for Sulfate Reducing Bacteria (SRB) and MIC forms of corrosion.
Known corrosive environments	High	Experience indicates that some soils and site conditions are corrosive regardless of testing. These include polluted sites, cinders, landfills, and peat bogs.
Soil temperature	Low	Changing soil temperature is of particular concern in northern climates where ground frost can cause movement. High temperatures can accelerate corrosion rates.
Groundwater levels	Medium	Groundwater levels need to be recorded including seasonal changes. Those caused for example by tidal variations can contribute to both soil instability and corrosive conditions. Constant high groundwater level can create anaerobic conditions.
Potential soil instabilities	Medium	Soil instabilities can arise from various causes, including, e.g., seismic movements and locations where a pipeline is on or adjacent to a slope or subject to manmade vibration, such as subway, rail line, and heavy vehicle traffic.

Source: USEPA (2009b).

TABLE 4.8 Indicators and Criticality from Environmental Data

variations in these systems. Some units provide data by downloading to mobile receivers that stop or pass nearby the monitors. Others are linked to automatic meter reading networks that transmit data to a base station or Website. Some systems are connected to water service lines, while others are connected to the top of valves.

4.7.6.2 In-Line Leak Detection

One of the most recent valuable leak detection developments has been "In-pipe Technology." Leaks can be detected by passing a hydrophone through the interior of the pipe to the point where the leak noise signal is detected. As the hydrophone travels through the pipe, it can only be a maximum of one pipe diameter away from the leak, so even small leaks can be detected. Commercially available tools include Sahara® and SmartBall®.

The SmartBall® is a free-swimming foam ball with an instrument filled with aluminum alloy core that records acoustic activity to identify leaks as it moves along the pipeline. It comprises a range of acoustic sensors, as well as an accelerometer, magnetometer, ultrasonic transmitter, and temperature sensors, which travel with the water flow down a pipe and detect leaks as it rolls. The sensors are encased in an aluminum alloy core with a power source and other electronic components. The core is encapsulated inside a protective outer foam shell or sphere. The outer foam shell provides additional surface area to propel the device and also eliminates the noise that the device might generate while traversing the pipeline. The diameter of the outer sphere depends on the pipe diameter and flow conditions. Figure 4.23 illustrates a SmartBall® traversing through a pipeline.

The SmartBall® is deployed into the water flow of a pipeline and captured at a downstream point. It continuously records acoustic data and emits an acoustic pulse every 3 seconds while the device

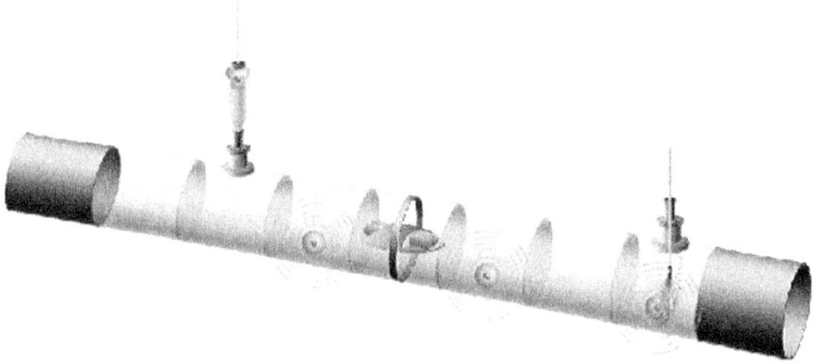

FIGURE 4.23 SmartBall® traversing through bends and valves in a pipeline. (*Source:* USEPA, 2009b.)

traverses the pipeline. A SmartBall® acoustic receiver, which is patented by Pure Technologies, is used to track the location of the device. The recorded acoustic data are analyzed to identify air pockets and leaks. Other sensory data are used to determine the location of air pockets and leaks. The severity of leaks is estimated by calibrated baseline data. Frequency analysis needs to be carried out to confirm that an acoustic anomaly is actually a leak.

The Sahara® system uses a hydrophone secured to an umbilical cable, which travels inside in-service water mains, to record leak noises. A locator signal can be tracked on the surface, enabling leaks to be marked for excavation and subsequent repair. Sahara® locates leaks through identifying the distinctive acoustic signals generated by leaks in the pipe wall, the joints, or steel welds. The magnitude of the leaks can also be estimated from the acoustic signal. Gas pockets in the pipeline are also detected by their unique acoustic signature. Figure 4.24 shows the Sahara® system in use.

The Sahara® sensors are launched into in-service water mains through a launching chamber that is mounted on a 2-in. (50 mm) or larger access point. A small parachute uses the flow of water to draw the sensor through the pipeline; alternatively, a preinstalled pull-tape can be used to draw the sensor through the line when no flow is available, such as on precommissioned pipelines. The sensor is joined to the surface of the control unit. The sensory data are displayed in real time.

Table 4.9 provides a summary of the two technologies described above.

4.7.6.3 Ground-Penetrating Radar

The GPR technology can be used for leak location. It uses electromagnetic wave propagation and scattering to locate and identify changes in the electrical and magnetic properties in the ground. It is the ability

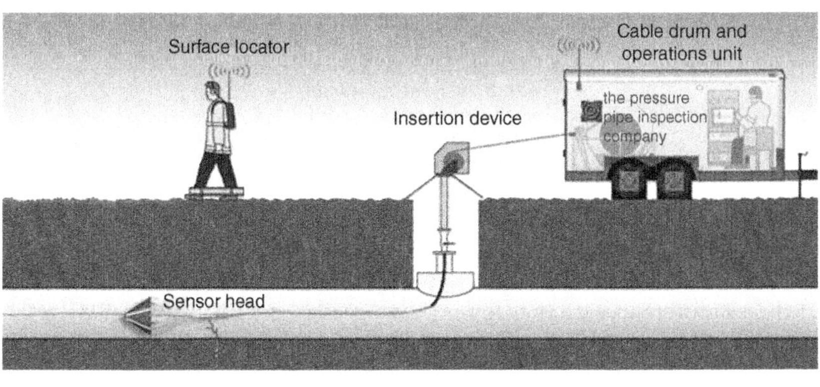

Figure 4.24 Sahara® System (*Source:* USEPA, 2009b.)

System	Typical Use	Advantage	Limitations
SmartBall®	• Detects leaks and air pockets in medium- and large-diameter (8 in. and greater) water pipes	• Can be used for any pipe material (concrete, steel, PVC, GRP, etc.). • Can be applied to detect air pockets and leaks on medium- and large-diameter pipe (>8 in.). • Can survey long pipelines with a single deployment. The total length of survey capacity depends on flow rates in the pipeline and battery life. The longest water line survey presently is 15 miles (25 km) under 2 ft/s flow. For higher flow rates, longer surveys could be performed. • Can detect very small noise disturbances along the pipeline. • Inspection is performed while a pipeline remains in service.	• The conventional SmartBall® cannot be used for pipelines with very high water pressure (>400 psi). • If the survey involves long pipe lengths, the surface sensor used for monitoring the pulses being emitted from the SmartBall® has to be moved along the pipe length. • The estimation of the leak magnitude is qualitative.
Sahara®	• Inspects in-service water mains for leaks, gas pockets, visible defects, and pipe wall thickness	• Can be used for in-service pipe inspection. • Can use existing 2-in. taps. • Sensitive to small leaks. • Surface tracking can map the pipeline under inspection. • Can be used for small mains (4 in.) and equally effective in large-diameter mains because of the proximity of the sensor to the leak. • Tether control allows withdrawal of the sensor when unexpected flow conditions are encountered.	• Intrusive technology • Requires access points at a frequency that is determined by bends and flow rates in the pipeline

Table 4.9 Comparison of SmartBall® and Sahara® Technology

to detect differences in the density and water content of soils that allows GPR to be used as a leak detector. However, as with all GPR investigations, there are limitations depending on soil conditions, surface pavement, and other underground structures in urban areas, and the use and data interpretation require a high level of skill and experience.

4.7.6.4 Radio-Frequency Interferometry

This technology uses a radio-frequency interferometer that transmits low power, ultra-high-frequency radio waves into the ground. The signal is reflected from the leaking water back to the antenna. The signal is amplified and processed by a sensitive interferometer system, which ignores signals that do not change with time but identifies rapidly changing signals that can be displayed to show strength and character.

4.7.6.5 Infrared Thermography

Infrared thermography is used to detect leaks in pipelines and the voids around them. High-resolution temperature measurements made by infrared cameras show variations in temperature in a pipe wall. The method is frequently used from aircraft flying over long and remote pipelines; as it is fast and provides instant feedback on site. Both the pressure-drop at a leak point and the plume of leaking fluid in the surrounding soil show measurable temperature changes. This method has been used for leak surveys on less accessible water mains.

4.8 Data Collection and Manipulation

4.8.1 Overview

The whole purpose of collecting data from inventories, records, observations, and inspections is to use the information to make an assessment of the current condition of the asset. The second aspect based on this assessment is to determine the likelihood of failure and predict the future structural condition and life expectancy. A third use is to prioritize inspection where additional data are needed.

The key to screening for likelihood of failure is to have a structured database or inventory that is constantly updated, can be interrogated, and records the full data on the pipeline, its surroundings, and its operation. There are programs of condition assessment developed for water utilities that provide guidance on developing and operating an appropriate database. If the database can be based on a GIS, it becomes more useful and effective. Various attempts to improve the reliability of life expectancy predictions have been developed. For example, when the extent and depth of pitting can be compared with a database of various types of pitting in pipes of a similar age and type, it becomes more meaningful. The Commonwealth Scientific and Industrial Research Organization of Australia has developed a failure model called pipeline asset and risk management system (PARMS) that takes into account variations in corrosion rate, the resistance as the pipe corrodes, and the applied service loads (USEPA, 2009b).

4.8.2 System Analysis

The PARMS planning uses two types of pipe deterioration models, (1) a nonhomogeneous Poisson-based model for high probability, low-consequence failure of pipes, requiring reactive renewal strategies (management of failure frequency) and (2) probabilistic/physical models for low-probability high-consequence failure of pipes requiring proactive renewal strategies (failure prevention). After calibration using the recorded history of failures, these models are used to forecast the number of failures expected in the planning period. The cost of failure is computed based on user input, including cost of repair, extent of the network that experiences interruption of service, penalties, and rebates. Based on future failure frequency and cost, as well as pipe replacement cost, candidate pipes for replacement are identified. These pipes are either replaced or the costs associated with their failure are reduced by installing additional isolation valves to reduce impact on the network.

Another example of water main data storage and management can be found in the City of New York. The New York City Department of Environmental Protection undertook nondestructive evaluation and development of an asset management system for wastewater force mains. They developed an asset inventory database first, including a standardized reporting tool. To develop a structural rating, a detailed finite element analysis was conducted on typical mains. Similar analysis can be implemented for water distribution mains. Table 4.10 defines the level of defect, and the recommended action.

Defects	Condition	Rating
Cracks, breaks, significant change in cross-section, and bending deflection >4 in. for CI, >0.06D for DI, and 0.1D for lined DI	*Failed:* Immediate action required	5
Greater than 75% loss of wall thickness at any cross-section, noticeable sag, or change in cross-section	*Severe damage:* Preventive action required	4
Wall thickness loss between 50% and 75%	*Moderate damage:* Preventive repair planned	3
Wall thickness loss between 25% and 50%	*Small damage:* Preventive repair on basis of remaining service life	2
Wall thickness loss between 0% and 25%	*Minimal damage:* No immediate action required	1

Source: USEPA (2009b).
Note: CI = cast iron, D = diameter, DI = ductile iron.

Table 4.10 Sample Structural Condition Ratings for Water Main b

This analysis can be used in ferrous pipes in the water network with the way the failure modes and indicators are represented and rated. The ultimate goal of analysis is to rate different failure modes and indicators, and recognize the planned action to rectify the problems. A current project undertaken by the National Research Council of Canada (NRC) in conjunction with the Water Research Foundation was "Dynamic Influences on the Deterioration Rates of Individual Water Mains" (Project #3052). The objective was to develop a model that considers both the static and dynamic factors impacting individual water mains. The pipe deterioration process is very complex with many static factors to consider like soil, pipe material, size, and age, as well as dynamic factors like climate, operation, and protection. The outcome is a model with analytical tool that covers dynamic factors for an individual water main (USEPA, 2009b).

4.8.3 Data Content for Water and Sewer Systems

Two types of raw data can be obtained from the inspection technologies described in the previous sections: visual-based technologies provide direct information about actual, observable distress indicators (cracks, delamination, etc.), while most other NDT technologies (Ultrasonic Testing, magnetic, etc.) provide signal patterns that require interpretation into distress indicators. In the latter case, signal interpretation is almost always proprietary knowledge, which concentrates on techniques and methods intended to interpret distress with inferential indicators into pipe condition ratings. Therefore, defect classification and rating, and pipe condition rating, are limited by the information provided by the inspection method, which is why it is important to selectively identify those that are relevant and that can be measured given available resources.

While the inspection and condition assessment of sewers started receiving attention in the late 1970s and the early 1980s, large water transmission mains started receiving attention relatively recently (the mid-1990s). In addition, large-diameter water transmission mains are inherently expensive components of the water supply system, and due to their high cost, the system often does not have enough redundancy to function while they are off-line for inspection. These are the main reasons why techniques and methods to interpret distress indicators into condition ratings started predominantly in sewer condition assessment and are in the process of evolving into the area of large water transmission mains. Some of the developers and/or providers of NDT services have developed their own methods to interpret distress indicators into condition ratings, but these methods are generally proprietary and often used for a specific NDT technology.

The full potential use of the inspection data generated by stated technologies is believed to emerge from the integration of asset-related datasets. The data integration can be two fold: (1) integration of new and heterogeneous sensing data; and (2) integration of existing data with the new data to leverage the performance of inspection and condition assessment process. First, the need to use different and heterogeneous pipeline inspection data arises from the fact that no single sensor can provide all relevant information about a pipe segment. Also, different pipe segments will require different combinations of sensor devices due to the limitations that pipe material, size, access, and charge level imply on the technology. Second, integration of existing data with the new data to leverage the performance of pipeline and its inspection and condition assessment process could be done by first providing prior probabilities of defect positions and types according to condition history. These probabilities could guide the location of inspection technology deployment and provide complementary information for inspection data analysis by reinforcing or negating the new inspection data.

As technology changes and different sensors are used to inspect pipe, it is important to provide means to integrate these different data so the condition history is not lost and trends in time and space may still be extracted from data. Data management (including integration) and analysis play a central role in this context because of the complexities that are likely to emerge from the heterogeneous nature of datasets, the diverse data types involved, from the presence of multimedia data, from the existence of multiple data sources, and from the large size of such datasets.

4.8.4 Point-Score Protocols for Water and Sewer Systems

Several protocols are available in the literature to record distress indicators in pipelines to translate them into condition rating. For example, WRc of UK, NRC of Canada, Water Services Association of Australia, and NASSCO provide these protocols. The WRc protocol is perhaps the most widely used protocol today. Both the WRc and the NRC protocols largely use a point-scoring approach, whereby each type of defect is assigned a score ("deduct value"). After a score is assigned to each observed defect, all of the scores are tallied and the totals are used to rate the condition of the pipe. While WRc uses a scale of 165 deduct values for condition rating, the NRC uses a 20-point scale. Tables 4.11 and 4.12 illustrate the comparison of point-score schemes and distress indicators of both protocols. It should be noted that both protocols contain separate scoring schemes for structural and operational observed defects. It is clear that while WRc uses a five-grade rating, the NRC uses six, from zero to five, where 0 = excellent, 1 = good (G), 2 = fair (F), 3 = poor (P), 4 = bad (B), and 5 = imminent collapse (IC).

The process of applying the protocol to real situations is not precise and is subjective. Often, two different evaluators will provide

Protocol	Condition States (Structural)					
	0 (E)	1 (G)	2 (F)	3 (P)	4 (B)	5 (IC)
WRc scores	Not Applicable(N/A)	<10	10–39	40–79	80–164	>165
NRC scores	0	1–4	5–9	10–14	15–19	20

Source: USEPA (2010).

Table 4.11 Comparison of Schemes of Two Point-Scoring Protocols

different scores to the same observed defects. Moreover, from Table 4.11, it can be seen that a score of 80 (WRc protocol) is in fact equivalent to 164, because both scores would translate to the same condition state of 4. Note that the point-score method described here was for sewer mains. Same can be implemented for water mains. Many water utilities use some type of point-score method, usually developed in house, for the evaluation of their water mains.

4.8.5 Data Sharing and Distribution

The relationship between the water and wastewater infrastructures continues to evolve and grow more, as recycling and reuse, conservation, and alternative development and wastewater management strategies evolve. This calls for a flexible, expandable simulation environment capable of simultaneous analysis of the infrastructure systems and their interdependency with other infrastructures.

The Water Infrastructure Simulation Environment (WISE) is an analytic framework supporting the evaluation of water and wastewater infrastructure in terms of both infrastructure-specific and interdependency issues. The WISE involves the integration of GIS with a wide range of infrastructure analysis tools, including industry standard hydraulic simulation engines, for example, EPANET[7] and Storm Water Management Model (SWMM), as well as Los Alamos National Laboratory interdependency simulation systems, such the Urban Infrastructure Suite and the Interdependent Energy Infrastructure Simulation System (IEISS). Key components in the WISE framework are ArcWISE, a GIS-based graphical user interface, and IEISS Water, a water infrastructure interdependency simulation capability within the IEISS.

[7]A public domain, water distribution system modeling software package developed by the United States Environmental Protection Agency.

Distress Indicator (Defect)	Distress Level	Unit	Scores	
			NRC	WRc
Longitudinal crack	• Light (up to 3 cracks, no leakage)	ft	10	33
	• Moderate (>3 cracks, leakage)	ft	16	131
Circumferential crack	• Light (up to 3 cracks, no leakage)	ft	10	33
	• Moderate (>3 cracks, leakage)	ft	16	131
Diagonal crack	• Light (up to 3 cracks, no leakage)	ft	10	N/A
	• Moderate (>3 cracks, leakage)	ft	16	N/A
	• Severe (multiple cracks, leakage)	ft	33	131
Longitudinal fracture	• Light (<10 mm)	ft	16	131
	• Moderate (10–25 mm or more than one)	ft	33	262
	• Severe (>25 mm)	ft	49	N/A
Circumferential fracture	• Light (<10 mm)	ft	16	131
	• Moderate (10–25 mm or more than one)	ft	33	262
	• Severe (>25 mm)	ft	49	N/A
Diagonal fracture	• Light (<10 mm)	ft	16	131
	• Moderate (10–25 mm or multiple)	ft	33	262
	• Severe (>25 mm)	ft	49	N/A
Deformation	• Light (<5% change in diameter)	ft	16	66
	• Moderate (5%–10% change in diameter)	ft	33	262
	• Severe (11%–25% change in diameter)	ft	49	541
Surface damage (spalling)	• Light	ft	10	16
	• Moderate	ft	33	66
	• Severe	ft	49	394
Joint displacement	• Light (<¼ pipe wall thickness)	Each	3	N/A
	• Moderate (¼ –½ pipe wall thickness)	Each	10	1
	• Severe (>½ pipe wall thickness)	Each	15	2
Broken pipe		Each	15	60
Collapse		Each	20	165

Source: USEPA (2010).

Table 4.12 Comparison of Distress Indicators of Two Point-Scoring Protocols

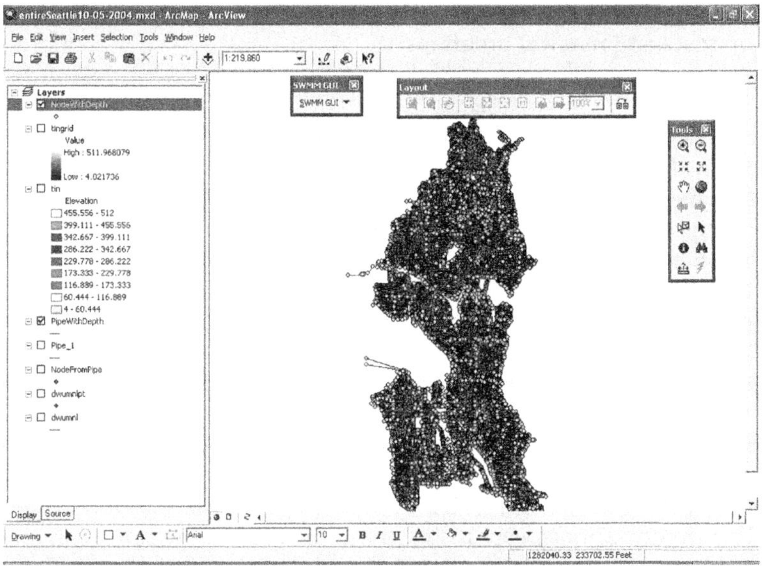

FIGURE 4.25 Example of ArcWISE application to sewerage system (*Source:* Los Alamos National Lab).

ArcWISE is a graphical user interface (GUI) built in ESRI's ArcGIS 9.0 (ESRI, 2004) with routines to read in GIS-based infrastructure datasets, as well as American Standard Code for Information Interchange (ASCII) text-based hydraulic simulation input files. From that point, the ArcWISE GUI can be used to improve attribution of network components, simplify networks, and visualize the system. Figure 4.25 displays the application of the ArcWISE GUI to a GIS sewerage dataset. This dataset contained approximately 60,000 nodes and pipes with roughly 20 percent of those components having poor attribution. The ArcWISE GUI facilitated the development of SWMM representation of these data while also correcting erroneous attribution in five and half hours.

An equally important variable in water distribution and sewage system simulation is the demand for service, that is, water demand and sewage production. The ArcWISE also contains modules to estimate those parameters based on customer geolocations, water use coefficients and time patterns, and population mobility.

The simulation of WISE can be used to visualize the interconnectivity between different systems, predict the outcome of incidents affecting the networks, measure the effects of disruptions in service, assess system robustness under varied future plans and forecasts, and identify components critical for the operation of the systems. All of these lead to the implementation of the "Level of Services" as defined in Chap. 2, on planning, prioritizing, and executing various upgrade and renewal programs.

4.9 System Integration and Operation

From the total asset management program (TAMP) perspective, it becomes obvious that a formal and holistic approach to infrastructure asset management is essential in order to provide services in the most cost-effective manner and demonstrate this to customers, investors, and other stakeholders (IIMM, 2006). The process of comprehensive asset management has gained increasing recognition within the water and wastewater industry, since such a formal approach provides information and analytical tools needed to manage existing assets more effectively and also plan for future needs. The flow of information with regard to integration and operation is illustrated in Fig. 4.26.

By implementing the TAMP concepts, utilities and other pipeline organizations responsible for capital infrastructure management can minimize the total cost of designing, acquiring, operating, maintaining, replacing, and disposing of capital assets over their useful lives while achieving desired service levels (GAO, 2004). Figure 4.26 illustrates the basic elements of such comprehensive asset management program and the relationships that form an integrated management system.

As described in Chap. 2, a typical "tactical plan" in a TAMP tracks associated tasks and data flow through each step of the plan. This has been graphically depicted in Fig. 4.27. The continuous cycle commences with data acquisition and proceeds in successive steps through

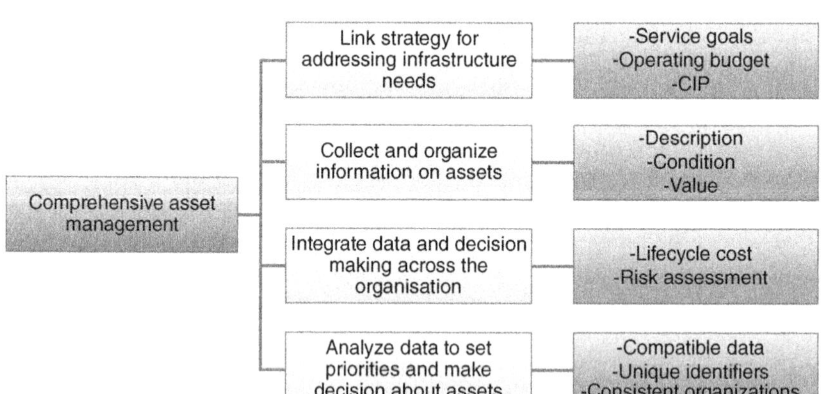

FIGURE 4.26 Elements of comprehensive asset management. (*Source:* GAO, 2004.)

data processing and storage prior to utilization in condition assessment, prioritization, budgeting, and planning tasks. Approved plans are implemented through work orders issued for Capital and Operations and Maintenance Program (O&M) projects, and upon completion of defined tasks, databases are updated to reflect new conditions.

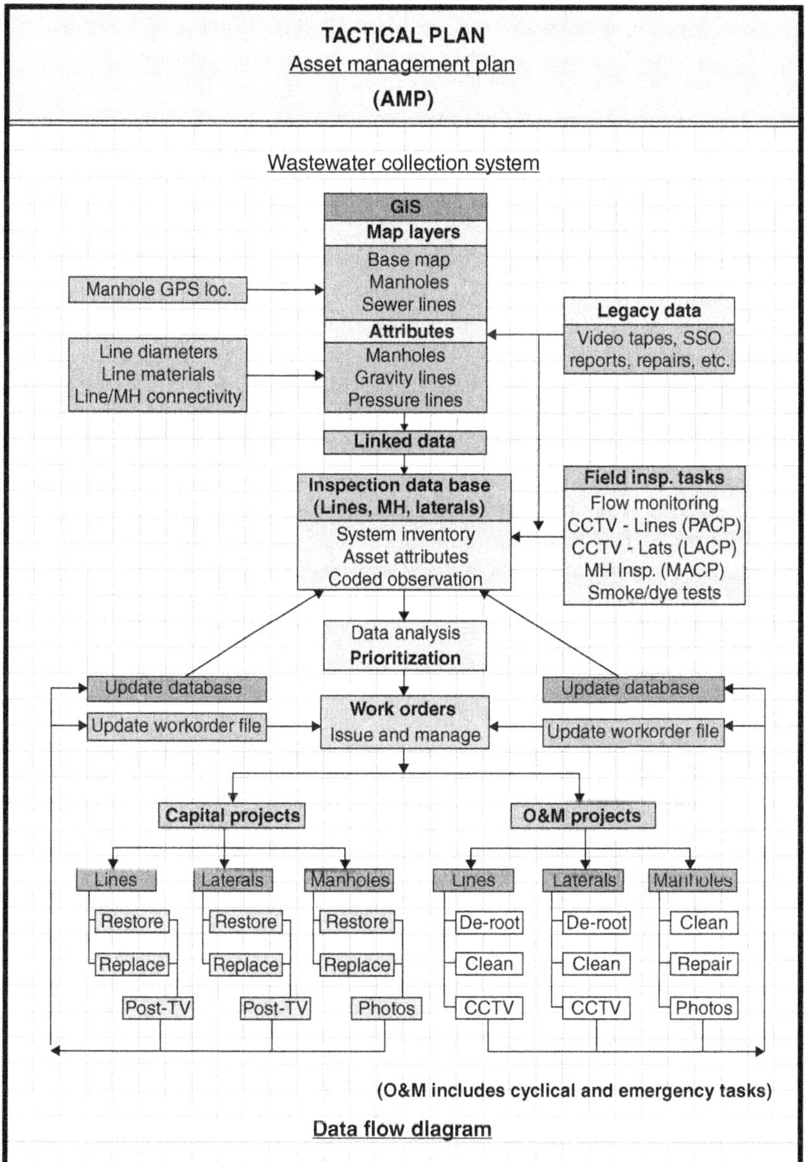

FIGURE 4.27 Tactical plan—data flow diagram.

The process of data acquisition, data processing, and data update is continuous, and new work orders are issued when justified by changed conditions identified through cyclical O&M and emergency efforts. Through this systematic approach, goals of the tactical underground infrastructure asset management plan are pursued, progress evaluated, and tactics revised to incorporate practices and processes optimized through experience and expanded knowledge.

The complex nature of tactical effort depicted in Fig. 4.25 will require support through comprehensive operations planning that includes not only procedural instructions but also training for personnel involved in plan implementation.

4.10 Chapter Summary

Condition assessment and life prediction modeling of the pipeline infrastructure provides the critical information needed to determine the condition of each pipe within the system and its estimated time to failure or remaining useful life. Critical gaps in the use of condition assessment as an asset management tool include lack of consistent, standard condition assessment protocols; methods for the systematic collection of data; and formal risk assessment methods to prioritize resources for maintenance and or/rehabilitation activities. Therefore, it is important to prioritize pipeline inspections so catastrophic failures are prevented. The level of data needed for day-to-day management is different from what is needed for making decisions on pipeline renewal and replacement. A detailed historical data is required for modeling purposes. There is also concern that creating large databases for condition assessment data would be overwhelming to pipeline owners. However, gathering data from a variety of inspection tools pipeline owners to make better decisions.

Pipeline owners should be knowledgeable and experienced in interpreting defects and distress indicators for their own networks. Providing a logical framework on recording the assets and the failures, using this information along with developed life expectancy models, and then applying observations and local knowledge could lead to significant improvements in pipeline performance. A proper condition assessment will prioritize pipeline renewal and replacement strategies.

A successful asset management program (AMP) will produce long-term rewards. Creating and implementing AMP requires strong leadership, persistence, patience, and knowledge. The first three characteristics are vital, but without knowledge it is doubtful whether they will be successful. Therefore, the personnel involved with AMP need to be informed and share "knowledge" applicable to every level of the pipeline asset management process.

4.11 Review Questions

1. Name main factors that have adversely impacted pipeline infrastructure.
2. Compare "reactive" with "proactive" strategy in repair and maintenance of pipeline infrastructure.
3. What is a "predictive" approach to pipeline renewal/ replacement needs, and what are its benefits?
4. Investigate and describe a "predictive models" to performance of a pipeline.
5. Name the decision making elements on pipe renewal through the assessment of pipeline physical conditions.
6. What are some parameters that have substantial impacts on pipe deterioration?
7. Briefly describe data collection, analysis, and management.
8. What is the process of asset registry?
9. Describe the GIS process.
10. Describe asset naming and numbering conventions.
11. Name several condition assessment tools and applications for wastewater systems.
12. Name several condition assessment tools and applications for drinking water systems.

CHAPTER 5

Pipe and Pipe Installation Considerations

5.1 Introduction

The term pipeline refers to a long line of connected segments of pipe, with pumps, valves, control devices, and other equipment or facilities needed for operating the system. Selection of pipe materials is an essential element in project planning, design, construction, life cycle cost analysis, asset management, and eventual renewal or replacement of the pipeline. The choice of a specific pipe has a major influence on the efficiency and longevity and is one of the most important elements for pipe selection. Long life-cycle pipes build asset value and provide economical service, minimize operation and maintenance costs and reduce impacts to consumers, environment, and the public. Availability of locally manufactured pipe appropriate for site and project conditions will reduce transportation and installation costs and improve sustainability considerations by reducing impacts to environment and carbon footprints. The longevity, corrosion resistance, and life expectancy of a pipeline are results of both pipe manufacturing process and best construction practices.

Pipeline construction has considerable uncertainties owing to the complex nature of the pipe-soil system. More complexity applies to trenchless methods when a machine is used to install the pipe blindly. However, trenchless technology only disturbs the ground in the area of pipe zone, but open-cut disturbs the soil from the ground surface.

Structural designs of pipelines employ models for the calculation of different behavioral factors, such as deflection, buckling, cracking, and so on, that are based on particular assumptions for interaction between pipe and soil. Since pipe and soil are typically designed as integral systems, it is crucial that the construction process achieves this design objective. Project plans and specifications provide the basic requirements for construction and installation. However, site conditions often vary from those anticipated during design, and the contractor and construction engineer must recognize these variations. Often, alternate or additional construction considerations are necessary to address unexpected or unspecified conditions.

5.2 Pipeline Construction Using Open-Cut Method

Pipeline installation using open-cut method involves mobilization of material, labor, and machinery to obtain, assemble, install, inspect, and test the pipeline system. Underground pipeline installation may seem relatively straightforward: (1) dig a trench, (2) lay the pipe in the trench, and (3) fill the trench back in. However, in reality, pipeline installation involves many important engineering and construction considerations. An underground pipeline system incorporates both properties of pipe and properties of soil to form a composite structure. Accordingly, structural designs of underground pipelines include specifications for the surrounding soil. The idea is for the system to be able to withstand the imposed demands through a composite action of the pipe-soil structure. If properly designed and constructed, this composite action is often advantageous and can be used to enhance the load-carrying capacity of pipelines. It follows that proper installation is extremely important to the performance of underground pipeline structures. Of particular importance are considerations for (1) proper selection and compaction of soils for foundation, bedding, embedment, and backfilling; (2) proper control over geometrical configurations of trench excavations (width, height, etc.); (3) control over appropriate soil cover; (4) control of line and grade of the pipe; and (5) control groundwater in the trench. In addition, appropriate care must be taken during transportation, handling, and installation to avoid cracks, gouges, buckling, and other forms of structural damage.

The physical mechanisms of failure modes of the pipe-soil structure depend upon properties of the pipe and the soils surrounding it. With reference to stiffness properties, pipes are broadly categorized as either rigid or flexible. Rigid pipes sustain applied loads by means of resistance against longitudinal and circumferential (ring) bending. Under maximum loading conditions, rigid pipes do not deform sufficiently enough to produce horizontal passive resistance from the soil surrounding the pipe (Young and Trott, 1984). Typical examples of rigid pipes are clay pipes and concrete pipes. On the other hand, flexible pipes are capable of deforming (without damage to the pipe) to the extent that the passive resistance of soils on the sides is mobilized providing additional support. American Society for Testing and Materials (ASTM) standards define flexible pipes as pipes that deflect more than 2 percent of their diameter without any sign of structural failure. Typical examples include ductile iron pipes and high-density polyethylene (HDPE) pipes.

Soils, on the other hand, require different properties for different sub-components of an underground installation (e.g., foundation, bedding, trench walls, etc.). Some of the important properties considered in design include soil stiffness (modulus), density, type of soil, and moisture content. Achieving the design properties in the field requires careful monitoring of construction activities along with testing and evaluation. The complexity of soil behavior introduces considerable uncertainty in defining and measuring of soil properties. The problem is augmented by spatial variability of natural soils combined with practical necessity to estimate the properties from very limited amount of sampling and testing. While designs use limited information measured at few strategic points, construction must be carried out over the entire spectrum of variability of soils. Coefficient of soil reaction (E') represents soil variability and is not directly measurable, but must be back calculated using observed pipe deflections which might be dependent on the pipe depth also. For this reason, E' is not a property of soil alone (like Young's modulus) but also a function of the pipe-soil system. This integral treatment of pipe and soil is common in current design standards. It follows that the outcome of design and construction of pipelines is not the pipe alone but the entire pipe-soil structure.

5.3 The Pipe-Soil System

As said previously, pipelines are designed as composite structures with pipe and soil forming an integral system to resist the applied loads. The pipe-soil structure generally comprises the pipe, the in situ soil in the foundation and the trench walls, and placed soils (for bedding and backfilling) at locations below and around the pipe. Ideally, installation of placed soils should achieve a state as close to the in situ state of soils as possible. However, this is practically impossible to achieve and appropriate compaction methods are needed. A typical layout of a trenched pipe installation and the associated terminology are illustrated in Figs. 5.1 and 5.2 and described in Table 5.1.

5.3.1 Rigid Pipes and Flexible Pipes

Broadly speaking, pipe materials fall into two categories: rigid and flexible. There are some pipe materials that may exhibit properties of both rigid and flexible pipes, such as asbestos cement (AC) pipe, and they may be identified as semi-rigid. Rigid pipes sustain applied loads by means of resistance against longitudinal and circumferential (ring) bending. Under maximum loading conditions, rigid pipes do not deform sufficiently enough to produce horizontal passive resistance from the soil surrounding the pipe. Typical examples of

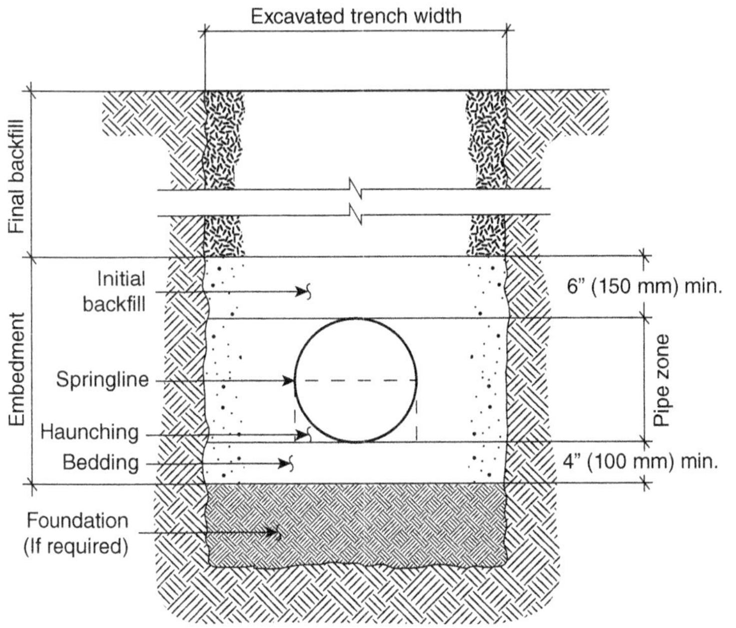

FIGURE 5.1 Typical cross-section of an underground pipeline installation for flexible pipe. (*Source:* ASTM D2321-11.)

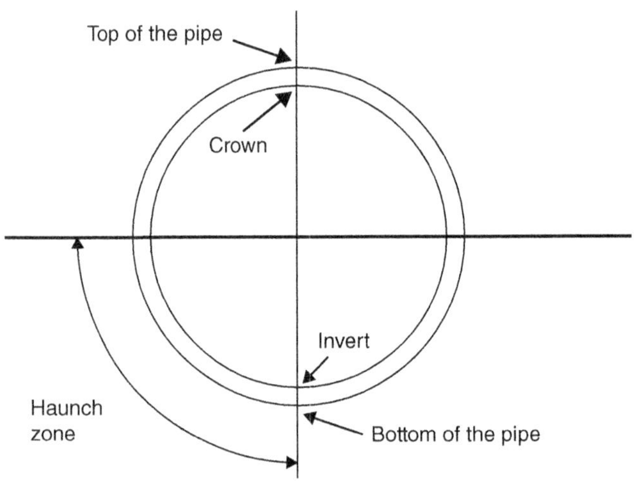

FIGURE 5.2 Pipe cross-section terminology.

Term	Description
Trench walls	The trench walls are comprised undisturbed in situ soils through which a trench is excavated. Depending upon the type of soil, they may be vertical or require sloping.
Bedding	Bedding is the soil placed on top of foundation and provides uniform support and grade for the pipe, except at pipe bells or large couplings where over-excavation is usually specified.
Foundation	The foundation may comprise undisturbed in situ soil or have imported soils to replace any unsuitable material at the bottom of the trench.
Embedment	Embedment is comprised material placed around the pipe providing a supporting structure. It consists of the bedding, haunching, and the initial backfill.
Haunch zone	Haunch zone is the area between the bottom of the pipe and the spring line. Backfill material in the haunch zone is critical in transferring forces in the lateral direction.
Springline	Springline is the horizontal centerline of the pipe.
Initial backfill	The initial backfill protects the pipe from final backfill placement. It typically begins at the springline of the pipe and continues about 6 to 12 in. on top of the pipe.
Final backfill	Material placed over the embedment up to the ground level.
Pipe zone	Depth of trench wall occupied by the pipe equal to the outside diameter (OD) of the pipe.
Trench width	Often specified in pipe design, this dimension must allow realistic side clearances to the outside diameter of the pipe, including belled ends and trench support systems. Excessive widths caused by careless excavation increase paving, loads on the pipe, quantities of earthwork and possibly the top width of the trench, affecting right-of-way, surface finishing, etc.

Source: Howard (2015).

Table 5.1 Pipeline Installation Terminology

rigid pipes are clay pipes and concrete pipes. On the other hand, flexi ble pipes are capable of deforming (without damage to the pipe) to the extent that the passive resistance of soils on the sides is mobilized providing additional support. The ASTM standards define flexible pipes as pipes that deflect more than 2 percent of their diameter without any sign of structural failure. Typical examples include ductile iron HDPE, steel, and polyvinyl chloride (PVC) pipes. Common terminology used to characterize properties of rigid and flexible pipes is strength and stiffness. While strength refers to the ability of rigid pipes in resisting loads and resulting stress in the pipe materials, stiffness refers to the ability of flexible pipes in resisting deflection. Pipes that overlap these two categories are sometimes referred to as

Rigid	Flexible
Concrete pipe	Steel pipe (SP)
Vitrified clay pipe (VCP)	Ductile iron pipe (DIP)
Prestressed concrete cylinder pipe (PCCP)	PVC pipe
Reinforced concrete pipe (RCP)	Polyethylene pipe (PE)
Bar-wrapped concrete cylinder pipe (BWCCP)	Fiberglass reinforced plastic pipe (FRP)
Asbestos-cement (AC) Pipe	
Polymer concrete pipe (PCP)	

TABLE 5.2 Examples of Rigid and Flexible Pipes

"semi-rigid" or "semi-flexible" or "intermediate" pipes. However, such distinction is seldom made in current design standards (Moser and Folkman, 2008). Examples of different types of rigid and flexible pipes are given in Table 5.2.

Rigid and flexible pipes differ in the way they transfer the applied loads to the surrounding soil structure. Figure 5.3 presents a simplified illustration of the load transfer mechanism for both types of pipes due to the vertical soil pressures. As can be seen in Fig. 5.3, rigid pipes sustain vertical loads by virtue of the material strength alone and with very little deflection. On the other hand, flexible pipes tend to deflect and use the horizontal passive resistance of the soil on the sides. As described in the following sections, this difference in behavior has important consequences in analysis, design, and installation of pipelines.

5.3.2 Soils

Soils are formed over several geologic periods through physical, chemical, and temperature effects (collectively known as weathering) on rocks and minerals. For engineering purposes, soils are often classified

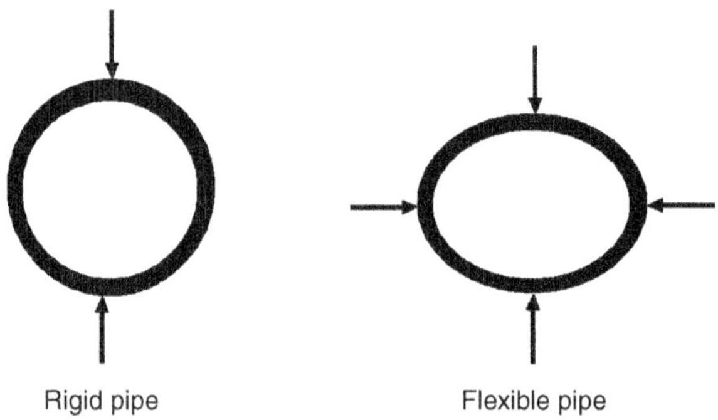

FIGURE 5.3 Load transfer mechanisms for rigid and flexible pipes.

as boulders, cobbles, gravel, sand, silt, and clay depending upon the particle sizes. Particle sizes for different soils can vary over a broad range with boulders having sizes in excess of 300 mm and clay having sizes less than 75 μm (Howard, 2015). This wide variation in particle sizes results in a wide variation of mechanical properties of soils. Furthermore, natural deposition resulting from weathering often entails a highly variable soil composition. Although, in localized areas, such as along a short pipe trench, the soil composition may appear more constant, the contractor must always be alert to changes in the characteristics of soils exposed during pipeline construction.

The first step in determining properties of soil is soil identification. ASTM D2487-11 describes a system for classification of soils for construction (see Table 5.3), which is a version of the Unified Soil Classification System (USCS). According to this system, four basic soil types are identified, viz., gravel, sand, silt, and clay. Boulders, cobbles, organic soils, such as peat, and so on, are ignored because they are generally not used in pipeline construction (Howard, 2015). Table 5.3 presents recommendations for installation and use of soils and aggregates for foundation and pipe embedment. These categories range from purely gravel, sand, silt, or clay to categories that are a combination of these basic types (e.g., silty sands, SM). In addition to providing an estimate of soil properties, such categorization gives the experienced contractor tools to compare a site with other locations and determine the required equipment and methods of construction (ASCE, 2009).

The desired properties of soil in pipeline construction are different from other types of construction owing to the integral nature of the pipe-soil structure. In most other types of construction, soils are expected to have enough bearing capacity to sustain the imposed dead weight of the structure. In pipeline construction, however, the weight of the pipeline is typically less than the weight of the earth displaced by it. Therefore, if trenchless technology methods were used, where the pipe is installed with minimal disturbance to the surrounding in situ soil, the only variable in design would be the strength of the pipe. So for trench installations restoring soil to its undisturbed state is desirable, although it may be practically impossible (Petroff, 1995).

5.3.3 Pipe-Soil Interaction

Whether a pipe is rigid or flexible, it has profound effect on the way in which it interacts with the surrounding soil. The interaction between pipe and soil influences the magnitude of loads exerted on the pipe and the manner in which the pipe transfers these loads to the surrounding soils. Calculation of loads exerted on underground pipelines can be traced back to the studies conducted by Anson Marston during the early part of the 20th century. The results were later

Soil Class[A]	Class I[B]	Class II	Class III	Class IV
General Recommendations and Restrictions	Acceptable and common where no migration is probable or when combined with a geotextile filter media. Suitable for use as a drainage blanket and under drain where adjacent material is suitably graded or when used with a geotextile filter fabric (see Eq. X1.8 of ASTM D-2321-11).	Where hydraulic gradient exists check gradation to minimize migration. Clean groups are suitable for use as a drainage blanket and underdrain (see Table 2 of ASTM D-2321-11). Uniform fine sands (SP) with more than 50% passing a #100 sieve (0.006 In., 0.15 mm) behave like silts and should be treated as Class IV soils.	Do not use where water conditions in trench prevent proper placement and compaction. Not recommended for use with pipes with stiffness of 9 psi or less.	Difficult to achieve high-soil stiffness. Do not use where water conditions in trench prevent proper placement and compaction. Not recommended for use with pipes with stiffness of 9 psi or less.
Foundation	Suitable as foundation and for replacing over-excavated and unstable trench bottom as restricted above.	Suitable as foundation and for replacing over-excavated and unstable trench bottom as restricted above. Install and compact in 12 in. (300 mm) maximum layers.	Suitable for replacing over-excavated trench bottom as restricted above. Install and compact in 6 in. (150 mm) maximum layers.	Suitable for replacing over-excavated trench bottom as restricted above. Install and compact in 6-in. (150 mm) maximum layers.
Pipe Embedment	Suitable as restricted above. Work material under pipe to provide uniform haunch support.	Suitable as restricted above. Work material under pipe to provide uniform haunch support.	Suitable as restricted above Difficult to place and compact in the haunch zone.	Suitable as restricted above. Difficult to place and compact in the haunch zone.
Embedment Compaction: Min Recommended Percent Compaction, SPD[D]	See Note[C]	85% (SW and SP soils) For GW and GP soils see Note[E]	90%	95%

Relative Compactive Effort Required to Achieve Minimum Percent Compaction	Low	Moderate	High	Very high
Compaction Methods	Vibration or impact	Vibration or impact	Impact	Impact
Required Moisture Control	None	None	Maintain near optimum to minimize compactive effort	Maintain near optimum to minimize compactive effort

[A] Class V materials are unsuitable as embedment. They may be used as final backfill as permitted by the engineer.

[B] Class I materials have higher stiffness than Class II materials, but data on specific soil stiffness of placed, uncompacted Class I materials can be taken equivalent to Class II materials compacted to 95% of maximum standard Proctor density (SPD95), and the soil stiffness of compacted Class I materials can be taken equivalent to Class II materials compacted to 100% of maximum standard Proctor denslly (SPD100). Even if placed uncompacted (i.e., dumped), Class I materials should always be worked into the haunch zone to assure complete placement.

[C] Suitable compaction typically achieved by dumped placement (i.e., uncompacted but worked into haunch zone to ensure complete placement).

[D] SPD is standard Proctor density as determined by Test Method D698.

[E] Place and compact GW and GP soils with at least two passes of compaction equipment.

Source: ASTM D-2321-11.

TABLE 5.3 Recommendations for Installation and Use of Soils and Aggregates for Foundation and Pipe Embedment

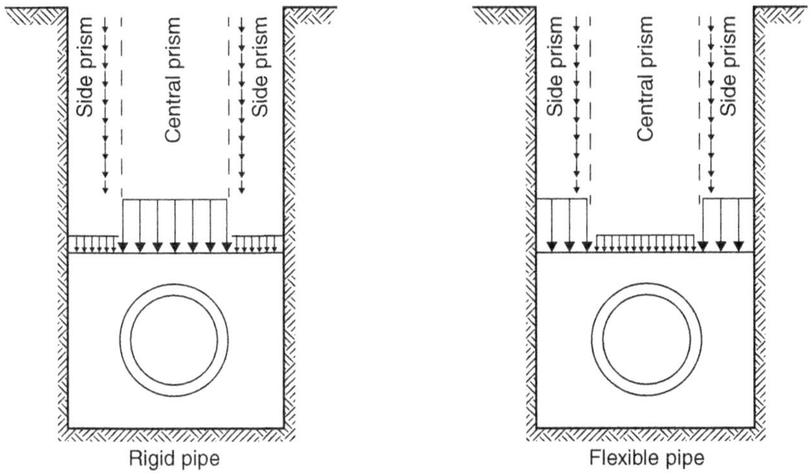

FIGURE 5.4 Trench load comparisons for rigid and flexible pipe. (*Source:* Najafi and Gokhale, 2005.)

expanded by Spangler and Watkins and are still in use today (Moser and Folkman, 2008). Figure 5.4 provides an illustration of the soil load distribution on rigid and flexible pipes. In the case of rigid pipes, the theory proposes that the soil in the side prism tends to settle relative to the central prism. This causes the pipe to assume full load of the central prism and portion of the load from the side prisms. In contrast, a flexible pipe tends to deflect, which results in a lowering of the pressure from the central prism.

5.3.3.1 Soils Arching Effect

Marston's theory (Spangler, 1982) proposes that the load due to the weight of the soil prism above an underground pipeline is modified by arching action. According to this theory, part of the soil weight is transferred between side prisms and the central prism resulting in either an increase or decrease in the effective weight of soil above the pipe. Arching effects could be classified as either positive or negative. Positive arching effects decrease the vertical pressure exerted on the pipeline, whereas negative arching effects increase the vertical pressure.

For pipelines installed in a trench, insufficiently compacted backfill and embedment material are more compressible than the adjacent native soil that has become well-compacted through natural consolidation. The more compressible backfill and embedment material has a tendency to consolidate and settle more than the native soils. As a result, some of the vertical soil load is transferred through shearing stresses between the side prisms and the central prism, and creates positive arching. The degree of transfer depends upon the type of backfill material and how well it is compacted. This arching action can be used advantageously for both rigid and flexible pipes by making the bedding immediately underneath the pipe more compressible than the adjacent bedding. However, care must be taken in order to avoid differential settlement of the pipe, especially for rigid pipes.

Assuming well-compacted soils in bedding and backfill, positive arching effects occur in the case of flexible pipes that deflect owing to their lower stiffness. This phenomenon is expected to occur naturally in trenchless installation of pipelines. The relative downward movement of the central prism within the trench mobilizes upward shearing stresses along the sides and creates an arching action that partially supports the soil column weight above the structure (see Fig. 5.5). In addition to this action, passive resistance of the soils adjacent to the pipe is mobilized and aids in transfer of loads. In contrast and again assuming well-compacted soils in bedding and backfill, negative arching occurs in the case of rigid pipes that do not deflect owing to their high stiffness. Owing to the relatively low stiffness of soils on the sides, shearing forces are transferred from the side prisms to the central prism increasing the effective vertical soil load on the pipe (see Fig. 5.5).

Installation conditions can have a significant effect on the soil loads acting on underground pipelines. In a long-term behavior study in Norway (Vaslestad et al., 1994), a 5.25-ft (1.6-m) diameter circular concrete culvert was constructed with imperfect trench condition (the compressible material was expanded polystyrene foam) with 46 ft (14 m) of embankment height. Vertical earth pressure directly above the culvert was only 25 percent of that due to the soil prism weight. Horizontal earth pressure at the mid elevation of the culvert, however, was 73 percent of that due to the soil column weight above that elevation. In the same research, the vertical pressure above a box culvert with 32.8 ft (10 m) backfill height, but with 1.7 ft (0.5 m) of

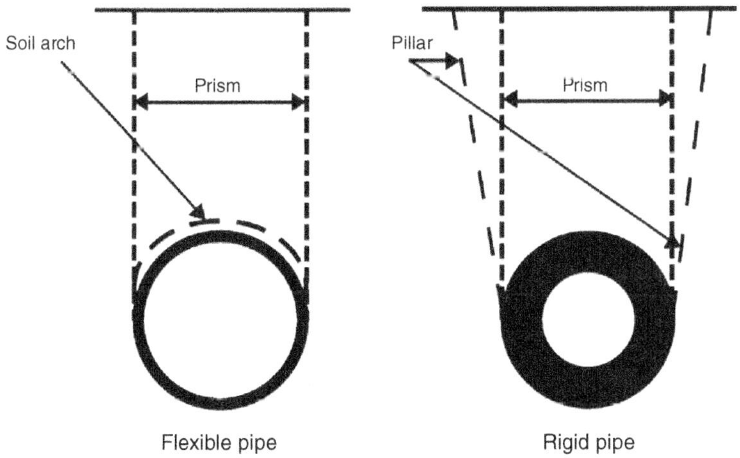

FIGURE 5.5 Arching effects for flexible and rigid pipes. (*Source:* Gabriel, 2006.)

expanded polystyrene foam placed immediately above the culvert, was 50 percent of the pressure due to the weight of the soil prism above. However, the vertical pressure above an identical box culvert, under the same embankment height without the foam, and with normally compacted backfill was about 120 percent of that due to the soil prism weight. These examples indicate that installation methods and culvert shape can strongly influence the magnitude and distribution of earth pressures on rigid culverts.

The magnitudes of the loads exerted on culverts depend on arching effects, which are the result of relative deformation of the backfill in a certain zone above the culverts. This deformation is related to both the soil and the structural stiffness. In the case of flexible culverts, culvert deformation results in arching effects, which reduce the vertical loading regardless of the installation method. Typically, the vertical earth pressure on flexible culverts is less than that due to the weight of the prism of soil above the culvert. For rigid culvert installation, arching effects can also be achieved by introducing compressible material into the backfill. In the case of positive projecting embankments, arching results in vertical earth pressures that are greater than that due to the weight of the prism of soil above the culvert. Thus, according to the Marston's theory, the vertical earth pressures on culverts are a function of the installation method, the soil and structural stiffness, the geometry of the structure, and the boundary condition with the natural ground. It is worth noting that even for flexible culverts, a load reduction is achieved as differential settlements transfer loads to the surrounding soil adjacent to the structure.

5.3.4 Behavior of Rigid Pipes

As noted earlier, rigid pipes are designed to transmit loads through their material strength. Since rigid pipes do not deflect appreciably, their design does not consider the horizontal passive resistance of the side soils. Therefore, the allowable load on rigid pipes depends only upon the strength of the pipe and the soil below the pipe. This is usually represented by the following relationship (Howard, 2015):

$$\text{Load on the pipe} = \frac{\text{Pipe strength} \times \text{Soil strength}}{\text{Safety factor}} \tag{5.1}$$

Typically, the strength of the material of a rigid pipe is determined in the laboratory by the three-edge bearing test (Fig. 5.6). The three-edge bearing strength is load per unit length required to cause either crushing or critical cracking of the pipe (Moser and Folkman, 2008).

Strength of soil refers to the amount of support on the bottom half of the pipe and depends upon the properties of soil and the contact area. Figure 5.7 illustrates two extremes of contact area available for pressure distribution below a rigid pipe (Howard, 2015). The point load is the worst possible support case, which gives the strength

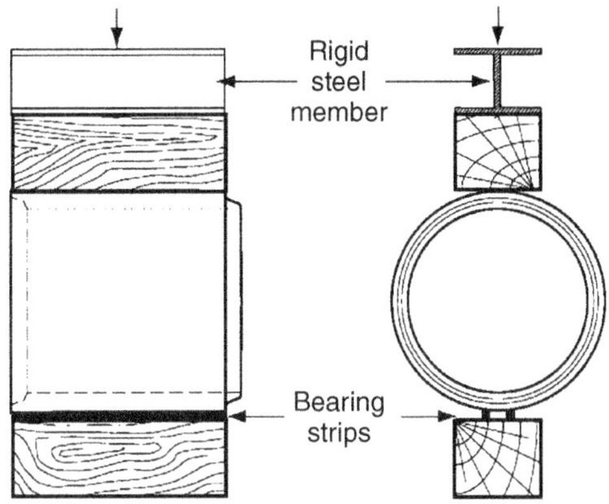

FIGURE 5.6 Three-edge bearing test. (*Source:* ACPA, 2000.)

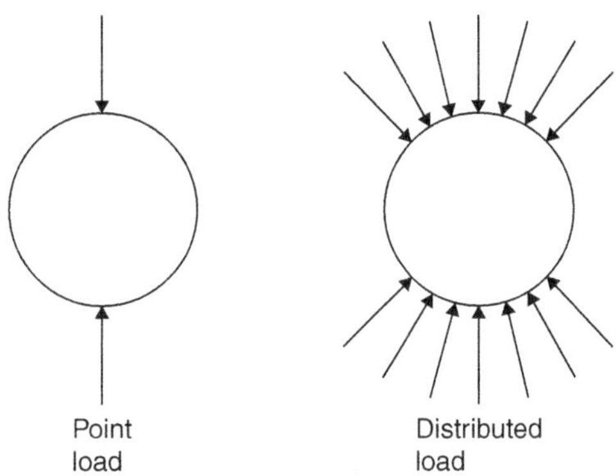

FIGURE 5.7 Load distribution on rigid pipes. (*Source:* Najafi, 2010.)

equal to the three-edge bearing test. For distributed load case, as for trenchless installation, the strength of the pipe may be enhanced due to uniform distribution of stress as shown in Fig. 5.7. Research has shown that the load required to cause failure in installation conditions is typically greater than the three-edge bearing strength owing to distribution of forces (Moser and Folkman, 2008). To ensure appropriate distribution, it is vitally important to achieve design densities in the haunch area of the pipe.

The ratio of field strength of a rigid pipe to the three-edge bearing strength is called a bedding factor and is given by following:

$$\text{Bedding factor} = \frac{\text{Field strength}}{\text{Three-edge bearing strength}} \tag{5.2}$$

Bedding factors are specified by different manufacturers for different types of pipes based on the placement methods and materials used.

5.3.5 Behavior of Flexible Pipes

Unlike rigid pipes, flexible pipes are designed to transmit part of the load to the side soils. As the load on the pipe increases, the vertical diameter of the pipe decreases and the horizontal diameter increases. The increase in horizontal diameter mobilizes the lateral resistance of the soils as shown in Fig. 5.8. Change in vertical or horizontal dimension of a pipe is usually represented as a percent change and is given by following:

$$\text{Percent deflection} = \frac{\text{Change in diameter}}{\text{Pipe diameter}} \times 100 \tag{5.3}$$

Consider the flexible steel pipe which is a perfect circle when it is laid on top of the bedding and no soil load has been placed. Steel is a linearly elastic (not viscoelastic) material. After backfilling, though, the steel pipe deflects. When first deflection takes place, two things happen. First, soil arching reduces the soil load on the steel pipe. So the load the pipe is resisting has decreased. Second, the material in the haunch zone has been further compacted by the expansion of the horizontal diameter. In other words, using Eq. (5.4), the numerator has decreased because of soil arching, and the denominator has increased because the soil's stiffness has increased due to compaction from the pipe's horizontal expansion.

$$\text{Pipe deflection} = \frac{\text{Load on the pipe}}{\text{Pipe stiffness} + \text{Soil stiffness}} \tag{5.4}$$

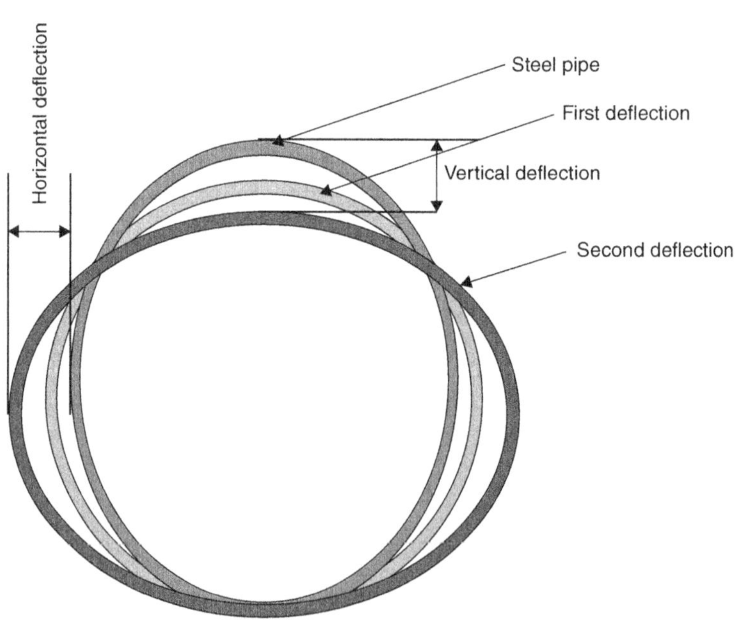

FIGURE 5.8 Load distribution on flexible pipes.

Nevertheless, the load on the pipe has not been reduced sufficiently yet. The pipe further deforms to the shape given by the second deflection. The load on the pipe reduces further due to soil arching. Soil stiffness increases because of greater density of the embedment material in the haunch.

Now the resistance of the combination of the pipe's stiffness and the soil's stiffness are great enough to prevent further deflection. That is how the pipe-soil system stabilizes.

5.4 Common Modes of Pipeline Failures

Underground pipeline failures can occur in several different modes and owing to different types of causes. Rajani (2001) presents three principal aspects of the physical failure mechanisms in pipelines.

1. Pipe properties, material type, pipe-soil interaction, and quality of installation.
2. Internal loads due to operational pressure and external loads due to soil overburden, traffic loads, frost loads, and third party interference.
3. Material deterioration due largely to external and internal chemical, biochemical, and electrochemical environment.

While the third aspect is usually relevant only in service conditions, the first two are very much applicable during and immediately after pipeline installation. Table 5.4 presents some of the factors that affect failure mechanisms in pipelines. For more information on pipeline deterioration, refer to Najafi (2005).

Pipe Factors	Environmental Factors
Type of pipe and pipe material	Defects during manufacturing
Location of the pipe	Damage during transportation, handling, and installation
Diameter	Soil loads (which depend on the type of soil, density, level of compaction, etc.)
Length	Point loads from projecting rocks and so on
Type of soil and embedment	Internal pressure loads
Joining method	Axial loads due to temperature, water hammer, and so on
Internal/external corrosion protection	Frost loads in soils
Wall thickness	Freezing and expansion of water
Depth of installation	Loads due to expansive soils
Bedding conditions	Third-party damage
Foundation conditions	Traffic loads

TABLE 5.4 Factors Affecting Pipeline Failures

Failure mechanisms for rigid pipes and flexible pipes differ in several respects. In general, rigid pipes fail in tension, and crack rather than deform if the imposed loads exceed the pipes inherent strength. Clarke (1968), reports the following major causes of failure in rigid pipelines:

- Inadequate load-carrying capacity of pipes
- Nonuniform bedding
- Inappropriate construction methods (e.g., excessive trench widths)
- Use of rigid jointing material resulting in a lack of axial flexibility and extensibility in pipeline
- Differential thermal deformation or moisture movements
- Differential settlement

Flexible pipes, in general, do not crack but fail by excessive deformation, buckling, or pipe flattening. Also, flexible pipes are more accommodating of faulty installation of embedment, bedding, or foundation because of their ability to deform. "However, improperly placed embedment material could lead to loss of side support, which is vital for flexible pipes and could result in over-deflection or flattening." Farshad (2006), reports the following major causes of failure in flexible pipeline.

- Fracture
- Buckling
- Weathering, color, and dimensional changes
- Voids, blisters, and delaminations
- Fatigue, corrosion
- Clogging of the pipe system

Specific mechanisms for failure of a pipeline also depend upon the type of material of the pipe. Structurally, a pipeline is said to have failed when the performance limits of its material have been reached. Although underground pipelines and the soil embedment around them are designed as an integral system, structural failure of the embedment soil leads to eventual failure of the pipe material. Therefore, the following discussion on failure mechanisms for rigid and flexible pipes focuses only on the failure modes of the pipe itself.

5.4.1 Failure Modes in Rigid Pipes

Rigid pipe failures occur when the performance limits for the pipe material are reached. Performance limits for rigid pipes may be categorized into:

- ring flexure,
- longitudinal flexure,
- shear,
- radial tension,
- longitudinal tension,
- cracking, and
- wall crushing.

Depending upon the specific pipe material actual mechanisms of failure may vary. For instance, radial tension due to hoop stresses may cause rupture of reinforcement in prestressed concrete cylinder pipe (ASCE, 2009). Ring flexure in concrete and clay pipes causes quarter point cracking as shown in Fig. 5.9. In cast iron pipe, longitudinal bending forces could cause circumferential failure of the pipe as shown in Fig. 5.10 (Makar, 2001).

Other mechanisms of failure include longitudinal cracking caused by pipe wall thinning and pressure surges in cast iron pipe as shown in Fig. 5.11 and spiral cracking of cast iron pipe as shown in Fig. 5.12 (Makar, 2001).

Ring flexure refers to cross-sectional bending of a pipe where the vertical diameter tends to decrease and horizontal diameter tends to increase. Owing to the high stiffness of rigid pipes, cracking at quarter points occurs before the pipe is able to deform (see Fig. 5.9). Longitudinal flexure is caused when a pipe is forced to behave as a beam resulting in axial bending stresses (see Fig. 5.13).

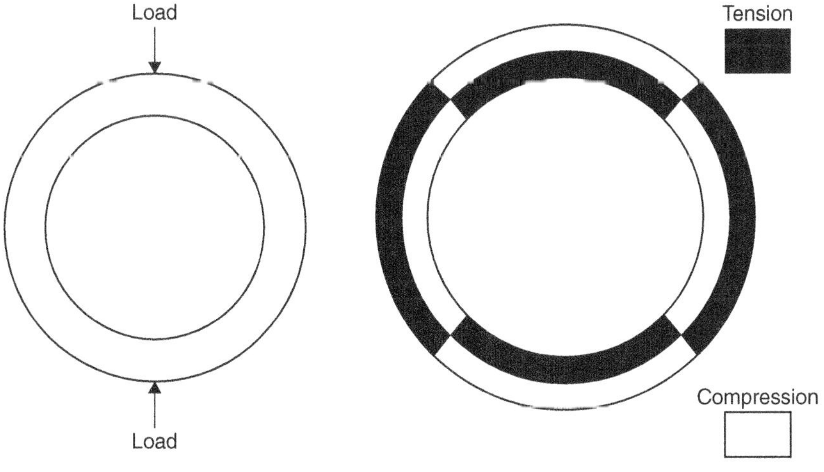

FIGURE 5.9 Zones of tension and compression in rigid pipes.

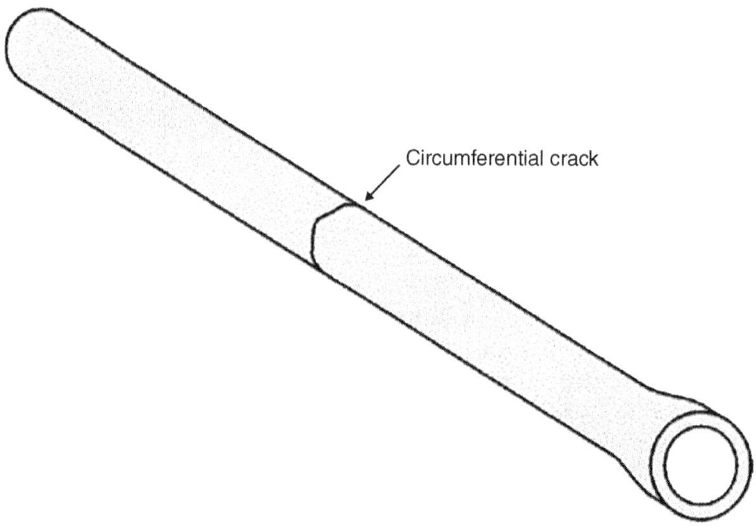

Figure 5.10 Circumferential failure of rigid cast iron pipe. (*Source:* Makar, 2000.)

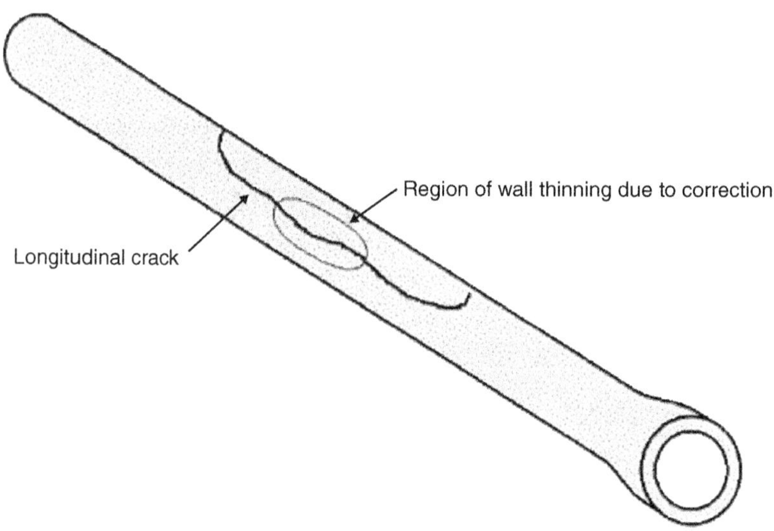

Figure 5.11 Longitudinal splitting of rigid cast iron pipe. (*Source:* Makar, 2000.)

Major causes for bending action include nonuniform bedding support, differential settlement, and ground movement due to external forces, such as earthquake, frost, expansive soils and so on (Moser and Folkman, 2008).

FIGURE 5.12 Spiral crack in rigid cast iron pipe. (*Source:* Makar, 2001.)

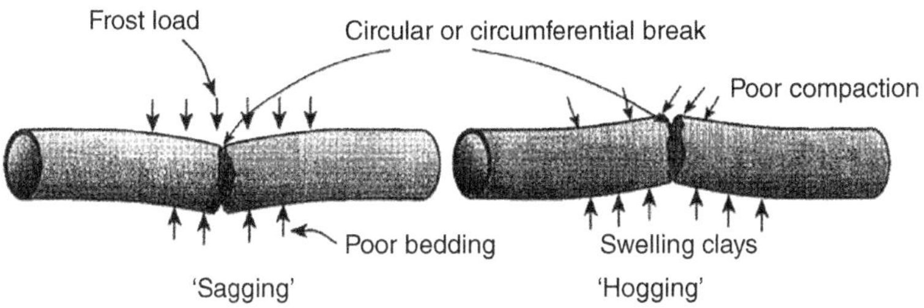

FIGURE 5.13 Bending or flexural failure in pipelines. (*Source:* Rajani et al., 1995.)

Some of the major causes of longitudinal bending or beam effect are (Moser and Folkman, 2008) as follows.

- Differential settlement of manholes or other structures that the pipe is connected to
- Nonuniform settlement or erosion of bedding
- Expansive soils
- Frost loading
- Nonuniform foundation
- Tree root growth pressure

Shear failure occurs in large diameter rigid pipelines owing to excessive shear stresses. Figure 5.14 illustrates an example of shearing of bell joints in large diameter (>18 in.) cast iron pipe. Longitudinal tension may be caused due to temperature effects or shrinkage of the embedment soil around the pipe (Moser, 2001).

Figure 5.14 Bell shearing in large diameter cast iron pipes. (*Source:* Makar, 2001.)

5.4.2 Failure Modes in Flexible Pipe

Flexible pipe failures occur when the performance limits for the pipe material are reached. Performance limits for flexible pipes refer to (Moser and Folkman, 2008)

- wall crushing,
- wall buckling,
- deflection,
- strain limit, and
- longitudinal stress.

5.4.2.1 Wall Crushing

Wall crushing occurs when the in-wall compressive stresses reach the yield stress or the ultimate stress of the pipe material. This type of failure occurs when pipe is embedded in very stiff material and subjected to very high loads (Massicotte, 2000). It is characterized by localized yielding at the springline of the pipe and usually occurs in smaller diameter pipes with thicker walls. Figure 5.15 provides an illustration of this type of failure. Normally, ring compression is the influential stressor causing this type of failure, although bending stresses also may cause wall crushing (Moser, 2001).

5.4.2.2 Wall Buckling

This type of failure occurs in pipes with low stiffness embedded in low stiffness soils. It is usually caused due to the pressure around the pipe

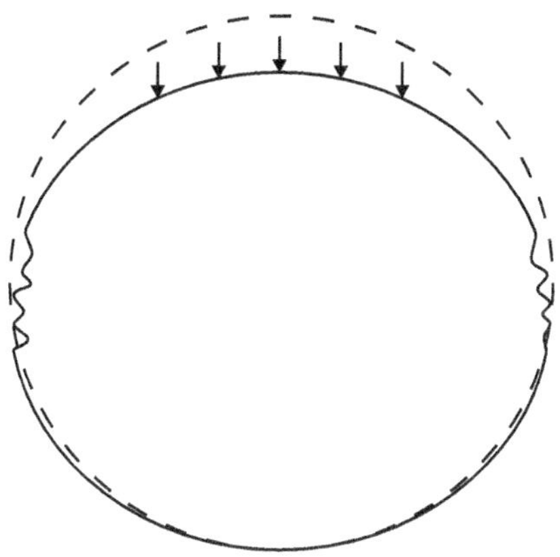

Figure 5.15 Wall crushing of flexible pipes. (*Source:* Moser and Folkman, 2008.)

or creation of vacuum inside (Moser, 2001). Design of large-diameter pipes may be governed by buckling, especially in conditions of high soil pressure and low stiffness soil (Reddy, 2002). Figure 5.16 gives an illustration of this phenomenon. Buckling is primarily dependent on the stiffness of the pipe, which in turn depends on the type of material and the pipe geometry (diameter, wall thickness, etc.). Resistance against buckling may be increased by stiffening the pipe walls by providing corrugations or by improving the soil stiffness. The soil stiffness may be improved by using a better embedment material, by providing better compaction, or both.

5.4.2.3 Deflection

Flexible pipes have relatively low bending stiffness when compared with rigid pipes. Owing to this fact, deflection is an important factor in

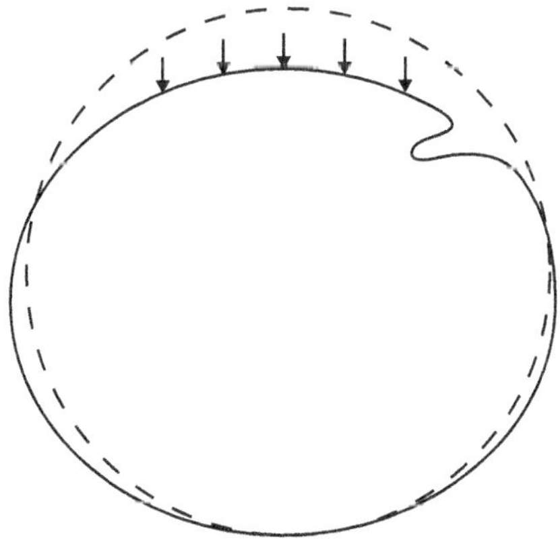

Figure 5.16 Wall buckling of flexible pipes. (*Source:* Moser and Folkman, 2008.)

design and installation of a flexible pipeline. Although flexible pipes are designed to deflect in order to initiate the passive soil resistance, deflection control is necessary to avoid the pipe from flattening or reversing its curvature. Deflection control plays an important role in the determination of quality of pipeline installation and can be a good indication of its long-term performance (Reddy, 2002). Deflection limits for flexible pipes usually range from 2 to 7.5 percent depending upon the pipe material (Howard, 2015). Typical deflection scenarios are illustrated in Figs. 5.17 and 5.18. The factors that affect pipe deflection are as follows:

- Pipe stiffness
- Soil resistance (type, density, modulus of elasticity, moisture content)
- Applied loads
- Trench configuration (geometry and embedment)
- Haunch support
- Construction stages

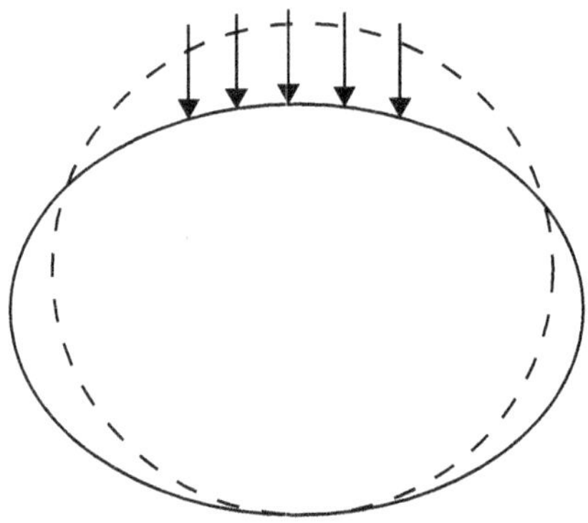

FIGURE 5.17 Ring deflection of a flexible pipe. (*Source:* Moser and Folkman, 2008.)

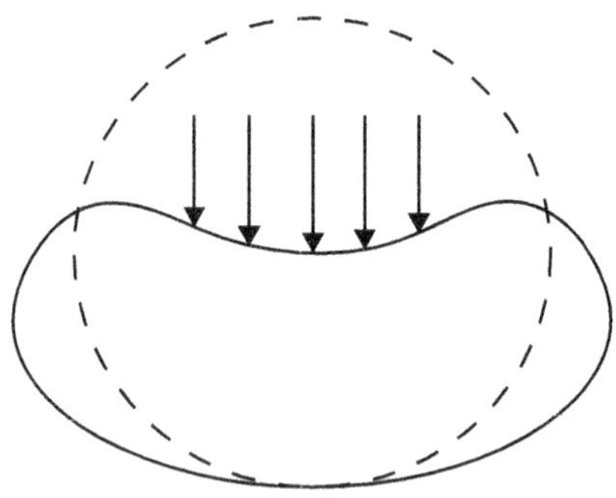

FIGURE 5.18 Reversal of curvature due to over-deflection in flexible pipes. (*Source:* Moser and Folkman, 2008.)

- Time
- Temperature
- Variability in construction procedures and in soil characteristics

5.4.2.4 Longitudinal Tensile Failure

In general, rigid or flexible pipe products are not designed to withstand longitudinal tension. Longitudinal failure in pipes can occur due to several factors that are given below (Moser and Folkman, 2008.).

- Thermal expansion or contraction
- Longitudinal bending
- Poisson's ratio (due to internal pressure)

The tendency of a pipe to move in the axial direction is resisted by the frictional forces of soil surrounding the pipe. This creates longitudinal stresses that can cause a circular break at weaker cross sections (cross sections weakened by corrosion or cracking). Figure 5.19 illustrates longitudinal tensile failure in pipes. Figure 5.20 illustrates the effect on pipe due to the loads.

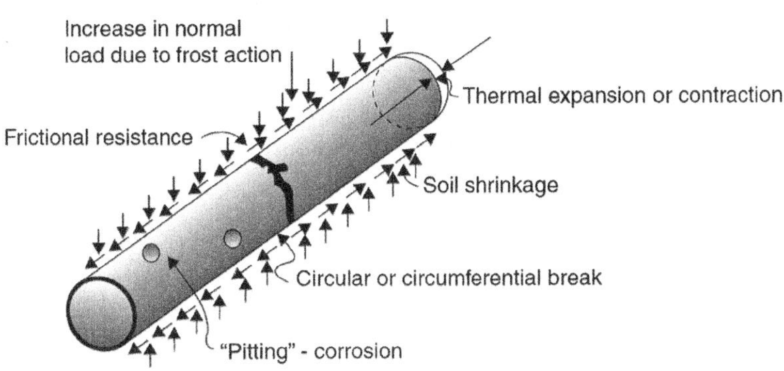

FIGURE 5.19 Longitudinal tensile failure in pipes. (*Source:* Rajani et al., 1995.)

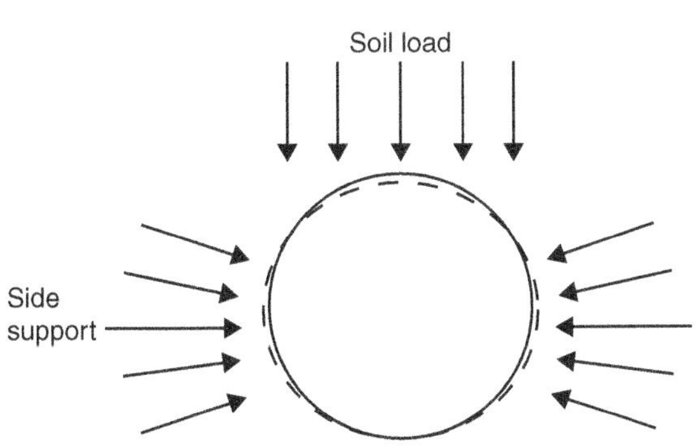

FIGURE 5.20 Side support for flexible pipes. (*Source:* NCSPA, 2000.)

5.5 Pipe Selection Considerations

Although economics is one of the engineering considerations during the design and selection of pipe material for a water or wastewater (sewage) piping system, there are several other factors that are as important. These considerations include the following:

- Type of fluid to be transported (potable water, wastewater, storm, sewer, water, oil, gas, and so on.)
- Construction conditions and methods used
- Life expectancy and related life cycle cost analysis
- Flow characteristics, such as corrosiveness and abrasion of fluid
- Ease of handling and installation of pipe
- Pipe's physical and chemical properties
- Availability of diameter sizes, pipe section lengths, and joints for trenchless technology
- Construction and operational stresses in the pipe. In many cases for trenchless technology projects, the construction stresses in the pipe exceeds the operational stresses.
- Location and pipe environment (inland, offshore, in-plant, corrosiveness of soil, etc.)
- Type of burial or support (underground, aboveground or elevated, underwater, etc.)
- Hydraulic properties of the pipe

There are some unique pipe properties that may be required in the planning and design of pipelines. These topics are discussed below in detail.

5.6 Cement-Based Pipes

Pipes in this category include concrete pipe (pressure and nonpressure types) as well as AC pipe. Both types incorporate Portland cement as the base material. Concrete pipes are designed with or without reinforcements and are used for both pressure and gravity applications. Asbestos-cement pipe is composed of a mixture of Portland cement, silica, and asbestos fiber, and was used in North America, mainly in potable water applications, though many sewer systems also have AC pipes in service. Asbestos-cement pipes have not been manufactured or installed in the United States since the late 1980s, but large quantities of the pipe continue to be in service in water systems at various locations. It is, therefore, important for engineers to have a working knowledge of the properties of this material. Except for AWWA C303 Bar-Wrapped Steel Cylinder Concrete Pipe, concrete and AC[1] pipes are rigid conduits and are therefore, designed to carry the full backfill loads following installation.

5.6.1 Concrete Pipes

Trenchless construction with concrete pipe was first performed by the Northern Pacific Railroad between 1896 and 1900, when the pipe was installed by jacking. The jacking of concrete pipe is performed both for pressure and gravity applications. Concrete pressure pipes are also installed by tunneling methods. For small-diameter pipe in short tunnels, such as those under a highway or a railroad, it is a common practice to slide the pipe through a liner followed by grouting the annular space.

5.6.1.1 Manufacturing

Concrete pipes can be made with any of the five types of Portland cement, though a low Tricalcium Aluminate (C3A) content cement is often selected simply to avoid the potential for degradation due to soil sulfates. There are five concrete pipe manufacturing processes used in North America, four of which use mechanical means to place and compact a dry concrete mix (zero slump with low water cement (w/c) ratios) into a form. A conventional wet mix and casting method is used in the fifth one. Each method is briefly discussed below.

1. *Centrifugal casting.* The form is rotated, acting as a centrifuge, while the concrete mix is fed into it. Vibration and compaction may also be employed to consolidate the concrete mix. The centrifuge promotes extraction of water from the mix. In this manufacturing process, the initial concrete mix may start out as a zero slump mix or not. Spinning the mix at high speed will drive excess water to the interior surface, resulting in a zero-slump, low water cement (w/c) ratio concrete after spinning.
2. *Dry casting.* Low-frequency, high-amplitude vibrations are used to distribute and compact the dry mix in the form in this process. The form is then removed and the pipe cured.
3. *Packerhead process.* In this process, a high-speed rotating device, the packerhead, acts as the inner form and compacts the concrete against the walls of the outer form. As the concrete is fed from the top of the form, the packerhead gradually moves up, forming the self-supporting uncured concrete.
4. *Tamping.* The concrete is mechanically compacted with oscillatory tampers between the walls of the outer form and an inner core. This method is declining in use.

[1]Asbestos concrete (AC) pipe may be considered as a "semi-rigid" pipe due to its thin wall.

5. *Wet casting.* This method is typically employed *only* for the manufacturing of pipes 36-in. and larger, typically using a concrete mix with higher water content and/or more plasticizing admixtures than the other four methods. Again, the pipe is made between the inner and outer walls of a form.

5.6.1.2 Types of Concrete Pipes

Below are five different types of concrete pipes made using Portland cement and used for pressure and gravity applications.

1. *Non-reinforced Concrete Pipe.* Used for nonpressure applications only, this pipe is manufactured in diameters of 4 in. through 36 in., in lengths up to 8 ft. Produced in three classes, its minimum strength requirement is the three-edge bearing per ASTM C-14. Since it is non-reinforced, structurally, it can only exhibit a brittle failure mode.
2. *Reinforced Concrete Pipe (RCP).* Typically produced to meet ASTM C-76, there are three types of reinforcements available for RCP: (1) welded wire reinforcing, (2) hot-rolled rod made of Grade 40 or 60 steel, and (3) cold-drawn steel wire made from hot-rolled rod. RCP is used in nonpressure and low pressure applications of up to 55 psi. There are five load-bearing capacities, ranging from 1,350 to 3,000 in units of lb/ft of diameter/ft of length known as "D-Load". Higher load capacities can be specially designed for unusual combinations of fill height or live load as per ASTM C-655. Variations of this pipe type are manufactured to AWWA C302 and AASHTO M170 standards, among others.
3. *Prestressed Concrete Cylinder Pipe (PCCP).* Often used in high-pressure applications, PCCP is a composite pipe of concrete and steel, capable of handling up to 500 psi. There are two types of PCCP: (1) lined cylinder pipe, with the concrete core inside the cylinder, and (2) embedded cylinder pipe, with the concrete core cast both inside and outside of the cylinder. A welded steel cylinder with joint rings attached at both ends exists in both types. After the core concrete is cast and cured, the pipe cores are wrapped with hard-drawn wire under high tensile stress, and then coated with cement mortar. Available diameters range from 16 to 144 in., though larger diameters have been manufactured for projects with sufficient lengths of pipe to justify the required manufacturing equipment. American Water Works Association (AWWA) C-301 is the governing manufacturing standard and AWWA C304 is the design standard. PCCP is designed to withstand both internal pressure and external load without the benefit of sidefill support.

4. *Reinforced Concrete Cylinder Pipe (RCCP).* This is similar to RCP, but includes a steel cylinder embedded in the pipe wall. Standard AWWA C-300 requires that the pipe be designed to withstand the combined, simultaneous loads from both internal and external pressure. Typically, available diameters range from 24 to 144 in., though larger diameters have been manufactured for projects with sufficient lengths of pipe to justify the required manufacturing equipment. Pressure applications are its main use.
5. *Bar-wrapped Steel-cylinder Concrete Pipe.* The steel cylinder is internally lined with a cement mortar lining. Once the lining is cured, a mild steel rod, under tension, is wrapped around the cylinder followed by the application of 1 in. cement mortar coating over the steel cylinder. Smaller-diameter pipes are considered rigid, whereas larger diameters behave as flexible conduits. AWWA C-303 is the standard governing its manufacturing and testing. Like the previous two types, bar-wrapped concrete pipe is also used in pressure applications.
6. *Polymer Concrete Pipe (PCP).* Polymer concrete originated more than 20 years ago in Germany. This type of pipe provides a corrosion-resistant concrete as needed for piping applications that require high concrete compressive strength and resistance to corrosive chemicals. These pipes are made by mixing a high strength, thermosetting resin with oven-dried aggregate to form a type of concrete. The resin within the mixture provides for bonding the aggregate, much like Portland cement does in traditional concrete pipes. Figure 5.21 illustrates typical sections of polymer concrete pipes.

The pipe sections are cast vertically between an inner and outer form and are vibrated for compaction. After the forms are removed, the section is heated in a kiln to finish curing the resin. These pipes can typically be used to carry highly aggressive wastes, for pipe jacking as they have very high compressive strengths (up to 17,000 psi), or for microtunneling. They can also be used for gravity flow or

FIGURE 5.21 Polymer concrete pipe. (*Source:* Meyer Polycrete Pipe.)

pressure applications, though pressure applications cannot allow pipe wall cracking since there is no cylinder to act as a water stop. Some manufacturers are also making polymer pipes in sizes appropriate for use as manholes. Also, semielliptical and circular-liner pipes have been developed for sliplining sewers. There are several ASTM standards for the product.

Polymer concrete pipes have several benefits, including high strength, corrosion resistance (they can be used in environments with pH ranges of 1 to 13), low wall roughness, and high abrasion resistance. The use of polymer concrete pipes is becoming more common in the following areas: direct bury, sliplining, jacking and microtunneling, tunneling, and above-the-ground applications. One of the main disadvantages in the past was the high cost associated with importing the pipe; its high cost relative to alternative materials limited its use to only niche markets where its superior qualities were needed. Another characteristic to be accounted for especially in the larger diameters is that, since it is non-reinforced, PCP can only exhibit a brittle failure mode if the external load bearing capacity is exceeded. However, since 2002, polymer concrete pipe has been manufactured in the United States under the brand name of Meyer Pipe. This is now enabling competition with products, such as glass-reinforced pipe (GRP), for jacking installations.

Fiberglass sleeve joints are made separately with elastomer sealing and spacing rings laminated into the sleeve. Factory fitting of the couplings to the end of the pipe is performed and pressure testing is done up to 35 psi. As an alternative to the couplings, stainless steel collars are also available.

5.6.1.3 Applicable Standards

Table 5.5 lists reinforced concrete pipe standards that are widely used in trenchless installations. Typical available diameter ranges are also given.

5.6.1.4 Joint Types

The concrete pipe industry has developed several different types of joints for the various types of pipes. Selection of the appropriate joint is based on the internal pressure as well as on the stringency of the

Pressure Standards	Gravity Standards	Available Diameters
ASTM C-361	–	12–108 in.
AWWA C-300	–	30–144 in.
AWWA C-301 and 302	–	12–144 in.
–	ASTM C-14, C-76, C-655	12–144 in.

Table 5.5 Concrete Pipe Standards

application for water tightness. Properly installed bell-and-spigot type joints can be completely water tight to 500 psi. These joints can be fabricated for trenchless methods, such as microtunneling and pipe jacking, that dictate that the joint be flush with the outside wall of the pipe barrel. For applications where some leakage can be tolerated[2], the tongue and groove joint or the modified tongue and groove joint is cost-effective. Table 5.6 shows joint types that may be used in trenchless construction. Also available are mastic or mortar joints, which are not well suited to prevent leakage except for transient hydrostatic heads below 5 psi. Above that pressure threshold, a gasketed joint meeting ASTM C-443 is probably appropriate. ASTM C-990 requires an in-plant hydrostatic test of 10 psi as "proof of design." Details of the joint are shown in Table 5.6.

5.6.1.5 Advantages and Limitations

Table 5.7 presents a summary of advantages and limitations of concrete gravity and pressure pipes. The widespread use of large-diameter concrete pipe in water and sewer applications is an indication of the material's acceptance at the municipal level. The vast selection of available diameters and pipe lengths is a convenience to the design engineer and contractor. On the pressure side, availability of various structural and pressure strengths also makes it convenient. The ability of manufacturers to make pipes capable of handling very high pressures in combination with high external loads in large diameters has resulted in its specification on some high-profile trenchless projects in recent years. Because a specialized work-crew is not needed; cost of labor for concrete pipe installation is not high. For trenchless construction, the high compressive strengths give it a definite advantage over some other pipe materials.

When used in open-cut construction, many concrete pipes installed in aggressive sanitary gravity applications, without the proper interior protection against hydrogen sulfide attack, have shown signs of early deterioration. For nonpressure concrete pipe, leakage through concrete joints and cracks in the pipe has been a constant source of inconvenience to municipalities. The issues of leaking joints and cracking pipe may be related, since leaking joints will result over time in the movement of soil from outside the pipe, reducing continuity of bedding support. In spite of the relative strength of the concrete pipe wall, concrete pipe's sensitivity to inconsistent bedding conditions in both shallow and deep installations has resulted in pipe failures by shear and beam breakage in many instances.

---

[2]It does not cost much more to require a true water-tight joint. These tongue and groove style joints will leak, the mastic fails over time, and so soil will slowly move from around the pipe and sometime in the future there will be a collapse of the soil around the pipe. Totally preventable with a good design.

Joint Type	Cross Section
Mastic filled tongue and groove joints (ASTM C-990) for low head applications. This is a poor joint for pressure applications, even for low pressures. Especially during joint deflections, which the pipeline will definitely experience over time, the joint will leak.	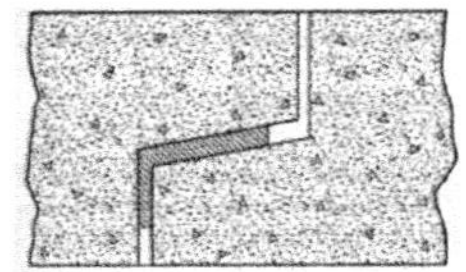
Compression type rubber gasket joints with internal bell and spigot with the gasket and joint meeting ASTM C-443 or C-361. Intended for use with pipes meeting the loading requirements of ASTM C-14, C-76, and C-655. Also a poor joint for pressure applications for the same reason stated above for mastic filled tongue and groove joints.	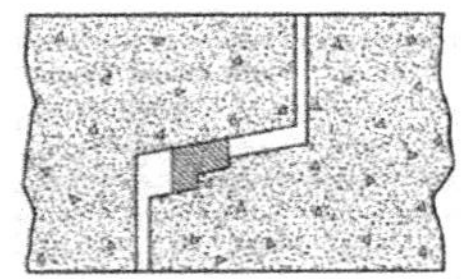
Steel end ring joint. This is a high pressure joint, intended for use with ASTM C-361. C-361, Sec. 8.1, permits three joint material combinations, "Joints shall utilize steel joint rings, steel bells and concrete spigots, or be formed entirely of concrete." Most RCP manufacturers can make an "all concrete" rubber gasketed joint that will pass a 72 psi test for "high pressure" performance category. This joint (all steel) is used for AWWA C300, AWWA C303, and AWWA C301. For C301 pipe, the outside of the pipe can be smoothed so the OD at the bell is the same as the OD of the pipe, so that pipe can be used in trenchless technology.	**AWWA C300** 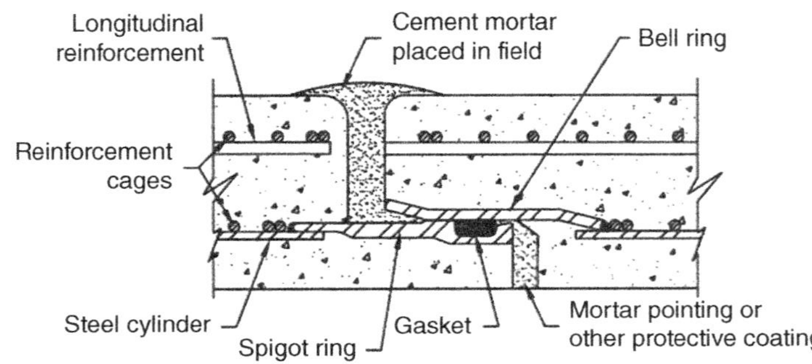

C300 Type pipe for pressure applications. Steel joint rings and, if required, testable Double Gasketed joints as shown on the detail. "Joints shall utilize Carnegie type steel joint rings, steel bells and steel." All RCCP manufacturers can make with rubber gasketed joint designed for 400+ psi for "high pressure" applications.	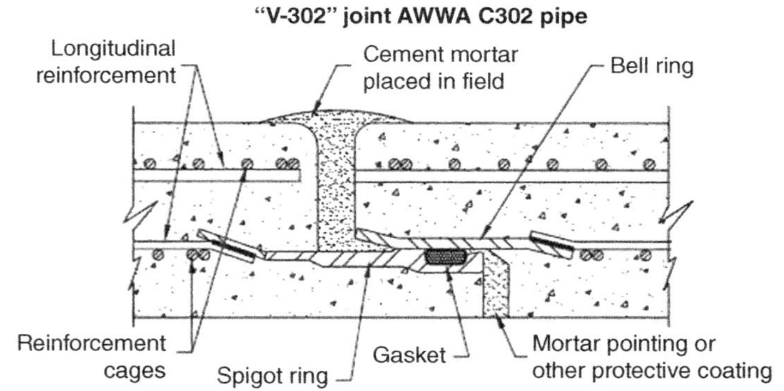

Table 5.6 Concrete Pipe Joints

Advantages	Limitations
Specialized work-crew not required for installation.	In open-cut construction, pipe is sensitive to bedding conditions, shear failure and beam breakage may occur. This may not be an issue on new construction with proper inspection. Any discontinuous bedding that would cause failure in concrete pipe would cause failure of any similar-type pipe.
Large selection of available nominal diameters and ease of assembly due to gasketed joints.	Handling and installation difficulty because of heavy weight except where weight would be advantageous because of flotation concerns. However, requires less soil support and thus less attention during installation as compared to flexible pipe.
Wide variety of pipe lengths available.	Susceptible to external corrosion in acidic soil environments, such as old garbage dumps or industry discharge oil pits.
Large selection of both structural and pressure strengths.	Requires internal protection due to H_2S attacks and internal microbiological-induced sulfuric acid corrosion at crown. A concern in sanitary applications only.
Low cost to install as pipe is designed for internal pressure and external load and due to rigidity, requires less support from compaction and side fill support. Relatively low cost of maintenance.	Generally more difficult to repair, particularly in cases of joint leakage or failure in pressure pipes.
Capability to withstand very high pressures without leak with a cylinder if properly installed.	Tendency to leak because of high pipe wall porosity and shrinkage cracking.
Ideal for pipe-jacking applications owing to high compressive strengths.	
Internal corrosion can be eliminated by using thermoplastic lining.	Somewhat lower abrasion resistance—Internal scouring can occur if solid content and flow velocities are high.
Potential for external sulfate corrosion may be reduced by use of Type V or other low C3A sulfate-resistant Portland cement.	In damaged pipe or in highly corrosive areas, reinforcements in PCCP may corrode or fail without little or no external evidence.

Table 5.7 Concrete Pipe Advantages and Limitations

For contractors, the heavy weight of concrete pipes can make it difficult to install when compared to some alternative materials. Concrete pipe is less sensitivite to bedding conditions in both shallow and deep installations as compared to flexible pipe types, over 85% of the pipe-soil system strength comes from the pipe. This is unlike flexible pipe where due to their low wall stiffness, only about 5 percent of the system's capacity is attributable to the pipe wall, thus requiring almost complete dependence on sidefill support as well as on quality of compaction and installation.

The trenchless installation, however, results in much more consistent bedding conditions in the circumferential direction. For trenchless installation of RCP, roughly 90% of the load bearing capacity of the soil/structure "system" is in the pipe wall.

In terms of "beam breakage," the statement is partially correct. Nonpressure, noncylider RCP is typically not designed for beam strength. The relatively short joint lengths in use allows that one length of pipe may move differentially with respect to the adjacent length relieving longitudinal stresses. The RCCP, on the other hand, with steel cylinder and its longer joint lengths, have sufficient longitudinal beam strength.

While exhibiting superior resistance to corrosion due to reinforcement encasement in high alkaline environment (concrete or mortar), certain aggressive corrosive environments have proven to be challenging to concrete pipe for both internal and external corrosion. For sanitary sewers, with hydrogen sulfide H_2S generation, many engineers and city authorities require concrete pipes to be internally lined to protect against hydrogen sulfides with PVC or HDPE. Studies suggest that unlined RCP for sanitary sewers can show signs of deterioration in less than 5 years after installation. Failures owing to H_2S attacks and internal microbiological-induced sulfuric acid corrosion at the crown of RCP have also been a topic of study in recent years. It is important to mention that steel and ductile iron pipe might be more sensitive to hydrogen sulfides as their mortar lining is much thinner than that of concrete pipe (i.e., only ½ in. for steel and 3/16 in. for DI pipe) and they are not recommended in aggressive sanitary sewer environment. In pressure applications, the loss of prestress due to damaged mortar coating or in extremely aggressive soils may result in in reinforcement wire breakage in PCCP which can lead to failures in high-pressure pipelines. Research to assess the risks of such failures has been undertaken by various engineers and entities. The preceding discussion about H_2S corrosion may not be an issue in storm sewer applications.

5.6.2 AC Pipe

With the asbestos-related litigations of the 1980s, bans were placed on all construction materials that used asbestos. Pipe was no exception.

In 1986, the United States Environmental Protection Agency (EPA or also called USEPA) published a proposed regulation on the commercial uses of asbestos in which a ban on the manufacturing and installation of AC pipe was proposed; this proposal was later carried out. Many AC manufacturing plants relocated to other countries, and production in the United States came to a complete stop. In the early 1990s, the ban on the pipe was lifted, but today AC pipe is no longer manufactured in the United States. Many countries of the world, including Mexico, continue to specify and install AC pipe for both water and sewer applications.

It is important to note that ingestion of asbestos fibers has not proven to be a health hazard. However, inhalation of the fibers can result in lung cancer. So, workers modifying AC pipe must be protected with masks and suits such that the fibers are not inhaled either at the work site or wherever else the worker's clothes and tools might release any airborne fibers.

Worldwide usage of AC pipe increased from 200,000 mi (miles) in the 1950s to 2 million mi in 1988. In the United States, it was estimated that by 1988, over 300,000 mi of AC pipe was in service in water systems. According to the Association of Asbestos Cement Product Producers, AC pipe is still used in the Southwestern United States but it must be imported from Mexico. However, this usage is certainly an anomaly. The hazards of asbestos inhalation are increased during removal of the pipe, so utilities have chosen to keep them in the ground. These pipes are regularly tapped and maintained, so a working knowledge of the material is helpful. There are no known instances of trenchless municipal piping construction with new AC pipe.

5.6.2.1 Manufacturing

Asbestos-cement water and sewer pipe is a fiber-reinforced cementitious product composed of an intimate mixture of Portland cement and silica. The controlled blending of these basic raw materials is built up on a rotating steel mandrel and then compacted with steel pressure rollers into a dense homogenous structure in which a strong bond is affected between the cement and the asbestos fibers. A smooth interior surface results from this process. Final curing of the product is done in an autoclave employing high-pressure steam for dimensional and chemical stability.

5.6.2.2 Joint Types

Asbestos-cement pipes were joined together by means of a gasket-joint coupling (also referred to as a double-bell coupling joint). The ends of two pieces of pipe were slipped into the gasketed coupling, effectively creating two watertight seals. Some manufacturers permitted angular movements of up to 5° at joints. Figure 5.22 shows a typical AC pipe joint.

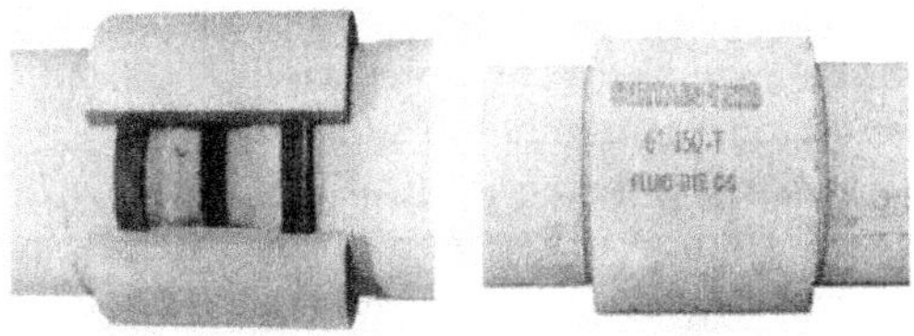

FIGURE 5.22 Cross-section of coupling joint in AC pipe.

5.6.2.3 Advantages and Limitations

Among its advantages, its long operational life, resistance to corrosion, light-weight in smaller diameters, and watertight joints may be mentioned. Known limitations of the pipe include a low flexural resistance as a whole, easily damaged by construction equipment because of its brittle nature, and low acid resistance. As concrete and cement corrode, the possibility of AC fibers getting released into the water in the pipe increases over time.

Significant research was done to study the effects of dissolved asbestos fibers in drinking water on human health. Asbestos-cement pipe manufacturers and the AWWA did not recommend the use of AC piping material with highly corrosive water, such as acidic water as it was more likely to attack AC piping products.

5.7 Vitrified Clay Pipe

The second group of rigid pipes among piping materials is VCP. Vitrified Clay is one of the oldest building materials with pipes dating back over 6,500 years. The use of VCP in sanitary sewer systems throughout the United States is yet to be matched by any other piping material. For two centuries, VCP was the only commercially available material capable of withstanding the chemically aggressive environments of sanitary sewers. Still today, the corrosion resistance properties of VCP are yet to be matched by any other piping material available. The earliest recorded use of clay pipe in the United States was in Washington, D.C. in 1815. Clay pipe is not used in pressure applications because of its inherently low tensile strength. Conversely, the compressive strength of vitrified clay averages 18,000 psi making it ideal for jacking applications.

Only 50 years ago, the sanitary sewer engineering community followed the philosophy that leakage of wastewater through pipe joints was a desirable methodology for effectively transporting suspended solids and reducing excessive flows within a sewer system. The accepted design theory at the time was: "dilution is the solution to pollution." Clay pipes manufactured during this time were supplied without factory applied compression joints as infiltration was considered a benefit. This philosophy soon changed as engineers realized the hazards posed by wastewater leakage to soils and groundwater sources. The cost of treating the excess effluent represented by the infiltration was a strong motivation for communities to adopt the change in philosophy.

The EPA's role in reducing infiltration or inflow (I/I) with the passing of several Congressional legislations, such as the Water Pollution Control Act, of 1972, and the Clean Water Act, 1977, were major factors in emphasis shifting to the requirement of watertight joints in sewer pipes. All modern-day clay pipes are equipped with factory applied flexible compression jointing systems. Whether on bell-and-spigot pipe or plain-end pipe, these leak-free joints provide resilience and flexibility to accommodate minor pipe movement.

All compression type jointing methods meet the requirements of ASTM-C425, *Standard Specification for Compression Joints for Vitrified Clay Pipe and Fittings*, which require that the joints do not leak. This standard was introduced in 1958 and became the norm nationwide in the early 1970s with the help of the Water Pollution Control Act of 1972. With the proper installation, a clay pipe sewer system will meet standard infiltration or exfiltration requirements. In trenchless construction, clay pipe's ability to withstand high compressive loads and external abrasion has resulted in its acceptance in pilot tube and other guided boring methods, pipe bursting, and other inline replacement methods (see Chap. 8), pipe jacking and microtunneling applications.

5.7.1 Manufacturing

The VCP is made of selected clay and shale that are aged to various degrees, and blended in specified combinations dependent on manufacturer and location. Large crushing wheels grind the clay in a heavy perforated metal pan until the finely ground clay passes through the perforations. The ground raw materials are mixed with water in a pug mill. The mixture is then forced through a vacuum, deairing chamber until a smooth, dense mixture forms. The mixture is extruded under extremely high pressures to form the pipe. After drying, the newly formed pipe is placed in kilns and heated to temperatures of approximately 2000°F. The finished pipe then undergoes a quality assurance/quality control (QA/QC) testing.

5.7.2 Applicable Standards

A relatively high minimum compressive strength of 7,000 psi (per ASTM C-1208) makes clay pipe appropriate for jacking and microtunneling installations. In 1994, ASTM C-1208 opened new doors for clay pipe in the trenchless construction and replacement (see Table 5.8).

5.7.3 Joint Types

The evolution of joints in clay pipe is an example of the impacts of continuous research, development, and innovation within a piping industry. To arrive at its present-day compression joints, clay pipes went

ASTM C12	Standard Practice for Installing Vitrified Clay Pipe Lines
ASTM C301	Standard Test Methods for Vitrified Clay Pipe
ASTM C425	Standard Specification for Compression Joints for Vitrified Clay Pipe and Fittings
ASTM C700	Standard Specification for Vitrified Clay Pipe, Extra Strength Standard Strength and Perforated
ASTM C828	Standard Test Method for Low-Pressure Air Test of Vitrified Clay Pipe Lines
ASTM C896	Standard Definitions of Terms Relating to Clay Products
ASTM C1091	Standard Test Method for Hydrostatic Infiltration Testing of Vitrified Clay Pipe Lines
ASTM C1208/ 1208M	Standard Specification for Vitrified Clay Pipe and Joints for Use in Jacking, Sliplining, Pipe Bursting, and Tunnels

Table 5.8 VCP Non-Pressure Standards for 4 to 42 in. Diameters

from having no joints to field-applied cement joints, to field-applied bitumastic joints to factory-installed cement mortar or bitumastic joints, and finally to the current compression seal joint. The molded polyurethane joint attached to the spigot end and a PVC collar, which is shrunk and rapidly fitted and sealed to the socket-end of rigid VCP barrel is shown in Fig. 5.23.

Figure 5.23 Innovative joints in clay pipe.

The jointing system in ASTM C-1208 clay jacking pipe is a precision ground recessed joint, ensuring dimensional accuracy and high end-bearing capacity. A stainless steel sleeve with elastomeric seals is used for jointing, as shown in Fig. 5.24*a* and *b*. The sealing element is compressed between bearing surfaces to promote watertight integrity. The ends of the pipe are fitted with a compression ring to distribute the jacking forces of installation.

Although not common in trenchless applications, the flexible compression joints conforming to ASTM C425 provide a tight joint

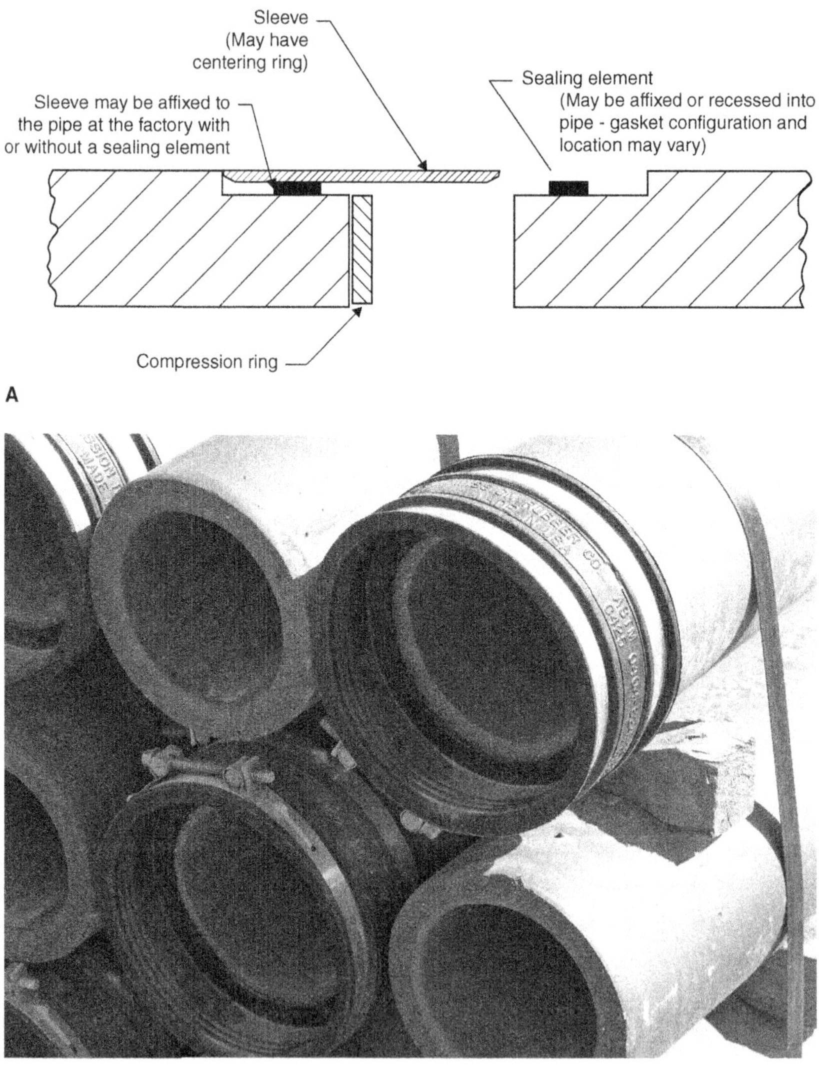

Figure 5.24 Jointing system in ASTM C-1208 clay jacking pipe.

whether the sewer is above or below groundwater. Flexible compression joints use polyester or polyurethane materials and are factory applied to both the bell and spigot of every pipe (see Fig. 5.25*a*). Another jointing system option is plain end pipe barrels joined with rubber compression couplings with stainless steel tightening bands (see Fig. 5.25*b*). These jointing systems allow the clay pipe to shift or

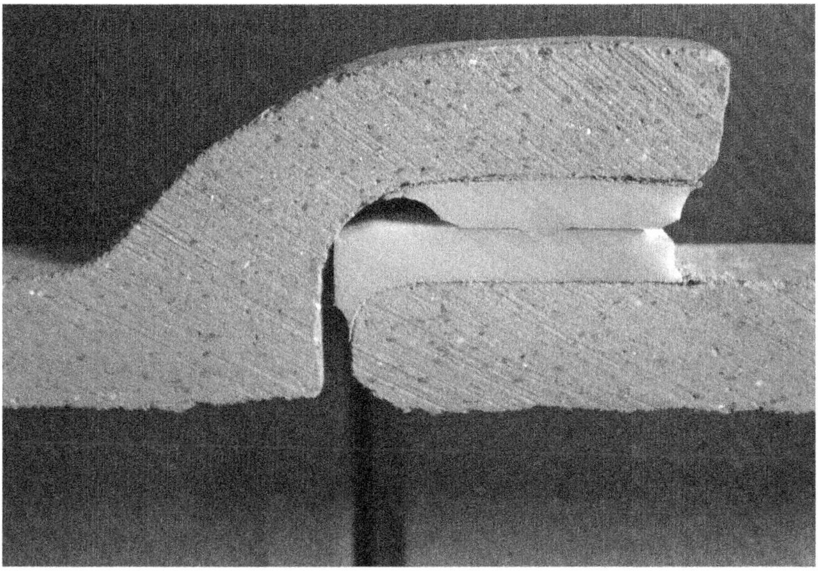

A

B

Figure 5.25 Flexible compression joints in ASTM C-425.

move with minor changes in earth conditions, preserving the leak-free joint without damaging the pipe. This provides a measure of forgiveness during installation and for the life of the line.

5.7.4 Advantages and Limitations

Table 5.9 summarizes some of the advantages and limitations of clay pipe. The inert nature of clay pipe is the reason behind its wide acceptance for use in sanitary sewer applications. Vitrified Clay is a fired ceramic with material properties unaffected by light, time or temperature making it the only corrosion resistant piping product where material degradation is not a concern over the service life. Resistance to both internal and external corrosion was its main advantage over other traditional piping materials such as concrete and cast iron. Clay pipe is also abrasion resistant, making it a suitable material in sewers with high solid content. It also permits various cleaning methods that would degrade or possibly damage other types of pipe. Its low sensitivity to temperature differentials prevents any significant expansion and contraction when buried in the ground. In recent years, the ability

Advantages	Limitations
Resistant to both internal and external corrosion.	Available for gravity applications only because of the characteristics of the pipe.
Proven history of long life of the pipe itself. The anticipated service life can be over 100 years if properly installed and tested with appropriate environmental conditions.	Sensitive to a firm and unyielding foundation beneath the pipe and bedding conditions—may be subject to shear and beam failure.
Factory applied flexible compression joints have been available since 1970.	Lack of a jointing system for pipe installed prior to 1970 may lead to leakage and root intrusion problems.
Ability to handle high compressive forces, making it ideal for jacking installations.	Short lengths, resulting in more pipe joints.
Abrasion resistant.	Only can be used for gravity pipe applications.
As a fired ceramic, its material properties are unaffected by light, time, or temperature.	
As a rigid conduit, VCP provides structural strength in the trench.	
Environmentally responsible pipe with natural environmentally responsible credentials of VCP as independently verified based on an ISO140001-compliant life cycle assessment.	

TABLE 5.9 Clay Pipe's Advantages and Limitations

of clay pipes to withstand external abrasion and relatively high compressive forces has led to its wide usage in trenchless applications, such as microtunneling and pipe jacking.

Although the inert characteristics of the material made it ideal for use in aggressive environments, such as sanitary sewers, the lack of effective joints associated with older pipelines had resulted in a drastic reduction of its use in sewer systems. Though innovation within the industry has led to better performing joints in recent years, the millions of feet of clay pipe installed in the ground decades ago are now responsible for I/I problems in sewer systems throughout the United States and Canada. The susceptibility to shear and beam breakage owing to poor foundation bedding conditions, ground movement, and nonflexible, field–applied joint had resulted in numerous leaks and cracks in a number of older sewer systems. The short lengths in which clay pipes are manufactured (usually 8 ft or less) increases the number of joints in a sewer main line, thus raising the chances of leakage through joints. Root intrusion through clay pipe joints has led municipalities to institute annual root control measures in older systems.

5.8 Plastics Pipes

The introduction of plastic pipes in the late 1950s in North America was the beginning of an industry that would eventually revolutionize the field of municipal water and sewer piping. Plastics are formed by the polymerization of molecules containing hydrogen and carbon. The three main types of plastics pipes widely in use in North America include PVC, medium density PE (MDPE) or HDPE, and GRP (also called fiberglass pipe). Polyvinyl chloride and PE fall into the group of "thermoplastics," while GRP is a "thermoset resin" pipe. The properties of thermoplastics for construction material applications are better appreciated with an understanding of "viscoelastics," discussed in the next section. Thermoset plastics are processed by a combination of chemicals and heat, and once formed, cannot be reshaped. Thermoplastics, such as PVC and PE, used in pipe manufacturing are referred to as "rigid plastics." The term rigid indicates that these materials do not contain any plasticizers, which would make them more ductile, and hence unsuitable for buried municipal piping applications. Both thermoplastic and thermoset pipes are considered to be flexible conduits and are designed accordingly.

In the field of municipal trenchless construction, PE has been the dominant piping material in the past decade. By butt-fusion of successive lengths of PE pipe, a long "jointless" conduit is created, which can be installed by trenchless methods such as horizontal directional drilling (HDD). PE is also used in open-trench construction. Traditionally, PVC has been the most dominant material for open-cut installations in the North American water and sewer markets because of its bell-and-spigot gasket joints, its light weight in smaller diameters, and ease to work with. In recent years, several manufacturers have also created proprietary PVC products with modified joints for trenchless installation. The GRP pipes are used in the United States for both pressure and gravity applications, though the latter is the more prominent use.

5.8.1 Properties of Viscoelastic Construction Materials

The PVC and PE are thermoplastics and thus viscoelastic materials. Viscoelastic materials exhibit elastic as well as viscous-like characteristics. A material that deforms under stress but regains its original shape and size when the load is removed is classified as elastic. Viscous materials, on the other hand, after being subjected to a deforming load, do not recover their original shape and size once the load is removed. In reality, all materials deviate from the linear relationship between stress and strain (Hooke's Law) at some point in various ways.

Defining the direct relationship between stress and strain, when a load is applied to a material, is the most common way to evaluate the strength and stiffness of that material. Graph A in Fig. 5.26 illustrates the linear relationship between stress and strain in elastic materials. In an ideal elastic material, strain returns to zero as soon as the material is unloaded, and the linear relationship is not typically time-dependent. But it should be noted that in all materials, this behavior is valid only up to a certain stress point, called the yield point, after which the strain in the material will increase dramatically by creep, before finally failing.

In the set of curves, B, in Fig. 5.26, it can be seen that the stress-strain relationship is somewhat different for viscoelastic materials than it is for elastic materials. Clearly, we no longer see a directly linear relationship between stress and strain, and the gradients of the

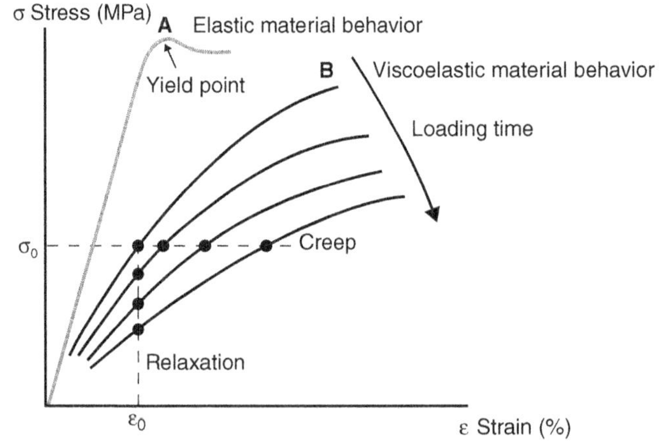

Figure 5.26 Stress-strain relationship in elastic and viscoelastic materials. (*Source:* Modified from Fig. 3.2.23 of Janson, 1999, p. 61.)

curves depend on the loading time. In other words, for a given stress level, the longer the loading time, the larger the strain reached. Creep is defined as continuing deformation (increasing strain) with time when the material is subjected to a constant stress. As a consequence of creep, failure of the material will occur after load is applied for a certain amount of time. So time dependency is a major factor to consider in viscoelastic material behavior. An important fact is that the time to failure is inversely proportional to the applied stress. In thermoplastic pressure pipe, it is, therefore, possible to find and apply a stress level that is low enough to ensure that the theoretical time to failure will surpass the design life of the pipeline.

In thermoplastic pipe applications, creep is prevented because the deflection of the pipe is kept constant, as is the case in buried PVC gravity (or pressure) pipe. Consequently, it can be seen from Fig. 5.26 that the initial stress decreases with time, and is referred to as the relaxation property of thermoplastic piping materials. These basic properties of viscoelastic materials, such as PVC and PE, enable engineers to design pipelines that ensure both structural integrity and the long-term design life of their municipal piping systems.

5.8.2 PVC Pipe

Polyvinyl chloride was discovered almost accidentally in the 19th century when German scientists, observing a newly created organic chemical gas, vinyl chloride (C_2H_3Cl), discovered that when it is exposed to sunlight, a chemical reaction took place, resulting in the creation of an off-white accumulation of solid material. Scientists had observed the first polymerization and creation of a new plastic material, PVC. In 1839, a technical paper was published detailing the observations of the process. In 1912, several decades after its accidental discovery, Fritz Klatte, another German, laid the groundwork for the technical production of PVC. The oldest known PVC pipe was manufactured and installed in the 1930s in World War II in Germany and continues to be in service today. The technology was brought to the United States following World War II, and by the mid-1950s ASTM groups were organized for plastic pipe standardization. In 1955, the total US shipments of thermoplastic pipe (PVC, PE, etc.) were estimated at about £40 million. By 2015, the annual use of PVC for pipe production had reached 6 billion pounds[3].

5.8.2.1 Manufacturing

A vinyl chloride monomer (molecule) comprises carbon, hydrogen, and chlorine, configured as shown in Fig. 5.27*a*. The PVC is obtained

[3]http://www.plasticsnews.com/article/20160128/NEWS/160129791/pp-sales-outperform-ps-eps-and-pvc-in-2015

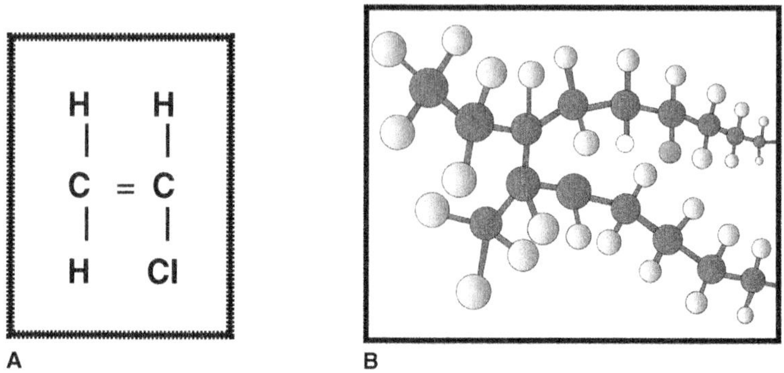

FIGURE 5.27 (a) Vinyl chloride molecule, (b) PVC chain.

by polymerization of single units of the vinyl chloride monomer, which join to create long chains and ultimately form PVC resin, Fig. 5.27*b*.

The PVC pipe is manufactured by blending PVC resin with stabilizers, pigments, lubricants, processing aids, and functional additives, and then heating this mixture to a temperature in the 400°F range. This causes the components to properly fuse and convert into a malleable state. In this molten form, the material is mechanically extruded into the pipe. Following the completion of the extrusion process, the pipe is allowed to cool, after which QA/QC testing is performed before the final delivery to the end user.

5.8.2.2 Types of PVC Pipe

There are three distinct types of PVC pipes manufactured in the world, each differentiated by either the way in which it is manufactured (which dictates the directional orientation of the molecules), or by the content of modifiers in its chemical formulation (which affect the ability of the pipe to withstand large impacts by absorption and dissipation of the energy). The term PVC is a generic designation, which includes PVC-U (unplasticized PVC), PVCO (molecularly oriented PVC), and PVC-M (modified PVC).

1. *PVC-U*. Unplasticized PVC is the most widely used piping material in water and sewer systems in North America. The molecular structure of PVC pipe is a random arrangement of long chain molecules, where molecular entanglement is prevalent throughout the length of the pipe. In general, the PVC molecules do not exhibit any definite directional orientation, and therefore a generally uniform strength prevails in both the radial (circumferential) and longitudinal directions. Testing has shown that the modulus of elasticity in 15-year-old PVC is only slightly higher in the longitudinal direction than

in the radial directions[4]. For simplification, the term PVC as used in this chapter will denote PVC-U, the conventional type of PVC most used in pipe manufacture. PVC pipes are manufactured for both pressure and gravity applications. All PVC pressure pipes (ASTM and AWWA standards) must meet the cell classification 12454, indicating a tensile strength of 7,000 psi, and a modulus of elasticity of 400,000 psi. PVC pressure pipes have a hydrostatic design basis (HDB) of 4,000 psi, per ASTM D-2837, to which a factor of safety of 2.0 is applied. This reduces the long-term design stress to 2,000 psi. The PVC gravity pipes are typically manufactured to cell class 12,454 or 12,364, where the latter has a tensile strength of 6,000 psi and a modulus of elasticity of at least 440,000 psi.

2. *PVC-O.* Molecularly oriented PVC is made in the United States by the expansion of PVC pipe made from conventional PVC resin; during the expansion process, the molecules become oriented in a generally radial or circumferential direction. This molecular reorientation increases the strength of the pipe in the hoop direction. Also, the resulting HDB is increased from 4,000 to 7,100 psi. Consequently, this stronger material can have a thinner wall than a conventional PVC pipe of the same pressure capacity. The two technologies available to manufacture PVC-O, batch (offline) and continuous (inline), are both used in the United States. There is one PVC-O manufacturer in the United States and another in Canada.

3. *PVC-M.* Modified PVC is produced by incorporation of additives or "impact modifiers" to enhance the toughness of the material. Resistance to fracture by absorption and dissipation of energy is an evidence of the toughness of the pipe material. Modified PVC is made and used mainly in Europe and Australia.

5.8.2.3 Solid-Wall and Profile-Wall PVC Pipes

The PVC pipes are available for pressure applications only as solid-wall pipes. For gravity applications, both solid-wall and profile-wall pipes are manufactured. Solid-wall pipe, as the name suggests, are made of a continuous wall of PVC of uniform thickness, as shown in Fig. 5.28*a*. Profile-wall pipe, on the other hand, is braced spirally or circumferentially with structural shapes, but provides a smooth-wall interior, as shown in Fig. 5.28*b*. Profile-wall pipes economize on the amount of material needed for fabrication; by altering the shape of

---

[4]PVC held under a constant pressure (water pipe) actually had the hoop strength increase compared to new pipe. For new pipe it is logical that the longitudinal strength may be slightly higher than the hoop because the molecules are likely slightly oriented in that direction as it is pulled off the extruder.

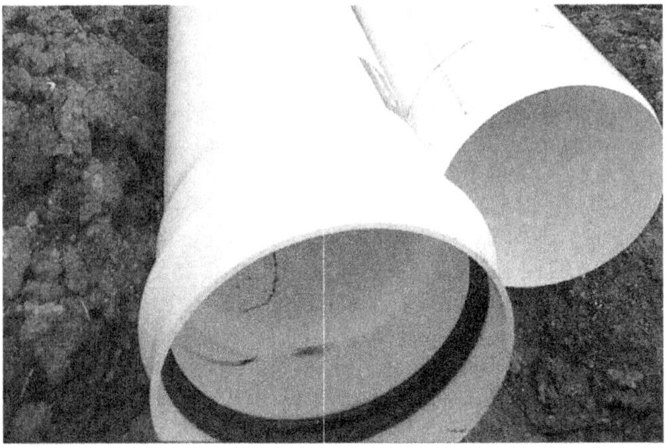

A

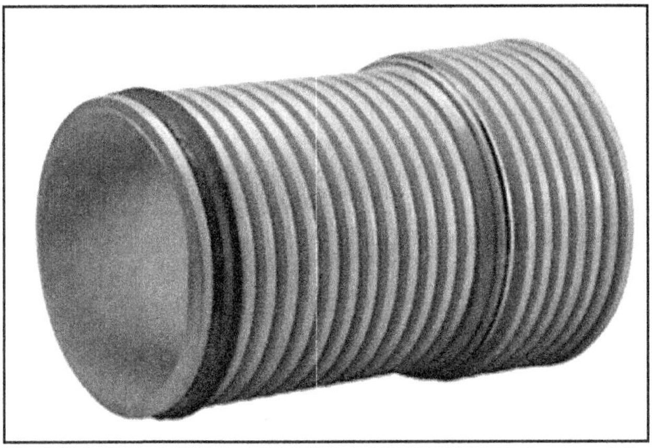

B

Figure 5.28 (a) Solid-wall PVC pipe. (b) Profile-wall PVC pipe.

the wall, the same stiffness as solid-wall pipe is achieved using less material. Profile-wall pipes generally fall into three categories—open profile (OP), closed profile (CP), and dual-wall corrugated profile (DWCP). The OP pipes have their rib-enforcements exposed on the outside of the pipe. The CP pipe make use of a CP that provides a continuous outer wall where the wall sections are hollow and are often described as an I-beam or honeycomb. The DWCP pipes have a smooth-wall waterway, braced circumferentially with an external corrugated wall.

From a design perspective, both solid-wall and profile-wall PVC gravity pipes are limited to a vertical deflection of 7.5 percent as per ASTM and Uni-Bell recommendations. However, tests must show that the structural integrity of the pipe is not compromised until a deflection in excess of 30% is achieved. The lower recommended design deflections are to ensure design flow is maintained and minimize surface deflections. Generally, solid-wall PVC pipes are used for open-cut installations as well as trenchless renewal methods such as sliplining and pipe replacement/bursting. Pipes used for trenchless technology have modified joints that facilitate their installation. For pressure pipe, AWWA recommends a deflection limit of 5 percent.

5.8.2.4 New PVC Pipe for Water Applications

Since the first PVC pipes were installed in the 1950s, some enhancements in the product were made to improve from a brittle material to a more flexible material. Recently a new PVC pipe (iPVC) has been developed in South Korea and currently is being tested in U.S. for possible use. iPVC pipe is currently manufactured in Korea with diameters 4, 6, 8, 10, and 12 in. and with dimension ratios of DR 14, 18, and 25. The iPVC pipe has been approved by the National Sanitation Foundation (NSF) under NSF/ANSI 61 for potable water applications (Venkatesh et al, 2016). Figure 5.29 illustrates an iPVC pipe installation in South Korea.

The manufacturing process of this pipe includes use of high strength resin, improved equipment and formulation method to optimize dispersion and distribution of raw materials, and improved extrusion methods to produce in high-processing temperature and high-melting viscosity. Table 5.10 presents some preliminary test results. Some of the differences in test results of materials were striking. For example, according to American Water Works Association (AWWA) standards, AWWA C900-07 and AWWA C909-09, the impact resistance requirement for PVC is 0.65 foot-pounds per inch (ft-lb). iPVC has an impact resistance of 1.26 ft-lb based on the testing conducted by PPI Systems Inc. This is 92% more than the AWWA requirement (Venkatesh et al, 2016).

FIGURE 5.29 iPVC pipe installation.

Tests	AWWA Requirements	iPVC
Impact Load (for 8-in. pipe)	100 foot-pounds	390 foot-pounds*
Tensile Strength at 73°F	7,000 psi	7,780 psi
Modulus of Elasticity	400,000 psi (Cell 5)	425,000 psi
Long Term Pressure	500 psi	500 psi
Hydrostatic proof test pressure 8-in. DR 18	600 psi	755 psi
Weight (pounds/foot) – 8-in. DR 18	11.5	10.6

*Izod impact value with TUP–A (sharp edge drop weight) for 4-in. iPVC pipe

(*Source:* Modified from Venkatesh et al., 2016, from testing conducted by American Water and Micobac Laboratories, Inc., funded by Water Research Foundation (WRF)

Table 5.10 Comparison of AWWA Requirements vs. Preliminary iPVC Pipe Results

5.8.2.5 PVC Applicable Standards and Products for Trenchless Technology

Although there are several widely used bell-and-spigot gasket-joint PVC pressure and gravity piping standards in North America written by several organizations, such as AWWA, ASTM, and Canadian Standards Association (CSA), products discussed in this chapter are specifically for trenchless applications. They are proprietary in nature in that they have uniquely designed jointing systems that enable the pipe to be pulled or pushed for various trenchless construction methods. Table 5.11 outlines details on each product along with pictures of joints. Also, there are several types of PVC piping products geared toward use in "trenchless renewal" market. Typically, these are solid-wall PVC pipes with proprietary joints. Information on these products is not included in this chapter but more details may be obtained from manufacturers or Uni-Bell PVC Pipe Association's Handbook of PVC Pipe.

1. Butt–Fused Joint *Fusible C-900, Fusible C-905, and Fusible PVC.* Until recently, PE was the only available thermoplastic pipe option that used butt-fused joints in the United States. In the late 2003, the water or wastewater industry saw the introduction of Fusible PVC, or Fusible C-900, or Fusible C-905. A first of its kind, this product combines a proprietary formulation with a unique fusion procedure that allows lengths of PVC pipe to be joined together in a continuous string for installation through a variety of methods, including HDD, sliplining (Chap. 7), pipe replacement/bursting (Chap. 8), and direct bury applications. Although Fusible

Pipe Type	Standard Compliance	Joint	Diameter
Fusible PVC	AWWA C-900/C-905		4 in. through 36 in.
TerraBrute	AWWA C-900		4 in. through 12 in.

Table 5.11 Trenchless PVC Products and Standards

Pipe Type	Standard Compliance	Joint	Diameter
Certa-Lok	AWWA C-900/C-905		4 in. through 24 in.
Internally Restrained PVC Pipe	AWWA C-900	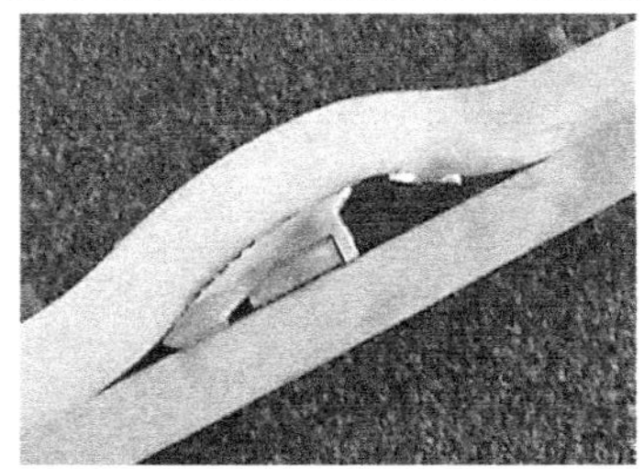	4 in. through 24 in.

Table 5.11 Trenchless PVC Products and Standards (*Continued*)

PVC is primarily for nondrinking water applications like recycled water, force-mains, gravity drains, and sewer applications, however, fusible C-900 or fusible C-905 are specifically for use in potable water systems. These pipes are manufactured to all requirements of AWWA C-900 and C-905, and are NSF certified. Surface scratches of up to 10 percent of the pipes' outside diameter are accepted, per AWWA, during pull-in of all PVC pipes. A major advantage of the fusible pipe is that those municipalities already using conventional PVC can now use the same material in trenchless construction, enabling the design of a "complete PVC system." It also allows municipal entities to inventory a single type of fittings and restraints for all situations. Tapping is performed in the same manner as gasket-joint PVC pipes.

2. *Pin and Groove Gasketed Joint TerraBrute.* This product meets AWWA C-900 standard requirements[5], but has a modified bell-and-spigot joint as shown in Table 5.11. Currently manufactured in diameters of up to 12 in., this pipe has been used in several buried and several non-buried pressure application in Canada. The bell-and-spigot joint modification increases the tensile load capacity of the pipe joint by a factor dependent on the diameter and the wall thickness of the pipe. Also, controlling the number of pins, pin diameter, depth of groove in the spigot, and wall thickness of the internal ring can optimize the tensile capacity of the joint. Installation involves insertion of the spigot into the bell, alignment of the "composite" pins with the groove on the spigot, and then hammering down of the pins. Once again, utilities in Canada have the ability to build complete PVC systems, even when portions of a project comprise trenchless construction.
3. *Spline–Lock Gasketed Joint Certa-Lok C-900/RJ.* This is another product conforming to AWWA C-900 and C-905 standards[6], with a proprietary joint type that makes the product ideal for trenchless installation via HDD. The pipe is also NSF certified for potable water usage. The Certa-Lok C-900/RJ is used in the new construction of water distribution or transmission lines or sewer force mains, or for new gravity sewer main installation. The joint has a groove machined in the pipe and in the coupling to allow the insertion of a flexible thermoplastic spline that provides a full 360° restrained joint with evenly distributed loading. Available diameters currently include

[5]The groove may not meet the required wall thickness.
[6]May not meet the required standard wall thickness.

4 in. through 24 in. The recent introduction of 16-in. pipe uses a fiberglass-based coupling joint system.

4. *Internally Restrained Gasketed Joint PVC Pipe.* This product also conforms to AWWA C-900. The joint is patented and is licensed to several PVC pipe and fabricated fitting manufacturers. The mechanism consists of a metal casing that sits adjacent to the Rieber gasket in the bell; the casing is molded into the "raceway" of the bell during pipe belling. A C-shaped grip ring with several rows of unidirectional serrations is manually inserted into the casing at the manufacturing facility. Both the casing and the grip ring are made of ductile iron that has been coated using an electro-coating process that achieves a uniform thickness and provides superior corrosion resistance. When the pipe arrives at a jobsite, the bell already contains the casing with the grip ring inserted in it, and no additional hardware is needed to provide a restrained joint. Internally restrained PVC pipe is suitable for HDD or for direct-bury applications that require a restrained joint. The brand names for the product are Bulldog, Diamond Lok 21, and Eagle Loc 900. Table 5.11 presents a summary of PVC pipe standards and jointing system.

5.8.2.6 Advantages and Limitations

The greatest contributor to the rapid adoption of PVC pipe in water and sewer systems is its inherent ability to withstand both internal and external corrosion. Vast amounts of corrodible piping materials are being replaced each year by PVC. In sewer systems, the completely leak-free joints are a definite advantage over traditional materials. Pipes with proprietary joints for trenchless installation have high tensile strengths, allowing for long lengths to be pulled in at a time. Abrasion resistance has been an advantage over alternative cementitious piping materials. Low internal friction enables the use of smaller diameter pipe in both pressure and gravity applications. An HDB, which is 2.5 times higher than alternative thermoplastic piping material, allows for thinner walls and therefore higher flows. A much smaller expansion coefficient than alternative thermoplastics significantly reduces expansion and contraction and concerns for open-cut, trenchless, and above-ground applications.

The sensitivity to temperatures on a long-term basis requires thermoplastics such as PVC to be derated if the pipe temperature is expected to exceed 73°F on a continuous basis. Pipe has a ultraviolet inhibitor that provides extended continuous exposure to high intensity UV from sunlight. Field testing of PVC pipe has shown a slightly reduced impact strength from UV in southern U.S. states, where UV is stronger. The PVC pipe cannot be deflected longitudinally as much as alternative thermoplastics. The minimum bending radius

Advantages	Limitations
Resistant to both internal and external corrosion.	Sensitive to operating temperature, must be derated in case of long-term operating temperatures with above 80°F and up to 140°F.
Gasket joints and fusible have an excellent track record of leak-free performance.	Sensitive to ultraviolet light if continuous long-term exposure to high intensity UV.
PVC is suitable in hydrocarbon contaminated soils.	Less longitudinal flexibility than alternative thermoplastic piping material.
All four restrained-joint PVC products have high tensile strengths for trenchless applications.	
High abrasion resistant for sewer applications.	
Low internal frictional resistance for both pressure and nonpressure applications.	Susceptible to chemical permeation in cases of gross contamination of some solvents.
At least 2.5 times stronger than other thermoplastic pipe (higher stiffness, higher HDB).	Susceptible to impact damage in cold temperatures.
Expansion is significantly lower than in alternative thermoplastic piping materials.	Tapping of fused PVC pipe must be done with extreme caution.

Table 5.12 Advantages and Limitations of PVC Pipe

for fusible PVC pipe is about 250 times the pipe diameter. The fusible PVC pipe might be susceptible than gasketed joint PVC to rapid crack propagation in the event a pressurized pipe is impacted or improperly tapped. Table 5.12 presents a summary of the advantages and limitations of PVC pipe.

5.8.3 PE Pipe

Polyethylene (PE) belongs to a group of thermoplastics known as polyolefins, materials made by polymerization of "olefin gases" including ethylene, propylene, and butylene.

Polyethylene was first discovered in 1898 by a German scientist, Hans von Pechmann, but the discovery was never commercialized. The PE was rediscovered in the United Kingdom in 1933 by the Imperial Chemical Company and later commercialized in 1939 to manufacture insulation for telephone and coaxial cables. The development of low-pressure reactors in the late 1950s greatly improved the commercial manufacturing of PE resins, and led to the commercial development of PE pipe in the 1960s. Early PE pipe use was mainly in the gas industry. Today, more than 90 percent of new and replacement gas distribution pipe in the North America is PE pipe. Other applications of PE pipe gradually spread to industrial and municipal systems. Since the development of HDPE materials in the 1970s, a significant use for PE is sliplining renewal of nonpressure gravity sanitary sewers. The development of ASTM and AWWA pressure piping standards led to the widespread use of PE pressure pipe in municipal applications in North America over the past decades. This is due to the recognition of the advantage of a corrosion resistant, fused pipeline in reducing water loss, and pipeline ruptures. Many municipalities that have already been specifying another thermoplastic material, PVC, are using PE in open-trench installations. Today, PE is the most widely used piping material for trenchless methods, such as HDD and inline replacement/bursting (Chap. 8), for new installation and renewal of municipal pressure and nonpressure piping systems.

ASTM classifies PE piping materials having specified density ranges as low-density (LDPE), MDPE, or HDPE. Higher density in PE provides greater hardness, stiffness, and tensile strength. The LDPE is generally not used for pressure piping, but due to its flexibility, it is used for pneumatic instrument controls. The MDPE pressure piping is predominantly used for gas distribution. The HDPE nonpressure piping materials are used for electrical and communications conduit and nonpressure corrugated and profile-wall pipes. The HDPE pressure piping materials are used with solid-wall pipes for gas, water, sewer force mains and industrial process piping, and for solid-wall nonpressure applications such as sanitary sewers and culverts. The advantages have been recognized by the nuclear industry where HDPE piping is now being used in safety water services where the reliability of non-corroding, electrically inert, and leak free piping system is critical.

Gas pipes are generally made from either MDPE or HDPE. Municipal water and sewer pipes and industrial pipes are generally made from HDPE. At 73°F, the pressure rating of HDPE is about 25 percent higher than MDPE pipes. In 2005, ASTM standards were modified to allow the introduction of an enhanced HDPE with significantly improved performance capabilities. Polyethylene materials are identified by code. The HDPE materials commonly used for municipal piping have a code of either PE3608 or PE4710. The PE4710 material has a pressure rating 25 percent higher than PE3608 material. Another advantage of the newly advanced PE4710 material is a dramatic increase in Environmental Stress Crack Resistance (ESCR). Some PENT tests on the advanced PE4710 material are exceeding 2,000 hours, a significant improvement over earlier generation resin formulations and providing a design service life of 100 years.

5.8.3.1 Manufacturing

The manufacturing process for HDPE pipe is plastic extrusion where molten PE resin extruded under pressure through specially designed extrusion machines and dies that form the melt into pipe. The pipe is drawn into cooling tanks to solidify and cool the pipe. The second method for producing large-diameter PE pipe is to extrude a wall profile onto a rotating mandrel. Wall profiles can have hollow sections to reduce the amount of material but maintain circumferential stiffness. Once the pipe section is created, the mandrel is collapsed and removed from the pipe's internal diameter (ID). The third method for corrugated profile-wall PE pipe is to extrude a thin wall pipe into a corrugator that has vacuum blocks that draw the pipe into a corrugated wall profile and cool it. Corrugated pipes can be produced with an ID liner for a smooth waterway, and a non-corrugated OD layer for a smooth OD. Solid-wall PE pipes are typically designed for internal pressure, but are used for pressure and nonpressure services. Spiral wound and corrugated profile-wall pipes are for nonpressure service.

PE resin is manufactured from the base ethylene molecule, see Fig. 5.30, a colorless gas composed of two double-bonded carbon atoms. Polymerization of the ethylene is performed with various catalysts, under heat and pressure, during which the double bond between the carbons is broken, resulting in the formation of a bond with another carbon atom. Polymerization produces long carbon-carbon chain molecules.

Copolymer PE pipes resins are formed by copolymerization with other monomers such as butene, hexene, and octane. Copolymerization produces a molecular structure with copolymer side chain branching. Side chain branching and molecule length significantly affect short-term and long-term properties, and have a lesser effect on mechanical properties such as density, ductility, and hardness. Homopolymers, resins manufactured without comonomers, are not used for PE piping.

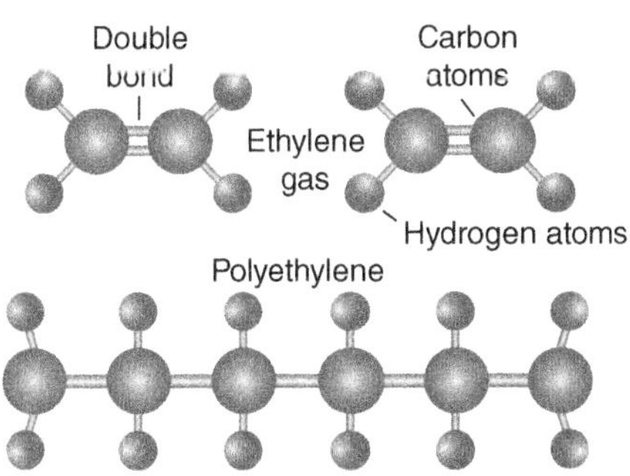

FIGURE 5.30 Ethylene molecule and polyethylene chain. (*Source:* Handbook of Polyethylene Pipe, Second Edition, 2008.)

	90°C (194°F) Test Temperature		
Categorization	Test Stress 2.48 MPa (360 psi) Time (h)	Test Stress 2.76 MPa (400 psi) Time (h)	Test Stress 3.10 MPa (450 psi) Time (h)
CC0	Unspecified	Unspecified	Unspecified
CC1	2700	1900	1200
CC2	7400	5100	3400
CC3	16200	11100	7400

TABLE 5.13 Minimum Log Average Test Times for Oxidative Resistance Classification

5.8.3.2 Engineering Properties

Polyethylene pipe materials are identified by categorizing physical property performance into numerically identified ranges in accordance with the ASTM D-3350. The resulting cell classification identifies cell ranges for density, melt flow rate, flexural modulus, tensile strength, slow crack growth (SCG) resistance and long-term hydrostatic strength. Higher numerical cell classifications represent higher cell property range values. UV resistance is identified with a letter code at the end of the cell classification. Standard specification ASTM D-3350 also includes additional minimum performance requirements for thermal stability, brittleness temperature, and tensile elongation. In 2014 an oxidative resistance classification table was added to D3350 which allows designers and owners the ability to quantify the level of oxidative resistance they may likely experience in demanding service conditions. Table 5.13 classifies the standards for materials as CC0 through CC3 with increasing numbers indicating increased performance.

PE pipe materials are generically identified by a material designation code that identifies the acronym for the type of plastic, ASTM D-3350 cell values for density and SCG resistance, and the hydrostatic design stress (HDS) rating for 73°F water in hundreds of psi with tens and units dropped. For example, PE3408 is a PE material having density and SCG resistance meeting ASTM D-3350 density cells 3 and 4, and a HDS rating of 800 psi. Typical material designation codes for PE pressure piping materials[7] are PE3408, PE3608, and PE4710. Table 5.14 provides material designation codes and typical ASTM D-3350 cell classifications. Figure 5.31 shows the HDPE pipe readied for pullback in a directional bore.

5.8.3.3 Applicable Standards

Environmental stress cracking may occur where the pipe is subjected to excessive contact stress due to improper embedment or excessive bending strain due to improper installation. Solid-wall HDPE pipes

[7]PE materials used for pressure piping are listed in PPI TR-4.

Material Designation Code	Typical ASTM D-3350 Cell Classifications
PE3408	345444C*
PE3608	345464C
PE4710	445474C or 445574C

*C identifies minimum 2% carbon black in the PE compound for UV resistance. For color products, E identifies color with UV resistance. C compounds are suitable for indefinite surface or above grade service. E compounds are sufficiently stabilized for storage, but must be buried.

Table 5.14 Material Designation Code and Cell Classification

Figure 5.31 HDPE pipe readied for pullback in a directional drilling.

manufactured to AWWA and ASTM standards for OD controlled pipe are typically used for trenchless installation applications. OD controlled pipes are required for butt-fusion joining that produces a continuous conduit ideal for pull-in installations such as HDD. Renewal methods such as PB also provide a good match for use of HDPE pipe. Testing has shown that surface scratches sustained during installation of up to 10 percent of the pipe's minimum wall thickness do not detrimentally affect short-term or long-term performance. Table 5.15 presents solid-wall HDPE pipe standard for trenchless.

Standards	Available Diameters
AWWA C-901	½–3 in.
AWWA C-906	4–63 in.
ASTM D-3035	½–65 in.
ASTM F-714	3 in.–63 in.

TABLE 5.15 Solid-Wall HDPE Pipe Standards for Trenchless

5.8.3.4 Joints

Butt-fusion of PE pipe is performed using commercially available fusion machines and performed in accordance with ASTM F-2620, "Heat Fusion Joining of Polyethylene Pipe and Fittings." The steps involve preparing ends of two pieces of pipe by planning the ends using special radial blades, heating the ends with a heat plate (after which the heat plate is removed), and finally fusing the ends together by application of pressure. Fusion joining parameters are specified in ASTM F-2620. For pressure applications, the internal bead has a negligible effect on flow, and turbulence from internal bead is similar to what occurs in a bell and spigot connection. The C-Factor used for fluid flow calculations was derived using empirical lab data with the internal bead in place. In gravity applications, some engineers believe that the turbulent flow created by the bead is helpful in keeping solids suspended in the flow; others feel it is detrimental to the flow. Though seldom required, internal and external beads (either or both) can be removed prior to installation without any effect on joint integrity.

Other types of joints in HDPE pipe include electrofusion and bell-and-spigot gasket joints used in some nonpressure profile-wall pipes.

5.8.3.5 Advantages and Limitations

The continuous "jointless" conduit that results from the butt-fusion of HDPE pipe makes it an ideal piping material for pull-in installations such as HDD and in-line replacement/bursting. The ability of HDPE pipe to withstand both the internal and external corrosion is advantageous for both water and sewer systems. Its high flexibility is a favorable characteristic during trenchless installations; there is no need for very long entry pits. A very low internal resistance to flow makes it a good material for both pressure and nonpressure gravity systems. In pressure-water systems, the expansive forces of freezing water do not cause the pipe to crack. In very cold temperatures, HDPE has a high resistance to failure by impact. The pipe also resists shatter-type or rapid crack-propagation failure, and the material is considered to be self-arresting by many.

In the early years of manufacturing, environmental stress cracking was a concern for HDPE pipe. In the past 25 years, higher quality resins have successfully overcome this problem in HDPE pressure pipes. The butt-fusion of HDPE necessitates the use of a skilled labor force. The fused joint is allowed to cool for an additional thirty minutes after removal from the fusion machine prior to installation. In the case of thick wall pipes (>2 in.) extra cooling may be required. The high sensitivity of HDPE to temperature differentials requires special design considerations for above grade pipelines. (Transitions from HDPE pressure pipe to gasket-joint pressure pipe may require a thrust anchor.) However, buried force-main lines require no special installation for thermal affects. Buried or sliplined gravity flow pipes generally require a wall anchor where they terminate in a manhole. At the same pressure rating, HDPE is a thicker-walled pipe when compared to many other pipe materials. The thicker wall effectively reduces the flow area within the pipe. Table 5.16 presents the advantages and limitations of HDPE.

Advantages	Limitations
Resistance to both internal and external corrosion. Low internal friction. Smooth interior.	Older PE materials are subject to environmental stress cracking due to improper embedment or excessive local bending. Newer materials have enhanced resistance.
Butt-fused joints effectively create a continuous "jointless" conduit.	Trained labor and special equipment required for butt-fusion.
Abrasion resistant. Use to convey sand and fly ash slurry.	Slightly smaller inside diameter than other pipes of the same outside diameter size.
High ductility and flexibility. Lightweight in smaller diameters. Typical minimum bend radius of 25 to 30 times pipe diameter.	Cannot be located with standard magnetic locating equipment unless buried with metallic wire or tape.
Excellent resistance to fatigue, water hammer, and repetitive surge pressures.	Sensitive to temperature differentials, resulting in significant expansion and contraction unless restrained by soil embedment even after burial.
May be repaired using mechanical couplings and saddles.	Lighter color products cannot be subjected to prolonged UV exposure such as unprotected storage, and installations must be buried.
Superior low temperature performance including high resistance to failure by impact, and does not easily crack under expansive forces of freezing water.	
Resists shatter-type or rapid crack-propagation failure.	

Table 5.16 Advantages and Limitations of HDPE

5.8.4 Glass-Reinforced Pipe

The third type of plastic pipe is GRP. Unlike PVC and HDPE, GRP is made of a thermoset material. The GRP was first manufactured in the United States in the 1950s as an alternative to corrosion-prone concrete and steel materials. Using a patented centrifugally cast manufacturing process; Perrault Fibercast Corporation of Oklahoma manufactured the first fiberglass reinforced polyester resin pipe. Traditionally used in industrial applications throughout the world, GRP is rapidly gaining market shares in North America for large-diameter municipal water and sewer applications. As a plastic material, GRP is designed as a flexible conduit, and is installed accordingly.

5.8.4.1 Manufacturing

Fiberglass composites are made of glass fiber reinforcements, thermosetting resins, and other additives, such as small aggregates, catalysts, hardeners, accelerators, and so on. Types of resin used include epoxy, polyester, and vinyl ester. The amount and orientation of the glass fibers increase the mechanical strength of the pipe. GRP is manufactured using one of the two methods:

1. *Centrifugal Casting*. This is the most widely used method of manufacture of GRP for municipal gravity sewer applications in North America. In this process, glass fiber reinforcements, resins, and aggregates are placed in a rotating steel mold. The centrifugal action of the mold removes air from the composite, resulting in a dense laminate free of voids. Material properties can be altered by varying aggregate content, resin type, and cure and reinforcement type, quantity, or orientation. The pipe is removed from the mold after heat curing.
2. *Filament Winding*. There are two distinct processes for filament winding—continuous and discontinuous. Discontinuous filament winding involves impregnating several glass reinforcing strands with a matrix resin and then the application of the wetted fibers to a mandrel under controlled tension in a predetermined pattern. Repeat of this process results in the desired wall thickness. In the discontinuous process, a standard length of mandrel is rotated, resin, glass, and aggregate are added to produce pipe that is helically reinforced. The process allows for the formation of a continuous bell-end, monolithic with the pipe wall. In the continuous process, an advancing mandrel causes the pipe to form. Continuous and chopped glass fiber rovings, resin, and aggregate are added to make the pipe. The pipe is then cured and cut into desired lengths.

5.8.4.2 Applicable Standards

There are currently three widely-used pressure and nonpressure piping standards for trenchless fiberglass pipe applications. Standard pressure

Pressure Standards	Nonpressure Standards	Available Diameters
–	ASTM D-3262	8–144 in.
ASTM D-3517	–	8–144 in.
AWWA C-950	–	1–156 in.

Table 5.17 GRP Standards for Trenchless Construction

classes for AWWA C-950 include 50, 100, 150, 200, 250, and over 250 psi. From the available pipe diameters, it can clearly be seen that GRP offers an alternative to other large-diameter traditional piping materials (see Table 5.17).

5.8.4.3 Joints

There are a variety of joints that have been developed over the years for GRP. The *AWWA M45* Fiberglass Design Manual (3rd Edition, 2014) lists many types of fiberglass pipe joints including the following types: coupling or bell-and-spigot joints, mechanically coupled joints, restrained coupling or bell-and-spigot joints, butt and wrap joints, bell-and-spigot with laminated overlay, bell-and-spigot adhesive joint, and flanged joints. Typically these joint systems are used for new pipeline construction in direct bury or aboveground installations. For renewal of existing pipeline systems, other flush profile joints are available to allow for capacity retention of the existing system.

Flowtite and Hobas are the leading manufacturers of municipal GRP pipe in the North America and have variations of the above-described joints. For new trenchless construction, pipe jacking in particular, they offer two separate types of joints: flush bell-spigot joint for gravity flow pipe jacking, and the flush filament-wound collar (FWC) coupling for pressure pipe jacking. A similar type of joint exists for use in the renewal of gravity pipe by sliplining.

In the flush bell-and-spigot joint type, as shown in Fig. 5.32, the sleeve is fitted to the pipe end, which has been machined down so that the joint outside the diameter is same as the pipe itself. An elastomeric

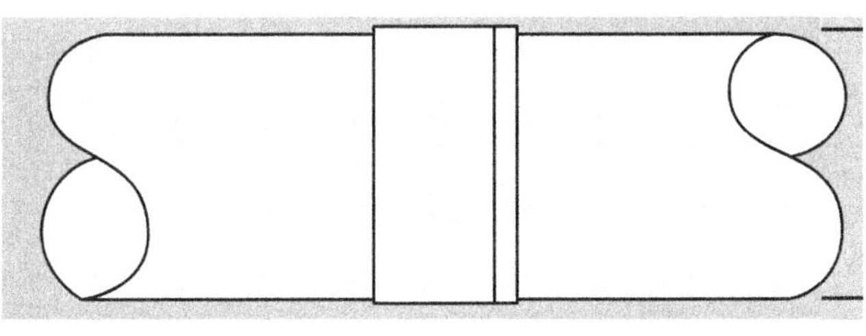

Figure 5.32 GRP flush bell-and-spigot joint for trenchless installation. (*Source:* Hobas Pipe USA.)

gasket, contained in a groove on the spigot end of the pipe, effectively seals the joint. The flush FWC coupling is a modified version of the FWC coupling [a filament-wound sleeve with an ethylene propylene diene monomer (EPDM) elastomeric membrane, the coupling is bonded to one end of the pipe during manufacturing].

5.8.4.4 Advantages and Limitations

Table 5.18 outlines some advantages and limitations of GRP. Excellent internal and external corrosion resistance in natural soils and corrosive wastewater and industrial applications has given GRP an advantage over traditional piping materials in large diameters. It also displays better abrasion resistance than cement-based pipes. It is significantly lighter than a concrete pipe of the same diameter. For pressure applications, a variety of available pressure classes makes it widely desirable to engineers and contractors.

GRP has high corrosion resistance and widely used in sanitary sewer applications where environment is highly corrosive to cementitious and ferrous materials. In order to GRP pipes to perform in these corrosive environments the long-term deflection of the pipe must be limited to 5%. For direct bury applications, this can be overcome during design by ensuring that the stresses are kept within a certain limit by analyzing the bedding and native soil conditions on the project. For trenchless sewer applications, the long-term deflection should also be review as part of the project design.

As with any pipe material, it is important not to damage the pipe during installation and renewal/replacement process. GRP pipes can be damaged due to severe impact forces. Although impact damage can occur to GRP pipe, it is also possible to repair this type of damage.

<table>
<tr><th>Advantages</th><th>Limitations</th></tr>
<tr><td>Good internal and external corrosion resistance in corrosive soils.</td><td>Possible to damage with severe impact.</td></tr>
<tr><td>Better abrasion resistance than cement-based pipes.</td><td rowspan="6">Resin selection is important for some chemicals.</td></tr>
<tr><td>Lightweight compared to alternative materials.</td></tr>
<tr><td>Various pressure classes available for pressure pipe.</td></tr>
<tr><td>Excellent hydraulics due to smooth interior.</td></tr>
<tr><td>Proven gasketed joint sealing systems for various installation and renewal/replacement methods.</td></tr>
<tr><td>Dimensionally stable for fluid service and weather exposure.</td></tr>
</table>

Table 5.18 GRP Advantages and Limitations

5.9 Metallic Pipes

Metallic pipes in present-day use include ductile iron pipe and steel pipe. The precursor to ductile iron was cast iron pipe. Metallic pipes have traditionally been used in pressure applications, but some areas of North America use them for gravity sewer applications also. Most metallic pipes used in the past were installed by open-cut methods. Recent acceptance of trenchless technologies has led to the manufacture of metallic pipes with joints suited for pull-in or jacking installations. Steel pipe with welded joints is a product of choice for large-diameter HDD applications (usually more than 22-in. diameter), horizontal auger boring (HAB), and pipe ramming (PR) projects. See specific chapters in this book for more information on these methods.

5.9.1 Ductile Iron Pipe

Ductile iron pipe was developed in the 1940s from gray cast iron by distributing the graphite into a spherical form instead of a flake form. This was achieved by the addition of inoculants such as magnesium to molten iron. It resulted in the ductile nature of the new pipe in addition to higher strength, impact resistance, and other improved properties. The commercial production of ductile iron pipe begun in 1955, and by the 1970s, it had almost completely replaced cast iron pipe in municipal applications. Since 1980, gray cast iron pipe for pressure pipe applications (not soil pipe) has not been produced in the United States.

The main use of ductile iron pipe is in water systems. However, the availability of thermoplastic and thermoset pipes and newer substitute materials represents some significant competition to ductile iron since the late 1960s. The newer materials are lighter (where this might be judged an advantage), in some cases less expensive, and perform many of the required functions within a distribution system. Above all, plastic pipes are inherently suited to withstand most forms of both internal and external corrosion that can be a concern to unprotected ductile iron pipe in some service environments.

However, corrosion problems for ductile iron pipe can be overcome by the use of appropriate corrosion protection. The protection methods include cement mortar or polymer linings for internal corrosion protection, maintenance of flow properties, and PE encasement per ANSI/AWWA C-105/A-21.5, or in uniquely aggressive environments cathodic protection, for external corrosion. A newer approach to iron pipe corrosion control, called the "Design Decision Model" (DDM), has been developed by the Ductile Iron Pipe Research Association (DIPRA). This method tailors the chosen corrosion protection method of the pipeline based on several factors, including substantial history, the corrosion potential of the specific environment involved, and the critical nature/priority of the specific pipeline involved.

In recent years, various manufacturers have developed restrained joints for ductile iron pipe; therefore, the industry is able to offer its products to the trenchless construction industry. Today, ductile iron pipe can be used for HDD, PB, casing and carrier pipe installation, and sliplining. The use of ductile iron pipe in sewer systems is less in comparison to its use in potable water distribution and transmission. Nevertheless, some municipal agencies use ductile iron pipe for their gravity flow applications. Flexible, restrained joint ductile iron pipes are particularly attractive for some HDD installations. In HDD applications, joints can rapidly be assembled in field conditions with minimum equipment or highly skilled technicians. The pipe also can be installed with limited access, and need not be bent (nor result in high bending stress/strain) when installed in a normal curved drillpath.

Ductile iron pipes are flexible conduits and are designed and installed accordingly. However, when internal cement-mortar lining is specified, the vertical deflection of the pipe is limited to 3 percent instead of 5 percent. Ductile iron pipes with flexible linings, such as proprietary ceramic epoxy and fusion bonded epoxy primer/heat-fused PE topcoat linings, are limited by ASTM A-716 and A-746 standards to 5% deflection.

5.9.1.1 Cast Iron Pipe

It is important to have an understanding of the predecessor of ductile iron pipe, the cast iron pipe. Cast iron pipe has been used throughout the world for many years in portable water systems. In the United States, cast iron pipe was introduced in the early 1800s. Also referred to as gray cast iron, it accounts for a very large portion of buried water piping material throughout North America even today. Cast iron is a very strong, but brittle material. Early unlined installations of cast iron pipe has lasted more than a century in some cases on account of the sheer thickness of the walls of the pipes. The thick walls played a sacrificial role as they slowly corroded over the years. Internal tuberculation in unlined cast iron pipe has caused severe hindrance to flow in most cases. Figure 5.33 presents a sample of internal tuberculation in unlined cast iron potable water pipe.

5.9.1.2 Manufacturing

Ductile iron pipe is centrifugally cast in metal molds precisely machined with a very slight taper from bell to spigot end. A head core which forms the bell and the anchoring configuration required for the sealing gasket and restrained joint mating surfaces is inserted into the bell end of the mold. These molds, with head core in place, are rotated while the molten iron is deposited at a controlled rate producing the required wall thickness. The casting machine rotates the metal mold at a speed sufficient to generate high gravity forces. The pipe begins to cool as high volumes of cooling water contained within the

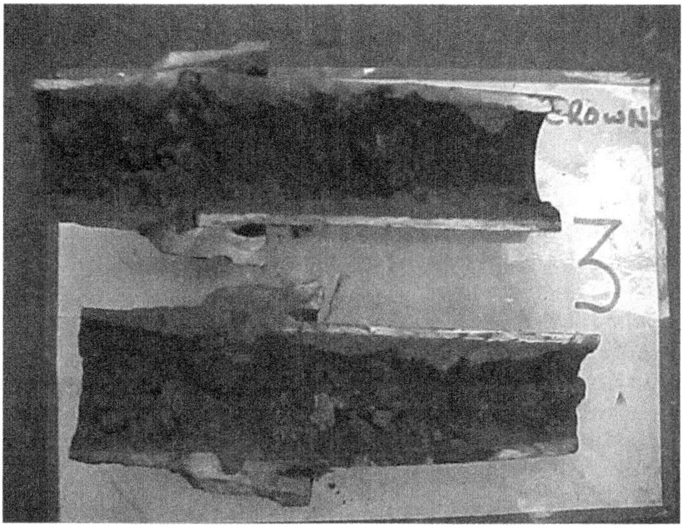

FIGURE 5.33 A sample of internal tuberculation in unlined cast iron potable water pipe. (*Source:* 3M Corrosion Protection Products.)

casting machine quenches the exterior of the spinning mold. The pipe then proceeds to the annealing ovens where it is reheated to a preset temperature, then cooled. This process reduces the carbides in the metal matrix. Carbides, when present form a very hard surface which will impact the elongation. Each pipe section is then hydrostatically tested to a minimum of 500 psi to prove the pipe's ductility. Finally, the lining and coating per the customers' requirements are added in separate inline manufacturing operations.

5.9.1.3 Applicable Standards

Table 5.19 lists a number of standards available for trenchless applications. It should be noted that these pipes can also be installed via open-cut methods. For trenchless pull-in installations, the joints must be restrained, as discussed in the following section. In recent years, ductile iron pipes with restrained joints have been used in trenchless construction methods (TCMs) as well as for trenchless renewal methods (TRMs).

Pressure Standards	Nonpressure Standards	Available Diameters
–	ASTM A-716	4–64 in.
–	ASTM A-746	
AWWA C-150/C-151	AWWA C-150/C-151	

TABLE 5.19 Applicable Ductile Iron Standards for Trenchless Technology

5.9.1.4 Joints

While restrained joints in ductile iron pipe have been available traditionally to accommodate the thrust forces acting on a pipeline, pipes with modern flexible restrained joints have been used in recent decades for various types of trenchless projects, both new constructions as well as renewals. Pipe bursting and HDD have been the most common applications of the flexible restrained joint ductile iron pipe. The joints are capable of withstanding internal pressure while providing effective joint restraint and joint flexibility up to 5 degrees. This combination of joint features provides the benefit of redistributing the tensile forces encountered during pull-in, virtually eliminating flexural tension due to bending. Pulling flexible restrained joints through a bore path is similar to pulling a chain that hinges every 18' or 20', as compared to pulling a solid rod or tube that must bend to negotiate the bore path.

Joints used for directional drilling or pipe bursting need to be boltless, flexible restrained with smooth contoured bells. Joints with bulky glands or flanges that may prevent the smooth flow of the drilling fluid/soil slurry over the joint should not be used. Since there are multiple proprietary designs for flexible restrained joints by various manufacturers, users should research the contoured bell designs and match the appropriate bell contour to the geology that the pipe will experience. Unfortunately, in a number of cases using HDD the selection of a pipe joint that had an abrupt change in the bell contour led to the pipe locking up during pipe pull back because the angular shape change in the bell contour snagged on a rock fragment. Figure 5.34 shows five proprietary restrained joints used for trenchless applications.

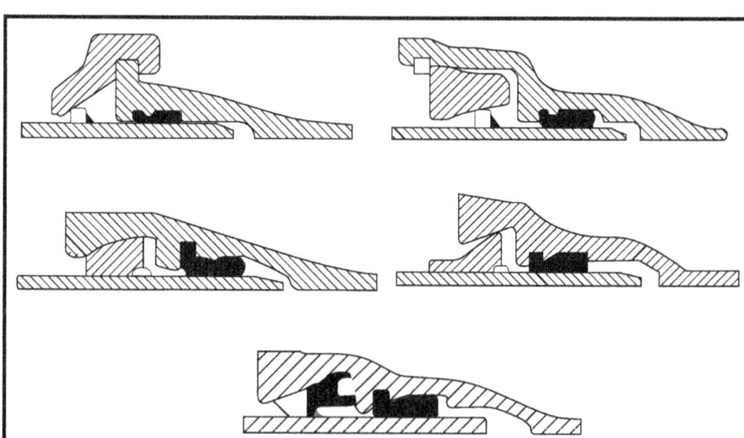

FIGURE 5.34 Proprietary ductile iron restrained joints used in trenchless construction.

Advantages	Limitations
Wide variety of internal and external corrosion protection systems is available.	Highly susceptible to corrosion, both internally and externally, unless protected.
Internal cement mortar lining prevents tuberculation and enhances hydraulic capability.	Not all available corrosion protection methods are effective.
Strong material, with high-load bearing strength, impact strength, and beam strength.	Internal cement mortar lining is easily damaged if struck with a backhoe.
Wide variety of joints enables various applications, including trenchless.	Cathodic protection may be cost prohibitive and is rarely used in municipal systems.
Available for both pressure and gravity applications.	Polyethylene encasement is damaged in improper installation.
Wide range of diameters and pressure classes available.	Heavyweight, resulting in high cost of labor.
Long laying lengths reduce joints in the system.	Lack of flexibility may make trenchless installations difficult.
Pipe is resistant to chemical permeation in contaminated areas.	Gaskets in the joints are vulnerable to chemical attack in contaminated soils.

Table 5.20 Advantages and Limitations of Ductile Iron Pipe

5.9.1.5 Advantages and Limitations

Table 5.20 summarizes the advantages and limitations of ductile iron pipe. The high-load bearing capacity, high impact strength, and high beam strength make it a sturdy piping material for pressure applications. There are many different types of joints, including restrained joints, which are used in trenchless applications. The long lengths of ductile iron pipe (20 ft) minimize the number of joints within a water or sewer system. The wide availability of various pressure classes and diameters makes it easily desirable to engineers and specifiers. The pipe itself is highly resistant to chemical permeation. Also, there are various internal and external corrosion protection systems available.

Ductile iron pipes with standard SBR gaskets are circa 99.9% impermeable to absorption and hydrocarbon or volatile hydrocarbon permeation, with the addition of special material gaskets make the system 100% impermeable, further minimize any concern of chemical attack or permeation into or out of piping systems. In soils that are uniquely aggressive to ductile iron pipe, external corrosion needs to be addressed as discussed in 5.10.1. The internal cement mortar lining prevents the deflection of the pipe beyond 3 percent, even though ductile iron is a flexible conduit.

5.9.2 Steel Pipe

Steel pipes, made from a versatile refinement of iron, have seen a wide range of usage for more than a century and a half. The development of high-strength steel pipes has made it possible to transport fluids such as natural gas, crude oil, and petroleum products over long distances. Initially, steel pipes had to be threaded together or riveted, which was difficult for large pipes, and they often leaked under high pressure. The development of electric arc welding machines in the 1920s made it possible to construct leakproof, high-pressure, large-diameter pipelines. One of the earliest steel water pipe installations in the United States, still in service today, was in San Francisco in 1863. Developments in technology have given way to riveted steel pipes evolving to the automatically welded steel pipes of today. Various other developments have resulted in the creation of different types of joints as well as effective mechanisms for prevention of corrosion, making steel more versatile for trenchless and open-trench applications.

In municipalities, steel pressure pipes are used today in large-diameter potable water transmission applications. In trenchless industry, steel pipes are used as in sliplining microtunneling, pipe jacking, tunneling, and piperamming, because of their high stiffness and compressive strengths.

5.9.2.1 Corrugated Steel Pipe

Corrugated steel pipes have been used for more than a century in gravity applications such as drainage and storm sewers. Though corrugated steel pipes have been used in some sanitary sewers, this is not the case today. Due to their relatively low compressive and tensile strengths, corrugated steel pipes are not used in trenchless or pressure applications. Therefore, a detailed discussion of corrugated steel piping products has been omitted in this chapter. The American Iron and Steel Institute's "Modern Sewer Design" is an excellent source for information on corrugated steel piping systems.

5.9.2.2 Manufacturing

Steel pipes used in municipal applications are manufactured by an automatic welding process. There are generally three types of steel pipe, each identified by the way in which it is manufactured:

1. *Rolled and Welded Pipe.* This is one of the oldest methods of steel pipe production, where plates of steel are rolled into cylindrical pipes, usually 6 to 12 ft in length, then welded in the circumferential and longitudinal directions. The pipes used in casing applications for trenchless technology are of this type. They are also used in other types of applications.
2. *Electric Resistance Welded (ERW) Pipe.* The ERW pipes are generally manufactured and used in smaller diameters up to 24 in.

Electric resistance welded is a single straight seam welding process where continuous coils of treated low carbon steel, called skelp, are shaped into cylindrical pipes by edge-forming, and then welded at the seam. These pipes can be manufactured in lengths of up to 100 ft. They are used in water systems as well as other industrial applications.

3. *Spiral Welded Pipe.* Starting with continuous rolls of steel similar to the type used for ERW pipe, the steel is fed into a machine and spirally wrapped against buttress rolls to form the pipe. The edges of the spiral pipe are then welded in and out by a double-submerged arc process. Spiral welded steel pipes are used in municipal water transmission applications in diameters of up to 156 in. Trenchless construction, such as HDD, have been used to install this type of pipe in the potable water systems.

5.9.2.3 Applicable Standards

ASTM A-139 Grade B is a common type of casing pipe used in the gravity trenchless construction. A pipe manufactured to the ASTM A-139 standard requires hydrostatic testing because it is often also used for medium internal pressure applications. As the use of the pipe for casing depends only on its structural capabilities, the hydrostatic test is not required. To exclude the hydrostatic test from the original standard, the casing pipe standard is referred to as ASTM A-139, Grade B (no hydro). The pressure pipe standard AWWA C-200 is used in large-diameter HDD projects for water transmission; in past years, some challenging projects in Texas, Washington, and Hawaii have used this steel pipe standard. Permalok™ steel pipe is manufactured as a proprietary mechanically interlocking joint type product for casing pipe and pressure trenchless applications. The ASTM standards to which the pipe must be compliant are listed in Table 5.21.

5.9.2.4 Joints

There are several types of welded joints available for steel pipe, each suited for a specific application.

Pressure Standards	Nonpressure Standards	Available Diameters
AWWA C-200 and ASTM A-139	ASTM A-139 Grade B (no hydro)	3–144 in. and higher
ASTM A-36, ASTM A-515, Grade 60, ASTM A-516 Grade 70 or ASTM A-572, and Grade 42 with Permalok™ T7 type joint	ASTM A-36, ASTM A-515, Grade 60, ASTM A516 Grade 70, or ASTM A-572, and Grade 42 with Permalok™ T5 type joint	

TABLE 5.21 Applicable Standards for Steel Pipe for Trenchless Construction

- Bell-and-spigot lap welded joints
- Butt-welded joints (single-V butt-welded, and double-V butt-welded)
- Butt strap welded joints
- Mitered lap-welded joints

There are also a number of non-welded joints.

- Mechanically interlocking joints
- Bell-and-spigot rubber gasket joints
- Harness joints
- Carnegie shape rubber gasket joints
- Mechanical couplings
- Split-sleeve mechanical coupling

For steel casing pipe used in boring and pipe jacking applications, it is important that there are no irregularities in alignment at the joints. If the casing is not straight, the ability of the contractor to keep the pipe in line and on grade is affected. Lap welded joint products are not recommended for jacking and boring installations. Though spiral welded pipes can be used for casing, it is a common practice to use a straight seam or seamless pipe. For jointing, both ends of the pipe can be beveled for welding, as in the single-V butt-welded joint (see Fig. 5.35). Another recommendation for achieving a good circumferential weld at the joint is to bevel one end of the casing pipe to a standard 37 degree bevel, and square cut the other end.

For casing and pressure pipe trenchless applications, a patented product line, Permalok™, offers two mechanical push-on joints. It is particularly attractive to the trenchless excavation industry because the joints are designed to be flush with the interior and exterior surfaces of the pipe (see Fig. 5.36). This joint type eliminates the time required for welding traditional-type steel pipes. Although using the Permalok™ pipe can be more than 2.5 times faster than welding, the pipe typically will cost more. Permalok™ joints are used for microtunneling (MT), HDD, PR, and HAB. Joint quality is further enhanced because the Permalok™ connector is consistently round,

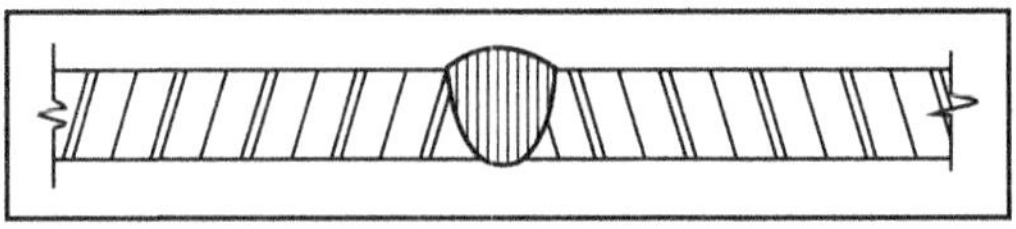

Figure 5.35 Single-V butt-welded joint.

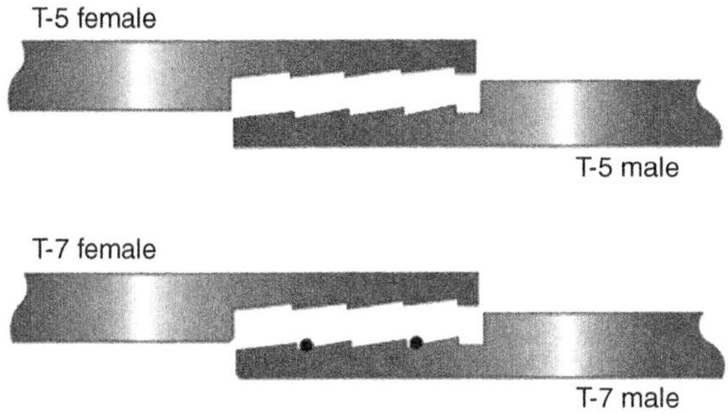

Figure 5.36 Interlocking jointing system. (*Source:* Permalok™.)

true, and perpendicular to the pipe axis. Its unique machined groove makes stabbing and aligning easy and quick. Also, the Permalok™ design, combined with the use of a sealant such as, room temperature vulcanization (RTV) Silicone, prevents leakage. Standard specifications, listed in Table 5.21, ensure that all pipe manufactured by Permalok™ meet guidelines. The T5 profile was patented for steel pipe casing applications and has been used in numerous trenchless construction projects nationwide. The T7 profile has been tested at pressures of up to 300 psi, and is used in trenchless pressure pipe applications. Figure 5.36 illustrates the Permalok™ interlocking jointing system.

5.9.2.5 Advantages and Limitations

Table 5.22 lists some of the advantages and limitations of steel pipe. With its high tensile strength, steel pipes are capable of handling high pressures in water transmission applications. The high compressive strength of steel pipe makes it a good material for jacking and boring. Assembly of non-weld joints is relatively easy owing to push-on bell-and-spigot gasket joints. Steel pipe has been known to adopt well in locations where ground movements occur. Good hydraulic properties of internally lined steel pipe make it ideal for use in municipal water systems. The availability of various methods and standards for both internal and external corrosion significantly prolongs the useful design life of the pipe.

Though various corrosion protection mechanisms are available, steel pipe is prone to corrosion attacks unless the protections are used. In large diameters, the pipe's ability to handle external pressure can be significantly expanded through the use of good backfill and embedment practices. For more information, refer to AWWA M11 "Steel Pipe—A Guide for Design and Installation," ASCE MOP No. 119 "Buried Flexible Steel Pipe Design and Structural Analysis," and AISI—SPFA "Welded Steel Pipe Design Manual."

Advantages	Limitations
Various standards and methods are available for internal and external corrosion protection.	When uncoated, prone to internal tuberculation and external corrosion, subject to electrolysis.
High tensile strength.	Use of internal and external corrosion protection raises price of the product.
High compressive strength.	Air-vacuum valves are necessary in large-diameter lines.
Easy to assemble, non-weld joints available.	
Adopts well to locations where soil movements occur.	Welding of joints requires skilled labor and is time consuming.
Good hydraulic properties when internally lined.	Special care required to ensure proper alignment at joint in welded pipe.
Resistance to external pressures in large-diameter sizes increases with good embedment efforts.	

Table 5.22 Advantages and Limitations of Steel Pipe

5.10 Corrosion Protection: General

There have been many approaches to corrosion protection of all kinds of pipes through the centuries. There are decades of practical experience with many applications and corrosion protection systems of iron and steel pipes, in particular. There are field and laboratory studies of gray and ductile iron pipes in widely divergent soil types and also in some notably very corrosive actual soil burial test sites, by DIPRA working in many cases in close conjunction with the utilities involved. Both unprotected iron and steel pipes will rapidly corrode in some soil environments, and in these environments suitable corrosion protection must be provided. It is also being discovered (in more recent applications of pipes that have not been around as long) that other piping materials, such as variously reinforced concrete, plastics, and composites, as well can also undergo forms of corrosion or environmental/stress related deterioration that are perhaps not now quite as obvious or known to many pipeline practitioners.

5.10.1 Corrosion Protection: Ductile Iron Pipe

The Ductile Iron Pipe Association (DIPRA) suggests that the majority of soils in the United States are not corrosive to ductile iron pipe. A 21st century approach to iron pipe corrosion control, called the "Design Decision Model" (or "DDM"), has been developed by DIPRA working in conjunction with the world's largest independent provider of corrosion protection services and equipment, as well as a noted statistician. This risk based model tailors the external corrosion protection method based on several factors including the corrosion potential of the specific environment, and the consequences of failure measured by the critical nature/priority of the specific pipeline involved. Based on this assessment of risk, the DDM suggests different methods for protection of ductile iron pipe that include:

1. Installing the pipe as-manufactured with its protective standard shop coating/annealing oxide system.
2. Encasing the pipe in polyethylene.
3. Encasing the pipe in polyethylene or encasing the pipe and providing bonded joints.
4. Encasing the pipe in polyethylene and providing bonded joints or providing life-extension cathodic protection currents, with or without encasement.
5. Cathodic protection.

Since 2015, domestic ductile iron pipe producers have also offered zinc coating for the exterior of ductile iron pipe. This coating has been an industry standard for improving the sustainability of ductile iron pipe manufactured outside the United States for more than 50 years.

Another innovation in corrosion control for ductile iron pipe that will add significantly to the sustainability of ductile iron pipe is a new polyethylene encasement, called V-Bio, an enhanced polyethylene encasement that targets anaerobic bacteria on the surface of the pipe and inhibits the formation of corrosion cells under the wrap. The availability of zinc coating in combination with V-Bio, in concert with the provisions outlined by the DDM for uniquely severe soil environments, are keys to long term sustainability. The addition of zinc and V-Bio should be considered for many trenchless installations of ductile iron pipe, particularly HDD where the depth of installation and the accessibility to the installed pipeline, would be difficult.

Developed in 1922, cement mortar lining has effectively eliminated internal corrosion in virtually all modern water pipelines that have been in service for many decades. Corrosion characterized by the formation of tubercles that reduce flow is common in older unlined systems, where many cast iron water mains nevertheless remain in service for over 100 years. Figure 5.33 presents a sample of internal tuberculation in unlined cast iron pipe. Many of these older lines have been rehabilitated by cleaning the lines and then applying a cement mortar lining or a NSF approved polymer lining. Cement lined ductile iron pipe offers a flow coefficient of 140.

Ductile iron pipe is also used for sanitary sewer applications, and primarily sewer force mains, utilizing standard cement mortar linings or special ceramic epoxy linings. In designing cement mortar lined sewer force mains in particular, it is important that the line remain full, particularly at high points in the pipeline, otherwise internal corrosion can quickly compromise the sustainability of ductile iron pipe. The potential for internal corrosion resulting from acid producing anaerobic bacteria must be protected by proper system design, maintenance through chemical treatment of effluent properties, and proper maintenance of air release valve for existing force mains. For new ductile iron pipe force mains, the addition of a polymer lining either just the high point and for some distance down slope in the pipeline, or for the entire force main, is not uncommon.

5.10.2 Corrosion Protection: Steel Pipe

There are a number of standards and processes, such as, AWWA C205, C222, and C214, that are the most common for both internal and external corrosion protection of steel pipe, including cement mortar lining, paints and polyurethane coatings and linings, tape coatings, coal tar enamel coatings, cement mortar coatings, and epoxy and polyurethane coatings and linings. Furthermore, cathodic protection is sometimes mandated by specifiers as an effective method of protection to extend the life of the steel pipe.

5.10.3 Corrosion of Concrete, Composite, and Plastic Pipes

All common piping materials can be deteriorated or corroded and designers should consult with manufacturers to learn pipe product limitations. It should be noted that sulfates and chlorides as well as low resistivity soils, in general, can corrode various concrete pipes, and gasoline contamination has been known to result in swelling and bursting of some buried plastic pipes, whereas such contamination will not burst metallic pipes.

The coverage of the topic of corrosion in this chapter is intended to make the reader aware of one of the significant problems being faced by pipeline industry today. These problems include corroding, deteriorating, and/or failing water and sewer lines. There are currently huge amounts of pipeline system underground, some of which are without effective corrosion protection or other provisions for adverse environmental conditions. Pipe leakage and failure in these unprotected pipelines can significantly influence rate of unaccounted-for-water and I/I in future. It is essential for engineers to be aware of the problems and available engineering solutions as they design and renew the pipeline systems of the future. Proper corrosion protection will ensure that the full potential of all piping materials for use in specific applications is realized in the long run.

5.11 Chapter Summary

Selection of pipe materials is an essential element in project planning, design, construction, life cycle cost analysis, asset management, and eventual renewal or replacement of the pipeline. The choice of a specific pipe has a major influence on the efficiency and longevity and is one of the most important elements for pipe planning. Long life-cycle pipes build asset value and provide economical service, minimize operation and maintenance costs and reduce impacts to consumers, environment, and the public.

5.12 Review Questions

1. Why is pipe material selection so important? Explain.
2. Describe rigid, semirigid, and flexible pipes.
3. How does the design process differ for rigid and flexible pipe?
4. Compare thermoset with thermoplastic materials.
5. Describe the advantages and the limitations of concrete pipe.
6. Describe the advantages and the limitations of clay pipe.
7. Describe the advantages and the limitations of PE pipe.
8. Describe the advantages and the limitations of PVC pipe.
9. Describe the advantages and the limitations of GRP pipe.
10. Describe the advantages and the limitations of ductile iron pipe.
11. Describe the advantages and the limitations of steel pipe.
12. Describe the steel pipe's mechanical jointing system.

CHAPTER 6

Cured-in-Place Pipe

6.1 Introduction

As a general note, this chapter uses the following definitions as per NASSCO guidelines:

1. Tube: Dry tube prior to wet out.
2. Liner: Combination of tube & resin. A wet out tube.
3. CIPP: The finished product/pipe.

Cured-in-place pipe (CIPP) installation is one of the most widely used methods of trenchless pipeline renewal for both structural and nonstructural purposes. The CIPP process involves a liquid thermoset resin-saturated material that is inserted into the existing pipeline by hydrostatic or air inversion, or by mechanically pulling-in with a winch and inflating. The material is typically heat-cured in place resulting in a CIPP product. Insituform introduced CIPP in the United Kingdom in 1971 and entered the US market with the first installation in 1976. In the 1980s, Ashimori Industries developed a method called hose lining basically for pressure pipe applications in the Japan gas industries. This increased number of available methods and provided many technical and economical benefits for utility owners through increased competition. Chapter 3 presented an overview of the CIPP design process. This chapter presents an overview of the CIPP method.

6.2 Site Compatibility and Applications

The CIPP can be used effectively for a wide range of applications that include sewers, gas pipelines, potable water pipelines, chemical and industrial pipelines, and pressure pipes. The physical properties of CIPP make it especially suitable for different types of pipe geometries, including straight pipes, pipes with bends, pipes with different cross-sectional geometries, pipes with varying cross-sections, pipes with lateral connections, and deformed and misaligned pipelines. However, several factors must be evaluated before choosing

CIPP for a project. Factors such as surface availability, congestion and type of use (residential, industrial, etc.), chemical composition of the fluid carried by the pipeline, number of laterals, number of manholes or access points, installation distance, renewal objectives, diameter, shape, ovality, previous point repairs, alignment, joint conditions, types of defects and structural capabilities of the existing pipe, height of watertable, and soil conditions around the existing pipe must be assessed before making a choice on the renewal system. The CIPP is also used for localized repairs in a wide range of systems. This is discussed in more detail in Chap. 14, where several localized repair techniques are described.

The possibility of negotiating bends depends on the installation and curing process of the various systems. Liners that are inverted during insertion can usually negotiate bends up to 90°. There are, however, limitations to the degree of bending that pulled-in CIPP, such as ultraviolet (UV)-cured liners, can manage. Bends present problems for liners that are pulled into place, because the liner must remain on the underlying protective foil and must under no circumstance become twisted. It is important to be aware that vertical creases may occur in the liner when negotiating sharp bends. Table 6.1 presents an overview of CIPP applications and possible limitations.

6.3 Impacts of Existing Pipeline Defects on CIPP

While cured-in-place pipe can be installed in a variety of shapes and configurations, levels of pipeline defect may dictate where the CIPP may not be able to effectively renew the existing pipe or when additional remedial work may require before CIPP installation.

The NASSCO Pipeline Assessment Certification Program (PACP) defines, in detail, each defect category and is a program designed to train CCTV operators to readily identify these defects in a common descriptive language uniform for the entire U.S. and Canada. The condition data identified by the CCTV operator are then compiled and through the use of a rating system can be assigned a priority for renewal. This condition assessment will allow the designer to select the most applicable technology to accomplish the renewal. Table 6.2 presents different defect types, defect descriptions, and conditions where defects will impact CIPP installations.

6.4 Main Characteristics

The primary components of CIPP are a flexible fabric tube and a thermosetting resin system. For typical CIPP applications, the resin is the primary structural component of the system. These resins generally fall into one of the following generic groups, each of which has

Pipeline Type	Suitable	Comments
Sewers	Yes	
Gas pipelines	No	See note A
Potable water pipelines	Yes	
Chemical or industrial pipelines	Yes	See note A
Straight pipelines	Yes	
Pipelines with bends	Yes	See note B
Circular pipes	Yes	
Noncircular pipes	Yes	See note C
Pipelines with varying cross sections	Yes	See note D
Pipelines with service lateral connections	Yes	See note E
Pipelines with deformations	Yes	See note F
Pressure pipelines	Yes	

Note:

A. It is applicable to hose liners. As each manufacturer uses different types of resin, reinforcement, and the like, the chemical resistance of CIPP varies with manufacturer and service provider. However, all systems using felt impregnation with polyester or other resins are suitable for aggressive wastewater. If special characteristics are required, individual manufacturers should be contacted to help finding the most suitable product for specific requirements with regard to chemical content, concentration, temperature, and so on.

B. The possibility of negotiating bends depends on the installation and curing processes of the various systems. Liners that are inverted during insertion can usually negotiate bends up to 90° depending on the radius of the bend. However, the number of bends that can be negotiated is limited by the pipeline size and geometry.

C. All the described methods are suitable for pipelines with circular, egg-shaped, oval, horseshoe, or arch cross-sections. However, for noncircular pipelines, customized design methods, such as finite element modeling may become necessary.

D. Some proprietary systems allow CIPP techniques to be used in pipelines with varying cross-sectional area.

E. Laterals can be reinstated using a robotic cutter. It is presently possible to reopen laterals on pipelines with a diameter of 3 in. or more. In person-entry systems, laterals are renewed manually. With CIPP systems, watertight connectors can be fitted between laterals and main lines. See Chap. 14 for localized repair methods.

F. The CIPP can be used where deformities are minor. However, the general advice when ovality is more than 10 percent is to evaluate each situation individually, complete the wall thickness design using design techniques other than ASTM F-1216, and then decide whether to rehabilitate with CIPP or to dig and make repairs. Although a liner can get through the host pipe deformities, all the deformities will be noticeable after installation, and additionally, there is a high probability for wrinkling in the CIPP in deformed areas.

Table 6.1 Overview of CIPP Applications and Possible Limitations

Defect Type	Defect Description	Possible CIPP Impact
Structural Defects (see Fig. 6.1)		
Cracks	A crack, in the pipe is visible but the pipe is still in place with no distortion.	Cracks typically will not reflect through the CIPP. This pipe condition is an ideal time to renew the pipe. CIPP installation will result in a well-shaped, round pipe within a pipe.
Fractured pipe	A crack that has become visibly open although the sections of pipeline pipe wall are still in place and not moved or displaced.	Fractures typically will not reflect through the CIPP. This pipe condition is still an ideal time to renew the old pipeline, resulting in a well-shaped pipe within a pipe.
Broken pipe	Pipe where pieces are noticeably displaced and have moved from their original position.	The CIPP can be installed when the pipe is broken but the displaced pipe sections will reflect in the final installed product.
Pipes with holes	Hole refers to a pipe which visibly has an unintentional hole in the pipe wall. Service connections are intentional holes in the pipe. A hole can be where the pipe has broken away and is missing.	The CIPP can be installed with a hole in the pipe. The pipe configuration will reflect in the final installed product much like a service connection dimple. Stabilization of the surrounding soil structure should be evaluated.
Deformed pipe	The pipe is damaged to the point that the original cross-section of the pipeline in noticeably altered.	The CIPP can be installed when the pipe is deformed but the deformation will reflect in the final installed product. If the crown of the pipe has developed a reverse curvature, the installed CIPP must be designed by a qualified engineer to determine the design requirements.
Collapsed pipe	A collapse is where deformation is so great there has been a complete loss of integrity of the pipeline with more than 40% of the cross-sectional area lost.	The CIPP should not be installed when the pipe collapse is greater than 40% of its cross-sectional area. A point excavation should be made to correct the collapsed section, and then the CIPP should be installed.
Joint problems	Joints can be offset, separated or angled.	The CIPP can be installed when the pipe is offset, separated or angled less than 40% of the cross-sectional area. The existing condition however, will reflect in the final installed product.

Pipe surface damage	Surface damage can appear in a number of ways. The surface can be rough, it can have aggregate missing or projecting, steel reinforcement visible and projecting and more.	Most surface damage will not prevent the proper installation of CIPP.
Missing pipe sections	A section of pipe is missing with only the dirt left in place.	Typically a preliner is installed first. Then the CIPP is installed inside the preliner. Grouting and stabilizing the surrounding soils should be evaluated.
Previous point repairs	An area of the existing pipeline system that was previously excavated and repaired with an undersized pipe section.	A CIPP liner can be installed but will significantly wrinkle through the undersized pipe section. The better alternative is to replace the undersized pipe with the correct size and then install the CIPP.
Brick pipe defects	Include missing bricks, missing mortar, and loss of crown key and dropped invert.	A CIPP liner can be installed but will take the shape of the old brick pipe. The alternative is to partially repair the existing pipe and restore its shape then installs the CIPP.
Operational and Maintenance Defects (see Fig. 6.2)		
Deposits & tuberculation	A multitude of different types of deposits may be found in an existing pipeline system.	All deposits should be removed from the existing pipe before installing the CIPP.
Tree roots	Roots are commonplace in various forms in a pipeline including fine roots, tap roots, medium roots and root balls.	All roots must be removed from the existing pipe before installing the CIPP.
Infiltration/ inflow (I & I)	Infiltration into a pipeline can range from a weeper, a dripper, a runner and a gusher.	CIPP can be installed effectively, with minor infiltration such as weeper, drippers and runners. Gushers should be sealed or a preliner used before installing the CIPP.
Obstructions in the pipe	Obstructions in the pipeline can vary from bricks and debris to pipes and cables installed through the pipeline.	All obstructions that will prevent the proper installation of the CIPP must be removed from the existing pipe.

(*Source:* NASSCO)

TABLE 6.2 Existing Pipe Defects and Possible CIPP Impacts

Defect Type	Defect Description	Possible CIPP Impact
Construction Features (see Fig. 6.3)		
Protruding service connections (hammer tap)	A protruding connection or hammer tap is a service connection installed, by a plumber into the main pipeline, usually by breaking a hole in the mainline and inserting the lateral pipe.	Protruding service connections should be cut back to within "-1/2" of the pipeline pipe wall before installing the CIPP. Lining over a long protruding connection may be possible, however, this will create a significant protrusion in the pipeline pipe which may cause a blockage.
Intruding joint seal	Joint seal material between two pipe sections has become loose and is protruding into the pipeline pipe.	All significantly intruding joint seals should be removed, from the existing pipe, before installing the CIPP.
Pipe shape change	This occurs when a pipeline changes shape either from round to oval or from one size to another without the existence of a manhole at the change location.	Pipeline pipe size changes can be accommodated with a custom designed tube. Some wrinkling and gaps behind the CIPP will occur at the transition point of the size change.
Pipe material change	This occurs when the pipeline pipe material is changed in mid-section without a manhole being constructed.	Pipeline material changes should not affect the installation of a CIPP unless the size also changes. Then a custom tube should be fabricated.
Pipe alignement change	During installation pipe joints may be deflected of brick pipe may be constructed to follow a curve in the road.	Small deflections in alignment of an existing pipe are typically not a concern when installing a CIPP.

TABLE 6.2 Existing Pipe Defects and Possible CIPP Impacts (*Continued*)

A

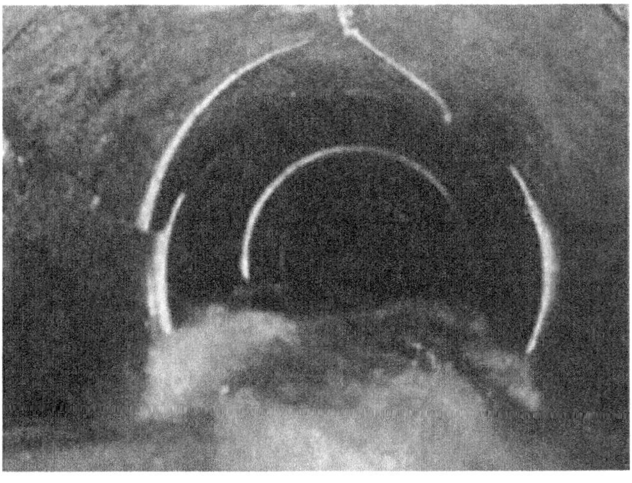

B

Figure 6.1 Structural defects. (a) Partially deteriorated, (b) Fully deteriorated (bottom of the pipe is up-buckled). (*Source:* Insituform Technologies.)

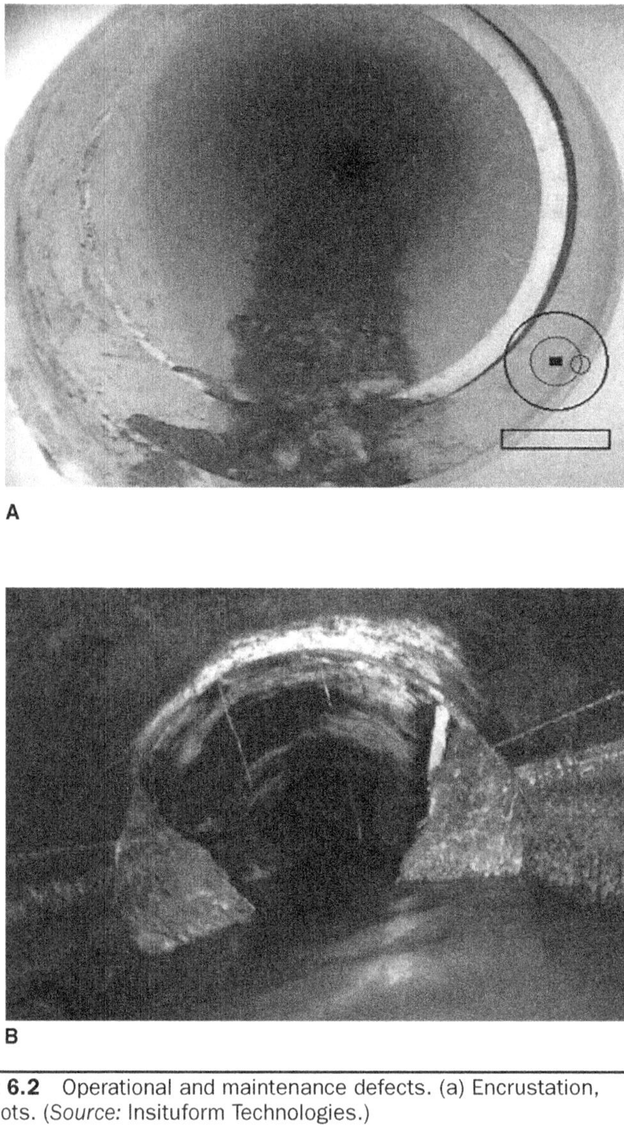

Figure 6.2 Operational and maintenance defects. (a) Encrustation, (b) Roots. (*Source:* Insituform Technologies.)

A

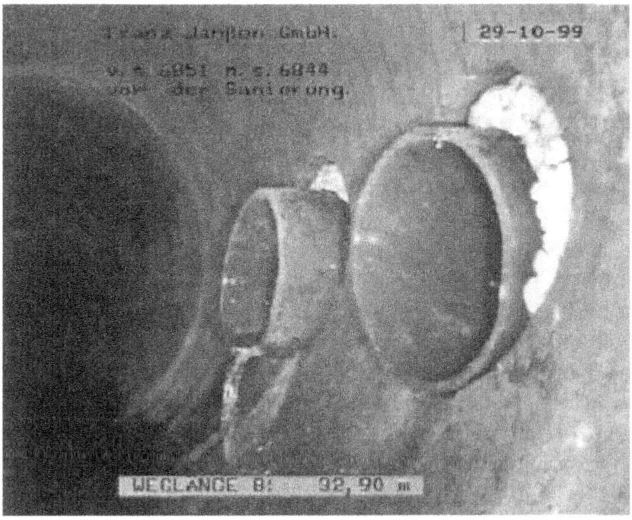

B

Figure 6.3 Construction defects. (a) Deposits, (b) Break-in, protruding laterals. (*Source:* Insituform Technologies.)

distinct chemical resistance and structural properties. The most common resin types used for CIPP applications are unsaturated polyester, vinyl ester, and epoxy.

All of above resins have distinct chemical resistance to fluid properties and domestic sewage.

Unsaturated polyester resins were originally selected for the first CIPP installations because of their chemical resistance to municipal sewage, good physical properties in CIPP composite, excellent working characteristics for CIPP installation procedures, and economic feasibility. Unsaturated polyester resins have remained the most widely used systems for the CIPP processes. Styrene, a primary ingredient used to manufacture the thermoset resin, is classified by the U.S. Department of Health and Human Services (HHS) as a "reasonably anticipated carcinogen." However, a pipeline renewal industry review has found that the levels of styrene exposure to workers and the public during CIPP installations are significantly lower than standards set by NIOSH (Occupational Safety and Health Administration)[1]. Once the CIPP is cured, it becomes essentially inert with no long-term styrene issues. To change to non-styrene resin, the pipeline owner must consider increased project costs. Most styrene complaints are related to odor, since some fumes enter the residences or businesses because of dry drain trap or through the foundation. It is important that CIPP contractors use good housekeeping practices and jobsite installation practices to prevent styrene from entering environment.

Vinyl ester and epoxy resin systems are used in industrial and pressure pipeline applications, where their special corrosion resistance at high pH, solvent resistance, and higher temperature performance are needed. These systems can be used in residential areas. However, it is typically not necessary and the costs will increase.

The primary function of the fabric tube is to carry and support the resins until it is installed and cured. This requires that the fabric tube withstand installation stresses with a controlled amount of stretch, but with enough flexibility to dimple at side connections and expand to fit the existing pipeline irregularities. The fabric tube material can be woven or nonwoven, with the most common material being a nonwoven, needled felt. Polyethylene, polypropylene, and polyurethane coatings are commonly used on the exterior, or interior, or both surfaces of the fabric tube to protect the resin during installation. The layers of the fabric tube can be seamless, as with some woven material, or longitudinally joined with stitching or heat bonding.

The primary differences between various CIPP systems are in the composition and structure of the tube, method of resin impregnation (by hand, by pressure or by vacuum), installation procedure, and curing process. Typical CIPP design factors are shown in Table 6.3. Refer to Chap. 3 for more information on CIPP design.

---

[1]Underground Construction Magazine, April 2015, p. 46.

A.	Fully or partially deteriorated existing pipe
B.	Groundwater elevation (if unknown, assume within 2 ft below ground surface)*
C.	Depth of cover at deepest point
D.	Unit weight of soil
E.	Modulus of soil stiffness
F.	Long-term modulus of elasticity of CIPP
G.	Live loads (HS20)
H.	Factor of safety against buckling (usually 2.0)
I.	Pipe ovality (less than 10%)

*With larger sewers, the groundwater elevation should be determined. Making assumptions, such as ft below grade for large diameters, can unnecessarily raise costs.

TABLE 6.3 Typical CIPP Design Parameters

6.5 Method Description

The CIPPs are installed either by inverting into the existing pipeline using water or air pressure, or by pulling in with a winch. The liner is then cured using hot water, steam, UV light, or at ambient temperature. The process involves several stages each of which is briefly described below.

6.5.1 Initial Setup: Cleaning, Televising, and Resin Impregnation (Wet-out Process)

First, the existing pipe to be lined is televised and cleaned. The distance between inspection manholes and access points is measured and all the damaged sections, appurtenances and service inlets are carefully noted. If an analysis of the existing pipe justifies use of the CIPP method, an order based on the actual length and diameter measurement is placed with the CIPP manufacturer, and the product is manufactured for insertion into the particular pipe. After the pipeline has been cleaned and televised, a flexible fabric tube, manufactured and cut according to specific project conditions, is impregnated with the designed resin and catalysts, to prepare it for insertion through a manhole or another convenient entry point. It is critical that the fabric tube is totally saturated to provide consistent finished physical properties required by design.

6.5.1.1 Tube Wet-out

Wet-out is the process of introducing the catalyzed resin liquid into the fabric tube. Wet-out of a tube can be accomplished several ways. The traditional wet-out is termed vacuum impregnation. Other systems that are equally as efficient are resin immersion and resin bath wet-out.

In the case of vacuum impregnation, once the resin and catalyst have been mixed, it is pumped into the fabric tube. Serial vacuum impregnation involves placing a vacuum suction attachment at predetermined intervals, evacuating the tube of air, allowing the resin to fill the evacuated tube area and patching the tube to seal the penetration of the vacuum device. Other impregnation techniques include resin bath immersion or utilizing gravity for the resin/catalyst mixture to fully impregnate the tube. Regardless of the method, *total saturation of the fabric tube is critical and mandatory.* Most tube coatings are transparent, which allows for a visual confirmation that the tube has been thoroughly wet-out with no apparent visible dry spots exists. During vacuum impregnation (wet-out) operations, the tube is fed through a set of rollers with a predesigned gap setting. The roller gap setting determines the volume of resin/catalyst mixture that is impregnated per unit length of wet-out tube. The roller gap setting is important in determining the final wall thickness of the CIPP. If the gap is not set wide enough, the tube will have insufficient resin and may be below design standards. If the gap setting is too wide, excess resin will be placed in the tube.

Figures 6.4 and 6.5 present the resin impregnation (wet-out) and vacuum impregnation process. Factory wet-out is the most common method for vacuum impregnating the resin/catalyst system into the fabric tube. In factory, the materials are stored in a controlled environment not subjected to weather or severe temperature changes. Typically, the resin is stored in large containers under controlled

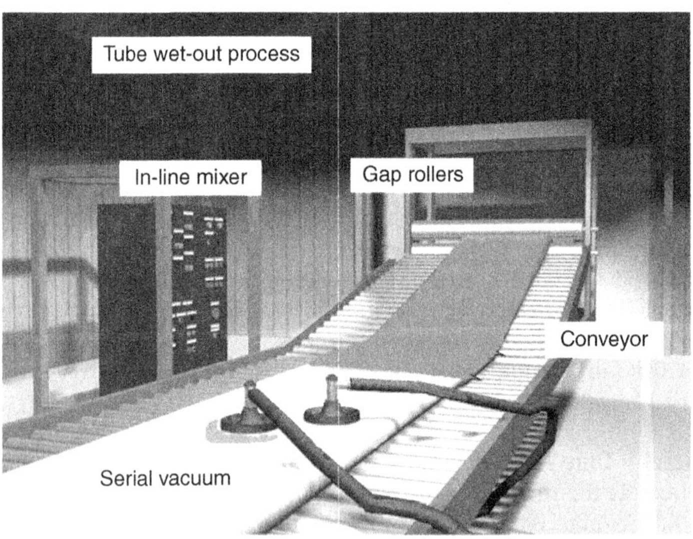

FIGURE 6.4 Resin impregnation (wet-out) process. (*Source:* Insituform Technologies.)

temperatures and the tube is allowed to reach required temperature under factory environment. The wet-out equipment is generally fixed in place and can be maintained on as-needed basis. Tubes can be wet-out using a static mixing system that regulates and combines the correct amounts of resin and catalyst for a proper wet-out of the tube. Once the tube is wet-out, it can be placed, on ice, into a refrigerated unit and delivered to the jobsite. The majority of CIPP tube wet-out is performed in a factory setting. Specially designed truck trailers make it possible to haul large diameter and long length tubes.

6.5.2 Installation

There are two basic methods of CIPP installation: the inversion method and the pull-in method. Specific variations of these techniques exist depending on the manufacturer.

6.5.2.1 Inversion Method

In the inversion method, the liner is clamped around an inversion ring or similar device and turned inside out (inverted) down the length of the pipe using air or water pressure. The liner is then forced by the air or water against the wall of the pipe to be rehabilitated.

1. *Water Inversion.* Water inversion has been used for over four decades and can be used for both small and large diameters. The liner is inverted by a water head or by a flow-through pressure device. When using water head, an inversion standpipe or down tube provides the necessary head to invert the liner into the host pipeline. For large-diameter pipes, the liner

Figure 6.5 CIPP vacuum impregnation process. (*Source:* Insituform Technologies.)

is inverted from an inversion ring or similar device which is typically located above grade. A rope is attached to the tail end of the tube to control the rate of inversion into the pipeline. Strengthening strips may be sandwiched between the layers of the tube to prevent longitudinal stretching during installation.

2. *Air Inversion.* Air inversion techniques can differ depending on which curing method is to be used. If the liner is to be cured with steam, it is inverted into the host pipe using air pressure. A guide cable can be attached, and often a rope is attached to the tail end to control the rate of inversion. When the liner is in place, pressure is maintained while the inversion air is replaced by steam for curing. Air inversion is particularly useful for steep sewers or sewers with significant elevation changes where inverting with water can lead to high water pressures at the lowest point of the sewer.

Water inversion and to a certain extent air inversion ensures that any water collected in pipelines with structural irregularities, such as sags, is forced out.

6.5.2.2 Pull-in Insertion

In the pull-in installation technique, liners may be pulled into place on protective membranes through the length of the pipe to be rehabilitated. The purpose of the membranes is to prevent the liner from being damaged when it is dragged over cracks, fractures, and corroded sections of the pipe. There is also a risk that loose pipe fragments will become dislodged when the membrane and liner are winched into pipe. It should be noted that tubes with an outer coating can be pulled into place without the membrane.

Then the liner is inflated using air and cured with steam or UV light. Winched insertion has limitations with regard to pipe diameter and length. The tube design prevents the liner from being damaged during installation.

6.5.2.3 Preliner Options

In some cases it may be necessary to install a preliner into the host pipeline to prevent washout or contamination of the resin, or to reduce friction during liner installation. A preliner may be required for heavily leaking pipelines, host pipes located under groundwater, in contaminated soils, and host pipes containing coatings that are not compatible with the tube and resin being installed. Preliners will remain in place and become part of the finished CIPP. Depending on the nature of the leakage or friction, two other options are available: grouting of the host pipeline to eliminate all infiltrating water, or installing a foil along the invert of the host pipe to reduce friction during pull-in of the liner.

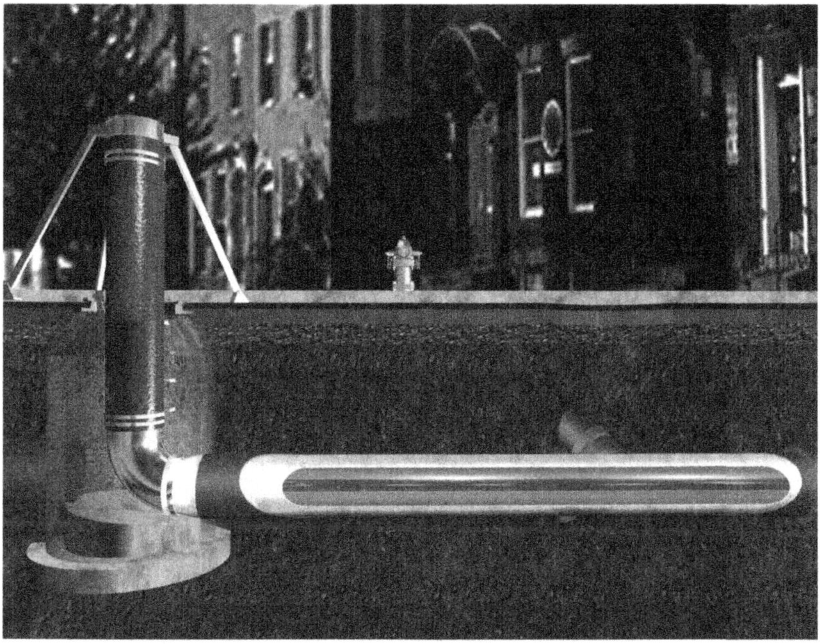

FIGURE 6.6 The liner curing process. (*Source:* Insituform Technologies.)

6.5.3 Curing

Curing of the resin is done by hot water, steam, UV light, or at ambient temperature. The specific choice of between these techniques should be determined before installation, because this may affect the choice of the installation process. Each of these installation methods is described in Fig. 6.6.

6.5.3.1 Water Curing

Hot water curing has been used for over four decades. The curing process can be documented, and continuous registration of water temperature ensures the correct laminate curing. It is also possible to control laminate cooling so that shrinkage and tensile stresses in the material can be minimized. Curing with water allows rehabilitation of long runs of large-diameter piping.

The main limitation of this process is that it can be slow and water used for curing is typically wasted. It is also necessary to consider the recipient of the large quantities of water used during the installation process.

6.5.3.2 Steam Curing

Curing with steam has been in use since the beginning of the 1990s. The main benefits of the method are that curing occurs rapidly and it can be used for steep sewers. Falls of up to 250 ft have been cured with steam. However, this method has the following disadvantages:

- It can be difficult to control the steam supply and the overheating can damage the CIPP.
- Cooling can occur suddenly, increasing the tensile stress in the laminate.
- It can be difficult to ensure complete curing in pipe sections containing structural irregularities or where groundwater infiltrates the pipe.
- When lining pipes of limited gradient or pipes with structural irregularities, it is important to be aware of the fact that "pools" of condensation can cause insufficient curing.

6.5.3.3 UV Light Curing

Curing with UV light has been used since the 1980s. The method provides rapid curing and is relatively simple to control. The curing process is controlled and documented by continuous registration of air pressure, laminate temperature, UV lamp output or light intensity, and curing rate.

With UV curing, cooling occurs continuously as the UV-light source gradually passes through the pipe, thus reducing tensile stresses in the liner. Overall, water and steam curing are more widely used.

6.6 Inspecting Installation of CIPP

A variety of project and site conditions can exist during the CIPP installations. Conditions can vary from wet to dry and from hot to cold environments. The inspector must document site and host pipe conditions and make sure that the CIPP is installed according to the contract and specifications.

The inspector must become thoroughly familiar with the contract documents for the project and most importantly the technical specifications. The technical specifications will guide the inspector about inspection areas, documentation requirements, test samples (identification, type, number of test samples required, location, maintenance and delivery, testing agency, documentation, and so on), to ensure that the CIPP is installed with specified quality. The specifications determine degree of inspection a specific project requires. Table 6.4 presents the various actions to be taken by the inspector on various operations before, during, and after installation of CIPP.

6.7 Pipe Plugging or Bypass Pumping

Bypass pumping of the existing fluid is required for the installation of CIPP. With most small diameter pipelines, plugging may be adequate but must be monitored on a regular basis to prevent flow backups.

Operation	Inspector Action
Liner material delivered on site	• Inspect the liner condition and verify that the delivered material is not damaged nor has prematurely exothermed before it is installed. • The liner wet-out report should be reviewed and matched to the tube material and resin, submitted by the contractor for the project. Record all information on the inspector report.
Sample homeowner notification	• Verify that homeowner notifications have been distributed as required in the contract documents and document on inspection report.
Pre-installation line flushing and cctv inspection of the pipe	• View CCTV inspection video of the existing pipe just prior to the installation of the liner by the contractor. Note condition and cleanliness of pipe.
Pipe plugging or bypassing	• Document the type of flow bypass provided by the contractor including size and number of pumps and the size of the discharge piping installed by the contractor, as specified or required.
Traffic control	• Document the traffic control provided by the contractor including detours, flagmen, and police details as required by the contract.
CIPP sample for testing of larger diameter	• Document that provisions, for CIPP samples, have been made by the contractor and record on inspection report.
Project odors	• Handout informational literature to affected properties and/or take a measurement of the styrene concentrations using the applicable detector tube testing device. Document all readings taken.
Monitoring the liner cure	• Maintain complete and accurate records of every step of the curing process and record all thermocouple temperatures from the start of the process to complete cool down of the liner.
CIPP samples	• Obtain samples, of the cured CIPP, directly from the contractor at the project site. • Send, cured CIPP samples, to an approved laboratory, maintaining a written chain of custody. • Review laboratory test results of the cured CIPP samples and compare to specified requirements in the contract documents. • Expedite testing and test results in a timely manner.
Service line reconnection	• Verify and document the condition of all service connections reinstated by the contractor.
Quality assurance & testing	• Visually inspect the installed CIPP and record all defects and any infiltration that is observed.
Defect documentation	• Record all defects encountered and have manufacturer recommend repair procedures.

(*Source:* NASSCO)

Table 6.4 Inspector Action on Various Operations

Operation	Inspector Action
Fit in existing pipe	• Check fit between the host pipe and the CIPP at the manhole and record on the inspection report.
Installation lengths	• Check the length of the CIPP. It should protrude several in. into each manhole or at a minimum be cut flush with the manhole wall. Document on inspection report.
Service connection	• Document the condition of each service connection and obtain manufacturers recommendation of specific repairs as required by the contract documents.
Physical property testing of cured CIPP	• Verify CIPP physical property compliance with the contract document requirements, through implementation of a chain of custody. • Verify water tightness of the CIPP in accordance with the specified requirements of the contract specifications and document on the inspection report.
Liner was possibly not properly or sufficiently iced down at the fold in the liner during transport in the refrigerated container (premature Exotherm)	• Cutout the exothermed section and use shorter length liner in another pipe if applicable. • Discard liner and furnish a new liner for installation.
Odors	• The resin component of the CIPP is typically a polyester styrene based material. The odor of styrene can typically be detected in extremely small quantities and may be detected at the manhole openings and sometimes it will migrate into a home if the home's trap is defective. The presence of styrene can be detected at less than one part per million (ppm). The threshold for continuous exposure to styrene in a confined factory setting is 50 ppm. When exposed to the atmosphere, styrene will dissipate very quickly. When an odor is detected it should normally dissipate in a matter of several hours and should not present a harmful environment for personnel in the vicinity of the installation. If styrene odors are detected in a residence, ventilate to eliminate the odors.

Table 6.4 Inspector Action on Various Operations (*Continued*)

Service connections in the residential systems typically do not require bypass pumping. The homeowners are notified usually a week or two in advance and again 24 hours in advance to refrain from the use of water while the CIPP installation process is underway. A small amount of water from the service connection typically will not affect the installation of the CIPP. In commercial and industrial applications it may be necessary to bypass flow from each individual service connection, since flow interruption may not be an option. In these cases, individual cleanouts are identified and a small pump is set-up directing the flow

Figure 6.7 Typical bypass pump setup for a large volume of existing flow. (*Source:* Insituform Technologies.)

to the mainline bypassing. In most cases, the project owner will require a pumping capacity that is equal to the anticipated peak flow (either recorded or anticipated) for the line section, plus 50 to 100 percent redundancy in the event the primary system fails to function. Most pump rental suppliers will determine bypass pumping suction and discharge configurations and setup requirements at the jobsite. However, pump operations for the duration of the project, are typically the responsibility of the general contractor. Figure 6.7 illustrates typical bypass pumping setup for a large volume of existing flow.

6.8 Quality Assurance and Testing

6.8.1 CIPP Inspection and Acceptance

The CIPP installation may be inspected visually if appropriate. Additionally, CCTV is commonly used for pre- and post-installation inspection purposes. Variations from true line and grade may be inherent with CIPP because of the conditions of the host pipe. However, certain conditions must be met, such as the new CIPP should not show any infiltration of groundwater. At the conclusion of the installation, all service connections must be accounted for and reconnected. Table 6.5 presents *sample* defect codes for CIPP evaluation.

6.8.2 Workmanship

The CIPP should be continuous over the entire length of an installation and be free of dry spots, lifts, and delaminations. If these conditions

CIPP Defects	Codes	CIPP Defects	Codes
Lining shrinkage	LS	Lining bubbles	LB
Lining wrinkles	LW	Lining fins or folds	LF
Foreign inclusions	FI	Lining delamination	LD
Lining cracks	LC	Hot dogs	HD
Dry spots	DS	Mis-cuts	MC
Lining wrinkles	LW	Annular space	AS
Lining elongation	LE	Lining thickness	LT
Lining dimples	LD	Lining lumps	LL
Under-cuts	UC	Overcuts	OC

TABLE 6.5 Sample Defect Codes for CIPP Evaluation

are present, the CIPP will be evaluated for its functionality to meet the applicable requirements of the contract documents. If the CIPP does not meet the requirements of the contract documents or the functions expected based on the technical specifications, the affected portions of CIPP may have to be repaired according to the contract provisions. Some project owners may apply a penalty in addition to the CIPP repair and/or replacement requirements (see following sections).

6.8.3 Quality Control Issues

6.8.3.1 Design Issues

The key design consideration in CIPP installation is the ability of the cured-in-place pipe to withstand buckling due to ground water and collapse due to soil and live loads. The resin and the tube make a composite material to determine the design thickness of the CIPP. There are generally two types of tubes—nonreinforced and reinforced. The reinforced tubes can be divided into fiberglass and carbon fiber laminates.

Loads cause the CIPP to creep reducing its capability to resist external pressures. Factors influencing the design of CIPP include physical properties of CIPP technology and site and project specific factors as shown in Table 6.6. The pipe wall thickness design is based on the anticipated effects on pipeline creep due to external loads over a 50-year period as specified by ASTM F1216-09, Eqs. X1.1 and X1.3. See Chap. 3 for sample CIPP design calculations.

6.8.3.2 Quality Issues

This section describes a *suggested* method for addressing CIPP quality issues. First, the project inspector will review CCTV post video within two weeks of submittal by the contractor. In the event the inspector

Condition of Pipe	Design Parameters	Liner Thickness
Partially deteriorated	Groundwater level increases	Increases
	Ovality increases	Increases
Fully deteriorated	Groundwater level increases	Increases
	Soil cover increases	Increases
	Ovality increases	Increases
	Soil modulus increases	Decreases

Table 6.6 Factors Affecting Design of CIPP

finds that a corresponding tolerance level has been exceeded, the inspector will flag the defect and handover the CCTV post video to the project manager representing the owner or the city engineer for secondary review. If the project manager agrees that the defect is in excess of acceptable tolerance levels, he or she will determine acceptable remedies to correct the defects at no extra cost to the project owner. If the contractor chooses not to correct the defects, a deductive penalty will be imposed against the contractor. Once the corrective work has been performed by the contractor to the satisfaction of the project manager and the inspector, full payment for the CIPP will be issued. In the event the contractor chooses to accept the penalty deduction stipulated in the contract rather than to correct the defect, a change order will be issued for the deduction and the remaining balance for the subject CIPP installation will be paid to the contractor.

6.8.3.2.1 Testing Issues CIPP samples must be tested based on the ASTM guidelines previously mentioned. If testing results fall short of required flexural modulus or thickness requirements, the required stiffness and actual stiffness of the CIPP samples will be compared. After comparison, if the average of actual stiffness calculated for the pipe is greater than 94 percent of the required stiffness, the pipe will be accepted but marked as noncompliant and a deductive penalty proportional to the stiffness deficiency will be assessed against the contractor. A change order will be issued for the deductive penalty, in proportion to the deficiency, and the remaining balance for the CIPP section will be paid to the contractor. If the average of actual stiffness for the CIPP section is less than 94 percent of the required stiffness, the CIPP section will not be accepted and the contractor will be required to install a second liner over the first liner to compensate for the deficiency, at no extra cost to the project owner. Full payment for the CIPP section will be issued to the contractor after the second liner is properly installed and accepted. Assuming an acceptable second liner is properly installed, no deductive penalty will be imposed against the contractor[2].

To identify every defective condition and to suggest every corrective measure or penalty is beyond scope of this book. This section is only intended as a guide to aid in identifying physical conditions of the installation, applying appropriate corrective measures and assigning deductive penalties for defective installations, which should all be included in the bid and contract documents. This way the contractors know, before the bidding and start of the project, what expected quality measures are, and possible outcomes if these quality measures are not met. This way, many quality problems, disputes and potential litigations may be avoided. The final decision for application, enforcement, and final acceptance will ultimately be made by the owner's project/construction manager or the city engineer on a case by case basis and as stipulated in the contract documents. Najafi (2010) presents a summary of *potential* CIPP defects, *suggested* acceptable tolerances, and *possible* repaired types that may be required. Figure 6.8 illustrates two potential CIPP defects.

6.8.4 Final Steps

After curing, the new pipe is cooled down. The ends of the cured pipe are then trimmed to form a smooth, seamless CIPP. A special cutting device or robot is used in conjunction with a closed-circuit television (CCTV) camera to reopen service connections, locating the dimples visually with the camera. The manhole openings and lateral cutouts are then sealed (if specified), and the final CCTV is provided to the customer if required.

Reopening the service connections can be done by person-entry for larger-diameter pipes or robotically for smaller diameters. Normally, a small dimple is left in the CIPP directly over each service connection, allowing them to be easily located and reopened. However, the number and locations of the service connections should be noted during the prelining televising process to ensure that all connections are reopened and to aid in locating those that are difficult to identify.

The new CIPP is of slightly smaller inside diameter fitted in configuration with the shape of the host pipe. The CIPP has a very

[2]The measured wall thickness has much more impact upon buckling resistance than the flexural modulus. Its value adjusts the buckling resistance by its cube power and not directly proportional. Design engineers and project owners should be cautioned that contractors may submit higher bids, or choose not to submit a bid, if the contract contains unjustified quality requirements, low tolerance expectations and possibilities of penalties or rework. Including or not including such language in the contract and justification of level of quality required needs to be evaluated based on project owner's needs, on the true project hydraulic, structural, and performance requirements, and must be evaluated in case by case basis using judgment of a knowledgeable professional engineer.

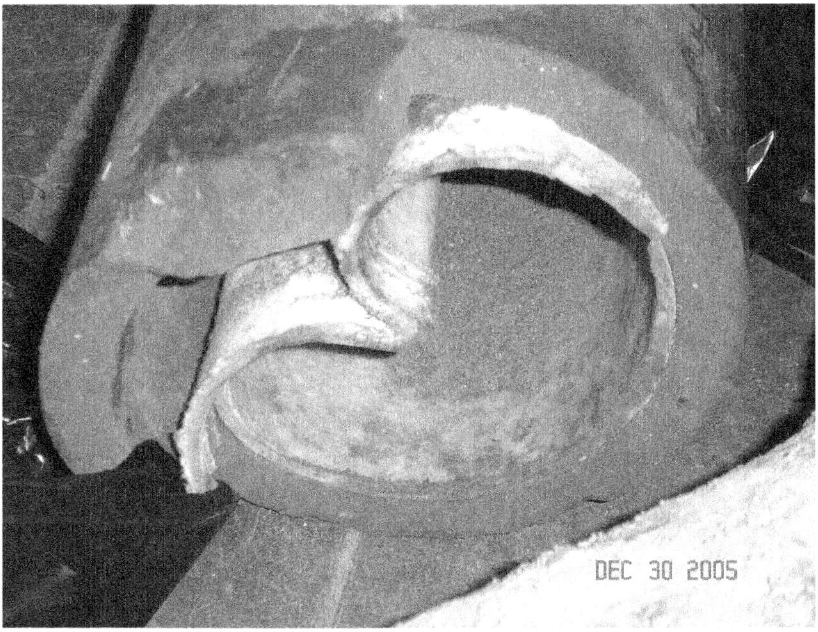

A

B

Figure 6.8 Lifts (CIPP defect).

smooth interior surface, which generally improves the flow capacity despite the slight decrease in diameter.

6.9 Major Advantages

- Grouting is not normally required because of the tight fit of the CIPP against the existing pipe except at manhole openings and at lateral openings if required.
- The CIPP has no joints and has a very smooth interior surface, which typically improves the flow capacity despite the slight decrease in diameter.
- Noncircular shapes can be accommodated often without a decrease in flow capacity.
- Lining is capable of accommodating bends and the existing pipe deformations. Most shapes of pipes can be lined.
- Entry is possible via existing manholes, access points, or through minor excavations. Remote-control internal lateral connection is possible.
- Although there is a small decrease in the existing pipe cross-section, the hydraulic capacity is typically improved because of a smooth CIPP surface (see Chap. 4).

6.10 Major Limitations

- The tube or the hose must be specially constructed for each project. (Some manufacturers provide stock materials to their installers for small diameters, such as 8, 10, 12, and 15 in.)
- The existing flow must be bypassed, rerouted or plugged during the installation process.
- Sealing may be required at liner ends and at lateral connections to prevent infiltration.
- The amount and the type of resin used are a contractor's function, and the owners should have technical specifications and inspection procedures to ensure proper resin quality and handling.
- The curing process must be carefully monitored, inspected, and tested to ensure obtaining the specified physical and chemical properties.
- The curing process may take 1 to 5 hours and styrene might be present with polyester and vinyl ester resin during the curing. On sanitary sewer jobs, the cure water is routinely discharged to the sanitary sewer after cooling to 100°F; however, water is removed from the jobsite if it is not a sanitary

sewer (storm sewer, culvert, etc.). The majority of work is in sanitary sewers where discharge of cure water is not a problem.

- Obstructions can inhibit the lining process.
- The cost of CIPP installations compared with other trenchless renewal methods can be a concern. Depending on the situation, the cost may be more than such methods as sliplining, thermoformed, and close-fit methods (used for pressure pipe only). However, also depending on the circumstances, CIPP may be the most cost-effective alternative. Therefore, it is generally recommended to evaluate the existing pipe problems and select the most cost-effective method for the conditions of the existing pipe and the long-term objectives of pipeline. Often it is difficult to disqualify CIPP during the planning or design phase simply because of cost. It may be better to bid the work with several acceptable specified products and let the low bid complete the work. With competing products of various advantages and disadvantages, an evaluated bid process can be used where price is not the only deciding factor.

6.11 Scheduling and Costs

6.11.1 Scheduling

Scheduling CIPP work is dependent upon many variables, including overall size of the project, project logistics, host pipe diameters, installation and cure methods, expected weather, resin system used, project environment (residential, industrial, commercial, busy streets, easements, etc.), and experience and skill of the installation crews. For a high-density residential project with 8 in. sewers, the installer may schedule two or even three CIPP installations per day per crew, whereas each installation for a large diameter interceptor may require an on-site wet-out with simultaneous installation followed by curing. Projects such as this could require 4 to 7 days to complete a single installation.

Scheduling sewer cleaning always presents a certain amount of unknowns. Whereas small diameter sewer cleaning is typically routine, there can be occurrences where difficulties are encountered, causing a delay in the CIPP installation. Heavy cleaning crews should be scheduled far enough in advance to compensate for these unknowns.

Cleaning large sewers is quite another matter. Having an estimate of how much debris is in the sewer helps the contractor schedule the cleaning operation and the subsequent CIPP operation. In some cases it may take several days to clean a reach of large diameter sewer. Since in many cases all bypass and/or flow diversion operations must be in operation during the cleaning process, all traffic control in place and all safety measures functional, cleaning large sewers can be a significant portion of the overall job cost.

If the crews are local, mobilization will be minimized. Some extra time may be allowed Monday morning to get setup for the week, and also for Friday or Friday afternoon for site clean-up, finishing small tasks, equipment maintenance, and mobilizing back to the shop. If the crews are from out-of-town, a day or more may be allowed for mobilization. Crews will be housed in local motels/hotels and may work for several weeks before returning to the home base. When the project is complete, another day or more will be required for demobilization.

An example of a productive project is one with a large number of small diameter installations all located within a small geographic area. When the crew only has to move immediately downstream for the next installation, two installations per day per crew should be attainable. In some cases, a crew may achieve three installations per day. If the installations are spread out over a large geographic area with significant driving times between installs, only one shot per day per crew may be feasible.

The productivity for residential small-diameter installations ultimately is dependent upon the ability to reconnect laterals. Any service connection that is closed by a CIPP installation must be reconnected in a timely manner. It is not desirable for a crew to install large amounts of small diameter CIPP per day if the lateral cutting team cannot keep up with the production. Highly productive CIPP installation crews may require two lateral cutting teams to ensure that all customers are back on line in a timely manner.

6.11.2 Costs

The three major cost components of a CIPP project are materials, labor, and equipment. Materials, principally the CIPP tube and resin, are typically a set cost for any CIPP project that can only be influenced by the installer's ability to negotiate tube and resin prices with the material suppliers. However, labor and equipment are quite another matter. The more efficient and productive the contractor can be, the less time is needed on the jobsite. Less time on the jobsite translates to lower labor and equipment costs.

CIPP installers can increase productivity by either performing multiple installations per day or installing greater CIPP lengths per shot. There is a direct proportional relationship in CIPP installation cost based on inversion lengths. Generally speaking, the longer the inversion length, the lower is the cost per foot. This is evident in the distribution of fixed costs over longer installation lengths.

Other cost components include subcontracts, small tools, rentals, meals and housing for the crews, and other incidental costs.

As a function of these costs, the contractor adds a margin or mark-up to cover overhead, profit, and risk. Overhead costs are typically set numbers for each contractor expressed as a percentage of sales. Because of the direct effect on bid pricing, contractors are continually looking for ways to keep their overhead costs at a minimum. Profit is of course desirable if the contractor wants to stay in business. Depending on how badly a contractor wants to secure the project, the amount placed in the bid for profit may vary.

The amount placed in the project pricing for risk is to cover any project unknowns or conditions that are known, thus causing the contractor to be apprehensive about possible scenarios during construction. Following are some areas for risk consideration:

1. *Prescriptive versus Performance Specifications:* Prescriptive specifications give step-by-step directions to the contractor describing techniques to be used to wet-out, install, and cure the CIPP. Performance specifications give general directions to the contractor but are quite specific regarding the end product desired. Contractors like performance specifications because they can use the installation techniques with which they have experience. Because they may not have experience with some of the requirements of the prescriptive specifications, a higher value of risk will be assigned, resulting in a higher final price.
2. *Materials Specified:* Materials that the contractor typically does not use may be specified. This is another prescriptive specification situation where the contractor will assign more risk and thus a higher price.
3. *Unknown Site Conditions:* Unknown site conditions where mitigating factors are not covered in the contract documents will increase risk. For example, no CCTV videos are available, and there is concern that some sewers may be collapsed. All point repairs are at the contractor's expense. A shaft is required for large CIPP installation, and the groundwater level at the time of the excavation is uncertain. Examples for this type of situation are almost endless.
4. *Known Site Conditions:* If a problem occurs during the project and excavation is not possible, that is a risky situation. For example, the sewer passes under a river and it is the only sewer the owner has that crosses the river. If something goes wrong, the contractor could incur large costs in correcting the situation. At times, extremely risky projects can be approached through a shared risk concept where both the owner and the contractor are at financial risk.

5. *Difficult Customers:* Some customers and/or their representatives are simply more difficult to deal with during a project. This generally leads to more meetings, negotiations, attorneys' time, letters back and forth, and so on throughout the project, all which tend to increase the contractor's costs. This leads to a higher risk, higher price situation.
6. *Penalty Clauses:* While some penalties for not meeting specified values are reasonable, some are not. If the contractor estimates that a certain percentage of the contract price will be deducted in penalties, that value can be added as risk during the bid or quote process.
7. *Removal of CIPP:* As with penalty clauses, there are circumstances where the CIPP should be removed because of quality issues. However, if the contract documents outline circumstances where the owner can ask that the CIPP be removed where other means of remediation should be acceptable, a red flag is raised for the contractor. Removal of a CIPP is difficult and expensive, and contractors will try to find other methods to remedy the situation. If CIPP installers determine that the contract documents are too onerous, many will not bid the work.
8. *Long Warranty Periods:* Many construction warranties are for one year. This does not mean that the product will only perform for 1 year, just that the contractor is responsible for 1 year. If longer warranty periods are specified, responsible contractors will purchase a bond for the warranty period, thus adding cost to the project. Public companies are required to take certain financial procedures when dealing with long warranties. Most deficiencies in CIPP work are apparent in the final inspection or become visible within 1 year.

Considering the many risk factors and variables that exist in CIPP projects, prices from project to project vary widely. Often project planners want to know how much a project will cost when little information is available. On some projects, the price for CIPP includes just the CIPP installation. Other projects may include other items as incidental to the CIPP installation, significantly raising the CIPP price per foot. In these cases, sewer cleaning, CCTV, lateral reconnections, restoration work, and so on may be considered incidental to the CIPP work.

One measure of CIPP price is \$/in-dia/ft. For example, a typical price for 8 in. CIPP is \$3/in-dia/ft, or \$24/ft. Depending on the project size, logistics and risk, prices for 8 in. CIPP may vary from \$2 to \$7/in-dia/ft. Although this range for 8 in. CIPP may seem extreme, the low and high values have recently been seen in the industry.

Table 6.7 and Fig. 6.9 present low and high values for the CIPP diameter technical envelop where the variance is due to project size,

Diameter, in.	Low*, $/dia-in.	Low Average†, $/dia-in.	Average‡, $/dia-in.	High§, $/dia-in.
5	2.77	4.35	5.05	8.54
8	1.58	2.21	3.92	5.99
10	1.74	2.13	4.11	7.10
12	1.82	2.12	4.09	5.84
15	1.72	2.19	3.50	3.98
18	2.19	2.42	4.78	7.5
21	2.22	2.55	4.82	10.02
24	1.95	2.44	5.03	10.92
27	2.04	3.41	5.55	11.35
30	2.08	2.89	5.88	12.32
35	2.50	3.05	4.52	5.82
42	2.35	4.09	5.34	5.51
48	3.03	3.23	5.50	11.35
54	3.55	4.32	5.42	5.53
50	3.03	3.44	4.77	5.00
72	4.87	5.20	5.45	5.70
84	4.93	5.34	5.34	5.89

*Single lowest bid of 5 CIPP contractors.
†Average of low bids of 5 CIPP contractors.
‡Average of average bids of 5 CIPP contractors.
§Single highest bid of 5 CIPP contractors.

Table 6.7 CIPP Sample Bid Prices

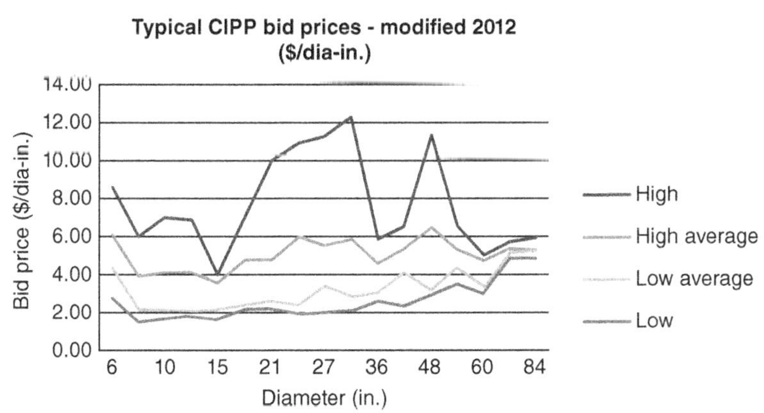

Figure 6.9 Sample CIPP average bid prices for different diameters.

logistics, difficulty, and risk. This table should be used as general guidance only and for gravity pipes and not for project estimates.

6.12 Chapter Summary

CIPP is one of the most common trenchless renewal methods. It is applicable in many pipe diameters, shapes, and applications. It has been successfully used in gravity applications, potable water pipelines, gas pipelines, and oil and industrial pipelines as well as road culverts and drainage structures. It is a method that requires high degree of supervision and inspection, as the product will be built at the jobsite, not at the plant and under controlled environment.

6.13 Review Questions

1. Describe history of CIPP.
2. Describe site compatibility and application of CIPP.
3. Explain impacts of existing pipe defects on CIPP.
4. Describe the 3 resin types.
5. List CIPP design parameters.
6. Describe the wet-out process.
7. Describe different CIPP installation methods.
8. What is the preliner?
9. Explain the three types of CIPP curing.
10. Describe CIPP inspection.
11. Describe bypassing.
12. Describe quality assurance and testing.
13. Explain quality control issues.
14. Explain CIPP advantages and limitations.
15. List some of the scheduling and cost factors.

CHAPTER 7

Sliplining

7.1 Introduction

Sliplining (SL) is one of the earliest and simplest forms of trenchless pipeline renewal methods and can be used for structural or nonstructural purposes. This technique has been used since the 1940s for renewal of deteriorated pipes. More than 70 years of experience has shown that sliplining is a proven and cost-effective renewal technique with all benefits of trenchless technology such as minimum disruption of service, surface traffic, and minimum property damage that would otherwise be caused by the open-cut method. Sliplining involves insertion of a new pipe of smaller diameter into the existing pipe and the grouting of the annular space between the existing pipe and the new pipe. The pipe materials most commonly used in sliplining are polyethylene (PE), glassfiber reinforced polyester (GRP), and PVC, although use of any other pipe material is possible.

7.2 Site Compatibility and Applications

Sliplining can generally be used, with limitations of technology, for application in potable water pipes, gravity sewers, chemical or industrial pipelines, and all pressure pipelines, although the predominate application is lining non-pressure sanitary sewers and pressure gas lines. Straight pipelines, pipelines with minor bends, and circular pipes, can be used with this method. It also has applications in, noncircular pipes, and pipelines with varying cross-section (see Chap. 9, Modified Sliplining).

While sliplining can be used to renew gravity pipelines; the decrease in existing pipe's cross-sectional area needs to be checked against any gain in flow because of better smoothness of the liner pipe surface or hydraulic characteristics. Also, it is usually necessary to excavate connections and disconnect them prior annulus grouting. Connection from inside the pipe is possible, although the process is more complicated when compared to cured-in-place-pipe (CIPP; see Chap. 6) and close-fit pipe renewal methods (see Chap. 11). For potable water pipes, approval of the relevant regulatory agency (such as NSF/ANSI 61 [NSF 2013]) is needed for all materials in contact with the potable water. In chemical or industrial pipelines, the liner pipe must be compatible with relevant chemicals. The liner pipe material's resistance to chemicals, high temperature conditions, and other adverse environmental conditions must also be evaluated.

Most bends cannot usually be negotiated, except for small angles (less than 6°). All bends add to the friction between the existing and new pipes during installation. This, as a result, reduces the length of the liner that can be pulled or pushed in without overstressing the pipe. Polyethylene, PVC and GRP pipes are available for noncircular sections, as described in Modified Sliplining (Chap. 9). For pipes with varying cross-sections, the liner pipe must be sized to minimum dimensions of the existing pipe, unless tapers are incorporated. Because of the high weight of large diameter liner pipe, it is unusual to pull a slipline pipe into a worker-entry (usually larger than 42 in.) pipe as a continuous string. Table 7.1 presents an overview of sliplining applications and possible limitations.

Pipeline Type	Suitable	Notes
Gravity sewers	Yes	A
Gas pipelines	Yes	
Potable water pipelines	Yes	B
Chemical or industrial pipelines	Yes	C
Straight pipelines	Yes	
Pipelines with bends	Yes	D
Circular pipes	Yes	
Noncircular pipes	No	
Pipelines with varying cross-sections	No	E
Pipelines with service lateral connections	Yes	F
Pipelines with deformations	Yes	G

A. Sliplining can be used to renew sewers, but it is not usually the first- renewal method for gravity pipelines less than 36 in. because of reduction in cross-sectional area of the existing pipe. In larger pipelines (36-in. and larger), segmental sliplining can be performed without bypass pumping.

B. Approval of the relevant regulatory agency and NSF/ANSI 61 is needed for all materials in contact with potable water.

C. The pipe material must be resistant to chemicals, extreme temperatures, and other requirements of industrial applications.

D. Normally, the sliplining method cannot accommodate bends more than 6°.

E. The new pipe must fit the smallest diameter in the existing pipeline unless adapters are built into the line.

F. For sewers, it is necessary to excavate lateral connections and disconnect them to the liner pipe before injecting grout. Connection from inside the pipe is a possibility, although the process is more complicated than for CIPP and closed-fit renewal methods. For potable waterlines, it is always necessary to excavate service connections.

G. Sliplining is not suitable for pipelines with considerable deformations.

Table 7.1 Overview of Sliplining Capabilities and Possible Limitations

7.3 Main Characteristics

Sliplining or pipe insertion can be used to install a flexible liner pipe of smaller diameter inside an existing pipe. Generally for pipes up to 24 in., the liner pipe outside diameter (OD) should be less than 90 percent of the existing pipe inside diameter (ID), to provide insertion clearance. For larger pipes, the liner pipe OD should be less than 95% of the existing pipe ID. Insertion is done by pulling or pushing the new liner pipe into the deteriorated existing pipeline. After installation and grouting of residual annulus, the new pipe forms a continuous, watertight pipe inside the existing one. Where the new pipe has to be laid to a specified grade, use of plastic or metal locators or spacers is essential. Spacers also help maintain the pipe location during annular grouting using a cementitious material or lightweight (cellular) grout. Flow can be used as ballast to help the liner pipe on the invert of existing pipe.

7.4 Sliplining Methods

Sliplining can be divided into two main categories: continuous and segmental. Each of these methods is explained in the following sections.

7.4.1 Continuous Sliplining

The continuous sliplining method (Figs. 7.1 and 7.2) involves accessing the deteriorated pipe at strategic points and inserting a high-density polyethylene (HDPE) or fusible PVC pipe (FPVCP), joined into a continuous line, through the existing pipe. This technique has been used to renew gravity sewers, sanitary force mains, water mains, outfall lines, gas mains, highway and drainage culverts, and other pipelines. It has been used to restore pipes as small as 1 in. (25.4 mm), with the maximum pipe diameter limited by the availability of new pipe material. Normally, the existing pipe must be empty or nearly empty during the liner pipe insertion.

This method can be used both for thin-walled and thick-walled liners. HDPE (or fusible PVC) solid-wall pipe is slipped into an existing pipeline after the joints are butt-fused (Figs. 7.1*a* and 7.2*a*). A guiding trench or an insertion pit is usually required. There is typically a slight-to-moderate loss of hydraulic capacity. Continuous sliplining leaves a residual annular space, which for gravity sewers, water mains, and other applications depending on the specific project conditions and type and condition of existing pipe and external loading, may require grouting. Excavation is required to reestablish each lateral connection.

A

B

Figure 7.1 Continuous sliplining with PE pipe. [*Source:* Plastic Pipe Institute (PPI).]

7.4.1.1 Polyethylene Pipe

One of the most common materials used for continuous sliplining is PE pipe. Its ability to be fusion welded into long jointless sections, which can be pulled into place, makes PE pipe a popular choice. Other reasons for the popularity of PE may be attributed to its flexibility and its abrasion resistant as well as its flexibility to negotiate minor bends during installation.

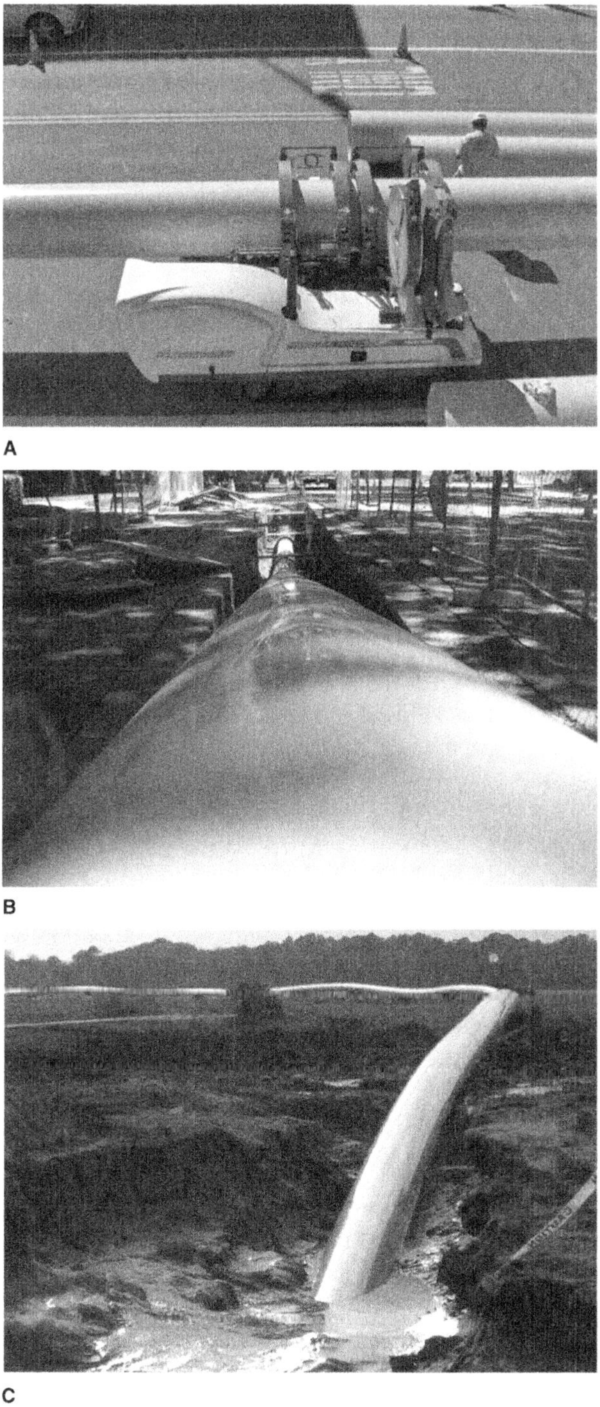

FIGURE 7.2 Continuous sliplining with FPVCP pipe. (*Source:* Underground Solutions.)

7.4.1.2 Fusible PVC Pipe

Fusible PVC pipe is commonly used for continuous sliplining. Its thermally butt-fused joints create a low-profile, fully restrained pipeline, leveraging the tensile capacity of PVC pipe material for long potential installations. The tensile properties of PVC also mean that FPVCP provides advantageous flow area and pressure class combinations in the constrained environment of a slipline installation compared to other pipe materials and joint configurations. PVC is a hard, abrasion-resistant material making it good for slipline installation. Insertion pits must be sized to allow for safe bending radius of the new pipe.

7.4.1.3 Other Pipe Materials

Additional pipes products with restrained joints such as PVC or ductile iron can also be pulled into the existing pipeline in a manner similar to that for PE pipe. These materials are usually installed a section (known as cartridge shot gun type) at a time, rather than being completely assembled prior to installation like PE or FPVCP. Flow bypassing is normally required for continuous sliplining. Due to high ring stiffness, for ductile iron pipe, grouting of the annular space is not required.

7.4.2 Segmental Sliplining

This method involves the use of short sections of pipe that incorporate a flush sleeve joint commonly used in microtunneling and pipe jacking methods or a low-profile bell-spigot. A number of plastic pipe products, such as GRP, PVC, PP, and PE, which include short-length sections with a variety of propriety smooth joints (both inside and outside), have been specially developed for sliplining sewers. This method is applicable for existing pipe diameters greater than 18 in.

In this type of sliplining, segments of the new pipe are assembled at access pits and pushed into the existing pipe. After the segmental pipe is installed, the annular space is grouted. The laterals are usually reconnected by excavation from outside. Figure 7.3 illustrates a segmental sliplining process.

7.4.2.1 Pipe Materials

It is possible to use any common pipe material that can be inserted into the existing pipe. To minimize the reduction in cross-sectional area of the existing pipe, the pipe must have smooth inside and outside joint. There is a wide variety of pipe products that meet these criteria. Most are intended for pipe jacking, microtunneling, or directional drilling; and their bell-less or low-profile bell configurations also make them well-suited for sliplining. It should be noted that the

A

B

FIGURE 7.3 Segmental sliplining. (*Source:* Hobas Pipe USA.)

main criterion is maximum ID of new pipe that fits inside the existing pipe. Assuming a flush joint is available in both cases, usually a thin-wall pipe with high stiffness can replace thicker wall pipe.

Pipe products such as vitrified clay, concrete, ductile iron, and centrifugally cast glassfiber reinforced polymer mortar (CCFRPM) pipe can also be pushed into the pipe from a relatively small access pit. These pipe products are pushed into the existing pipe using bucket of an excavator or jacking equipment. Flow bypassing is not normally required because the existing pipe remains open during the insertion process. In addition, because of the high bearing capacity of rigid pipe products, grouting of the annular space is not as critical as it is for flexible plastic pipes; however, grouting is still normally done to lock the pipe in place and to maintain the grade. Refer to Chap. 5 for more information on flexible and rigid pipes and properties of different pipe materials.

7.5 Installation

During the sliplining process, the pipe is pulled or pushed into the existing pipe from a launch pit to a reception pit or existing manhole. Installation procedure is slightly different for each of the two methods of sliplining (i.e., continuous or segmental). However, certain typical tasks common to all of these methods are listed below:

- Inspecting the existing pipe
- Cleaning and clearing the existing pipe from any obstruction
- Providing access to the existing pipe
- Joining (by butt fusion or gasketed bell and spigot) lengths of PE, GRP, PVC and possibly other pipe materials
- Inserting the liner pipe and positioning it inside the existing pipe
- Stabilizing by grouting the annular space (if required)
- Lateral connections
- Construction of terminal connections

7.5.1 Installation: Continuous Sliplining

As discussed previously, PE or PVC pipes can be fused together to form long segments. This fusing may be done either aboveground or in the launch pit. Fusion aboveground may require long launch pits because of the limitations on bending set by the minimum permissible radius of curvature for PE or PVC pipes. This is especially true for deeply buried and large diameter existing pipes. Fusion in the launch pit allows the pit to be shorter, but the rate of installation is limited by the time taken to fuse pipe segments and cool welds. The cooling phase in particular is important for service life of the new pipeline. This is because too short cooling times weaken the strength of the new pipe during installation and in the long term.

Beads are formed internally and externally on both FPVCP and PE pipes during fusion welding. In the case of sewer pipelines, both internal and external beads can be removed before new pipe insertion, but usually is not required as it is a difficult task. In the case of water pipelines, internal beads are not usually removed because they have little consequence to the operation or hydraulics of the new pipeline. In addition, debeading water lines introduces the risk of contamination from the use of debeading tools.

When liner pipes are pulled into position, it is connected to towing or pulling or head. The pulling head grips the new pipe and transfers the pulling force from the winch cable or pulling device to the liner pipe. The pulling head should provide a secure connection without producing high levels of localized tensile stress. In some systems, the end of the liner pipe is closed off to prevent soil or other debris from entering the liner pipe. This is especially important in the case of potable water pipes.

A breakaway connector can be fitted between the winch cable and pulling head to avoid subjecting the PE pipe to excessive stress. Such connectors can be adjusted to separate at loads lower than the permissible load on the new pipe. A broken connector is usually preferable to the liner pipe being damaged with possibility of subsequent service failure. The use of breakaway connectors also induces operators to avoid using excessive pulling force during the new pipe insertion.

7.5.1.1 Installation by Pulling

Small-diameter sliplining pipes can often be pulled into position with the aid of a sleeve made of diamond-shaped mesh. The sleeve is fastened to the winch cable and pulled over to the end of the liner pipe. During insertion, the sleeve tightens and grips the pipe end, which can then be pulled into position within the existing pipe. Short sections of small piping can be pulled into place manually, but most pipes must be pulled into place by the winch cable. The winch must provide a steady, progressive pull without snatching or uncontrollably varying the force. The winch must be positioned and the cable carefully guided. This is done through placing additional pulleys in the access pit, manhole or reception pit to ensure that the cable has an unobstructed path and does not rub against any part of the launch pit or the existing pipe.

7.5.1.2 Installation by Pushing

Numerous types of manually or hydraulically powered machines are available for pushing new liner pipes into existing ones. Some are designed to operate from launch pit, whereas others are positioned on the ground immediately behind the pit. The machine grips the new liner pipe and pushes it forward into the existing pipe. The grip mechanism is then released and returned to the starting position, and the process is repeated.

7.5.2 Installation: Segmental Sliplining

The exact segmental sliplining installation procedure is dependent on the new pipe size and material used, for example, whether reinforced pipe materials or plastics are used. Pipe segments are prefabricated sections specifically designed for each project. They are constructed from two pieces, joined longitudinally and circumferentially, and can be made from glass reinforced cement (GRC), GRP, polymer concrete (PC), and concrete. Units are assembled manually, prior to grouting of the annular gap. Individual lining units are passed into the pipeline via manholes or special access shafts. For worker-entry pipelines, the lining begins at the furthermost point, from where the units are positioned and joined. Each existing pipe segment is renewed, and then grouted before proceeding to the next existing pipe section. For non-worker-entry pipes, the segments are pushed or pulled into existing pipeline until entire length has been lined. Excavation is required to restore lateral connections.

7.5.3 Connecting Service Laterals

Excavation is normally necessary to connect laterals and branch lines while sliplining gravity sewers. To connect internally, openings can be cut in the liner pipes before injecting grout, and laterals and branch lines can be sealed off using inflatable bags. Sealing the laterals and branch lines off would prevent any grout from being forced into them. However, the complexity of such operations only justifies their use in situations where outside access via excavation is difficult or impossible. Moreover, it is only possible to use this procedure in large-diameter pipes and with low or no existing flow.

Usually connections must be excavated and branch lines disconnected before grout is injected. Connecting laterals to liner pipes is carried out in the same way as for new installations. Special connectors must be used to connect the new junction to the existing branch. Where waterlines are concerned, excavation is always necessary for connecting service lines or for making any other type of connection to the existing pipe.

7.5.4 Grouting

Sliplining systems in which the liner pipe bonds to the existing pipe to form a composite pipeline, and systems in which the new pipe only functions as a permanent mold for the annular grout require the use of cementitious grout with a minimum compressive strength of 1.5 to 3 psi. Liner pipes that are held in place by the existing pipe, but do not need to bond to it, only require a grouting material capable of transferring loads from one pipe to the other. The strength of some of the injection grouts used for this purpose corresponds to that of stiff clay.

Most sewer pipes are now grouted using a foamed (or cellular) material, which is lighter (lower density) than water, so even very large diameters can be grouted in one lift—avoiding floatation by using the flow as a ballast.

Injection grout made of ordinary Portland cement and pulverized fly ash is commonly used, although numerous special grouting materials are available. One of these is a grout with extremely low viscosity that is capable of flowing through the annular space by gravity or minimum pressure and which sets within 20 minutes. One benefit of these grouts is that they allow stage grouting to proceed more rapidly than with conventional materials.

The forces applied to liner pipes during grout injection are often greater than those experienced during normal service. Failures owing to grout pressure and floatation forces are related to the weight of the grout displaced by the liner pipe (i.e., the volume of the liner pipe multiplied by the unit weight of the grout) rather than the weight of the grout in the annular space.

It is a common practice to fill up the liner with water during grouting as this helps counteract the floatation forces and resist external pressure. Even so, as the relative density of most grouts is higher than 1.0, it may still be necessary to inject grout in stages. This is especially true for large gravity sewers, where the gradient of the sewer is critical and floatation is unacceptable.

Grout is normally injected through a standpipe under gravity pressure. This method results in a very low grout pressure compared to pressurize grouting. In many cases, the maximum pressure is less than 5 psi; however, it is important that the grout pressure does not exceed the pipe's allowable external pressure.

7.6 Design Considerations

Pipes used for sliplining are generally, but not always, similar type and specification to those used for new construction. PE and PVC pipes are usually aimed at applications where internal pressure is the main criterion, and the design of PE slipliners in pressure pipes should follow the same principles as for new pipes. The structural design of slipliners in pressure pipes should follow the same principles as for new pressure pipes. Thin-walled (nonstructural) liners may occasionally be used, provided the existing pipe has sufficient load bearing capacity, and that complete grout fill of the annulus can be achieved so that the new pipe is fully supported. This is often difficult to guarantee, and thin-walled liner pipes are therefore not favored for basic sliplining, although they are frequently used in close-fit method described in Chap. 11.

Annulus grouting may not be required when renewing pressure pipes, but is usually necessary for gravity pipelines to increase the ring stiffness of the new liner pipe. Slipliners in sewers are usually designed to be restrained by the existing pipe and the annulus grout, but do not form a bond with the existing pipe wall. In such cases, the grout acts only as filler, and does not require high structural strength. Systems that rely on the existing pipe for structural support are sometimes known as *interactive or semi-structural lining* techniques. Because of the relatively low flexural modulus of the liner pipe, thick-walled pipes may be needed to withstand high external loading. This may be a significant factor with gravity pipes, which are laid at considerable depths or are subjected to high vehicle loadings. In such cases, it may be more economical to design the liner pipe as a permanent support for high-strength grout, rather than to increase the wall thickness of the liner pipe itself. In this type of lining system, the grout is the main structural element. In all cases, the liner pipe must be designed to withstand not only the internal and external forces in service, but also the loads during installation—particularly winching forces and grout pressure. In summary, design considerations should include:

- Hydraulic analysis for flow capacity
- Insertion forces capacity of liner pipe
- Analysis of curves and inflection points (PIs)[1]
- Stiffness to avoid buckling during grouting
- Long-term external and internal loading conditions

7.7 Sliplining Gaslines

Several techniques are developed to allow the insertion of a new PE pipeline into existing gas mains or lateral service lines without interrupting the gas supply. These methods generally rely on gas flowing through the annular space between the existing and new pipelines during installation, and so take advantage of a reduction in pipe diameter. This may be acceptable in the case of existing mains originally designed for gas pressures lower than gas pressures currently available. It is outside the scope of this book to describe the many proprietary systems for live insertion. Strict and detailed safety procedures must be laid down for installation, and the following section is intended *only* as a general guide to the basic principles.

Renewal systems are available for low- and medium-pressure gas mains. The first stage is to isolate the section of the main to be renewed by bypassing at one or both ends of the isolated section. The new PE pipe is then fed through seals attached to the existing main at the entry pit and is pushed using pneumatic or hydraulic machines through the entire length of main to be renewed. Typical insertion lengths are between 300 to more than 1,500 ft.

[1]A point on a curve at which the curve changes from being concave downward to concave upward, or vice versa.

There are many installation techniques. In the simplest method, the new PE pipe is passed through seals in the exit pit, and can then be connected either to the existing pipe or to a new, generally higher pressure, network. In all methods, during the installation, the annular space between the existing pipe and the new pipe is utilized to maintain gas supply to customers. In order to facilitate the transfer gas flow to the new PE pipe, polyurethane foam is injected into the annular space to stop the flow of gas. This process would allow the existing main to be closed and gas flow diverted to new pipe.

Gas mains from 3 to 18 in. in diameter are commonly relined using the above method. In fact much larger insertions have been done. In 2008, Missouri Gas Energy inserted 19,000 ft of 20-in. DR 6.5 HDPE pipe into an existing 24-in. CI pipe. Insertion was accomplished in a number of pulls. For the renewal of gas service pipes, a technique is available that allows the existing gas meter position to be maintained by enabling the insertion of PE pipe through a 90° elbow, around a tee, or through a number of long-radius bends. After removal of the meter and the main stopcock, the line-blowing assembly is fitted to the service connection at the meter position. Air is blown through the existing service pipe to remove any loose rust. The pipe receiver, bend, and standpipe are fitted to the service, and air is allowed quickly into the pipe to blow a line through to the far end. This is then used to pull back the winch cable, and the winch is fitted to the top of the pipe receiver. A short length of the PE pipe is winched through to remove any further rust or encrustation. Full-length pipe is installed by using the winch in combination with a pushing force applied manually from the other end. A pressure test is conducted after a brief period to allow the pipe to recover from any stretching. The technique can also be modified for the renewal of water services.

A method of live insertion for gas service pipes has been developed in which a new PE pipe is pushed into an existing cast iron or other service line through a sealing system attached to the existing pipe, either inside the customer's premise or by means of a small excavation outside the building. No excavation is necessary at the service connection with the gas main in the road or street. The annular space between the existing and the new pipes is filled with a permanent sealant, which is prevented from entering the mains system by a type of nose cone fitted to the leading end of the PE pipe. The system is available for services from 1 to 2 in. diameter.

7.8 Advantages

- As a renewal method, sliplining does not require investment in costly specialized equipment.
- Jacking pipes and fittings, as used for trenchless construction methods (TCM), can also be used for sliplining.

- Sliplining is a conceptually simple technique, which can be applied to either pressure or gravity pipelines.
- Sliplining can be used for structural or nonstructural purposes.
- Live insertion (with existing flow) is possible (for some materials and systems).
- It is much less expensive than installation of a new pipe with open-cut method particularly as burial depth increases.

7.9 Limitations

- The most important limitation of sliplining is the reduction of pipe diameter inherent in the system. It is therefore necessary to establish whether the use of sliplining is compatible with the capacity requirements of the renewed pipeline. However, a loss of cross-sectional area does not proportionally represent loss of capacity because of improved hydraulic characteristics of the liner pipe. In some cases, flow capacity is increased despite the decrease in diameter.
- Where manhole access is not possible, a pit excavation is required for access during installation process.
- For lateral connections, open-cut excavation is required.
- Grouting is generally recommended.
- Generally cannot pass through and/or seal at significant inflection points (PIs).

A number of resources are available that give detailed design equations and recommendations for using PE pipe. These include ASTM F585, "Standard Practices for Insertion of Flexible Polyethylene Pipe into Existing Sewers" and the Plastic Pipe Institute's Handbook of Polyethylene Pipe (2008).

7.10 Case Study

A recent evaluation of a sliplining installation in 1997 showed effectiveness and reliability of this method to renew an old pipeline. A 30-in. ID HDPE rated to 100 psi (DR 17) with surge pressure of 60 psi, with fused butt joints, was installed in a demonstration project. Special flanges for connections to existing bar-wrapped concrete pressure pipe as shown in Fig. 7.4 are provided.

Provisions were included in the contract documents to perform an assessment of the existing 30-in. HDPE water main. The criterion used for the HDPE assessment was the following:

- Evaluate performance of joints by visual inspection for leaks at butt-fused joints and special flanged connection to the concrete pipe.

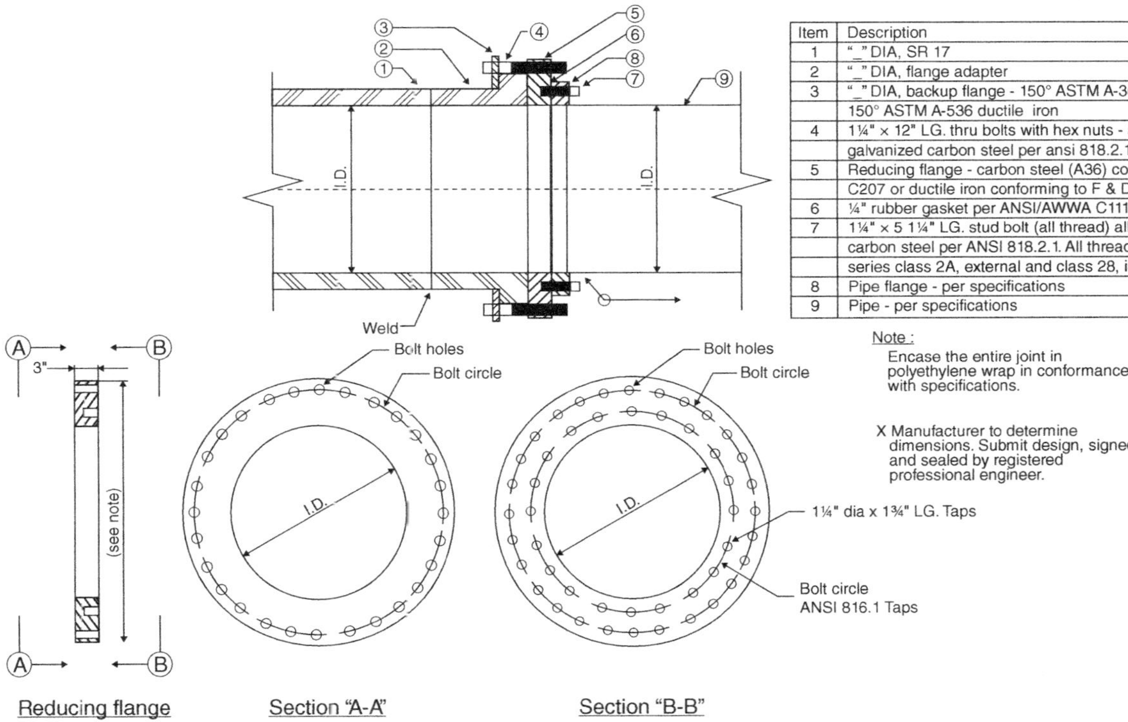

Item	Description
1	"_" DIA, SR 17
2	"_" DIA, flange adapter
3	"_" DIA, backup flange - 150° ASTM A-36 carbon steel or
	150° ASTM A-536 ductile iron
4	1¼" × 12" LG. thru bolts with hex nuts - alloy steel or
	galvanized carbon steel per ansi 818.2.1.
5	Reducing flange - carbon steel (A36) conforming to AWWA
	C207 or ductile iron conforming to F & D ANSI/AWWA A21.10/C110
6	¼" rubber gasket per ANSI/AWWA C111/A21.11
7	1¼" × 5 1¼" LG. stud bolt (all thread) alloy steel or
	carbon steel per ANSI 818.2.1. All threads are coarse-thread
	series class 2A, external and class 28, internal
8	Pipe flange - per specifications
9	Pipe - per specifications

FIGURE 7.4 Special flanges for connections to existing bar-rapped concrete pressure pipe.

FIGURE 7.5 45° bend.

- Evaluate performance of HDPE material characteristics by direct measurement of the OD of the pipe to quantify ballooning or elongation due to thrust at bends. This was done by verifying that deformations in pipe material (if any) do not exceed allowable stress for HDPE material.

Two locations were identified as critical areas to be representative of the total length of HDPE installed: the 45° bend and the flanged connection to concrete pressure pipe (see Figs. 7.5 and 7.6).

In 2011, 14 years later, the HDPE water main was exposed at each of above locations for visual evaluation of the pipe's performance. No leaks were observed in the butt-fused joints or at the flanged connection. Also, the OD of the HDPE water main was measured at each of the exposed locations to determine if ballooning or elongation of the

FIGURE 7.6 Flanged connection.

FIGURE 7.7 Outside diameter measurement.

HDPE line had occurred as a result of thrust (Fig. 7.7). The OD of the pipe matched the original OD prior to installation. However, if these phenomena occur simultaneously, they may have offsetting effects that are difficult to quantify individually. Based on the results of this limited assessment, the HDPE water main at each of the referenced locations was performing adequately under the current operating conditions.

7.10.1 Sliplining a 42-in. PCCP Pipeline

During the early morning hours of June 8, 2004, an existing 42-in. water main in Houston (Texas) suffered a catastrophic failure. This failure occurred in a pipe section of prestressed concrete cylinder pipe (PCCP) installed in 1991–1992. The area served by the pump station consisted of approximately 75 square miles. Several residential properties were flooded as a result of the failure. A failure assessment to design the repair for the water line was performed.

This particular PCCP was composed of a concrete core, prestressing wires wrapped around a thin steel cylinder, and mortar coating. Figure 7.7 shows a cross-section of typical lined PCCP. Immediately following the rupture, a visual inspection of the failed water line and testing of mortar and wire samples revealed the internal cracking appeared consistently at invert, crown, and spring lines for a distance of approximately 25,000 ft (see Figs. 7.8 and 7.9). The visual inspection was accompanied by acoustic testing to determine if the mortar lining was delaminating from the steel cylinder; however, the lining did not exhibit any noticeable delaminations.

Several pipe sections were excavated to investigate how the internal cracking correlated to overall pipe condition. Each piece excavated

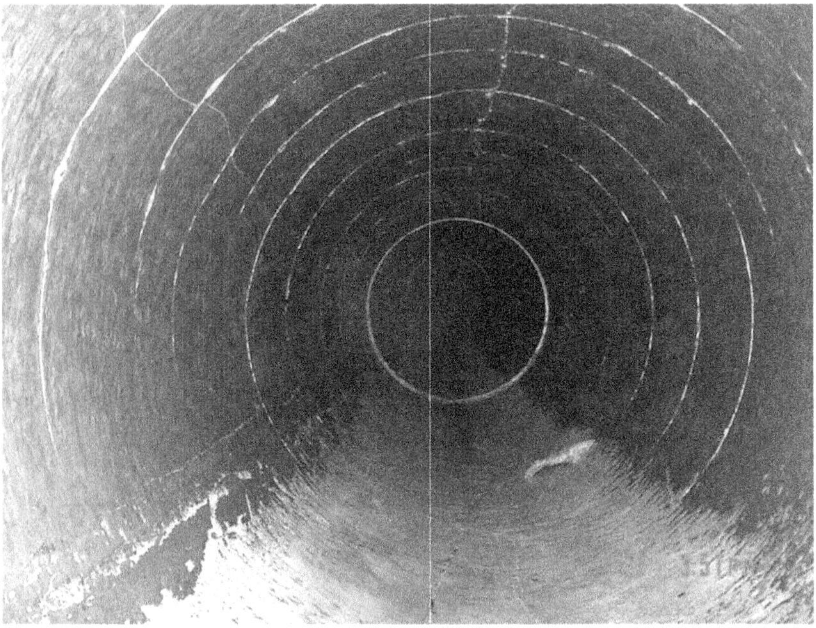

A

B

FIGURE 7.8 Internal cracks.

FIGURE 7.9 External cracks.

showed signs of external cracking, approximately in line with the internal cracks, indicating the cracks extended through the entire pipe wall. Limited wire testing performed on sample revealed no signs of hydrogen embrittlement or other obvious manufacturing defects. Figure 7.10 illustrates failed pipe section removal.

FIGURE 7.10 Failed pipe section removal.

It was determined that the repeated rapid closure of an emergency shut-off valve within the plant probably created pressure transients which could have exceeded the pipe design and caused the cracking. Hydraulic modeling of this closure scenario performed after the failure indicated that pressures in the pipe may have surged above the maximum design of the pipe of 60 psi for a total surge pressure of approximately 300 psi.

Because this pipeline was the only source of surface water to the area, groundwater wells had to be brought online and were required to operate at full capacity to meet the water demand during the outage. Hydraulic modeling of the area suggested that groundwater wells alone were not sufficient to meet the upcoming summer peak water demands in the area. Therefore, the city was faced with the task of renewing the 42-in. water line and safely returning it to service before the summer peak demands. This schedule left only 10 months to complete both design and construction.

7.10.1.1 Design Options

Several options, including replacing the pipeline, were evaluated for restoring surface water to the pump station. Sliplining the existing pipeline with HDPE meant the existing water line could be used as a primary liner, reducing disruption to pavement and traffic. The limitations of this method included requirements that flow into the pump station be shut down during the duration of construction, and the capacity of the pipeline would be reduced to approximately two-thirds of its original capacity.

Sliplining was deemed to be a viable method due to the specific conditions of the existing pipeline and the project site. At the time of the existing pipeline construction, there was little development in the area and the project site was relatively flat, consistent vertical and horizontal alignments, with few bends. Also, only three interconnections existed along the failed portion of the line.

7.10.1.2 Other Considerations

Hydraulic modeling showed that current water demand in the area required a pipe with an approximately 36-in. ID, which was achievable by sliplining HDPE. However, to achieve the future projected water needs, an additional parallel water line was planned.

7.10.1.3 Pipe Design

It was determined that the existing concrete pipe would continue to degrade over time. As a result, it was decided that the HDPE pipe should be designed to withstand the full internal pressure and overburden weight without any support from the existing pipeline. A DR 17 HDPE pipe for a working pressure of 100 psi and a surge pressure of 60 psi was selected.

A DR 17 HDPE pipe has a wall thickness of approximately 2.3 in. with an OD of approximately 39.37 in., leaving a small room to maneuver within the existing 42-in. PCCP. Although the pipe could have been pulled in a straight run of up to 4,000 ft, it could only be pulled through deflections of up to 5°. Excavation pits were therefore required at multiple areas where horizontal or vertical bends exceeded 5°. Pits were also required where valves, fittings and connections were installed, and to remove existing butterfly valves.

The slipline pipe was only 6% smaller than the host pipe with a 42-in. typical inside dimension. Based on internal inspections, the existing concrete pipe did have a smooth, regular inside surface, however, workmanship of the interior field-grouted joints occasionally left mortar buildup at the joints, reducing the inside dimensions.

To mitigate the risk of getting the HDPE pipe damaged or stuck in the existing pipe, the contractor was required to pull a test mandrel consisting of a 40-ft-long section of HDPE prior to pulling fused sections of the HDPE slipline pipe.

7.10.1.4 Construction

During construction, the key to pulling the pipe was a pre-installation inspection—during design. Based on this inspection, the design should account for the majority of pipe ID issues, and the required number of pits and fittings. The 40-ft-long test section of HDPE which served as a mandrel was a good indicator of the ease of pipe insertion. One additional access pit was required along the project and was determined when the test mandrel got stuck. The actual HDPE water main never became stuck during pulling. During installation, the contractor was able to overcome friction between the existing pipe and the sliplined pipe by applying a viscous bentonite material along the invert of the existing pipe prior to insertion. Figure 7.11 shows a butt-fused section of HDPE pipe. Figure 7.12 illustrates construction pits showing the tight fit of sliplined installation. Figure 7.13 illustrates grouting of annulus space.

High-density polyethylene pipe had to be fused together in sections equal to the distance between proposed insertion pits. The fused sections ranged in length from 1,700 linear feet to 300 linear feet. To reduce impact to traffic and the local businesses, the longer sections were fused in approximately 500-ft runs to avoid closure of cross-traffic through median openings at intersections. Once each 500-ft segment was prepared for a particular run, these segments were fused in a matter of hours and the entire run was pulled in a single day. The other major obstacles encountered during construction were traffic control, access to businesses when long runs of pipe were butt-fused along the right-of-way.

7.10.1.5 Case Study Summary

As a result of the city's willingness to evaluate new products, the trial installation of HDPE provided an effective alternative. Sliplining failed 42-in. PCCP ultimately saved the city between $2 million and $4 million considering other options. The contractor was able to

FIGURE 7.11 Butt-fused sections of the HDPE sliplined pipe.

FIGURE 7.12 Construction pits showing the tight fit of sliplined installation.

FIGURE 7.13 Grouting of annulus space.

install approximately 6,250 ft of pipe per month, and completed 25,000 ft of installation of the HDPE sections within 4 months of notice to proceed. Although the use of HDPE may not be cost effective in all applications, the use of this material for this specific project provided a critical option which allowed the city to save both time and money.

7.11 Chapter Summary

Sliplining is a cost-effective option that needs to be considered for renewing pipelines which require a cost-effective structural option. This method can be used for both gravity and pressure pipelines. Several pipe materials are available with this method, including HDPE, FPVC, and GRP. The limitation of this method is reducing the cross-sectional area of the existing pipeline which must be evaluated with a better hydraulic capacity of the new sliplined pipe. The case studies provided effectiveness of sliplining.

7.12 Review Questions

1. What is sliplining?
2. In what conditions sliplining is suitable?
3. What types of pipe materials can be used in sliplining?
4. Describe grouting process in sliplining method and its objectives.

5. List advantages and limitations of sliplining.
6. Describe the two main sliplining methods.
7. How the lateral connections are reinstated in sliplining? Explain.
8. What type of sliplining requires grouting? Explain.
9. Discuss sliplining design considerations.
10. Discuss sliplining method for gas pipeline applications.

CHAPTER 8

In-line Replacement and Pipe Bursting

8.1 Introduction

In-line replacement (ILR) and pipe bursting (PB) technology originated in the 1970s and can be used to replace water, wastewater, and gas pipelines. This method is most cost-effective if a new pipe with a larger diameter is required. The ILR methods can be used to replace virtually all types of pipe materials such as concrete, clay, steel, ductile iron, cast iron, asbestos cement, PVC, and polyethylene (PE). In certain site and project conditions, diameters of up to 48 in. and drive lengths of 1,500 ft have been completed with pipe bursting. The new pipe can be PE (most common); PVC, ductile iron, glassfiber reinforced polyester mortar, and vitrified clay. The ILR can be divided into two main categories of pipe bursting (PB) and pipe removal (PR). Each of these categories is further divided into different methods (see Fig. 8.1). The basic difference among these systems is in the source of energy and the method of breaking the existing pipe and some consequent differences in operation that are described in the following sections. The selection of a specific replacement method depends on soil conditions, groundwater conditions, degree of upsizing required, type of new pipe, original construction method of the existing pipeline, depth of the pipeline, availability of experienced contractors, and so on.

8.1.1 Method Applicability

It is important to choose the best ILR method for specific conditions of a project. Such parameters as the site space available, the chemical composition of the fluid carried by the pipeline, the pressure and the temperature in the pipeline, number of service laterals, number of manholes, number of bends, type of existing pipe deterioration, etc.,

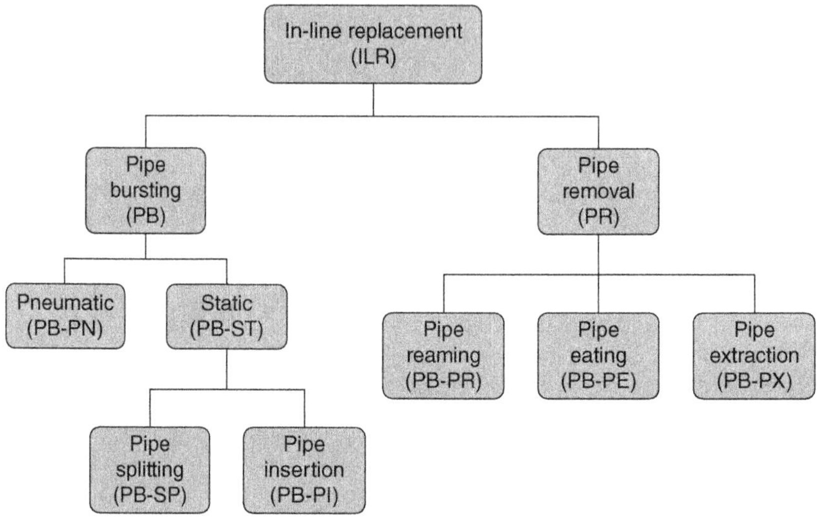

FIGURE 8.1 Categories of in-line replacement methods.

must be assessed from project to project. Table 8.1 provides an overview of the applications and possible limitations for ILR methods.

In-line replacement techniques allow the existing pipeline to be upsized while removing displaced joints and realigning the pipeline.

Pipeline Type	Suitable	Comments
Sewers	Yes	
Gas pipelines	Yes	Note A
Potable water pipelines	Yes	
Chemical/industrial pipelines	Yes	
Straight pipelines	Yes	
Pipelines with bends	Yes	Note B
Circular pipes	Yes	
Noncircular (deformed or oval) pipes	Yes	Note C
Pipelines with varying cross sections	Yes	Note D
Pipelines with lateral connections	Yes	
Pipelines with deformations	Yes	Note E
Pressure pipelines	Yes	

A. Grooves in the pipe surface must be documented and gas pockets removed.
B. Depending on whether bends less than 20°.
C. Become circular after replacement.
D. The new pipe will have same diameter.
E. Deformities are removed.

TABLE 8.1 Overview of In-line Replacement Applications and Possible Limitations

Moreover, the method does not require much of the preparatory work, such as, pipeline cleaning, grinding of intruding lateral connections, root cutting, cutting of displaced joints, required by other renewal methods. However, subsurface conditions must be thoroughly investigated for suitability of ILR methods. For example, when PB is used below the watertable or when the existing pipeline is surrounded by sand, the pneumatic pipe burster can create a vacuum that prevents progress. This problem can be solved by lowering the watertable or by using other pipe-bursting methods, such as static heads which advance solely on high pulling forces without the use of pneumatic actions.

8.1.2 Range of Applications

In-line replacement can be applied on a wide range of pipe sizes and types, in a variety of soil and site conditions. The maximum pipe size for PB is 48 in., although larger pipes have been replaced. It is expected that bursting of large pipes (e.g., 80 in.) will also be possible with larger equipment in the future. The most common PB is size-for-size, or upsizing the diameter of the existing pipe (e.g., 8 to 12 in.). Large upsizing requires more energy and causes more ground disturbance and can slow the replacement operation, and therefore, need more engineering evaluation for method and site applicability.

The existing pipes suitable for PB are typically brittle materials, such as vitrified clay, cast iron, plain concrete, asbestos, or plastics (PVC or PE). Reinforced concrete pipe (RCP) can also be replaced. Ductile iron and steel pipes are not ideal for pipe bursting, but can be replaced using pipe removal (pipe reaming).

The bursting length is usually between 300 and 400 ft, which is a typical distance between sewer manholes, nut drive lengths of up to 1,500 ft have been achieved. A long burst requires more powerful equipment. Pipe reaming replacement has been conducted for drive lengths of 1,300 ft

8.1.3 Unfavorable Conditions

Unfavorable site conditions for PB are (1) expansive clay soils, (2) obstructions such as collapsed pipes, (3) metallic point repairs that reinforce the existing pipe with ductile material, (4) concrete envelopment, and (5) adjacent pipes or utility lines that are very close to the pipe being replaced. Some obstructions do not necessarily prevent the pipe bursting and pipe removal operations, and sometimes can be removed with localized excavations. Limitations for ILR and PB applicability are in three forms: (1) economic aspects (its comparative costs versus cost of open-cut replacement), (2) safety aspects (potential damage to the neighboring utilities and structures), and (3) technical aspects (the ability to provide sufficient energy to complete the replacement operation).

Pipe bursting operation may create outward ground displacements adjacent to the pipe alignment. The bursting operation can cause ground heave or settlement above or at some distance from the pipe alignment. However, ground displacements dissipate rapidly away from the bursting operation. The most critical conditions for ground displacement are when: the pipe to be burst is shallow, upsizing for large-diameter pipes are used, and when deteriorated existing utilities are present within a distance of twice or three times diameter of the existing pipe.

Pneumatic PB may create noticeable ground vibrations on the surface above bursting operation. Unless PB is carried out at very close distances, it is unlikely to damage existing utilities or nearby structures. As a rule of thumb, the head should not pass closer than 2.5 ft from buried pipes and 8 ft from sensitive surface structures. If distances are less than these, special measures must be taken to protect the existing structures, such as excavating the crossing point to relieve stress on the existing pipe surroundings.

For most soil conditions, it is simply necessary to provide the required power to burst pipe, displace the soil and pull the replacement pipe. Longer bursts can be accomplished in favorable ground conditions. Bentonite or polymer lubrication may be injected into the annular space behind the bursting head to help keep the annular space open and to reduce the frictional drag on the replacement pipe.

8.2 Construction Considerations

8.2.1 Service Laterals

Service connections (sewer, gas, and water) to existing pipeline can create problems during bursting. This is regardless of whether these connections have been excavated prior to bursting or not. The services are usually excavated prior to bursting to provide temporary bypass service and to protect the services during the bursting operation. If the connections are not excavated prior to bursting, they can be damaged even at some distance from the connection. However, if the connections are excavated prior to bursting, a change in the profile of the replacement pipe may be created due to reduced resistance to upward movement of the replacement pipe at lateral point. This problem can be reduced by excavating beneath the pipe, as well as above the pipe, at the connection. For the reaming or pipe eating processes that have less impacts on the service lines, a wye with a riser to the surface can be installed on the service line about 2 to 3 ft from the main, and a plug between the main and the wye inserted prior to bursting. During the replacement operation the riser can be monitored and after replacement the connection to the main can be excavated and reconnected.

8.2.2 Insertion and Reception Pits

For sewer applications, insertion and reception pits are usually excavated in place of manholes. For replacement of gas and water lines, service pits can be expanded and used for the insertion and reception pit. Pit walls, if more than 5-ft deep, must be sloped or shored in accordance with Occupational Safety and Health Administration (OSHA) standards. For all static rod and cable pull machines, the bursting machine should be properly braced to resist the horizontal force necessary for the bursting operation. This may require the pit or manhole wall to have a thrust block with proper structural capabilities. Inadequate thrust block structural capacity to resist the forces can cause wall deformation that may result in soil failure and surface heave near the wall.

Different PB systems have different requirements in terms of the space required in the reception pit. Some systems may be able to operate within the existing manholes and others may need to excavate a pit for the pulling frame. The insertion pit must be in sufficient size to allow the pipe to be inserted. For continuous HDPE and PVC pipes, this means that the pipe must be able to be fed from the surface into the existing pipe alignment without overstressing the pipe in bending. Pipe manufacturers' guidelines must be followed for minimum bending radius.

8.2.3 Bursting Operation

The bursting of the existing pipe can be performed without stoppage if the replacement pipe is fused and a winch with continuous cable is used. When rigid rods are used, the bursting operation temporarily halts, for each rod sections to be unthreaded and removed from the reception pit. In addition, when the pipe is installed in segments, the preparation of each successive pipe segment also interrupts the operation.

8.2.4 Reconnection of Service

The reconnection of service, sealing of the annular space at the manhole location, or backfilling of the insertion pit for the new PE pipe must be delayed for the manufacturer's recommended time, but normally not less than 4 hours. This period allows for PE pipe shrinkage due to cooling and pipe relaxation owing to the tensile stresses induced in the pipe during installation. Following the relaxation period, the annular space in the manhole wall may be sealed. Sealing is extended a minimum of 8 in. into a manhole wall in such a manner as to form a smooth, watertight joint. Ensuring a proper bond between the PVC or PE replacement pipe and the new manhole wall joint is critical.

Service connections can be reconnected with specially designed fittings by various methods. The saddles, made of a material compatible with that of new pipe, are connected to create a leak-free joint. Different types of fused saddles (electrofusion saddles, conventional fusion saddles) are installed in accordance with manufacturer's recommended procedures. Connection of new service laterals to the pipe also can be accomplished by compression-fit service or mechanical connections. After testing and inspection to ensure that the service connections meets specifications, the pipeline is returned to service.

8.3 Pipe Bursting Methods

The main method of ILR is PB which will be discussed in detail in this chapter (Fig. 8.2). Pipe bursting is a trenchless replacement method in which an existing pipe is replaced either by percussion action (pneumatic method) or by splitting (static method), applied by a bursting head. At the same time, a new pipe of the same or larger diameter is pulled in replacing the existing pipe. Usually, the back-end of the bursting head is larger or is connected to an expander and then to the new pipe, and the front-end is connected to a winch cable (pneumatic method) or pulling rod (static method). Pipe bursting can be used to replace sewer, water, gas, and other pipelines, and is used particularly to upsize the capacity of existing pipelines, or where existing pipelines are in such poor condition that renewal is impossible and replacement is required. Pipe bursting is also used in situations where it is not possible to use open-cut techniques on the entire section of a pipeline, or where open-cut method is inconvenient for local residents and disruptive traffic. The PB industry changed significantly in 2005 with the expiration of British Gas/Advantica pipe bursting patents, opening the market to new manufacturers and allowing contractors to perform PB without paying licensing fees.

Figure 8.2 Pipe bursting process. (*Source:* Hammerhead Trenchless Equipment.)

Pipe bursting is widely used for the replacement of deteriorated pipes with a new pipe of the same or larger diameter. It is an economic pipe replacement alternative that reduces impact to social life and traffic disruptions as well as reduces damage to pavement, sidewalks, driveways, green areas, surrounding structures, etc., when compared to the open-cut technique. This method is especially cost-effective if the existing pipe is in located urban areas, is out of capacity, and deep. Replacing an existing pipe with a larger one is termed upsizing. One-size upsizing refers to replacing the existing pipe with a pipe one standard size larger, for example, replacing 8-in. pipe with a 10-in. size. Similarly, two-size upsizing is replacement of the existing pipe with a pipe two standard sizes larger, for example, replacing 8-in. pipe with a 12-in. size.

8.3.1 Pneumatic Pipe Bursting (PB-PN)

In the pneumatic pipe bursting, the bursting head is a cone-shaped soil displacement hammer (see Fig. 8.3). The bursting head is driven by compressed air, and operated at a rate of 180 to 580 blows/min. The percussive action of the bursting head is similar to hammering a nail into a wall, where each impact pushes the nail a small distance farther into the wall. In a like manner, the bursting head creates a small fracture with every stroke, and thus continuously cracks and breaks the existing pipe. The percussive action of the bursting head is combined with the tension from the winch cable, which is inserted through the existing pipe and attached to the front of the bursting head. The winch cable keeps the bursting head pressed against the existing pipe wall, and pulls the new pipe behind the head. The air

FIGURE 8.3 The pneumatic pipe bursting method. (*Source:* Hammerhead Trenchless Equipment.)

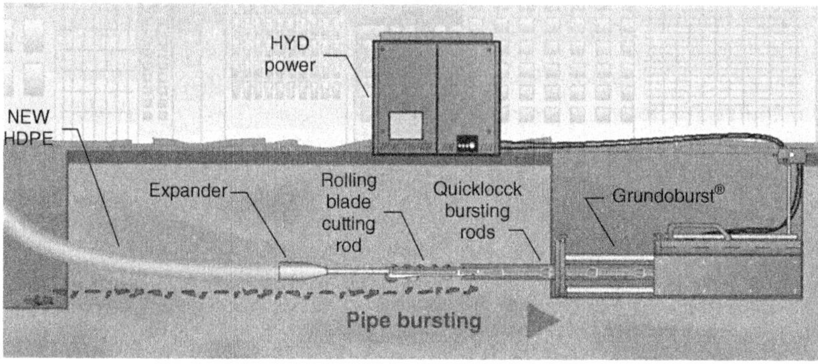

FIGURE 8.4 Static pipe bursting method. (*Source:* TT Technologies.)

pressure required for the percussion is supplied from the air compressor through a hose, which is inserted through the new pipe and connected to the rear of the bursting tool. The air compressor and the winch are kept at constant pressure and tension values, respectively.

The expander can be front-end (attached to the front of the hammer) for pipes smaller than 12 in. or back-end (attached to the back of the hammer) for pipes larger than 12 in. The front-end expander allows withdrawing the hammer through the PE pipe after removing the expander from the existing manhole or access pit without damaging the manhole. The bursting starts once (1) the head is attached to the new pipe, (2) the winch cable is inserted through the existing pipe and attached to the head, and (3) the air compressor and the winch are set at constant pressure and tension values. The process continues with little operator intervention until the head reaches the pulling pit or manhole at which point it is separated from the PE Pipe.

8.3.2 Static Pipe Bursting (PB-ST)

Another common method of PB is the static pull system. In the static pull system, the force for breaking the existing pipe comes only from pulling the bursting head forward. Either a rod assembly or a winch cable pulls the head (Fig. 8.4). The rod assembly or winch cable is inserted through the existing pipe and attached to the front of the bursting head at the insertion pit. The tensile force applied to the bursting head must overcome the frictional force and also breaking force of existing pipe and is therefore significant. The cone-shaped bursting head transfers this horizontal pulling force into a radial force, which breaks the existing pipe and provides a space for the new pipe. If a rod assembly is used for pulling, the bursting process is done in consecutive sequences, rather than continuously. Prior to bursting, the segmented rods are inserted into the existing pipe from the reception pit. The rods are only a few feet long, and during insertion, they are threaded together to reach the bursting head at the insertion pit. At the insertion pit, they are attached to the front end of the bursting head, and the new pipe is connected to its rear end. In each sequence during the bursting, the hydraulic unit in the reception pit pulls the rods for the length of individual rods, and the rods are separated from the rest of the rod assembly as they reach the reception pit. If a winch cable is used instead of rods, the pulling process can be continuous. However, a typical cable system does not transmit as a large pulling force to the bursting head as a rod assembly.

8.3.3 Pipe Splitting (PB-PS)

"Pipe splitting" is a form of static pipe-bursting method used for pipes that are not brittle, such as steel and ductile iron gas mains (Fig 8.5). Instead of the bursting head, the system uses a splitter, which cuts the existing pipe along one line on the bottom and opens it out, rather than fracturing it. The splitter is pulled through the existing pipe by either a wire rope or steel rods. The splitting system consists of one or more of three parts: (1) a pair of rotary slitter wheels, which make the first cut, (2) a hardened sail blade on the underside of the splitter, which follows, and (3) an expander, whose conical shape and off-centered alignment force the split pipe to expand and unwrap. The unwrapping of the pipe is smooth, without generating hoop stresses or longitudinal bending in the pipe walls that could cause high pulling forces and jamming. The splitting and unwrapping of the existing pipe creates a hole immediately behind the splitter and large enough to allow the new pipe to be pulled in. The existing pipe dislocates to a position above the hole and over the new pipe, thus protecting the new pipe from damage.

8.3.4 Pipe Insertion (PB-PI)

The pipe insertion system (Fig. 8.6) pushes or jacks a new pipe into the existing and deteriorated pipe. This system uses the columnar strength of segmented bell-less jacking pipe to advance the lead train through the existing pipe. The lead train consists of five sections:

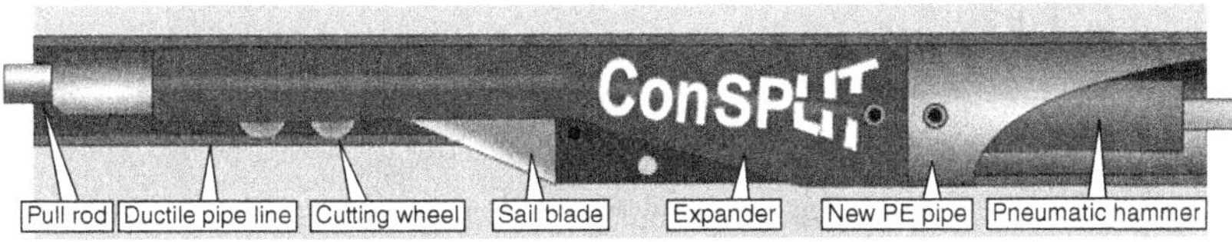

Figure 8.5 Pipe splitting system. (*Source:* ConSplit.)

the lead, a heavy steel guide pipe, which maintains the alignment within the center of the existing pipe; the cracker that fractures the existing pipe; the cone expander that radially expands the fractured line into the surrounding soil; the front jack, a hydraulic cylinder that provides axial thrust to the (above) penetration or compaction pieces; and the pipe adapter that provides mating surfaces—linking the new pipe to the front jack.

The last section, the pipe adapter, is fitted with a lubricant injection port where the lubricant (polymer and/or bentonite) can be injected into the annular space surrounding the new replacement pipe. The introduction of a lubricant allows for the efficient replacement of the existing pipelines even in soft sticky clays or wet sands.

Dual flexible hose sections, which transport lubricant and hydraulic fluid to the front train, are fed through each new pipe section. Each new hose section is connected to the previous sections and to the operator's control panel with quick-disconnect couplings. Using the new pipe as a support column, the front jack advances the lead train into the existing pipe independent of the advance of the new pipe column. The new pipe is jacked behind the lead train, piece by piece, by the jacking frame (in the entry or insertion pit). The primary jacking frame applies the required thrust to advance the new pipe column (as the front jack is retracted). Instrumentation and controls at the operator's control panel (at the jacking frame) allow the operator to feel his way through the existing pipe as the new pipe column and front train are inch-wormed into the existing pipe. Upon completion of the pipe replacement, the lead train is disassembled easily inside a typical 4-ft diameter receiving manhole and the new pipe is jacked into its final position. Pipe insertion technique is illustrated in Fig. 8.6.

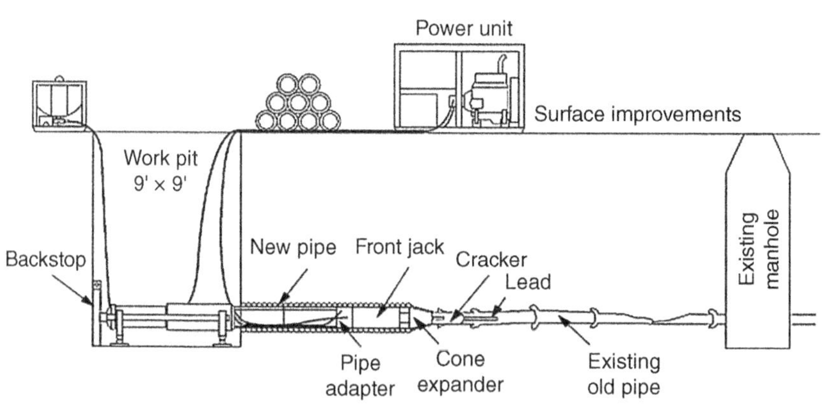

FIGURE 8.6 Pipe insertion. (*Source:* Tenbusch Insertion Method.)

8.4 Pipe Removal Methods

Pipe removal is a replacement technique, which breaks and removes the existing pipe fragments rather than displacing them into the surrounding ground as the case with pipe bursting. Pipe removal is further divided into three methods, as pipe reaming, pipe eating, and pipe ejection or extraction. The most common form of PR is pipe reaming. Pipe eating and pipe ejection have rarely been used in the United States.

8.4.1 Pipe Reaming (PR-PR)

Pipe reaming is a pipe replacement technique that uses a horizontal directional drilling (HDD) machine (see Najafi (2013) for more information on HDD) with minor modifications. After pushing the drill rods through the old pipeline and connecting the rods to a special reamer (Fig. 8.7), the new PE pipe string is attached to the reamer via a swivel and towing head. As the drill rig rotates and simultaneously pulls back, the existing pipe is ground and replaced by the new PE pipe. Removal of the existing pipe is accomplished by mixing the spoil with drilling fluid and transferring it to an exit pit for removal via a vacuum truck. Horizontal directional drilling contractors or utility contractors that use an HDD rig can add inexpensively modified reamers of various types for different pipe materials and

FIGURE 8.7 Pipe reaming method. (*Source:* Nowak Pipe Reaming.)

ground conditions. Pipe reaming is limited to non-metallic pipeline replacement. The surrounding environmental conditions (groundwater, sand, rock, concrete encasement, etc.) that prohibit other procedures are not obstacles to successful installations.

Pipe reaming method can remove existing vitrified clay pipe (VCP), polyvinyl chloride (PVC), asbestos cement (AC), or nonreinforced concrete pipe and simultaneous replacement with an HDPE or PVC pipe of equal or larger diameter. Dependent on the project and site conditions, double diameter replacement is the practical maximum upsize. Removal of the existing pipe is accomplished by back reaming using a directional drilling machine with a cutter head that breaks the existing pipe to small pieces, which are carried along with excavated soil by the drilling fluid to an extraction point.

High density polyethylene (HDPE), restrained joint or fusible PVC, and ductile iron pipes provide the new pipe that will satisfy performance as well as installation requirements. The horizontal directional drilling (HDD) equipment requirements for the pipe reaming system generally serve two purposes for the contractors. The purchase of HDD equipment provides the contractor with not only the ability to enter the trenchless market with the pipe reaming system, but also to enter the HDD market using the same equipment. A modified backreaming tool is the only special tool required that may be modified at the contractor's shop or purchased from a manufacturer. Whether or not the prospective user owns a directional drill, the initial investment can be readily amortized.

8.4.2 Pipe Eating (PR-PE)

Pipe eating uses a modified microtunneling machine, which has been specially adapted to allow the existing pipeline to be broken up, and removed. The MTBM has a larger diameter than the existing pipeline (see Najafi (2013) for a detailed description of the microtunneling method). The microtunnel boring machine (MTBM) crushes the existing defective pipeline and removes the particles by the circulating slurry system. A new pipe is simultaneously installed by jacking it behind the microtunneling machine. The new pipe may follow the line of the existing pipe on the entire length, or may cross the elevation of the existing pipe on a limited segment only. The microtunneling system is remotely-controlled and is guided with a laser line from the drive pit. During the operation, the MTBM removes or eats whatever is in its way, the existing pipe or the ground. The MTBM has a cutterhead and a shield section. The cutterhead has cutting teeth and rollers that cut the pipe, and cutting arrangements close to the edge of the shield cut the ground to the required diameter to take the new pipe. The cutterhead is cone-shaped, which puts the existing pipe material into tension and thus reduces the heavy wear of cutting teeth. The shield section carries the cutterhead and its hydraulic motor system. The MTBM is launched from a drive pit, where a thrust frame is located. The jacking frame provides a thrust force that is applied to the new pipe to push the MTBM forward.

8.4.3 Pipe Ejection or Extraction (PR-PX)

Pipe ejection or extraction is a technique that removes the existing pipe as a whole from the ground, by pushing or pulling it toward a reception pit where it is broken and taken out. Pipe ejection (a modified version of pipe jacking) and pipe extraction (modified static pipe bursting) are PR systems, in which the unbroken existing pipe is removed from the ground, while the new pipe is simultaneously installed. The existing pipe is broken into pieces only as it completely exits out of the ground. These techniques are applicable only for existing pipes with sufficient remaining thrust capacity to withstand the push or pull forces. These methods are used in projects with shorter replacement drives to avoid high frictional resistance.

In pipe ejection, the jacking frame in the insertion pit pushes out the existing pipe by jacking the new pipe segments forward. The new pipe is placed against the existing pipe, and as the new pipe is jacked, the existing pipe is pushed toward a reception pit or manhole. At the reception pit, the existing pipe is broken into pieces and removed. The jacking frame and the insertion pit are sized to fit the length of individual new pipe sections.

In pipe extraction, the new pipe is pulled into place of the existing pipe. An extraction machine is placed into reception pit, and the replacement pipe is fed from the entry or insertion pit. A pulling device (a pulling rod assembly) is inserted through the existing pipe, and attached to the extraction machine at one end, and a tool assembly, which is connected to the new pipe, at the other end. The tool assembly consists of a centralizing device, pull plates, and a cylindrical expander or plug, which allows the method to handle size-for-size replacement or upsizing.

8.5 Pipe Materials

8.5.1 Existing (Old) Pipe Material

In most bursting applications, the existing pipe is composed of a rigid material such as VCP, cast iron, plain concrete, asbestos, or some plastics. Reinforced concrete pipe has been successfully replaced when it was not heavily reinforced or if it was substantially deteriorated. The diameter of the existing pipe typically ranges from 2 to 30 in., although bursting of pipes of larger diameter is occurring more frequently. A segment of 300- to 400-ft is a typical length for bursting, although longer drives have been completed using powerful bursting systems. However, if there have been point repairs on the existing pipe, particularly using ductile iron; the overall process becomes difficult.

8.5.2 Replacement (New) Pipe Material

High- and medium-density PE (HDPE and MDPE) are the most common replacement pipes for pipe-bursting applications[1]. The new PE4710 pipe provides better performance and greater strength capability than the previous PE3608/3408 pipe. The main advantages of PE pipe are its continuity, flexibility, and versatility. The continuity, which is obtained by butt fusing together long segments in the field, reduces the possibility of stopping the process. Up to 6 in. diameter, the pipe may also be supplied in continuous lengths on a reel. The flexibility allows bending the pipe for curved insertion into the pit and inside the existing pipe. In addition, it is a versatile material that meets the other requirements for gas, water, and wastewater pipelines. The relatively smooth interior surface reduces the friction between the fluid flow and the pipe wall, which allows higher flow speed and increased flow capacity. The smooth exterior surface also reduces the friction between the soil and the pipe, facilitating the pulling operation. In general, PE pipe does not significantly degrade and does not support bacteriological growth. The relatively higher thermal expansion coefficients represent the main disadvantage of PE pipes, requiring proper installation and restraint. Refer to Chap. 5 for more information on different pipe materials.

The internal surface of PE pipe is smoother than that of the concrete or clay pipe. For gravity applications, Chezy-Manning equation (Eq. 8.1) demonstrates that the flow capacity of the PE pipe is 44 percent greater than those of the concrete or clay pipes assuming the internal diameter for the old clay or concrete pipe equals that of the replacement PE pipe.

$$Q = \frac{1.49}{n} A(r_H)^{2/3} \sqrt{S} \tag{8.1}$$

where Q is flow quantity (cfs), n is Manning roughness coefficient (unitless), A is internal cross-sectional area of the pipe (ft^2), r_H is hydraulic radius (ft), S is the slope of the energy line (unitless), which is parallel to the water surface and pipe invert if the flow is uniform. The n value for clay or concrete pipes ranges between 0.012 and 0.015, in comparison to approximately 0.009 for PE; the ratio corresponds to an average increase of 44 percent for the latter material.

In addition to PE and FPVC, other new pipe materials are DI, VCP, and RCP which can be pushed and/or pulled. Rigid pipes (RCP and VCP, see Chap. 5) cannot be assembled into a single pipe string prior to the bursting, which allows a continuous pulling operation.

[1]Use of restrained joint PVC with static systems or fusible PVC (FPVC) pipe with both static and pneumatic system have increased recently, and should be considered as viable options. The PVC pipe provides larger inside diameter for the same DR, which provides more flow capacity.

However, they can be jacked into position behind the bursting head or maintained in compression by towing them via a cap connected to the cable or rod that passes through the pipes. Therefore, the static pull system is the only bursting system that can be used with the rigid pipes. The joints of these pipes must be designed for trenchless installations to be smooth inside and out.

8.6 Pipe Bursting versus Other Alternatives

For repair and replacement, conventional techniques require open-cut excavation to expose and replace the pipe. As described in Chap. 6, CIPP can be used to structurally renew an existing pipe. The main advantage of CIPP over PB is reducing the need for surface excavation for access pits and reconnection of service laterals.

Pipe bursting is more cost effective compared to CIPP when

- there are few lateral connections to be reconnected with the replacement section;
- the existing pipe is structurally deteriorated; and
- additional flow capacity is needed.

For pressure applications, a 41 percent increase in the inside pipe diameter doubles the cross-sectional area of the pipe and consequently doubles the flow capacity of the pipe. For gravity applications, after some algebraic manipulation to the above-mentioned Chezy-Manning equation, it can be shown that a 15 percent and 32 percent increase in the inside diameter of the pipe combined with the smoother pipe surface, can produce a 100 percent and 200 percent increase in the flow capacity, respectively.

Pipe bursting has substantial advantages over open-cut replacements; it is much faster, more efficient, and more cost effective than open cut, especially in sewer applications due to the large depths at which gravity sewer pipes are typically installed. For open-cut, sewer pipelines require shoring, spoil removal and hauling embedment materials, and possible dewatering, which substantially increase the cost of open-cut replacement. The increased existing pipeline depth has a minimal effect on the cost per foot for pipe bursting, as shown in Fig. 8.8. Specific studies carried out in the United States have shown that pipe-bursting cost savings are as high as 44 percent with an average savings of 25 percent, compared to open-cut. These cost savings could be much greater if social costs are considered, such as if the existing pipe is deep (more than 10 ft), surface area is developed (pavement, green areas, and heavy traffic), underground conditions include unstable soil conditions and high watertable (see Najafi and Gokhale (2005) for social costs details). Furthermore, open-cut excavation can potentially cause significant damage to nearby buildings and structures.

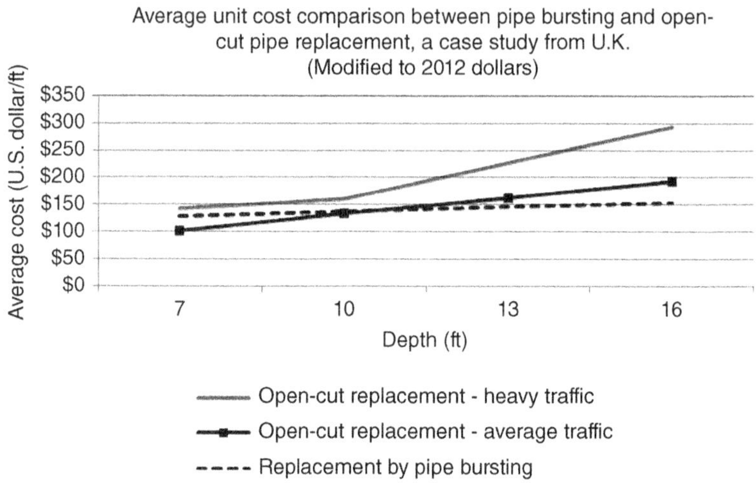

FIGURE 8.8 Cost comparison between pipe bursting and open-cut replacement. (*Source:* ASCE, 2007.)

As stated above, in addition to the direct cost benefits of PB over open-cut, pipe bursting has several indirect cost savings, including less traffic disturbance, road or lane closing, shorter time for replacement, less business interruptions, and less environmental and social impacts. Pipe bursting creates minimal interference with other utilities and fewer safety hazards (for both operators and the public) due to reduced open excavation.

8.7 Pipe-Bursting Project Classification

The American Society of Civil Engineers (ASCE, 2007) and the International Pipe Bursting Association (IPBA, 2012) classify PB projects into three categories, in terms of difficulty: class A—routine; class B—moderately difficult to challenging; and class C—challenging to extremely challenging, as shown in Table 8.2. Note that the degree of difficulty increases as more than one of the above criteria applies (ASCE, 2007).

8.8 Advantages and Limitations

Typical PB replacement length is between 300 ft and 500 ft; however, in favorable conditions, longer drives have been completed successfully. The size of pipes being burst typically ranges from 2 to 30 in., although larger size pipes can be burst. Pipe bursting is commonly performed size-for-size and one-size upsize above the diameter of the existing pipe. Larger upsizes (up to three pipe sizes)

Criteria	A	B	C
Degree of difficulty	Minimal	Moderate	Challenging
Depth	<12 ft	12–7 ft	>7 ft
Existing pipe ID (in.)	2–12	12–7	20–36
New pipe diameter	Size for size	Single upsize	Double/triple upsize
Burst length	<350 ft	350–450 ft	>450 ft
Trench width	Relatively wide trench compared to upsized diameter	Trench width less than 4 in., wider than upsize diameter	Moderately compressible soils outside trench (medium to dense sand, medium to stiff clay)
Soil	Compressible soils outside trench (soft clay, loose sand)	Moderately compressible soils outside trench (medium dense to dense sand, medium to stiff clay)	Constricted trench geometry (width less than or equal to outside diameter of burst head)

Source: IPBA (2012).

Table 8.2 Pipe-Bursting Classification

have been successful, but the larger the pipe upsizing, the more energy is required and greater ground movement will be experienced. It is important to pay close attention to the project surroundings, depth of installation, and soil conditions when replacing an existing pipe, especially in unfavorable conditions, such as expansive clay soils, pipe repairs made with ductile material, collapsed pipe, concrete encasement, sleeves, and adjacent utility lines. Pipe bursting has the following specific limitations:

- Insertion and pulling shafts are required, especially for larger bursts.
- Excavation for lateral connections is required.
- Expansive clay soils may cause difficulties for bursting existing pipe.
- A collapsed (existing) pipe segment may require excavation at that point to allow the insertion of pulling cable or rod and to repair pipe sag.
- Point repairs with ductile material can interfere with the bursting and replacement process.

- If the existing sewer pipeline is significantly out of line and grade, the new line will also tend to be out of line and grade, although some corrections of localized sags are possible.

8.9 Preliminary Surveying

To achieve the best possible result of pipe-bursting techniques, it is important to carry out detailed preliminary surveying to determine, at the very least, the following:

- Location and type of any previous point repairs.
- Soil conditions around the existing pipe—this can be the type of embedment for existing pipe and surrounding soil conditions.
- Location of nearby utilities.
- Survey of nearby surface structures and documenting their conditions.
- The conditions of the existing pipeline—using closed-circuit television (CCTV) inspection or similar.
- The need for cleaning based on CCTV inspection.
- Location of valves and appurtenances.
- Pipeline length and depth.
- Pipeline dimensions, deformations, and displaced joints and bends.
- Groundwater depth.
- Soil and live (traffic) loads on the pipe.
- Pipe service conditions (type of flow, pressures, etc.).
- Service lateral information, including locations, dimensions, unused service lines, active service lines, and intruding service line connections.
- Existing access point or manhole conditions.

8.10 Preparatory Work

Once necessary preliminary surveying has been completed and replacement is about to commence, there are several routine preparatory work that should be carried out, such as following:

- Planned bypass pumping, its duration, and the number of customers affected. Bypassing potentially can be eliminated if the bursting job for a section of 300-400 ft can be performed in one day using static PB and a prechlorinated new pipe in accordance with local and state regulations is used. In this case, the consumers must be notified to stop their service for the expected duration of bursting operation. It should be noted that in areas near hospitals, nursing homes, fire departments, schools, and police facilities, etc., the bypass pumping may be required. The contractor may need a risk assessment and contingency plan just in case if the bursting duration due to operational and installation difficulties exceed the expected duration, so temporary service to customers can be reinstated by bypassing.

- Planned traffic diversions and their advertising.
- Pipeline shutoff (timings, duration, and number of laterals affected).
- Sizes and location of entry and exit pits.
- Data logging and marking of lateral connections.

Usually, the preparatory work required will depend on what the pipeline is designed to carry, and whether it is a sewer or pressure pipeline. For a sewer, for example, preparatory work would include the following.

- Excavation of laterals and their disconnection from the main, for example, by placing vertical supports in the excavations
- Breaking up manhole floors so that the bursting head and PE pipe can pass through without raising the pipeline profile
- Closing off/sealing unused laterals

8.11 Design Considerations

Pipe bursting projects can be divided into three phases: pre-design, design, and construction. The pre-design phase includes collecting information about the existing pipeline, investigating the alternative solutions, and verifying that PB is the best solution. The design phase requires investigating the conditions of the existing pipe and its embedment, nearby utilities and structures, determining pit locations, bypass pumping requirements, developing detailed drawings and specifications, confirming that PB is the optimum solution, and selecting the appropriate bursting system. The construction phase includes submittals, bypass pumping, pit construction and shoring, implementation of the bursting system, installation of lateral connections, and restoration including testing and disinfection (for water applications).

8.11.1 Pre-design Phase

During the pre-design phase, the ability to influence the cost of the project is the highest, and the cost of project modification is the lowest, as shown in Fig. 8.9. This is especially true for small jobs where the contractor's cost savings (from design modification) are minimal,

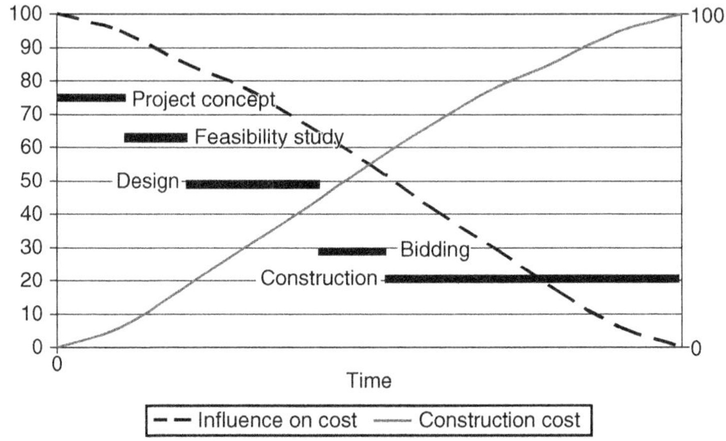

FIGURE 8.9 Ability to influence construction cost over time. (*Source:* Project Management Institute Inc., 2004.)

and the benefits do not justify the risk of being responsible for the redesign and its consequences. Therefore, effort invested in this phase will pay dividends later.

The pre-design phase includes collecting information about the old and new pipelines. The designer determines the maximum flow requirements for the future design life of the pipeline (considering the future economic development and population growth trends), and then calculates the diameter of the new pipe. This phase also includes investigating potential solutions and collecting relevant information to perform an initial (tentative) evaluation of the various solutions. For example, possible alternatives may include lining the existing pipe, replacing the pipeline via open-cut, replacing it by pipe bursting or pipe removal, and so on.

Although open-cut is typically the specified method of construction for some pipeline projects, PB is often a viable alternative. The process may therefore include preparation and submittal of two (or more) initial bids: one based on open-cut and the second a value-engineering proposal based on pipe bursting. In order to avoid possible increased overall project costs due to risk and contingency factors, it is important that the initial bidding information is as complete as practical for all considered alternatives, including both open-cut and bursting. This procedure will help ensure that competition is maintained and the risk and impact of changes is reduced, as was illustrated in Fig. 8.9. If PB is determined to be a likely or possible solution, the design team proceeds to the design phase.

8.11.2 Design Phase

The design phase starts with collecting additional information regarding the existing pipeline, including current flow volume for bypass pumping, lateral connections, embedment width and compaction levels, and manhole locations. This phase also includes locating nearby utilities, investigating soil and embedment material, and developing risk assessment plans. The feasibility of PB as the optimal solution may need to be re-evaluated in light of the new collected information. The designer completes this phase with the development of detailed drawings and specifications, and completes (final) bid documents, which includes a listing of required submittals. The drawings should provide all relevant information concerning the existing pipeline and environment, such as diameter and material, plan view and profile, nearby utilities and structures (crossing and parallel), repair clamps, concrete encasement, fittings, and so on. Some of this information may be collected by means of a CCTV system or equivalent.

8.11.3 Utility Survey

The presence and nature of surrounding utilities will have a significant impact on the success of the pipe-bursting operation, and the design engineer must, therefore, identify and locate these existing utilities as accurately as possible. However, the "exact" location of these utilities will be identified during the construction phase through visual locating, such as vacuum potholing. The identification of nearby utilities is critical for the following reasons:

- The presence of nearby utilities may eliminate PB as a viable construction method.
- Existing utilities may affect the location of insertion and pulling/jacking pits.
- The risk of causing significant damage to existing utilities.
- Protection of existing utilities from the ground movement due to bursting operation.
- Risk of injuries and fatalities to the workers and public by avoiding damage to existing utilities.
- Number of utilities to be exposed, to be properly reflected in contractor's bid.

An on-site investigation should indicate the locations of many utilities; for example, sewer manholes indicate the presence of a sewer line and fire hydrants indicate the presence of a water line, and so on. The engineer should contact the one-call center for information on existing utilities in the area, review the available as-built drawings from the different utility owners, and ideally consider geographic information system data (if available), utility maps, and conduct surface and subsurface investigations to superimpose these utilities on plans and profiles.

8.11.4 Investigation of Existing Pipe and Site Conditions

Investigation of the existing pipe condition assists in selecting the suitable renewal/replacement technique and provides the exact location of the lateral connections. The nature and condition of the existing pipe may render PB as an unsuitable solution. The presence of sags in the pipeline may require correction prior to bursting. The host pipe (diameter, material, and conditions) and the diameter of the new pipe guide the contractor in the selection of an appropriate type of bursting system, size, and accessories, and should therefore, be reflected in the bidding process and construction phase. The site conditions and surface features may affect the locations of the insertion and pulling pits, staging area for fused pipe, traffic control planes, and footprint for the required bursting system components.

8.11.5 Insertion and Pulling Shaft/Pit Requirements

When planning for pit locations, the engineer identifies locations where excavation is required to replace manholes, valves, lateral connections, or fittings. These locations also serve as insertion or pulling pits. However, if excavation at the manhole location is not necessary or feasible, pit excavation at other locations may be considered. In selecting the location of these pits, the engineer has to consider the following issues:

- Sufficient staging area for the fused replacement pipe to avoid blocking driveways and intersecting roads.
- Pit length should be sufficient to allow alignment of the bursting head with old pipeline and for the PE pipe to bend safely from ground surface to entry point.
- Space for the construction equipment, including backhoe, loader, crane, and so on.
- Nearby flow bypass discharge location or space to lay bypass lines without blocking driveways and intersecting roads.
- Traffic control around pits.
- Soil borings close to proposed pits.
- Discharge locations for dewatering, if necessary.
- Ability to use same pit to insert or pull pipes more than once.

In general, the engineer recommends locations for the insertion and pulling pits but leaves the final determination to the contractor (through a submittal approval process), subject to minimizing excavation and disturbance to the surrounding environment.

8.11.6 Geotechnical Investigation Requirements

The soil and subsurface investigation includes collecting the necessary technical information to properly designing the project. This process assists the contractor in submitting a proper bid by selecting the appropriate bursting system (type and size), shoring of the pulling and insertion pits, dewatering system, compacting backfill material, and so on, thereby increasing the chance of success during the construction phase of the project. Thus, if the original soil borings (during the original pipe installation) are available, they should be reviewed and provided as part of the supplemental information available to the bidders. The determination of the trench geometry and backfill material and compaction is important for the designer and contractor.

The soil investigation activities comprise performing soil borings, standard penetration tests, groundwater level determinations, original trench geometry investigation, and native soil and existing pipe embedment material classifications. If the presence of washouts or voids around the existing pipe is suspected, a ground penetrating radar survey may assist in determining the locations and magnitudes of these voids. Special attention should be given to the presence of major difficulties that may render PB not feasible, such as the presence of rock, hard cemented dense soils, very soft or loose soils, reinforced concrete encasement, very narrow trench in hard soils or rock, or ductile point repairs. If contaminated soil is suspected, the type and extent of contamination should be identified and indicated in the contract documents. The contractor should be requested to take the necessary measures to handle and dispose of this contaminated soil.

The soil around the pipe (embedment and native soil) must be compressible in order to absorb the diametric expansion. Compressible soils are ideal because the outward ground displacements will be limited to an area surrounding the pipe alignment. Soils with long "standup time" allow the overcut (expanded hole) to remain open for most of the bursting operation, thus reducing the friction force between the soil and the pipe. The overcut reduces the required pulling forces and axial stress on the new pipe during installation. Original pipe embedment material is the most suitable soil for bursting followed by (increasing difficulty) compressible clay, loose cobble, beach and running sand, densely compacted clay, and sandstone. Somewhat less favorable ground conditions for PB also include soils below the water-table and expandable soils. Special soils, such as highly expansive clay soils or collapsible soils, may also cause problems.

In very soft or loose soils, significant ground movement may take place causing sags in the new pipeline and damage to nearby structures. In severe situations, the soil particles migrate to the interior of the existing pipe, converting the bursting operation into a piercing operation.

Although there may be some plausible advantages in performing PB below the groundwater table, such as possible reduced friction between the pipe and soil, as well as reduced effective soil pressure on the pipe due to the lower buoyant weight of the soil above, this situation requires dewatering of shafts or pits.

8.11.7 Maximum Allowable Operating Pressure

For pressure applications, such as water, gas, and force mains, the maximum allowable pressure should be determined based on the maximum surge pressure that pipe will be subjected to and the maximum operating pressure for the pipe. The PE pipe should be designed to withstand the maximum allowable operating and surge pressures according to conventional design procedures, such as provided in *the Plastics Pipe Institute Handbook of Polyethylene Pipe* (PPI, 2008). DR 17[2] is typically used for bursting pressure or gravity pipe unless a higher pressure rating is required. For relatively shallow bursting runs, where high tensile and vertical forces are not expected, DR 21 may be used.

8.12 Risk Assessment Plan

It is important to pay close attention to the project surroundings (surface and subsurface conditions) to identify unfavorable conditions and significant risks. These conditions require extra attention in order to ensure the safety of the construction personnel and public, as well as surrounding facilities and infrastructure. Most underground and pipeline construction projects entail some risks due to unknown subsurface conditions. The risks associated with PB include damage to nearby facilities, failure to complete the project using the selected technology, and time and/or budget overrun. The risk of damage to nearby utilities, buried structures, and pavement increases under adverse soil conditions, improper construction techniques, design errors, inaccurate location of utilities, and so on. There are also risks associated with flow bypass, dewatering, shoring, if inappropriate procedures are employed. The lists of additional risks that may halt or imped the bursting operation and/or create problems include the following:

- Settlement at insertion/pulling pits if the density of the existing pipe embedment materials exceeds that of native soil.
- Bursting through sharp curves.
- Concrete encasement or steel point repair of existing pipe.
- Excessive bursting lengths.
- Damage to new (replacement) pipe from sharp edge or fragments of existing pipe being burst/split.
- Damage to laterals from bursting of main line.

---

[2]The DR, or dimension ratio, of the pipe is the outer diameter divided by the wall thickness.

- The presence of rock under the existing pipe may create a 'bump' in the replacement pipe.
- Collapsed replacement pipe.

Projects within class C (challenging to extremely challenging as indicated in Table 8.2) must be carefully examined in terms of required forces and ground displacements. In addition, due to the enlargement of the hole and accompanying compaction of surrounding soil, the depth of the existing pipe affects the extent of ground displacement above the pipe. If the pipe is shallow, damage to the pavement may take place. Saw cutting the pavement prior to bursting might be advisable. If the existing pipe is below the groundwater table (GWT), the difficulties increase. Insertion and pulling pits or shafts grow larger and more complex as the depth increases.

If there are unacceptable sags in the existing sewer line, these sags must be corrected before bursting. The sags can be corrected by local excavation, surface grouting, or grouting from within the pipe. Some reduction of sag magnitude may be expected (without corrective measures) from the bursting operation, but the extent to which the problem is corrected depends on the relative stiffness of the soil below the sagging section. Alternatively, the pull or insertion shafts or pits can sometimes be located over the sag area, which allows the problem to be corrected.

If there is erosion of the soil around the pipe, the bursting head and the following PE pipe will tend to deviate toward the void or lower density region. Similarly, if there is a hard soil layer or rock close to the pipe, the bursting head will tend to displace toward the softer soil. In shallow conditions, the ground will deviate mostly upward toward the ground surface. If the conditions change substantially along the length of the burst, this may cause some change in the grade and/or alignment of the pipe. When the grade is critical, these possibilities should be considered.

In general, pipe-bursting operations can be performed successfully and safely if site and project conditions are known before bursting and appropriate measures are taken to address these conditions. There are well-known solutions to all of the above mentioned risks and problems, and experienced project engineers or construction managers identify these risks and develop a risk management plan to address these specific risks for their project. This plan includes quantification of the likelihood of the identified events and their associated impact or degree of damage; it also includes measures to eliminate, mitigate, transfer, or undertake these risks. One of the general measures to mitigate the project risks is building and maintaining cooperative relationships among owners, engineers, contractors, equipment manufacturers, and pipe suppliers. The identification and development of a realistic plan to manage and share risks appropriately is an important part of effectively communicating responsibilities, defining roles, and building a strong team.

8.12.1 Ground Movement Associated with Pipe Bursting

The pipe-bursting process creates a cavity in the soil around the pipe through which the new pipe is pulled. This cavity creates a compression plastic zone around the new pipe outlined by an elastic zone, as shown in Fig. 8.10. The magnitude of the compression and the dimensions of these zones correlate with the amount of upsizing, the diameter of the pipe, and the type of soil (Atalah, 1998). Based on investigations of the ground movements and vibrations associated with bursting small-diameter pipes in soft soils (Atalah, 1998) and with large-diameter pipes in rock conditions (Atalah, 2004)[3], guidelines have been developed for safe separation distances from existing nearby utilities, structures, and pavement (see Fig. 8.11).

Figure 8.11 compares the peak particle velocity (PPV) of the soil versus the distance from the source of the vibration for different types of construction equipment, including pneumatic pipe-bursting results based on the aforementioned investigations (i.e., small-diameter PB in

[3]The investigation of the large diameter bursting in rock conditions included upsizing 24-in. diameter RCP pipes by less than 50 percent. The studies of the small diameter bursting in soft soils included upsizing 8-in. and 10-in. diameter VCP by less than 30 percent.

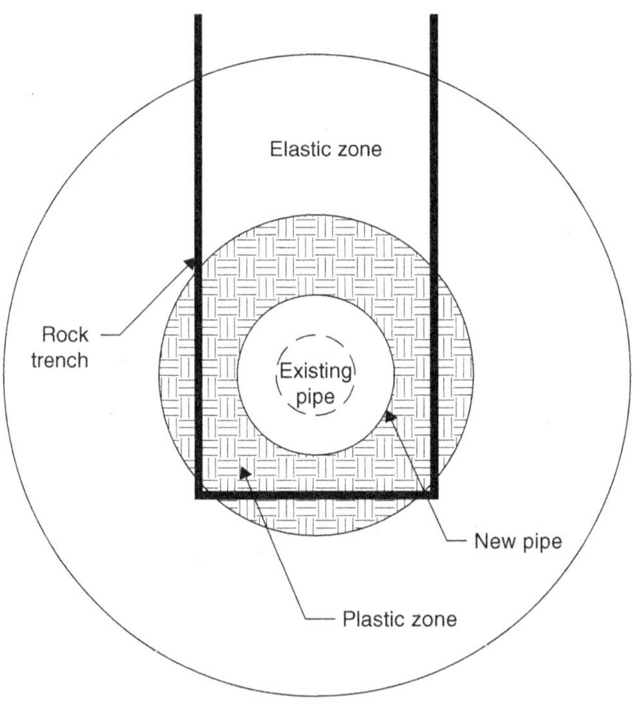

FIGURE 8.10 Cavity expansion and the plastic and elastic zones.

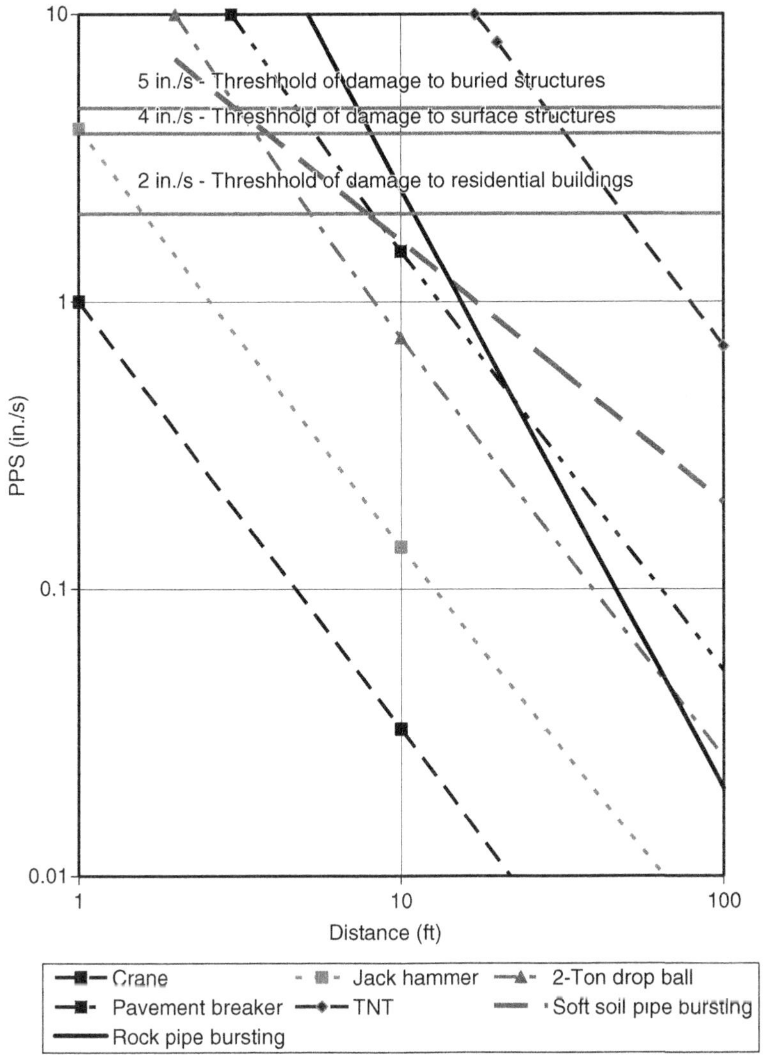

FIGURE 8.11 The attenuation lines versus distance for different construction equipment.

soft soils and large-diameter bursting in rock conditions.). The PPV is the velocity of soil particles as they vibrate due to the construction activities. In all cases, there is a strong correlation between the distance from the construction tool and the level of vibration. The vibrations due to bursting are seen to rapidly decline to levels that do not cause damage to buildings. For structurally sound residential buildings, minimum separations of 11 ft and 8 ft are recommended for large-diameter bursting in rock conditions and small-diameter bursting in soft soils, respectively. Corresponding minimum separations of 8 ft and 4 ft are recommended from structurally sound commercial structures. Considering pipes as "buried structures," the statistical analysis as reflected in Fig. 8.11 suggests that 7½ ft represents a safe separation distance for pipe-bursting operations for either small- or large-diameter pipes, in various soils, with reduced spacing possible for small sizes. Belowground pipelines include deep main lines installed in the right of way, which are usually far from the residential or commercial buildings.

8.13 Plans and Specifications

The contract documents typically include the contract agreement, general conditions, special conditions, project plans, specifications, geotechnical report, and CCTV records. The plans and specifications for pipe-bursting projects should have all the required information for typical open-cut water or wastewater pipeline projects plus the information listed in this section. The drawings should provide information about the existing site conditions and the required construction work. Description of site constraints (i.e., work hours, noise, etc.) and the procedures to review the CCTV data should be listed in the notes section in the drawings. The plans may also include information to show erosion and sediment control requirements, flow bypassing plans, and service connection and reinstatement details. The plans should typically include:

- limits of work; horizontal and vertical control references.
- topography and survey points of existing structures.
- boundaries, easements, and rights-of-ways.
- existing utilities, sizes, locations, and pipe materials.
- the verification requirements for existing utilities.
- plan and profile of the design alignment.
- existing point repairs, encasement, sleeves, and so on.
- construction easement and the allowable work areas around the insertion and pulling pits.
- details for lateral connections and connections to the rest of the network.
- restoration plans.
- traffic control plans.
- existing flow measurements for bypass pumping (Najafi, 2007).
- quality control/quality assurance (QA/QC) measures.

The technical specifications supplement the drawings in communicating the project requirements. Information to be included in the technical specifications should include the following:

- General
  - Minimum contractor qualifications.
  - Permit matrix and responsibilities.
  - Safety requirements with focus on confined space entry, flow bypass, and shoring.
  - Scheduling requirements and construction sequence.
  - Submittals.
- Pipe and manhole materials
  - Standards and tolerances for pipe material, wall thickness and class, and testing and certification requirements.
  - Construction installation instructions for pipe joining and handling.
  - Fittings, appurtenances, and connection-adaptors.
  - Acceptable material performance criteria and tests.
- Construction considerations
  - Flow bypassing, downtime limits, and service reinstatement requirements.
  - Spill and emergency response plans.
  - Traffic control requirements.
  - Erosion and sement control requirements.
  - Existing conditions documentation (e.g., photographs, videos, and interviews).
  - Protection plan for existing structure and utility (ground movement monitoring).
  - Accuracy requirements of the installed pipe.
  - Daily construction monitoring reports.
  - Field testing and follow-up requirements for pipe joining, pipe leakage, disinfection, backfill, and so on.
  - Site restoration and spoil material disposal requirements (Najafi, 2007).

8.13.1 Submittals

The submittals for pipe-bursting projects usually include site layout plans, sequence of bursting, shoring design for all the excavations, bypass pumping plan, manufacturers' specifications of the selected bursting system and its components, dewatering plan, new pipe material, lateral connections material and plans, site layout plans, contingency plan, and so on. The site layout plans should show the location of the insertion and pulling shafts, dimensions of shafts, traffic flow, safety and communication plan, storage space to store and lay the new pipe, and so on. Site restoration and cleanup plans should also be included in the submittals.

8.13.2 QC/QA Issues

The project specifications should state the QC/QC measures required to ensure that the project is executed according to the contract specifications. In addition to the QA/QC measures usually specified for a traditional open-cut project, there are several measures specific to pipe-bursting operations. These measures can take the form of tests, certifications, inspection procedures, and so on. Extensive listing of the relevant required submittals, careful preparation of the submittals, and careful review and approval of the submittals are significant steps in the QC/QA program. The QC/QA program states the performance criteria for the new pipeline and the acceptable tolerance from these criteria. For example, the invert of the new pipe should not deviate from the invert of the existing pipe by more than a specified number of inches, the depth of sags in the line should not exceed one inch, and the difference in the vertical and horizontal dimensions of the new pipe diameter (ovalling) should not exceed 2 percent.

The QC program should state how these performance criteria will be measured, tested, and checked. Some methods for verifying performance are post bursting CCTV inspection, pressure testing, and mandrel test. A surface and subsurface displacement-monitoring program can be outlined in the specifications along with the acceptable amount of ground movements. Certifications from the manufacturers of the bursting system, replacement pipe, and other material may be required to verify that they meet the contract specifications, based on tests or past projects conducted by the manufacturer or a third party. For challenging projects, the presence of a representative from bursting system manufacturer at the jobsite may be required. The owner's QA program should ensure that the field and management teams of the contractor have the knowledge and the experience needed to complete the project successfully and are able to respond appropriately to unforeseen problems.

It is critical that the contractor ensures that the replacement pipe meets the specifications, before, during, and after bursting. Adhering to the QA/QC plans during the pipe manufacturing and shipping to the site, along with proper unloading of the pipe, reduces risk of pipe failures. Pipe fusion operations should be performed by well trained (preferably certified) workers under appropriate supervision to reduce the risk of future problems when repair would be difficult and costly. For pressure application, the PE pipe should be inspected and pressure tested prior to bursting, as well as after completion of the installation.

8.13.3 Dispute Resolution Mechanisms

The contract should include clauses addressing the possibility of different site conditions and unforeseen conditions that allow contract time and amount adjustment if the conditions at site "materially" differ from the conditions expected and indicated in the bid documents. These clauses facilitate resolving disputes efficiently and quickly without negative impacts on the project. If site conditions are significantly different from those described in the contract documents, and the contractor or owner can show that the different conditions impacted the work, the contract value and duration should be adjusted accordingly. Conducting the proper surface and subsurface investigations, outlined earlier in this chapter, should minimize the occurrence of project disputes and possibility of work stoppage during the pipe bursting operations.

8.13.4 Permitting Issues

Permits from all affected parties should be secured before the start of the bursting operation. Some of these permits could be secured by the owner, and the rest should be secured by the contractor. The responsibilities for obtaining the individual permits should be outlined in the specifications and stated on the drawings. Permits to burst under the road and to modify the regular traffic flow according to the project traffic control plan should be secured from department of transportation (DOT), local government, or municipality. If the pipe crosses beneath a runway, taxiway, drainage ditch, irrigation channel or canal, and railroad track, permits should be secured from the appropriate agencies. Communications should take place well before start of bursting operation to inform residents and businesses about road closures, night or weekend work, service disruptions, driveway blockings, and so on.

8.13.5 Typical Bidding Form

In addition to providing the owner with the total price of the project for comparisons, the bid form should provide the contractor and the owner with a mechanism for fair pricing and payment system based on the progress during construction. The unit prices in the form can also be used to resolve disputes amicably. It is recommended that the bursting operation be measured in linear footage and segmented by classification or sections from manhole to manhole, or from insertion to pulling pits. Segmentation by section or bursting class provides the owner and the contractor with a better pricing mechanism and helps reducing future disputes. Lateral connections should be measured by number, indicated as separate bid items, and segmented by depth. Cleaning, testing, and post inspection (CCTV) of the new pipe should be a separate bid item. Bypass pumping can be priced on daily basis or as a lump sum measured by each pipe segment.

8.14 Construction Considerations

Upon awarding the contract, a notice to proceed is issued. At this stage, the contractor usually takes the following steps:

- Preconstruction survey
- Cleaning or pigging of pipeline, if necessary
- The CCTV inspection, if necessary
- Excavations at services for temporary bypass
- Setting up temporary bypass or connections to customers
- Excavation of insertion and pulling pits
- Fusion of PE pipe sections, or joining other pipe materials, as necessary
- Setting up the winch or hydraulic pulling unit and insertion of pulling cable or pulling rods inside the existing pipe
- Installation of hoses through the PE pipe to attach to bursting head (air supply hoses for pneumatic systems), as appropriate
- Connection of bursting head to pulling cable or rod
- Pipe bursting and replacement with new pipe
- Removal of bursting head and hoses from the pipe
- Post installation inspection
- Pipeline chlorination if it is not pre-chlorinated (for water mains)
- Reconnection of services and reinstating manhole connections
- Site restoration

Butt fusion of PE replacement pipe is typically carried out prior to the bursting operation, so that all fused joints possibly can be chlorinated (for water lines), and tested. The new pipe should not be dragged over the ground, and possibly pipe rollers should be used for both insertion and transportation of the new pipe. The ends of water or gas pipes should be capped to prevent contaminants entry.

8.14.1 Typical Pipe Bursting Operation Layout

The first step in planning the pipe bursting operation is the optimization of the locations of the insertion and pulling pits by using the insertion pits to insert the new pipe in two (opposite) directions. This optimization reduces the amount of excavation, mobilization, and demobilization efforts. These pits should preferably be located at manholes or lateral connections in sewer lines and at fire hydrants or gate valves in water applications. The length of the segment between the insertion and pulling pits should not generate friction forces that exceed the capabilities of the bursting system and the tensile strength of the new pipe.

The next step is to ensure that the area around every shaft is sufficient for safe operation of the necessary equipment and work staging. The insertion pit contains a flat section and sloped section; the flat portion must be sufficiently long to allow alignment of the centerline of the bursting head with that of the existing pipe. The slopped section must be sufficiently long to allow the PE pipe to bend without damage to the pipe (i.e., accommodate the manufacturer's minimum bending radius requirements.). PE pipes can typically be cold bent to a radius of 25 to 30 times the OD of the pipe, depending on the DR value. For example, for a 7-in. HDPE pipe with a DR of 17, the minimum horizontal length of the insertion shaft is 12 times the diameter of the new pipe (7 ft) plus a sloped length of 2.5 times the depth of the pit, as shown in Fig. 8.12.

The width of the pit depends on the pipe diameter and required working space around the pipe. The pulling pit must be large enough to allow for operation of the winch or pull-back device, along with removal of the bursting head. Due to the flexibility of PE pipe, the outside laydown area of the pipe prior to insertion does not necessarily have to be in line with the existing pipe.

Acceptable arrangements for traffic control, based on departments of transportation (DOT) and local government regulations, and for arranging the fused PE pipe, with minimum inconvenience to nearby residents and businesses, must be carefully considered. The flow bypass pumping and pipe layout should also be planned. If it is necessary to dewater, safe and proper flow discharge plans are required. The contractor has presumably previously submitted the jobsite layout plan that reflects the intended method of construction, and addresses the above considerations. The contractor does not start the bursting operation before the engineer or inspector has reviewed and approved the jobsite

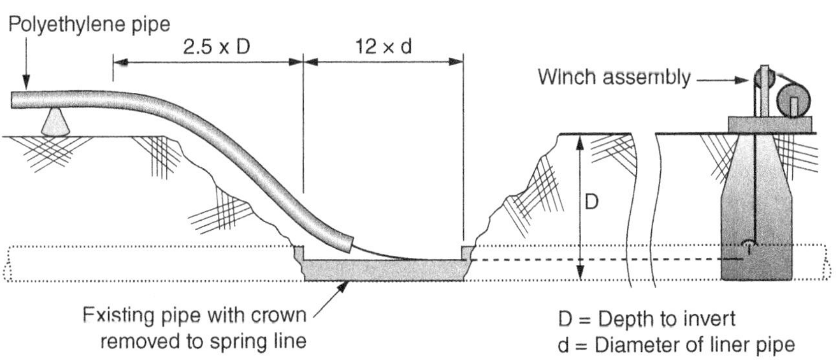

FIGURE 8.12 Insertion shaft dimensions for HDPE pipe with DR 17. (*Source:* PPI, 2008.)

layout plan. The site inspector enforces the adherence to this plan unless there is a valid reason for deviation, and that has been approved by the engineer or owner. If contaminated soil is excavated, the contractor must take the necessary measures for handling and disposal.

8.14.2 Shoring the Insertion and Pulling Shafts/Pits

Proper shoring of the insertion and pulling pits is essential for the safety of the workers and the stability of the surrounding environment. The trench shoring or bracing must be constructed to comply with Occupational Safety and Health Administration (OSHA) standards. There are several available means of shoring these shafts: trench box, soldier pile and wood lagging, steel sheet piles, corrugated pipes, and so on. In addition, if space is available, another option is to slope the sides of the shaft to provide stability. The judgment and the supervision of a competent person (as defined by OSHA) or a qualified geotechnical engineer is required to ensure the shoring is safe.

In the pulling pits (or manhole), the static bursting machine will thrust against the wall on one side. This side must be capable of withstanding the corresponding pressures. In static bursting applications, the pulling forces can be a few hundred tons, so the contractor should possibly construct a thrust block to distribute the force over a reasonably large area. The thrust block, shoring, and soil behind the shoring must be able to withstand the stresses from the pulling system. The passive earth pressure of the soil must exceed the stresses generated by the pulling system with an acceptable factor of safety.

8.14.3 Matching System Components to Reduce Risk of Failure

One of the most critical activities prior to initiating the bursting operation is to ensure that the system has sufficient power to burst the existing pipe segment between the insertion pit to the pulling pit. The system must be able to burst the existing pipe as well as overcome the friction between the soil and the outside surface of the new pipe with a reasonable margin of safety to account for unforeseen repair sleeves, clamps, and so on. In general, the bursting system components should be appropriately matched to the project; for example, the winch capacity should be compatible with the bursting head size and the conditions of the job.

The contractor should adhere to the sizing guidelines stated in the operations manual issued by the bursting system manufacturer to match the system with the needs of the job. The manufacturer should be consulted if there is any doubts regarding the adequacy of the system for that specific run under those particular conditions (soil, depth, type of existing pipe, etc.). Lubrication of the outside surface of the pipe with polymer or bentonite (depending on the type of soil) can reduce the coefficient of friction between the pipe and the soil, and consequently reduce the required pulling force.

8.14.4 Existing Nearby Utilities

The contractor must perform its due diligence to identify, locate, and verify the nearby underground utilities prior to digging the pits and initiating the bursting operation. This includes contacting the local one-call center to request representatives of the utilities to locate their existing lines. Then the contractor should verify the exact location and depth of these utilities via potholing and/or vacuum excavation. Manual excavation may be required within a few inches from the existing utilities to avoid damage. Vacuum excavation is an excellent tool to expose utilities with minimum surface excavation and minimum risk to the existing utilities.

Underground utilities that are in moderately good condition are unlikely to be damaged by vibrations at distances of greater than 2½ ft from the bursting head for small sizes (less than 12 in. in diameter), in soft soil, which is a typical application for pneumatic pipe bursting operations (Atalah, 1998). The safe distance for large-diameter bursting (up to 24 in.) is about 7½ ft. These guidelines are generally consistent with that reported by Rogers (1995), indicating that ground displacements are unlikely to cause problems at distances greater than 2 to 3 diameters from the pipe alignment. Utilities that are closer to the bursting head than these distances should be exposed prior to bursting so the vibration from the bursting operation would be isolated or reduced prior to reaching the utility in question.

It should be noted that it is possible to specify static pipe bursting equipment when near existing utilities due to the lack of vibrations inherent with static pipe bursting. Also, after a nearby utility is exposed, it is a good practice to excavate under the utility to allow relief to the utility as the bursting head passes by.

8.14.5 Bypass Pumping Considerations

One of the objectives of the bursting team (owner, engineer, contractor, etc.) is to minimize customers' service interruptions for water and gas applications, and continuation of flow for sewer applications. The key for achieving this objective is the bypass pumping system. For water applications, the system should be able to deliver the needed flow volume with the specified pressure to the customers. For gravity applications, the system should be able to adequately pump the upstream flow and discharge it to the manhole downstream of the run being burst. The plan should ensure that the bypass system has adequate pumping capacity to handle the flow with emergency backup pumps to ensure no interruption to existing services. The bypass pipes and fittings should have sufficient strength to withstand the surge water pressures. Contractual arrangements between the owner and the contractor should be made regarding third party damage due to the disconnection and reconnection of the water lines without fault of the contractor. In critical high flow situations, it is prudent to have a second primary pump on standby in the case of failure of the first pump, and it sometimes is necessary to have automatic controls in place to switch the pumps in case of failure of the primary pump.

8.14.6 Dewatering Considerations

The pulling and the insertion pits should be dry during installation. Therefore, if rain is expected or the pipe invert is slightly below the groundwater table (GWT) in clay soils, installation of a dewatering sump pump at one corner of the pit is required. Ditches at the perimeter of the pit, sloped towards the sump pump location, lined with filter fabric, and filled with gravel, may be necessary to direct the water towards the sump pump. If the pipe invert is significantly below the GWT in sandy or silty soils, a more elaborate dewatering system is recommended, such as wellpoint system, deep wells, or a larger sump pump system. As the water level is drawn down, soil particles tend to travel with the water toward the dewatering system undermining utilities and structures. Therefore, as it is the case with every dewatering system, the contractor should take all necessary measures to prevent the migration of the soil particles from underneath nearby buildings and utilities. Since the discharge flow volume in this case is expected to be large, a suitable discharge in compliance with the local requirements is necessary. If a sump pump is used, preliminary treatment of flow to reduce the sediments may be necessary before discharging into storm water streams.

8.14.7 Ground Movements Monitoring Program

The safety of nearby buildings and structures is paramount, as in the case of deep open-cut installations. The safety of nearby structures can be compromised if these structures are subjected to excessive ground movements or vibration. For extremely challenging pipe bursting operations (class C, Table 8.2), preconstruction survey and monitoring of ground movement is advisable. A preconstruction survey of all the nearby buildings and structures that documents all existing cracks, cosmetic problems, and structural deficiencies is recommended prior to any work on site. The ground elevations of carefully planned control points (on nearby buildings and on the ground surface) around the insertion and pulling pits and pipe alignment should be surveyed and monitored prior to bursting, during bursting, and after bursting. These preconstruction surveys and elevation monitoring can significantly reduce the risk of unwarranted claims to the contractor and the owner.

8.14.8 Pipe Connection to the Manhole

The thermal elasticity of the PE material causes changes in the pipe length; for example, 1 in. change in length per 100-ft of pipe for each 10°F temperature change. Therefore, in extreme hot or cold weather conditions when there is significant difference between the temperature of the deep soil and the ambient air temperature, it is recommended to allow the pipe to rest for 12 to 24 hours prior to tie-ins. Also when pipe has been pulled to a significant portion of its allowable tensile load, it may be prudent to let the pipe rest as well before connecting to other pipes, fittings, manholes, and lateral connection. This allows the pipe to recover from any stretching that may have occurred during bursting.

In most pipe bursting applications, the sewer line is old and deteriorated, including the manholes along the line. In most cases, it is, therefore, economical on a life-cycle cost basis, to replace the original deteriorated manholes and use their locations as pulling or insertion pits. As shown in Fig. 8.13, when existing manholes are replaced with new ones, connections to PE pipe can be made using flexible rubber manhole connectors called "boots." A pipe clamp is used to tighten the boot around the PE pipe for new manholes, as shown in Fig. 8.14.

If the original manhole is in reasonable conditions and it is judged to be capable of providing adequate future service, the manhole bench is removed and the pipe opening is enlarged to allow the passage of the bursting head. Expandable urethane grout and oakum can be used to create a seal between the existing pipe opening and the new PE pipe, as shown in Fig. 8.15. The compression ability of seal allows PE pipe movement.

Frequently, the inlets and outlets of the manhole are damaged during the bursting operation, such that the resulting inlet or outlet is no longer round. A low shrink polymer cement grout may then be used to repair the damage. To obtain a good seal to the PE pipe, a special PVC fitting with bell end and sand texture outer surface (as shown in Fig. 8.15) may be used. The grout bonds to the manhole and the rough sanded surface of the PVC fitting provides a good seal. The gasket between the PVC fitting and the PE pipe allows the

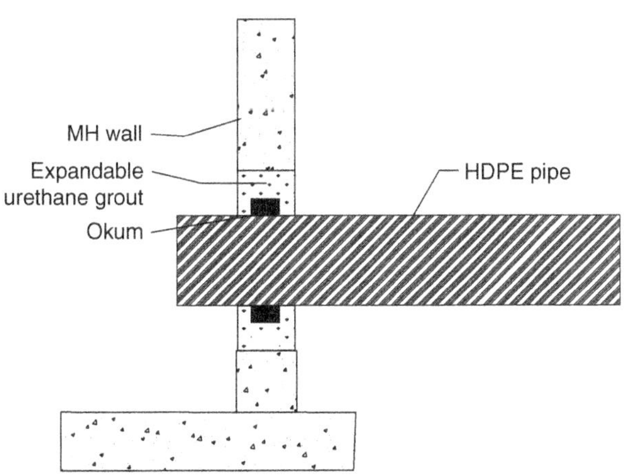

FIGURE 8.13 Connecting PE pipe to original manhole.

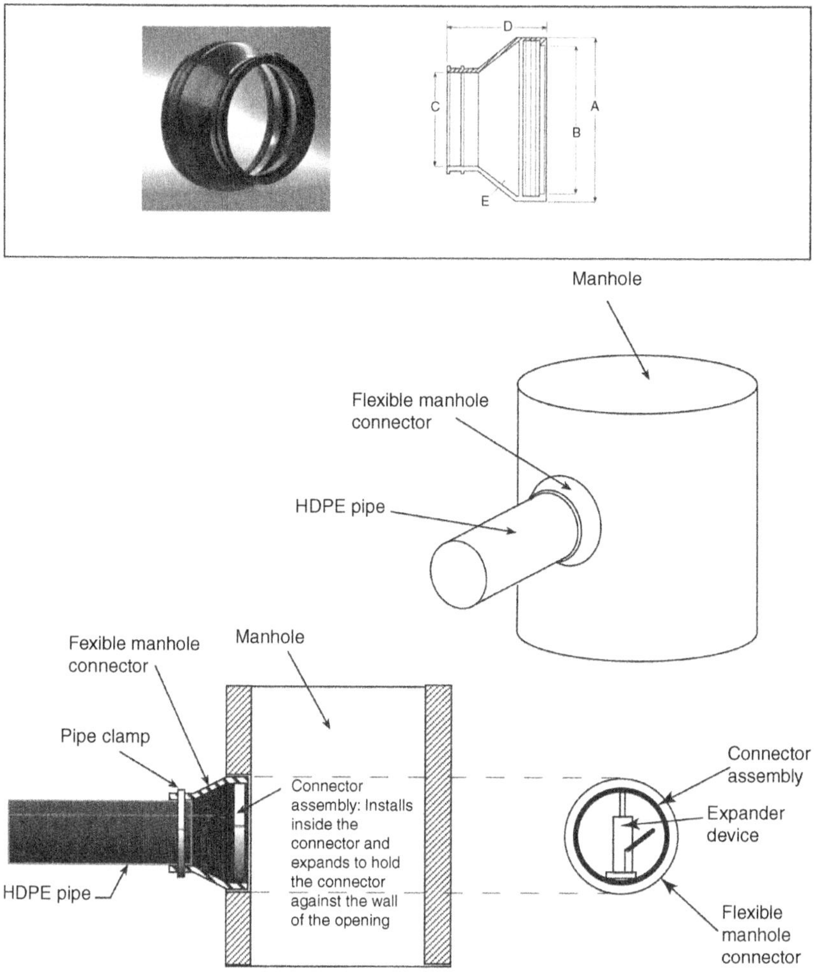

FIGURE 8.14 Connecting PE pipe to new concrete manhole.

PE pipe to move if expansion or contraction occurs. The PVC fitting requires PE pipes with DR of 21 or lower.

8.14.9 Pipe Connection to Other Pipes

PE pipes are joined to other PE fittings by heat fusion or mechanical fittings. They are joined to other material by means of compression fittings, flanges, or other qualified transition fittings (PPI, 2008).

8.14.10 Pipe-Bursting Water Mains

The most common materials for existing water mains are cast iron, DI, and PVC. All these three pipe can be replaced by PB but each requires

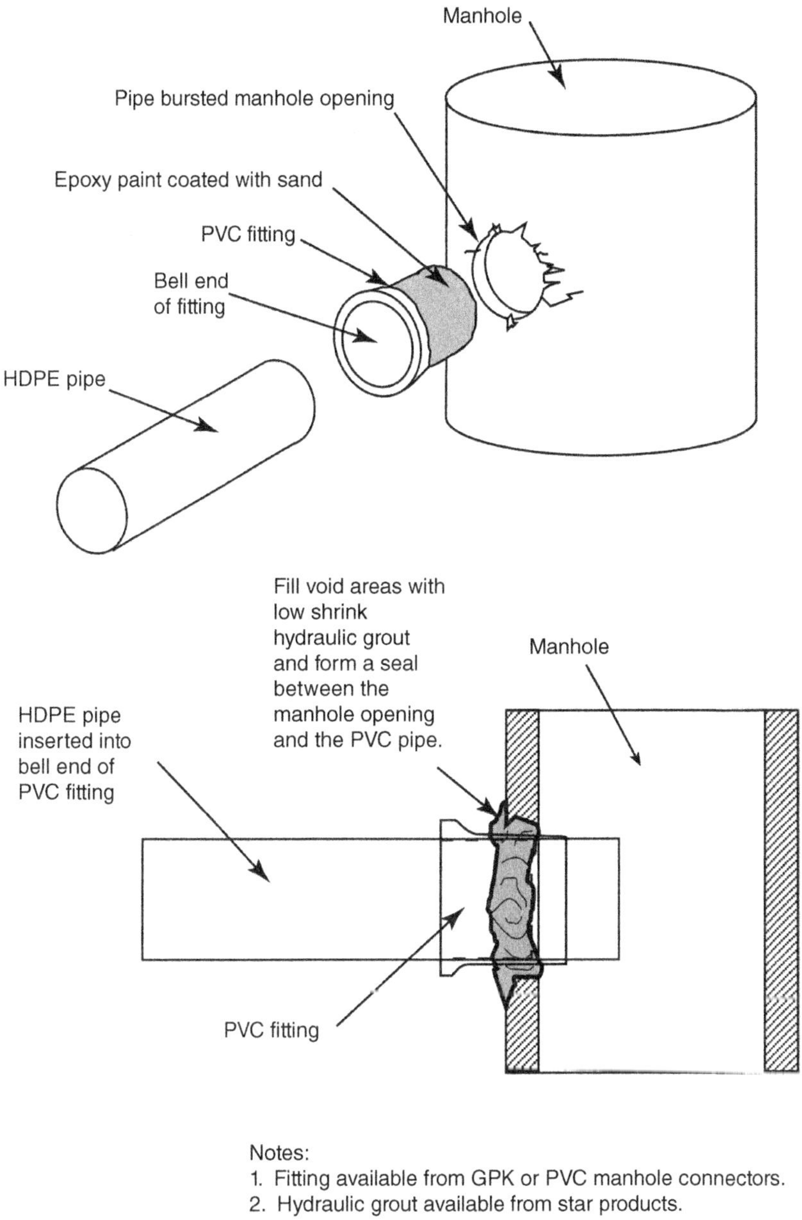

FIGURE 8.15 Connecting PE pipe to original manhole with damaged inlets/outlet.

a different piping burst approach. Cast iron pipe is a relatively brittle material, and therefore, the basic pipe bursting system is sufficient to shatter the pipe. The PVC pipes are not brittle and require multi-blade cutting accessories in front of the bursting head to facilitate cutting the

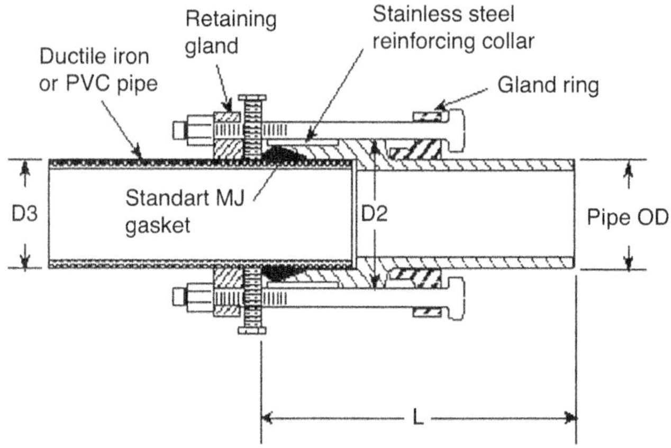

FIGURE 8.16 Connection to existing PVC or DI pipes using female MJ adapter.

pipe. Ductile iron pipe also is not brittle; and therefore, pipe splitting is the most suitable bursting system.

Connections between the replacement PE pipe and remaining segments of non-PE pipe, including valves, are accomplished using mechanical joint (MJ) adapters, which are butt fused to the PE pipe. A gland ring is then used to make a restrained connection between the two types of pipes. Figure 8.16 shows connections to PVC or DI pipes made using a female MJ connector, which is butt fused to the PE pipe. This connection provides a restrained connection (Fig. 8.16).

During installation, a temporary above-ground PE bypass is usually installed to continue to supply water to the community, while the main line is under construction. Figure 8.17 shows an example of above-ground PE bypass pipe.

FIGURE 8.17 An example of above-ground PE bypass pipe.

The PE pipe can be pre-chlorinated prior to bursting to reduce the overall installation time and inconvenience to the home/business owners (see case study in Sec. 8.26). A trace wire usually is required on all PE water pipes.

8.14.11 Service Connections

The service (lateral) connections stated in the submittal list should explain the proposed material and connection procedures. Video inspection of the original sewer line normally provides the location of service connections. In replacing water and gas lines, metal detectors can often be used. Standard practice is to locate and expose services prior to pipe bursting. Service connections can be made with Inserta Tee, specially designed fusion fittings, or strap-on saddles. Figure 8.18 shows an electrofusion saddle with a cutter attached, by which it is convenient to hot tap PE pipelines for water or gas applications.

For sewer applications, a "window" is cut in the PE pipe wall, and then one of the above fittings is used to connect the new PE pipe to the lateral connection. The Inserta Tee connection is a three piece service fitting consisting of a PVC hub, rubber sleeve, and stainless steel band, as shown in Fig. 8.19.

The Inserta Tee is compression fit into the cored wall of a sewer mainline and requires no special tooling. Inserta Tees are designed to connect 4 in. through 15 in. services to solid wall, profile, closed profile, and corrugated pipes. The PE lateral connection options include fusing a lateral HPDE pipe to the main line and placing an electrofusion sewer saddle. Fusing a lateral PE to the main PE line requires curved iron that allows heating the ends and exposed surfaces of both pipes. This connection requires a highly skilled fusion worker,

Figure 8.18 Electrofusion saddle with cutter attached.

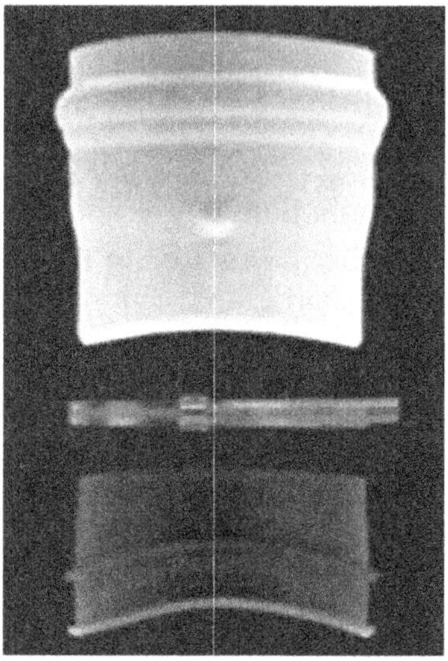

FIGURE 8.19 Inserta Tee fittings for sewer lateral connections.

because it is usually made in restricted muddy space. An electrofusion saddle is mounted on the opening for fusion with the main line, as shown in Fig. 8.18. It is important to ensure that all exposed surfaces are clean and maintained in an acceptable condition for the fusion operation. Strap-on saddles use a PE or PVC saddle that is lined with a rubber layer; the saddle is strapped around the main line using a stainless steel band, (Fig. 8.20). After testing and inspecting the line and the connections, the excavation is backfilled and the line returned to service. Additional details for gravity and pressure service connections are presented in the *Handbook of Polyethylene Pipe* (PPI, 2008).

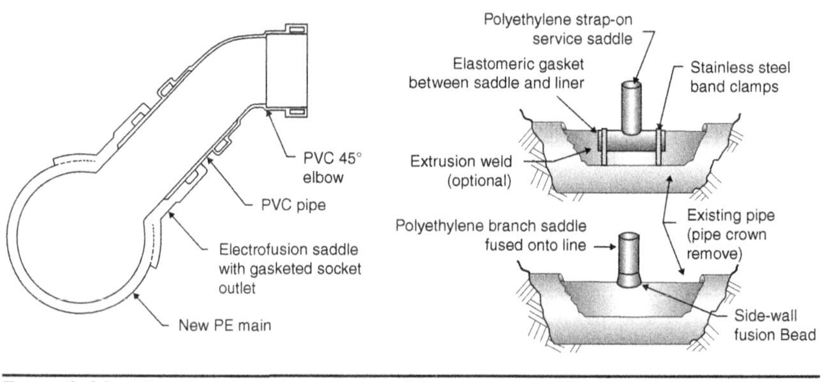

FIGURE 8.20 Pipe fusion and strap-on saddle sewer lateral connections.

8.15 Groves on the Outside Surface of the Pipe

A common misconception regarding PB is that the pipe fragments from the original pipe can damage the PE pipe as pulled into the cavity. British Gas Company conducted a detailed study on bursting cast iron pipes and concluded that there is no significant damage to the HDPE replacement pipe. Related supporting research, conducted at Louisiana Tech University and Bowling Green State University, indicates any possible groves are very shallow and narrow when bursting clay, asbestos, and concrete pipes. The widest groove was 0.07 in. across and the deepest groove was 0.03 in., as shown in Fig. 8.21. Cast iron, clay, concrete, and asbestos pipes generally shatter without producing sharp shards that may potentially produce significant damage to HDPE pipe. An exception to this rule occurs when attempting to significantly upsize ductile iron pipe. It is, therefore, recommended to limit PE replacement pipe to size-on-size bursting, or a single upsize, when bursting DI pipe. If larger upsize of DI pipe is required, PE pipes with a protective outside surface, similar to that shown in Fig. 8.22, may be used.

If the PE replacement pipe will be dragged for a long distance over rough surface or pavement, the contractor can reduce the risk of scratching the pipe by placing it on rollers or cut-outs of available (scrap) PE or PVC pipe, to keep the pipe off the rough surface. Sometimes, it is necessary for the contractor to push down on the replacement PE pipe with the bucket of the excavator to ensure that the PE

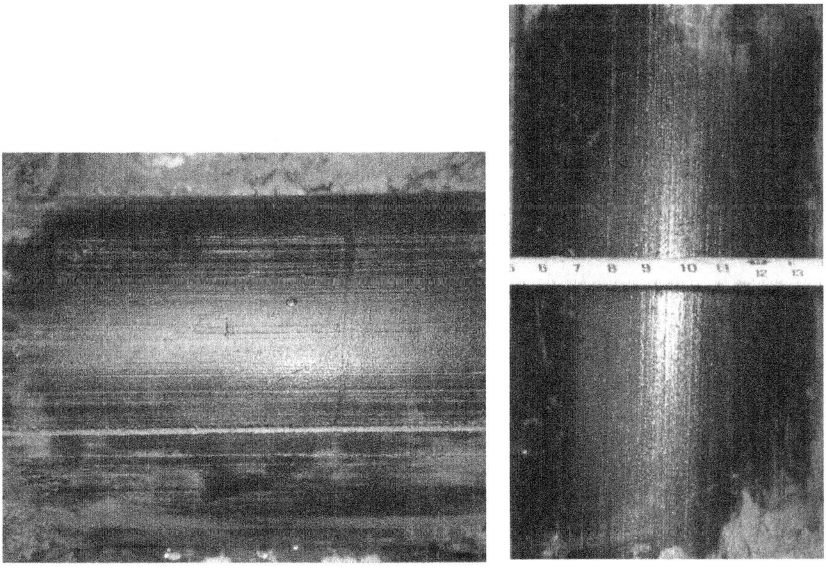

A B

FIGURE 8.21 Grooves on the outside surface of the PE pipes after bursting.

FIGURE 8.22 Protecting the PE pipe from shards.

pipe is properly aligned with the original pipe at the entry point in the insertion pit (Fig. 8.20), in the event the shaft is not of sufficient length. Wheels, similar to those shown in Fig. 8.23 (*a* and *b*), reduce the risk of grooves or scratches on the PE pipe. When the head end of the pipe reaches the pulling pit, the pipe should be inspected for surface damage. Surface scratches or defects in excess of 10 percent of the wall thickness should be rejected.

8.16 As-built Drawings

As-built drawings are usually required for any underground utility construction, including pipe bursting projects. The bursting contractor should mark the new line manholes, and ancillary structure information on a copy of the plans, marked and dedicated as new as-built. The contractor should document on these plans any changes to the original layout of the underground utilities and structures that occurred during the construction phase. For example, rerouting any utility due to the excavation of the pits, reconfiguration of other utilities required for bypass pumping, and so on should be marked on new as-built drawings. These changes, shown on new as-built drawings, should be verified and used to update the corresponding certified record electronic files for the locality.

A

B

FIGURE 8.23 Protecting the PE pipe from grooves using welded wheels on the bucket.

8.17 Contingency Plan

An important submittal is the contractor's contingency plan. Most contractors have contingency plans that include planned corrective actions if certain events take place. Pipe-bursting projects require specific additions to their standard contingency plan. Some of these specific events that are unique to PB and should be addressed include the following:

- Excessive ground movements or vibrations
- Slow bursting progress or stuck bursting head
- Problems with bypass system and with diverting and reconnecting the services to the customers
- Damage to existing waterline, gas line, sewer line, power cable, and so on
- Dewatering problem in insertion or pulling pits or at lateral connection pit

8.18 Safety Considerations

Standard safety procedures, as followed in typical open-cut construction, should also be followed in bursting projects. In addition, the workers should understand the components of the bursting system and how they function, with special attention to the moving parts in the system. The workers should be trained and equipped with the necessary tools for confined space entry; as workers may operate in live sewers during flow bypass and diversion, which typically takes place within a manhole. The static machine pushes against a thrust block that (along with the soil behind it) must withstand the forces applied by static bursting machine. The stability of the soil behind the thrust block should not be compromised. During the flow bypass, the upstream pipe will be plugged; these plugs should be braced and preferably remotely inflated and deflated. Prior to bursting, the contractor must ensure that there is no unforeseen gas line or power line close to bursting head.

8.19 Completion Work

After installation, it is often necessary to carry out completion work. This may, for example, take the form of the following:

- Leak testing
- Customer information following service line reinstatement
- Manhole finishing to ensure continued smooth flow
- Final CCTV inspection to document the completed replacement

The purpose of completion work is to ensure that tasks over and above actual pipeline replacement are performed and documented in accordance with the contractor's own quality system, and satisfaction of the pipeline owner.

8.20 Replacing Laterals

Pipe bursting can be used to replace service laterals in the same way as for mains. When replacing a lateral, the pipe is usually inserted from a convenient point on the property or by establishing a new boundary manhole. In this case, pipe bursting is performed toward the mains connection. Replaced laterals are connected to the mains using saddle connectors. Depending on its making, these can be fastened in various ways, including bolts, stainless steel collars, or electrofusion (see Chap. 13 for more information on lateral renewal).

8.21 Pipe Bursting with Segmental Pipes

Basically, PB with segmental pipes is performed in the same way as PB with long, fusion-joined pipe strings. However, with segmental method, only one pipe section (such as 8-ft long) is inserted at a time, after which the process is stopped and a new pipe section is connected and pulled into the existing pipeline. The benefit of this method is that the launch pit will be shorter. The drawback is that the method is more time consuming because of the many joints that must be fused and cooled or otherwise connected.

8.22 Quality Assurance and Documentation

A final quality review compares the documented work performed with the requirements, and descriptions of the contractor and any supervising authority. A final review should always be conducted, its purpose being to confirm that work done is in compliance with agreements made.

When a pipe-bursting project is completed, the contractor provides the pipeline owner with documentation of the quality of the work performed in the form of a quality manual. In this respect, it is important that during the tendering phase, the pipeline owner ensures that such documentation can be prepared and submitted on completion of the renovation work. The quality manual should at least contain or state the following:

- Declaration stating which products/installation methods have been used.
- How preparatory work and preliminary surveying have been performed.
- Information on pipe material handling.

- Availability and use of installation manuals.
- How supplementary work and final QA have been carried out and documented.
- As-built drawings.
- How deviations in relation to the instruction manual have been dealt with.
- Complaint processing.

8.23 Cost Estimating

Estimating pipe bursting projects for bidding purposes should be as detailed, methodical, and systematic as for open-cut installations. For each run, the contractor should estimate the labor, material, and equipment necessary for excavation and shoring of shafts or pits, shaft bottom stabilization (concrete or gravel), bursting system set up, pipe fusion, lateral connection excavation, bypass pumping, bursting operation, service reconnection, shaft backfill, surface restoration, and potential dewatering.

As was shown in Fig. 8.8, the cost per foot of pipe-bursting installations is less than that of open-cut in unfavorable (average or light traffic) situations. Figure 8.8 also showed that the cost is less than that of open-cut in favorable (heavy traffic) conditions if the depth of cut is more than 10 ft. Table 8.3 shows the unit cost from several previous pipe-bursting projects with different sizes and upsize percentages.

Based on a survey of bidding prices from several municipalities and contractors, the Trenchless Technology Center[4] has developed Fig. 8.24 (modified to 2012 prices) for PB for size-to-size replacement. The bid prices ranges for upsizing less than 20% and upsizing larger than 20%, using pipe-bursting replacement of different diameters pipes, are shown in Fig. 8.25.

[4]Louisiana Tech University.

Project No.	Existing to New Pipe Information	Length (ft)	Overall Cost/ft ($)
1	6-in. VCP to 8-in. HDPE	8500	102
2	8-in. concrete to 12-in. HDPE	350	254
3	8-in. concrete to 14-in. HDPE	700	273
4	10-in. PVC to 16-in. HDPE	520	293
5	12-in. AC Pipe to 14-in. HDPE	2650	203
6	24-in. RCP to 14-in. VCP	521	483

TABLE 8.3 Sample Units Costs for Various Pipe-Bursting Projects (2012 Prices)

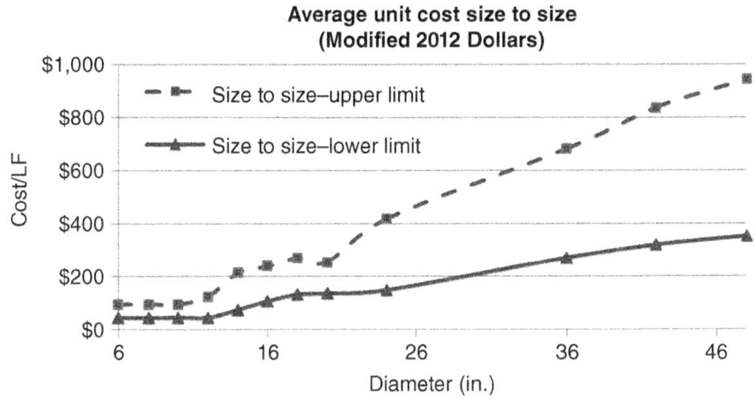

FIGURE 8.24 Bid price ranges for size-to-size replacement for different pipe diameters.

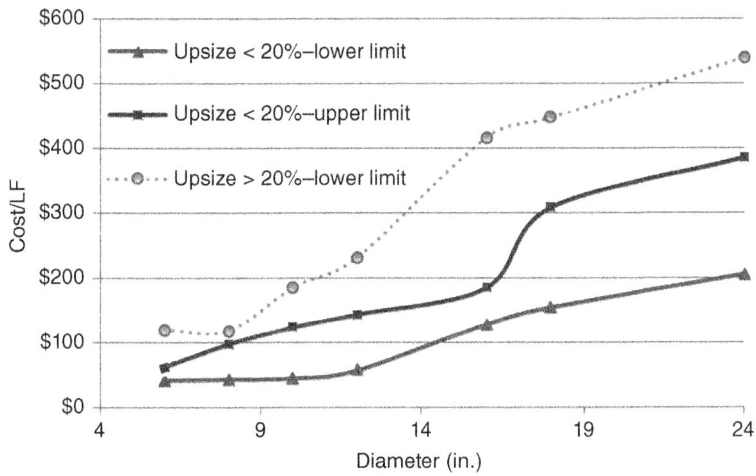

FIGURE 8.25 The bid price ranges for upsizing less or more than 20 percent.

8.24 Potential Problems

Some of the causes of such problems as excess pipe sag, relatively large soil displacement, and inadequate protection of nearby utilities, inappropriate bursting system, unforeseen obstacles, and unexpected site restrictions are described in this section.

If, prior to initiating the bursting operation, the existing pipeline is determined to display excessive sag, this condition may be repaired using local open-cut method. However, if such a condition was discovered after initiating bursting, the contractor can repair the sag by exposing the location and adjusting the soil beneath the pipe. Replacement of a section of pipe may be necessary at this excavation. If direct excavation at the location is not feasible, grouting or otherwise stabilizing the soil beneath the pipe, while simultaneously reducing the sag, may be possible.

If excessive ground movement is anticipated close to an existing structure, a ground movement and vibrations monitoring plan should be developed. If problematic levels are observed, a slower rate of bursting is mandated. If the ground movement remains excessive, the bursting operation should be halted, until an analysis of the causes and corrective options is conducted. If there is a gas line, waterline, or sewer line that is particularly close to the bursting head and at risk of damage, exposure of the line significantly reduces risk of damage. The corresponding excavation should be performed using means that do not damage the line, such as vacuum or manual excavation. If the existing pipe is shallow and there is a high risk of damaging the surface pavement, saw cutting the pavement prior to bursting prevents the spreading of the damage to the rest of the pavement. After the operation is completed, the pavement over the replacement can be repaired or replaced. Table 8.4 presents some of the common problems and possible solutions.

8.25 Pipe Design

Table 8.2 indicated the "classification" of the pipe-bursting operation based on various parameters characterizing the planned operation, indicating the relative difficulty of successfully completing the overall installation. However, it is useful to be able to independently analyze a particular application to verify the compatibility of the considered pipe strength (i.e., wall thickness) based on the length and depth of the proposed installation. General guidelines ("rules of thumb") for this purpose have been developed, considering both the pull strength and the resistance to collapse for HDPE pipe, and are presented in detail in Najafi (2010).

8.26 Case Study: Asbestos Pipe Replacement[5]

Asbestos cement (AC) pipe comprises a significant portion of the water distribution systems in many North American (U.S. and Canada) utilities. It is estimated that about 12 to 15 percent of North American water distribution systems are AC pipes. In some utilities, as high as 85 percent of their water mains are AC pipes (Hu et al., 2013). While it is difficult to accurately measure how much AC pipe remains in the

[5]Modified from Edward Alan Ambler, P.E., LEED AP, Published at *Underground Construction Magazine*, September and October 2015.

Problem	Description and Possible Solution
Damage to the new product pipe, during installation.	Damages can occur from improper handling, exceeding the bending radius at the insertion pit, abrasion from handling or dragging above ground, exceeding the safe pull load of the pipe during installation, and from broken fragments of the existing pipe.
Unanticipated geotechnical conditions.	Changes in soil conditions can cause challenges and change the rate of insertion or forces required for expansion. The geotechnical report should include a cross representation of the entire work zone with particular attention given to the soils in and around the existing trench.
Narrow trench geometry.	Knowledge of the geometry of the original trench construction is an important consideration prior to undertaking a pipe-bursting operation and selection of the pipe-bursting tools. The resistance to expansion forces can add considerable forces to the operation and can cause the bursting head to displace unevenly.
Contaminated ground.	Contaminated soils require special permitting or disposal. This can affect pipe-bursting operations and schedules. Also, the new pipe material being installed may not be tolerant to conditions found in the ground (permeation problem for water pipes).
Loss of lubricants.	There is potential that the lubricants may be lost into a void or cavity, or other geotechnical condition cause the fluid to escape from annulus. This could cause excessive installation forces or exceed maximum safe pull force. Special consideration should be given on sanitary sewer installations to prevent lubricants migrating through lateral connections to inside businesses and residential areas.
Excessive pit wall movement during a static pull.	The pull loads required for a static installation require shoring of the face of the receiving pit. If insufficient sheet piling or thrust block is used, it may stop or delay operation.
Breaking or shattering the original pipe in running soil, below the groundwater table.	This may result in the pipe filled with soils. After verifying that the bursting head did not damage any nearby water line, the site should be dewatered.
The soil around the pipe prone to flowing or running, possibly leading to excessive friction and associated elevated pull forces; (see Sec. 8.11.6).	Lubrication of outside surface of the pipe with suitable lubricant is an effective way to reduce required pulling force.

TABLE 8.4 Common Pipe-Bursting Problems

Problem	Description and Possible Solution
Certain components or accessories of the system (for example, the winch, air compressor, hydraulic components, cutting accessories in front of bursting head, etc.) are undersized or not compatible with the application.	The addition of appropriate accessories in front of the bursting head can cut PVC fittings, ductile clamps or fittings, and so on, reducing the potential of stopping, or slowing the bursting operation.
The bursting system does not have sufficient power relative to the application (upsize percentage, large diameter, length, etc.).	If the problem becomes evident shortly after the start of the bursting, the bursting system should be upgraded with one that has more power. If the problem occurs when the unit is close to the pulling shaft, toward the end of the run, the operation should be continued until the bursting head reaches the pulling shaft. At that point, the system may be upgraded for performing the next run, unless it is determined that an isolated problem, such as that due to a repair ductile clamp or a fitting, and unlikely to occur again. As an alternative to employing a more powerful bursting system, the length of the runs may be reduced. If the problem arises in the middle of the bursting, at a location where excavation is feasible, the bursting head should be exposed by a new shaft and the system upgraded. The introduced shaft can serve as an insertion shaft for the remainder of the run.

Table 8.4 Common Pipe-Bursting Problems (*Continued*)

ground and in what condition, there is an estimated 630,000 miles of AC pipe in North America (Von Aspern, 2009).

Asbestos fibers were used in combination with cement to manufacture pipes over 100 years ago in Genoa, Italy. Between 1906 and 1913, Genoa Company combined asbestos fibers with cement to produce a reinforced pipe that could handle high pressures. It was first brought to North America in 1929 when the Johns-Manville Corporation used an AC pipe manufacturing machine. AC pipe was a common choice for potable water main construction during the 1940s, through 1960s. The installation of AC pipe was curtailed in the early 1980s due to health concerns associated with the AC manufacturing process (Hu et al., 2013). The U.S. Environmental Protection Agency (USEPA) issued a complete ban on all asbestos-containing products in 1979. However, the U.S. Fifth Circuit Court of Appeals overturned the ban. The court later reinforced the USEPA's responsibility to regulate asbestos (Von Aspern et al., 2012).

The city of Casselberry is a medium size town in suburban Orlando that is considered to be 95 percent developed. Much of the city has been developed for many decades with a significant amount of the development occurring between 1950 and 1980. This time frame occurs with the increased popularity of installing asbestos cement (AC) pipe within the United States.

Prior to 2009, the city was appropriating $300,000 per year to replace existing potable water mains throughout the city, which replaced approximately one mile per year. The city owns and maintains 215 miles of potable water main in its distribution network including 95 miles of AC pipe. The assumed 50-year service life of the existing asbestos cement pipe is over and the current replacement schedule was not sustainable, requiring 215 years to replace the entire distribution network (Ambler et al., 2014). Then the city received a substantial grant from the Florida Department of Environmental Protection through the American Recovery and Reinvestment Act to embark on a project focused on renewal of 35 miles of asbestos cement pipe. This allowed Casselberry to make dramatic headway into positive renewal of the city's buried pipeline infrastructure.

Renewal of the existing AC pipe is governed under the National Emissions Standards for Hazardous Air Pollutant (NESHAP), a subsection of the Clean Air Act. U.S. Congress established much of the basic structure of the Clean Air Act in 1970 and made major revisions in 1977s and 1990s. Modifications to the Clean Air Act require Congress to approve any changes. NESHAP governs renewal of existing AC pipe line by establishing limits on the release of asbestos fibers during handling of the existing AC pipes.

NESHAP provides definition for the distinction of asbestos-containing materials, (ACMs). Friable or regulated ACM (RACM) is defined as any material containing more than one percent asbestos that, when dry, can be crumbled, pulverized or reduced to powder by hand pressure. Non-friable ACM is any material containing more than 1 percent asbestos that, when dry, cannot be crumbled, pulverized or reduced to powder by hand pressure.

8.26.1 Two Categories of non-friable ACM

EPA defines two categories of non-friable ACM, Category I and Category II non-friable ACM. Category I non-friable ACM is any asbestos-containing packing, gasket, resilient floor covering or asphalt roofing product which contains more than 1 percent asbestos. Category II non-friable ACM is any material, excluding Category I non-friable ACM, containing more than 1 percent asbestos that, when dry, cannot be crumbled, pulverized, or reduced to powder by hand pressure (Sec. 61.141) (Ambler et al., 2012). This essentially means asbestos products that can be crumbled by hand pressure to release asbestos fibers are considered to be hazardous.

NESHAP provides exemptions from its regulations based on the quantity of ACM. For AC pipe, the quantity threshold is 260 ft regardless of diameter, in one calendar year. Other exemptions from NESHAP or clarifications of its requirements for AC pipe have been provided by interpretive letters in response to questions posed to EPA (EPA, 1990). These clarifications state that any AC pipe is exempt from NESHAP if it is removed intact and disposed in a waste disposal site, capped and abandoned in place or grout-filled and abandoned. If the existing AC pipeline undergoes pipe bursting, it becomes subject to NESHAP provisions (Ambler et al., 2012). The Water Research Foundation (WRF, *ww.waterrf.org*) data collected on the city of Casselberry AC pipe bursting project as summarized in WRF Report #4465 has been the most data collected on an AC pipe bursting to date (Matthews et al., 2015).

EPA has mainly left enforcement and interpretation of NESHAP's applicability to AC pipe bursting up to the state regulatory agencies. A survey of 50 state asbestos regulatory agencies conducted by Battelle (*www.battelle.org*) found that majority of states adhere to NESHAP regulations and concluded that any process that makes asbestos fibers friable should be regulated and requires licensed contractors. According to WRF data, most utilities preferred to abandon AC pipe in-place when possible, or replace it by excavating. Although other methods of AC pipe renewal exist, such as CIPP, SIPP, pipe reaming as well as pipe bursting, utilities are hesitant to employ them based on their understanding and interpretation of NESHAP and state regulations (Matthews et al., 2015).

The extensive sampling described in WRF Project #4465 indicates that PB does not release asbestos fibers into the air creating a hazardous environment for pipe workers, residents and others within the working area (Matthews et al., 2015).

8.26.2 Asbestos Concerns

As NESHAP has defined, the areas directly on top of AC pipelines that have been pipe burst and remain in place are considered to be in-active hazardous waste sites. These sites are required to be identified as in-active hazardous waste site on the property deed. However, public right-of-ways do not maintain a property deed for this process to be implicitly met. This conflict brought industry and Casselberry project leaders to Washington D.C. to meet with top EPA staff to discuss PB and the applicability of NESHAP to pipe burst AC pipe in 2010. EPA officials embraced the environmental, social and economic benefits of pipe bursting AC pipe and understood the risks of asbestos exposure due to pipe bursting would be mitigated over traditional open-cut method. While PB was met with a positive response, modification of the existing NESHAP regulations would require an Act of Congress to complete. EPA officials recommended industry representatives present the EPA Administrator with an "Administrator Approved Alternate" process that can cover AC pipe bursting (Ambler et al., 2012).

Some regulators, utility providers and engineers still have concerns that the remaining pipe fragments will be excavated and expose public with the risk of asbestos inhalation. Potential future excavation of the fragments of AC pipe will be limited to utility crossings, unintended resident excavation but primarily, future renewal of the new HDPE pipeline. Subsequently, the remaining fragments of AC pipe would closely surround the new HDPE pipeline and excavators would stop digging once the HDPE pipe was discovered (Ambler et al., 2014).

If another utility company or underground excavator were required to perform excavation in the vicinity of the remaining AC pipe fragments, they would likely only be doing so where the proposed utility crosses the remaining AC pipe fragments, not parallel the remaining AC pipe fragments. Typically, any excavation along the existing pipe and AC pipe fragments would be less than the 260 linear feet exemption previously established in NESHAP (Ambler et al., 2014).

8.26.3 Future Handling

Utility owners performing AC pipe bursting projects must acknowledge the future handling requirements for the AC pipe fragments remaining in the ground. Emergency repair procedures performed on newer HDPE pipe will typically disturb less than 260 linear feet of the remaining AC pipe fragments. If the utility owner chooses to replace the new HDPE pipe in its entirety, they will have to acknowledge appropriate handling of the remaining AC pipe fragments. Significant release of asbestos fibers from the remaining AC pipe fragments can occur through excavation and crushing of the remaining pipe fragments. Once the AC pipe fragments remain in place surrounding the replacement pipe, the risk of exposing people to asbestos fibers is mitigated (Ambler et al., 2014). AC pipe bursting should be recognized as the preferential method of replacing existing AC pipelines since the majority of the force executed on the existing pipeline occurs underground.

8.26.4 Case Study Summary

This case study presented some characteristics of AC pipe, quantities of AC pipe installed in the U.S., and an overview of regulations governing working with AC pipe. The city of Casselberry has performed AC pipe bursting to replace 35 miles of AC pipe while meeting all existing regulations and minimizing risk to its workers and the general public. The city has elected to move forward to progressive utility distribution system renewal while leaving the fragments of AC pipe in the ground.

8.27 Chapter Summary

This chapter presented general background as well as basic planning, design, and construction considerations for in-line replacement and pipe bursting methods, primarily based on the use of PE pipe. Although other pipe materials may be employed for pipe bursting, PE pipe is most commonly used material for this application, due to its continuity, versatility, and flexibility. In comparison to other renewal methods, including conventional open-cut construction, PB is often the method of choice when the existing pipe has few service lateral connections, is significantly deteriorated and, most importantly, "requires additional capacity." This chapter discussed the predesign, design, and construction phases of pipe bursting.

8.28 Review Questions

1. Describe ILR and name its different methods.
2. Describe the different variations of PB method.
3. Compare pipe eating and pipe reaming methods.
4. In what conditions would you recommend PB method?
5. In what conditions would you recommend pipe reaming method?
6. What are the construction considerations for PB method? Explain.
7. Describe the effects of ground conditions on pipe bursting.
8. Describe how surrounding utilities may affect PB process.
9. Discuss the applicability of existing pipe materials for ILR process.

CHAPTER 9
Modified Sliplining

9.1 Introduction

Most pipelines are of small diameter, and majority of the renewal techniques have been directed toward these small, circular pipes. However, for pipelines with diameters larger than 48 in., and noncircular shapes, special methods are needed. Such pipelines constitute only a fraction of the entire pipeline system, and they are found only in larger towns and cities. The damage found in these pipelines is generally the same as that found in small pipelines, such as cracks and fractures or the corrosion of concrete or masonry, often as a result of aggressive wastewater. These large-diameter pipelines are often located at great depths in the center of towns and cities. Open-cut replacement can involve overwhelming social costs because of the enormous impact such work has on local residents, businesses, and traffic.

9.1.1 Background

In 1979, the Water Research Centre (WRc) in the UK was commissioned by the UK government and several other companies to undertake a 5-year research in methods and materials for structural renewal of large-diameter noncircular sewers. The $18 million project was completed in 1984, and the WRc's manual on sewerage rehabilitation design was published. This manual has provisions for three design types: I to III. Type I design assumes bonding of a new pipe with the existing pipe. This design type relies on the composite action created between the host sewer, the sewer lining, and the bonding agent that holds them together (normally a structural grout). Type I designs take into account vertical pressures resulting from road loadings and overburden to depths of up to approximately 26 ft. Type I designs do not consider the effects of external hydraulic head or groundwater.

Type I designs rely on the uniform transfer of loads from the existing pipe to the sewer lining through the grout. Therefore, it is of primary importance that the grout is of sufficient strength and has enough resistance to shear at the liner and sewer pipe wall interfaces. Foam or low-strength grouts are not suitable for Type I designs.

Type II designs, similar to Type I designs, are intended for use with repair systems that utilize some form of internal lining. In a Type II design, the liner thickness is calculated such that it is capable of withstanding external hydraulic loading conditions. This design type does not take into account the effects of overburden or road loading. Type II liners are designed as a "standalone" system using formulas derived from Timoshenko's equations for flexible pipes. The composite action created by bonding of the liner and host sewer is not considered.

Type III designs are intended for use where external groundwater is not present (or likely to be present) and where corrosion resistance or improvement in flow capacity is the main renewal objective. Type III designs are considered nonstructural.

Where a combination of Type I and Type II designs is used, the design is referred to as a Type II design with a Type I check. In these cases, the required liner thickness is calculated using Type II design formulas, with this thickness used in Type I design calculations. The liner system is then modified until it meets factor of safety requirements for both design types.

9.2 Primary Characteristics

Many systems for lining large-diameter pipes are currently available. This chapter presents fiberglass liners, also called glassfiber reinforced polyester panels, manufactured as pipe segments or panels. Glassfiber liners can be manufactured to fit any size and/or shape. The main structure of the liner and the makeup of the liner wall incorporate a corrosion barrier on the inside surface followed by a layer of preformed, powder bound, chopped strand glass mat consolidated with isophthalic resin.

The pipes are made of various combinations of fiberglass mats, glass fibers, polyester, vinylester, epoxy resin, and sometimes silica sand. When the pipes are machine made, a 2- to 3-mm-thick fiberglass mat is laid first and then impregnated with polyester resin to form a corrosion-resistant layer. Subsequently, the mold is rotated while polyester-soaked fibers are spun on. Fine silica sand is sometimes applied at the same time as the fibers. The process is continued until the required wall thickness is achieved.

"Channeline" is a brand name which is handmade in molds and can, therefore, be produced either as complete pipes or as pipe segments. This system also allows the production of asymmetrical profiles or profiles that have negative curvature for part of the profile–as in the case of profiles with floor channels. As in other liner systems, the type of resin and fiberglass used is chosen on the basis of requirements for strength and resistance to chemicals and extreme temperatures.

FIGURE 9.1 Glassfiber panels.

"SPR" is another brand using a spiral-wound trenchless pipeline renewal process designed for use in medium- to large-diameter pipelines, typically 18 in. and larger. SPR can be used in worker-entry and non-worker-entry applications. SPR utilizes steel-reinforced interlocking PVC profile strips grouted in place with an engineered specified cementitious grout. The installation equipment can be utilized via standard access points without or with minimum surface excavation. SPR can also be installed in vertical applications, such as wet wells, access shafts, and other large-diameter structures. Figure 9.1 presents an egg-shaped fiberglass liner and Fig. 9.2 illustrates the spiral-wound process.

FIGURE 9.2 Spiral-wound lining process. (*Source:* SPR.)

9.2.1 Project Planning

Panel liners can be designed and constructed as self-supporting, flexible pipes. Strength can be achieved by sandwiching together the existing pipe, the panel liner, and the grout filling the annular space. Different design criteria are used in different parts of the world. Where there are no national standards, guidelines from the WRc in the UK, as described previously, are often used.

In addition to static calculations, the hydraulic capacity of the pipeline should be investigated and calculated. It should be mentioned that despite new liner pipe reducing pipe diameter, the resulting smooth surface would usually increase hydraulic capacity in relation to the existing pipeline.

9.2.2 Modified Sliplining Applications

As for other renewal projects, it is important to choose the best method for each individual project. The available space, the chemical composition of the fluid carried by the pipeline, the pressure and temperature in the pipeline, the number of laterals, the number of manholes, and so on, must be assessed from project to project. Table 9.1 presents an overview of applications and possible limitations for the modified sliplining method.

Pipeline Type	Suitable	Comments
Sewers	Yes	See note A
Gas pipelines	No	
Potable water pipelines	No	
Chemical/industrial sewers	Yes	See note B
Pipelines with bends	Yes	
Circular pipes	Yes	
Noncircular pipes	Yes	
Pipelines with varying cross-sections	Yes	See note C
Pipelines with lateral connections	Yes	
Pipelines with deformations	Yes	See note C
Pressure pipelines	No	

A. The method is suitable for the manhole-to-manhole renewal of sewer lines with diameters larger than 48 in.

B. Each project must be individually assessed to find the most suitable renewal technique. The available space, the chemical composition of the fluid carried by the pipeline, the pressure and temperature in the pipeline, the number of laterals, the number of manholes, and so on must be assessed from project to project.

C. There may be a significant diameter reduction for pipes with varying cross sections and deformations.

TABLE 9.1 Overview of Modified Sliplining Applications for Different Conditions

9.3 Preliminary Inspection

To achieve the best possible result of pipeline renewal using modified sliplining method, it is important to carry out a detailed preliminary survey to determine, at very least, the following:

- The condition of the existing pipeline–using manual or closed-circuit television (CCTV) inspection.
- The need for cleaning.
- Pipeline length and depth.
- Pipeline dimensions, deformations, and displaced joints.
- Watertable depth.
- Location of laterals, including dimensions, unused service lines, and intruding service line connections.
- Traffic loads.
- Other pipeline data and manhole conditions.
- Location and sizes of access pits, if necessary.

9.4 Preparatory Work

Once necessary preliminary surveying has been completed and renewal is about to commence, there are several routines for the way in which preparatory work should be carried out in order to achieve effective pipeline renewal. These include

- Any planned bypass pumping, including its duration and the number of consumers affected
- Planned traffic diversions
- Pipeline shut-off (timing, duration, and number of laterals affected)
- Grinding of intruding service line connections
- Measurement of internal diameter
- Grinding or cutting away of scattered deposits
- Water jetting/root cutting, if necessary
- Closing/blocking off unused service lines
- Restoration of the surface and access shafts/pits
- Sealing leaks to prevent groundwater infiltration

Renewing large-diameter pipelines requires detailed planning and close collaboration between the pipeline owner, contractor, and the consultant. In the case of storm sewers, the method has the advantage that flow bypassing is not always necessary. Flooding usually does not damage work in progress; however, safety of workers inside the pipe must be considered.

9.5 Safety Issues

During the installation, certain safety and work environment requirements, according to local, regional and national regulations, must be met. Most important of all, Occupational Safety and Health Administration's safety regulations for confined-space entry must be addressed by a competent person. The existing pipeline must be effectively closed off upstream (and possibly also downstream) to ensure that flow cannot enter. If this is accomplished with inflatable sealing plugs, the plugs must be securely held in place, with chains for example, and the flow pressure should be constantly monitored. Two plugs should be positioned upstream as an extra precaution. The pumps must be constantly monitored to ensure that the level of wastewater behind the plugs does not become so high that it subjects the plugs to excessive loads.

Adequate light must be supplied and ventilation be provided. A gas detector must be used continuously to monitor oxygen levels and to warn workers the presence of poisonous or explosive gasses.

There must be escape routes from the pipeline in both directions and there must always be a watchman in the launch pit who is continuously in touch with workers in the pipeline. There must be a telephone and emergency equipment on site, and workers must be instructed in emergency procedures. As is the case with all other sewer work, personnel must, of course, be vaccinated.

Workers need to manually position each prefabricated fiberglass pipe segment. The pipe sections are inserted into the existing pipeline through manholes and access chambers or via launch pits if necessary. Point source or localized installations are also possible. Most liners are machine made as centrifugal-cast or spun sections. Modified sliplining liners (panels) usually have tongue-and-groove joints that are sealed with either rubber sealing rings or polyurethane or epoxy filler. Liner segments can have specially developed longitudinal joints that are sealed with epoxy filler.

Drive lengths of up to 3,000 ft can be installed from a manhole or an access point. However, ventilation and escape routes must be provided at short intervals. A small work site should be established at the ground level. This should be cordoned off to ensure the safety of workers and passersby. There should be room for necessary installation equipment. A crane or excavator should be positioned above the launch pit to lower panels to the pipeline entrance. There should also be a compressor, a generator as well as tool storage and personnel facilities.

The panels are transported to their position in the pipeline. Work is started at the opposite end of the pipeline so as to finish in the entry-pit. Packing pieces are wedged between the panel liner and the existing pipe as installation progresses. If there is room, structural

FIGURE 9.3 Filling the annular space with grout.

irregularities can be evened out to a certain extent if the dimensions of the panels and existing pipe are large enough to allow adjustment in the positioning of individual panels.

Once the panels have been positioned throughout the length of the pipeline, the annular space is filled with structural grout (Fig. 9.3). Care must be taken during the pumping process to prevent the panels from being subjected to excessive external pressure by checking the pump pressures for floatation forces that the panels will be subjected. In this respect, it should be noted that cement grouting creates greater floatation than water because of its higher density. Foamed concrete, which has lower density, can restrict floatation forces. The possibility of using foamed concrete is, however, limited if the cement grout is to contribute to the pipe strength, that is, foaming reduces the compressive strength of concrete. Another possibility for reducing floatation forces is to inject the cement grout in two or three stages while allowing it to set between successive injections.

Changes in pipe diameter can be negotiated with prefabricated transition panels. Such panels are often expensive because a mold must be made for each individual irregularity. Another possibility is to cut the panels to be installed on either side of the irregularity so that they fit and then connect the two with a matrix made by hand onsite. Bends are similarly negotiated using prefabricated segments or panels cut to fit.

As for other renewal methods, service line connections are reopened once the panels are in place. Watertight connections can be established with filler, fiberglass matrix, or transition profiles, or a combination of these methods. Depending on various factors, including the diameter and length of the pipeline, the weight of the panels and the number of connections, advance rates of 20–100 ft per day can be expected. Figure 9.4 illustrates the modified sliplining installation

A

B

Figure 9.4 Modified sliplining installation process.

process. Modified sliplining may be advantageous to segmental sliplining where a whole pipe section is installed for the following reasons:

- Existing pipeline access may be limited to manholes where there is only room for inserting pipe segments. In such circumstances, segments must be connected as installation progresses inside the existing pipe.
- Pipe transport from the factory to the renewal site can be expensive and problematic where large-diameter pipelines are concerned. These can, therefore, be delivered in segments and subsequently assembled either in the pipeline itself or—if the access conditions to the pipeline allow—aboveground.

9.6 Relining with Steel Pipe

Similar to sliplining, steel "split-can liners," designed to required operational parameters, are fully re-rounded once inside the existing prestressed concrete cylinder pipe (PCCP), then longitudinally welded (see Fig. 9.5). Adjoining re-rounded pipe section joints are welded together, and lining is applied in the field after installation. The annular space, which is typically no more than 2-in. is grouted. When complete, the existing PCCP is fully structurally renewed without with minimum loss of hydraulic capacity.

A

B

FIGURE 9.5 Relining with steel pipe. (*Source:* Northwest Pipe.)

C

D

FIGURE 9.5 Relining with steel pipe. (*Continued*)

9.7 Project Delivery

After installation, it is often necessary to carry out the completion work. This may, for example, take the following steps:

- Connecting manholes and removing end seals.
- Reinstating service laterals.

E

FIGURE 9.5 Relining with steel pipe. (*Continued*)

- Conducting final inspection.
- Conducting leak testing.
- Notifying residents and businesses about work completion.

9.8 Advantages

- Modified sliplining can be used in any worker-entry diameter and any shape (circular, oval, vertically sided, symmetrical, or nonsymmetrical).
- Modified sliplining is applicable for both corrosion control and structural purposes.
- Modified sliplining can be installed in restricted flow conditions, which in many cases eliminates the need for bypassing the flow.
- Modified sliplining can be used for renewal of culverts.
- Modified sliplining can be used for long drive length where access is limited.
- Nonstandard shapes up to 21 ft in diameter are possible.
- There might be an improved hydraulic capacity by providing a lower coefficient of friction.

- Modified sliplining (panel liner) is abrasion resistant and chemical resistant.
- The new liner provides structural support after the grouting process.
- It is possible place panels, according to engineering specifications.
- Annular space does not have to be equidistant from sides of the existing pipe.
- Frames can be differentially positioned along the existing pipe to allow restoration of original slope.

9.9 Limitations

- Modified sliplining can be used only in worker-entry pipes.
- Significant reduction in pipe cross-sectional area is possible.
- Access shafts may be required.
- The annular space must be grouted.

9.10 Chapter Summary

The modified sliplining is a cost-effective method for large diameter (more than 48 in.) gravity pipelines. The major benefit of this method is its flexibility with odd-shaped pipes and sewer lines. There will be reduction in cross-sectional area of the pipe, which must be compared with any improvements due to a better coefficient of roughness of the lined pipe. This method is applicable specifically for structural applications, where the existing pipe is deteriorated or additional loads are placed on the pipe.

9.11 Review Question

1. A Midwest city has a large combined sewer that runs under a busy downtown street. The sewer generally follows an old creek bed and is horseshoe shaped, measuring 12 ft wide and 12 ft high. Manholes with 30-in. openings are located approximately every 800 ft. The sewer is constructed of stone blocks, and some of the blocks have become dislodged, falling into the sewer. Also, the sewer is leaking heavily from infiltration. Because of these reasons, the sewer is in need of rehabilitation to stop leaks and restore and maintain structural capability. Also, since the city has a program to disconnect storm water piping from this sewer, capacity is not a problem. What type of rehabilitation product should be used for this sewer?

CHAPTER 10

Spray-in-Place Pipe

10.1 Introduction

Spray-in-place pipe (SIPP) or linings and coatings[1] using spray-on methods have been used to protect and renew water and sewer infrastructure (pipelines, tanks, reservoirs, clarifiers, primary and secondary retention and treatment basins, pump stations, diversion boxes, manholes, and other structures) for decades. Shotcrete, an air-assisted spray-on SIPP method for cementitious products, was developed at the dawn of the ninth century in Allentown, Pennsylvania, and became accepted as a construction method in 1910. Today, high-tech polymer SIPPs and composite SIPP methods are used to restore, protect, and renew a wide range of concrete, masonry, and steel structures.

The principal objective of SIPP is to apply a monolithic application that inhibits further deterioration. Deterioration is relevant to the existing pipeline materials and application as well as environmental conditions. In water mains, it is characterized by tuberculation, scale buildup, and corrosion that can reduce pipeline's flow capability. SIPP methods are effective at eliminating infiltration while providing containment in both sewage collection systems and water distribution pipelines. In corrosive sewer environments, pipe crowns, pump stations, and manholes can lose an inch or more of concrete in less than a year. SIPPs can mitigate further degradation and, if needed, can structurally enhance and renew severely damaged structures.

10.2 Material Type and Purpose

SIPPs may be applied to renew and protect aging pipelines or to protect new pipelines to increase their service life. The primary materials used for SIPPs generally fall into two broad categories of cementitious and polymers.

---

[1]Lining is *generally* the term used for spraying inside the pipe and coating is the term for spraying outside of the pipe. This chapter is focused on SIPP or linings, as they are used to renew and protect existing and deteriorated pipelines.

10.2.1 Cementitious

Cementitious materials are used in both water and sewer applications primarily for its most basic characteristic—cost effectiveness. In water applications, cement mortar SIPP is most common and serves two main functions: the alkalinity of the cement inhibits corrosion (especially in case of an iron pipe), and the relatively smooth internal surface reduces hydraulic roughness and improves flow characteristics. It should be noted that cement mortar SIPP is also applied to many new cast iron and ductile iron water pipes to inhibit corrosion. In this application, the SIPP does not fulfill a structural function; rather, it reduces the rate at which the pipe will deteriorate. This SIPP is not appropriate for pipes which leak, or where corrosion has reduced the wall thickness significantly. It is also important to apply sufficient thickness of mortar to create the alkaline environment at the mortar or iron interface for iron water pipes. As with steel reinforcement in concrete structures, inadequate cover to the metal will allow the onset of corrosion, which will cause the mortar to crack and spall.

In sewer applications, cementitious SIPPs comprise various base materials such as Portland and calcium aluminate. These materials are used to line underground structures to prevent infiltration and rebuild structures deteriorated because of age and erosion from infiltration or inflow (I/I). Calcium aluminate cements are frequently used to provide some sulfide resistance in rehabilitating sewer structures where mild corrosion is evident. Cementitious materials are also frequently used as a basecoat material for polymer SIPPs in more corrosive sewer environments.

10.2.2 Polymers

The Clean Water Act in the United States has introduced efforts to reduce unwanted I/I into the collection systems. Therefore, more concentrated sewage without heavy metal and industrial pollutants results in much more aggressive microbiological induced corrosion problems. Accelerated corrosion problems in today's infrastructure, as well as greater awareness of corrosion issues, have led to the development and the use of SIPPs to increase infrastructure life expectancies. As a result of inherent safety, performance, and quick return-to-service attributes, solvent free 100 percent solids epoxy, urethane, and urea SIPPs have proven to have unique advantages for successful renewal and corrosion protection. These thermoset SIPPs are essentially inert plastics when cured and therefore resistant to most corrosive elements found in these environments. They have longer life expectancies when properly applied by professionally trained applicators using specialized spray equipment to achieve optimal performance as defined by the product manufacturers. Because these products contain no solvents, they require no evaporative process to cure and do not emit destructive volatile organic compounds (VOCs) found in many SIPPs used in above grade SIPPs. Polymer formulations having 100 percent solids (little or no VOC content) are safer to use in confined spaces and can also be formulated for thicker, structure enhancing applications. Such products now dominate the underground infrastructure protection industry because many can cure and bond to concrete, brick, steel, and cast iron, in damp underground environments while providing necessary protection from the aggressive sewer environment.

Compared with cementitious products, most polymers deliver superior chemical resistance in the most severe corrosive environments, although manufacturers should be consulted for specific recommendations. Polymers can be formulated to be structural or nonstructural[2], including extraordinary long-term strengths, flexibility, and elongation. Most polymers used in underground applications have been formulated for "ultra-high build" thicknesses of 40 to 250 mils[3] single-coat capability towing to the jagged and deep profiles found on new and deteriorated concrete surfaces. Polymer SIPPs are often used to topcoat cement mortar SIPPs in severe duty environments.

10.2.3 Epoxies

The use of epoxy linings to protect and renew water mains originated in the United Kingdom during the 1970s and has been used successfully primarily as a protective barrier lining to prevent corrosion of cast iron pipe. Epoxies have longer curing and return to service times than polyuria and polyurethane; as a 16-hour return to service is common. Epoxies have the best moisture tolerance making them ideal for penetrating into porous substrates in damp underground environments and developing the necessary adhesion for long-term service. Epoxies are generally rigid and can be formulated with high tensile and flexural strengths for use in structural renewal or with lower strengths but excellent chemical resistance to be used as a protective (nonstructural) barrier. Unlike other polymers, epoxies generally do not contain toxins, which is important when considering potable water applications. Epoxies are more surface and moisture tolerant, which provides improved adhesion to less-than-adequately prepared substrates.

10.2.4 Polyurea and Polyurethane

Polyurea and polyurethane, are made up of two components: an isocyanate (base) and a polyol (activator). The resultant material can be formulated in numerous ways, including variations in physical properties with the two primary variants being elastomeric and rigid.

---

[2]*Structural, semi-structural,* and *nonstructural* are terms that must be verified with manufacturers through short- and long-term testing results. Performance of materials in field applications also depends on quality of installation and substrate preparation and cleaning.

[3]Thousandth of an inch.

Polyurea can be formulated to obtain extremely fast cure times. Polyurea has a cure time of 1 hour, offering the potential to renew a section of existing pipeline and put it back to service in 1 day, thereby significantly improving productivity and possibly eliminating bypassing costs.

Elastomeric polyurethanes have been used extensively in larger tanks and treatment structures such as clarifiers and digesters. To date, these elastomeric polyurethanes have been utilized only for corrosion protection, and their suitability in the water pipe renewal market will remain in this category. Only the development of rigid polyurethanes made structural renewal of pipelines possible. Both elastomeric and rigid polyurethanes offer extremely short cure times with the ability to return the pipe back to service within 30 to 60 minutes. Therefore, as with polyureas, construction efficiency is greatly enhanced. The flexible formulations can be used in nonstructural (protective) applications (Fig. 10.1), whereas the rigid products supply structural enhancement to deteriorated structures and corrosion protection.

10.2.5 Polyesters

Polyesters use styrenated resins, which are rigid and more moisture sensitive and can require more extensive ventilation procedures in

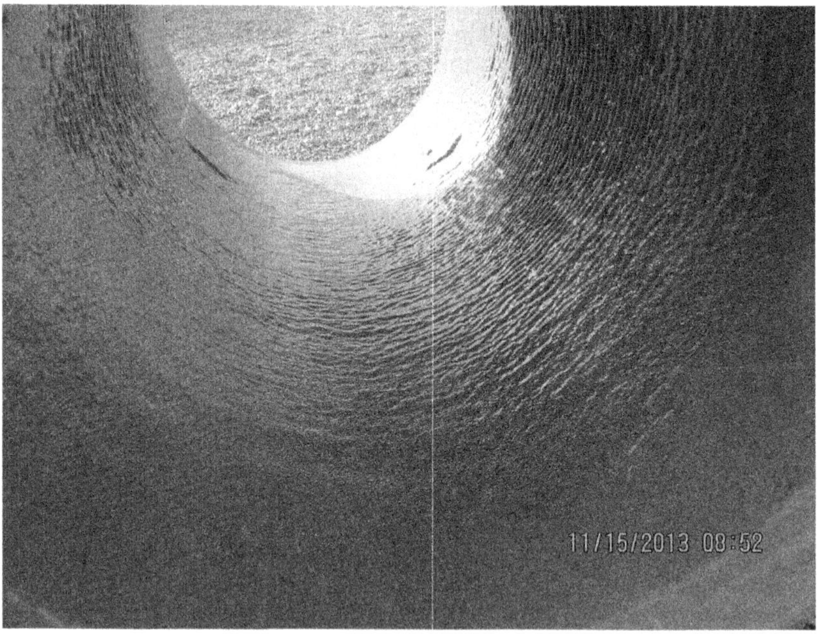

Figure 10.1 Flexible polyurethane SIPP.

confined spaces. The resin must provide the same expansion and contraction rates as the substrate after curing. There are very few polyester formulations that are used as independent SIPPs in underground application. Those available require intense field chemistry and trained installers. Polyesters are frequently used in CIPP SIPP systems.

10.3 Installation Process

10.3.1 Ideal Properties

In conjunction with chemical resistance and monolithic coverage, adhesion is generally regarded as a required attribute of SIPPs. Other attributes vary greatly between polymer and cementitious SIPPs, the most prevalent types used in underground infrastructure protection and renewal. Some may be excellent for bridging moving cracks in concrete structures but may have low chemical resistance owing to inherently higher porosity; others may exhibit excellent long-term strength but poor adhesion in damp environments. As for any renewal method, true project needs should be evaluated and matched with proven product attributes.

Moisture can weaken a SIPP's curing process as well as its ability to bond to the host structure. Although moisture is relatively easy to mitigate in above-surface structures, it cannot be completely avoided in below grade, especially in concrete and masonry structures. Therefore, a SIPP with high moisture tolerance offers an adhesion advantage for below-grade SIPP projects. Epoxies can generally be formulated to offer the best moisture tolerance, although some urethanes and urea also offer moderate tolerance or require the use of an epoxy primer. Other attributes to analyze include structural enhancement, permeability, chemical resistance, quick return-to-service, maintenance, and visibility.

10.3.2 Site Investigation and Preparation

Every pipe-renewal project is unique and requires a careful site investigation before it can begin. A successful project may require surface and subsurface investigations. Trenchless pipe renewal methods require that the project owners make available all necessary information so that one can reasonably anticipate obstacles that might be encountered during the SIPP operation. In the planning phase of the project, surface and subsurface survey information provides assistance in determining the suitability of the pipe SIPP method for the renewal of deteriorated water pipes. Accurate information of this sort would lead to reduction of installation problems, reworks, change orders, and associated social costs. The investigation of a SIPP site includes evaluation of the number and condition of service laterals, easement restrictions, potholing locations and regulations, and environmental concerns. A surface survey may include the following steps:

- Work-area requirement
- Location of all existing utilities
- Utilities and structures adjacent to proposed pothole locations
- Surface features, descriptions and layouts of roadways, sidewalks, and so on
- Listing of visible subsurface utility landmarks, such as fire hydrants, valves, and so on

It is essential to determine all valve and hydrants locations along the length of the pipe to be lined. It is recommended that contractor or installer physically check all the dimensions supplied by the project owner in the form of drawings, videos, and reports to ensure accuracy. Once surface and subsurface surveys are completed, thickness and lengths of the SIPPs to be installed can be determined.

Proper site inspection is important for selection of appropriate SIPP parameters and any modifications in design thickness. Because of difficulties in inspecting underground water pipes, there are possibilities of unanticipated obstacles, such as the presence of debris, silt, and so on, which sometimes make the task of SIPP difficult.

10.3.3 Pipe Inspection

The main objective of pipe inspection is to examine the condition of the existing pipe before and after liner installation. Closed-circuit television (CCTV) inspection is a method of choice to inspect the interior of water pipes and also to evaluate installed liner quality. Valves or water hydrants maybe used as insertion points for these recording devices.

The purpose of a preSIPP inspection is to assess pipe condition and ensure a successful liner installation. This inspection provides a good idea of the degree of cleaning required to prepare an existing pipe before the start of a SIPP operation. A preSIPP inspection may also reveal the need for other forms of water pipe preparations, such as removal of projecting service lateral connections. The state or presence of the following items may be revealed during a preSIPP inspection:

- Leaking valves and ferrules
- Leaking stop taps
- Dropped joints
- Protruding ferrules
- Structural failures (cracks, holes, and so on)
- Recleaning requirements

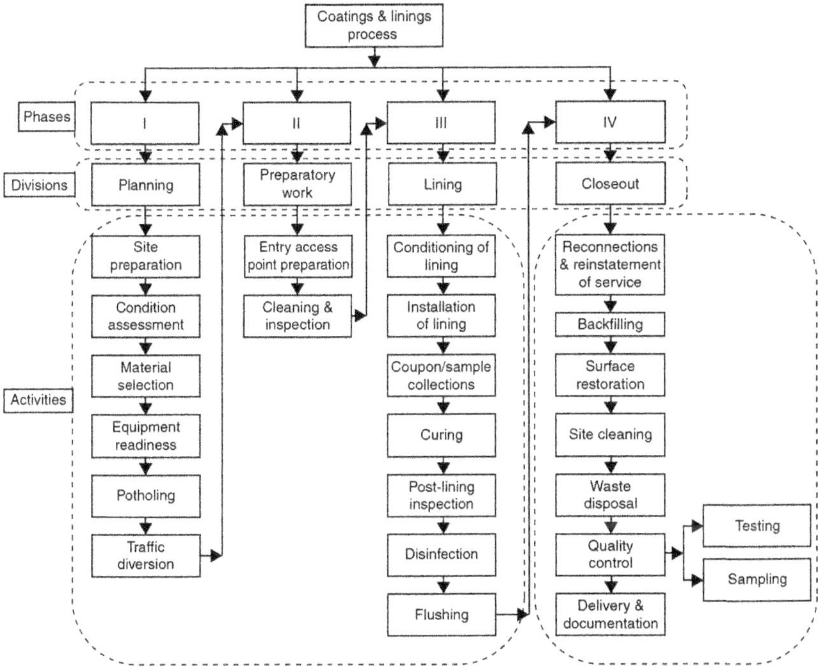

FIGURE 10.2 Pipe assessment method selection guide. [*Source:* Adapted from American Water Works Association (AWWA), Workshop on Condition Assessment Inspection Devices for Water Transmission Mains.]

- Debris
- Standing water

Figure 10.2 presents phases of SIPP installation method (Najafi, 2010).

10.3.4 Water Pipe Cleaning Methods

Cleaning of old water pipes is prioritized based on the age of the pipeline and the frequency of problems it encounters and must be routinely scheduled. In most of the water pipes, sediments accumulate and biofilms develop, increasing the risk of color, taste, and odor problems, along with the chance of coliform regrowth.

Cleaning must be performed for the areas to be coated. Before isolating and cleaning a section of pipeline, water flow must be stopped or bypassed. The section of pipe will then be emptied and cleaned. The cleaning technique chosen depends on the pipe material, and previous SIPP and entry point locations. It should be noted that if the structural condition of the existing pipe is poor, and the pipe wall is thin, then some cleaning methods may result in damage to the pipe section being cleaned.

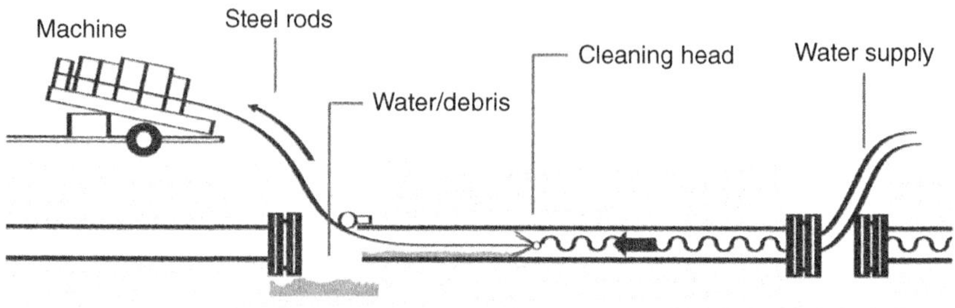

Figure 10.3 Conventional cleaning method using a steel rod. (*Source:* 3M Water Infrastructure.)

Figure 10.3 illustrates a conventional cleaning method using a steel rod. Table 10.1 presents a summary of water pipe cleaning methods.

10.3.4.1 New Cleaning Method

The patented "Tomahawk System" launches abrasives stones of various calibers into a high-volume, low-pressure air stream, removing all internal corrosion, sediments, deposits and old coatings (i.e., tar, cement) from drinking water pipes (see Fig. 10.4). Following initial axial cleaning with larger caliber stone, the process then uses an airborne camera to radially deflect smaller stone, cleaning critical areas such as service connections. Tomahawk provides a prepared, dry host quality for customers. pipe surface for best-liner bond. The airborne "Scout" camera also creates localized vacuum and turbulence which draws water and debris from service connections and joints. This "vaculation" operation allows the operator to detect leaking services, valves and expose visible anomalies such as cracks and holes in real time. The process is enclosed, waterless, dust free and typically takes two hours per street. The addition of airborne lining (adding a polymeric coating to the airstream following cleaning) allows this technology to renew buried water pipes with a bonded liner.

10.3.5 Surface Preparation

The application of any SIPP requires diligent surface preparation. It is necessary to clean and profile the substrate prior to application. Also, most protective materials require that all active infiltration be stopped prior to application and some require a dry surface to attain long-term adhesion to the host structure. SIPP systems can be either machine or hand applied. Small-diameter (<36 in. internal diameter) pipe applications are generally carried out using spincast equipment, a rotating head that is winched through the pipeline. Large-diameter applications can also be robotically applied or, more likely, installed with handheld spray equipment, similar to what is used for SIPP diversion boxes, pump station wet wells, manholes, and similar underground structures where worker entry is possible.

Method	Procedures
Drag scrapers	Designed to remove hard deposits and nodules. This apparatus is made up of a number of spring steel scrapers mounted on a central shaft. These scrapers remove deposits as they are winched through the pipe. The central shaft is fitted with a towing eye at each end to allow the scraper to be pulled back if necessary. These are available for all pipe sizes ranging from 2 in. (50 mm) to 12 in. (600 mm).
Power boring	Hydraulically powered device utilized to remove tuberculation and encrustation from the water pipes. The device is a rack feed boring machine, a compact, diesel-powered unit using hydraulic pressure to rotate steel cutter blades at 750 rpm through the pipe (see Fig. 10.4).
High pressure water jetting	High-pressure water jets are pulled through the pipe, so high-velocity water blasts deposits and films from the pipe surface.
Poly pigging	This is similar to soft swab pigging (see below) except that it differs in its construction. For softer scales, the pig is simply foam covered by a plastic shell, while for harder scales abrasive materials are adhered to the pig's exterior (see Fig. 10.5).
Air scouring	Volumes of air and water are alternately introduced to the pipe through fire hydrants, creating turbulence that scours film, and lifts and transports sediment.
Rodding	Uses an engine and a drive unit with sectional rods. As blades rotate, they break up the deposits and loosen the debris.
Balling	A threaded rubber cleaning ball that spins and scrubs the interior of the pipe as flow increases in the pipelines.
Flushing	Introduction of alternatively high flow rate of water into the pipes, removing floatables, sands, and grids.
Soft swab pigging	Foam plug is forced through the pipe using water.
Chemical cleaning	Acid is circulated through a closed system, dissolving mineral scale, biological growth, and corrosion byproducts. Finally, the pipe is flushed and disinfected. Some common chemicals used are H_2S gas, bio-acids, digester, enzymes, catalysts, hydroxides, and neutralizers.
Scooter	This is a hydraulic method. A round, rubber-rimmed, and hinged metal shield is mounted on the steel framework on small wheels, which works as a plug to build a headwater. This action scours the inner walls of the pipeline.

Source: Adapted from AWWA, M28-Rehabilitation of Water Mains.

TABLE 10.1 Water Pipe Cleaning Methods

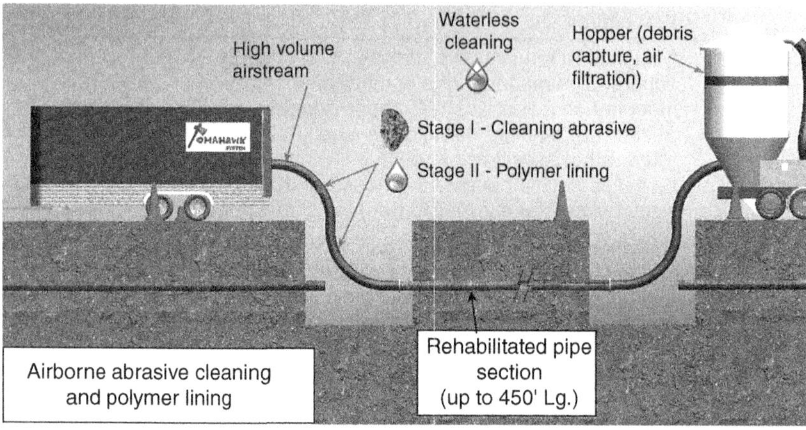

FIGURE 10.4 Tomahawk Cleaning System.

The objective of surface preparation is to provide a clean, sound surface with adequate profile and porosity that is essential to ensure proper adhesion. Proper site inspection is important to be able to specify the right SIPP product and appropriate thickness. Because of the "out-of-sight, out-of-mind" nature of underground infrastructure, there are always unanticipated difficulties, ranging from active I/I to the presence of various contaminants, which makes the task of SIPP application rather difficult hence the strong need for professional, experienced applicators. There are, however, additional resources available to assist both specifier and trained SIPP applicator in selecting the right surface preparation method(s). In particular, the National Association of Corrosion Engineers (NACE), the Society for Protective SIPPs (SSPC), and the International Concrete Repair Institute (ICRI) have prepared extensive guidelines, including visual guides that aid in proper selection of SIPPs, surface preparation, and application methods.

Cleaning is the first and most commonly inadequately performed task before SIPP application. For cleaning in worker-entry structures, mechanical abrasion is preferred whenever practical, but rust, latent concrete, and other surface contaminants can generally be removed by low-to-high pressure water cleaning, abrasive blasting, shot blasting, and power or hand tooling. For small or hard-to-reach places, hand grinders and wire brushing may be used. If oil, grease, or other hydrocarbon deposits have contaminated the surface, steam, hot-water blasting, or other methods may be necessary to clean and decontaminate the surface chemically.

Non-worker-entry pipe cleaning techniques can include robotic low and high pressure water jetting, scraping, pigging, rack-feed borers, and mechanically driven devices, such as cutters and chain flails. There is often a balance to be drawn between removing all traces of corrosion and avoiding damage to the pipe wall, and some of the more aggressive techniques should be used with caution. Pipe scrapers are designed to remove hard deposits and nodules when winched through a pipe, and consist of a number of spring steel blades mounted on a central shaft. A towing eye is fitted to each end of the shaft, allowing the tool to be pulled back, if necessary. Wire brush pigs comprise two circular wire brushes on a central shaft with a towing eye at each end, and are used to remove loose deposits and dust prior to SIPP. They may also be used to remove debris loosened by a pipe scraper. Cleaning pigs are available in a wide range of types, and are usually molded from hard resin with an abrasive outer layer. Some have carbide studs around the barrel to remove hard deposits. They are normally driven through the main by water pressure, and can travel distances of several kilometers in continuous pipelines. In a heavily encrusted pipe, pigging is carried out in stages using pigs of increasing size. Foam pigs are generally pushed through a pipeline by air or water pressure, but versions are available that can be pulled through with a towing rope. They are generally used to remove dust or fluids from pipes of any material, and are also suitable for line drying.

Surface preparation procedures strongly depend on the type of surface being prepared and the profile necessary for the SIPP being applied. Some recommended surface preparation procedures for different materials are discussed below. It is very important to understand that the performance standard for cleaning is paramount; how that is accomplished will be up to the contractor who may have to use a myriad of different techniques to achieve the desired standard.

10.3.5.1 Concrete and Masonry

All concrete and masonry surfaces must be sound, clean, and dust and contaminant free. Existing SIPPs that have failed because of incompatibility or debonding must be completely removed. Any form of release, curing compounds, toppings, waxes, oils, greases, and the like must be removed prior to application.

Cleaning for concrete or masonry materials can generally be accomplished with low-pressure water cleaning using equipment capable of 5000 psi at 4 gal/min (NACE No. 5/SSPC-SP 12 and NACE No. 6/SSPC-SP 13). Mechanical methods, such as high-pressure water cleaning, water jetting, abrasive blasting, shotblasting, grinding, or scarifying, may be used to remove previous SIPPs, laitance, and disintegrated material. Detergent cleaning and hot-water blasting may be necessary to remove oil and grease from the substrate. Chemical cleaning, such as acid etching with muriatic acid, can be used in certain situations. However, care must be taken to remove all the residual acid prior to application of any SIPP and this method is typically not effective on vertical or overhead surfaces. Following cleaning, the surfaces should be prepared to the profile as recommended by the SIPP manufacturer. This may require more aggressive mechanical methods. Typically an ICRI CSP-4 or CSP-5 (ICRI Guideline No. 03732) profile is recommended to be achieved for most high-build SIPPs. Whatever methods are selected for use, it must be ensured that a uniform, sound, and clean surface without excessive damage is obtained.

10.3.5.2 Steel

Steel structures should be prepared to a white or near white metal (SSPC SP-5 or SP-10) based upon its intended service environment and manufacturer recommendations for that specific SIPP product. Prior to blasting, all surfaces should be decontaminated should oils, grease, soluble salts, or other contaminants be present. All scale, deposits, weld splatter, and soluble salts should be removed, and all rough and sharp edges rounded off prior to SIPP. Surface preparation is generally accomplished by dry abrasive blast or power tool cleaning. Vacuum sweep can be used to remove the remaining dust and debris. SIPP must be applied quickly to avoid flash rust or contamination of prepared surfaces.

10.3.6 Repair Materials

Repair and patching is necessary for final surface preparation, especially on concrete and masonry substrates. The procedures used for this are also dependent on the type of substrate and profile created during the cleaning process.

10.3.6.1 Concrete and Masonry

Where necessary, voids should be filled and jagged surfaces profiled with repair materials compatible with the SIPP. Any area exhibiting movement or cracking owing to expansion and contraction should be grouted and patched according to appropriate crack repair or expansion joint procedure provided by the manufacturer. All surfaces that show exposed structural steel, spalling greater than ¾ in. deep or cracks greater than 3/8 in. wide, should be patched using a quick-setting, high-strength cement mortar or high-build, nonsagging epoxy grout after initial cleaning. All concrete that is not sound or has been damaged by chemical exposure should be removed. If, in areas to be patched, reinforcement is missing and radial cracking from a spall site exists, steel replacement should be considered by the project engineer.

In masonry structures where loss of mortar has created gaps greater than ¼ in. in diameter between the bricks or blocks, the voids can generally be filled using a compatible quick-setting, high-strength cement mortar (see Fig. 10.5). Whenever structural integrity is questioned, a basecoat (½ in. or greater) of high-strength cement mortar, or additional epoxy SIPP or grout should be considered.

A

B

FIGURE 10.5 Rack feed boring. (*Source:* 3M Water Infrastructure.)

Most repair materials require a minimum cure time and certain environmental conditions which should be observed to allow the material to obtain optimum physical properties. Following cure, the repair areas should be cleaned to remove laitance and any unadhered materials. Repair materials may also need to be profiled to promote adhesion of the topcoat. Inspectors may refer to ASTM D4258 and D4259, NACE No. 6, SSPC SP-13 and ICRI 03732 standards as guidelines for the performance and inspection of concrete surface preparation.

10.3.6.2 Steel

Steel surfaces should be thoroughly inspected, and when necessary, ultrasonically tested to detect thin spots in the structure, which may need reinforcement. A fiberglass fabric patch can be applied whenever corrosion or erosion has removed the safety factor of steel (generally greater than 50 percent of the original thickness). Fiberglass fabric may be rolled into applied resin or chopped glass spray applied with the resin. Additional polymer coats and fiberglass layers may be applied to achieve greater thickness, fill remaining voids, cover exposed fibers, or more strength.

10.3.7 Equipment and Process

SIPP systems range widely in their applications. In each application, the material is applied in one or more layers to the interior of a structure that has been adequately cleaned and prepared. Most SIPP systems provide for mechanical adhesion to the prepared substrate and chemical adhesion to each successive layer of SIPP. The system can be used to coat the entire or a portion of the structure.

SIPP systems are applied by both machine and hand. With water pipelines (non-worker entry), the principal objective is to remove scale and corrosion, and then apply a SIPP that further inhibits deterioration and seals minor leaks. As most mains are structurally sound, the most common techniques to improve flow and prolong pipe life are to renew the existing pipes with cement mortar or polymeric resins, applied by a robotic spraying machine that is winched through the pipe at a constant predetermined rate. The AWWA standards for cement mortar SIPP and the Manual 28 for rehabilitation of water mains provide detailed procedures.

Sanitary sewer applications are different from water pipelines. Development work has been carried out on spray-on SIPP techniques for non-worker-entry sewers, but so far no such method has achieved commercial prominence. This may be partly because of the different requirements of sewer renewal, where the aim is usually to increase the pipe's resistance to the external loading rather than to prevent corrosion, and partly because of the practical difficulties of ensuring that infiltration, and the resulting wet areas or standing water, to the sewer is completely stopped while the material is being applied and cured.

In the United States, spray-on SIPP is seldom used in gas mains, although in some countries it is used extensively in gas service pipes.

The following covers various application methods as related to the SIPP types used in underground applications:

10.3.7.1 Cementitious

1. *Non-Worker-Entry Pipes.* Cement mortars are applied to the inside wall of non-worker-entry sewers to form SIPPs with limited structural integrity. Application is generally carried out by a spraying machine which is either fed through hoses from the surface, or, particularly in larger pipes, may have its own hopper containing premixed mortar (see Fig. 10.6). Forward speed control of the machine is important to produce a consistent thickness of mortar. Trowelling follows spray application, which may be carried out by rotating spatulas fitted to the spraying machine, or sometimes by a simple tubular shield of the required internal diameter, which is pulled through behind the machine. Whatever system is used, it is essential to centralize the equipment within the existing pipe so that the SIPP is of constant thickness around the whole perimeter.

   Process for non-worker-entry pipelines:

   - The pipeline is cleaned to remove encrustation and protrusions.
   - Remote inspection of the main is carried out to ensure no infiltration and a relatively clean surface for adhesion of the SIPP.

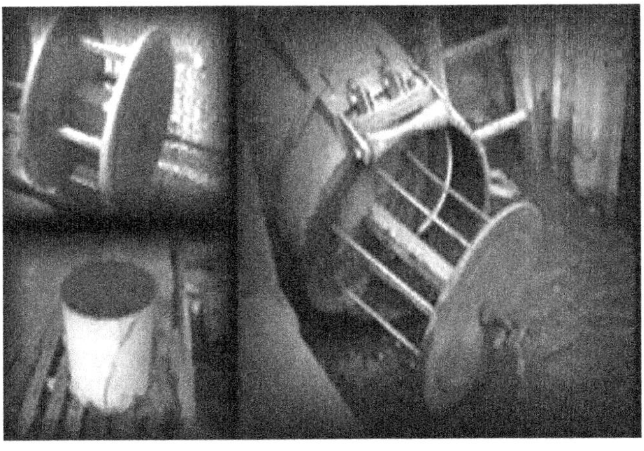

Figure 10.6 Pigging method. (*Source:* 3M Water Infrastructure.)

- The cement is mixed onsite and pumped though lines to a spinner head.
- The spinner head is winched through the main at a required speed.
- The cement is allowed to set.
- A uniform 4-mm SIPP of cement is achieved.
- Ordinary Portland cement or a blend containing blast furnace slag is used for SIPP.

2. *Worker-Entry Pipelines.* For worker-entry pipes, spraying of gunite (dry concrete) or shotcrete (wet concrete) is possible. It is essentially a stabilization technique for old brickwork and masonry pipelines. The sprayed concrete forms a new inner skin which enhances strength, reduces leakage, and prevents further loss of mortar. The concrete is trowelled after application to produce a smooth surface. Reinforcement may be fixed to the existing pipe wall prior to applying the concrete, adding to the structural strength, and creating a new reinforced concrete pipe with the old pipeline. The design of the concrete mix is important to ensure full penetration and encapsulation of the reinforcement. Precast concrete units are commonly used to line the invert, especially if the flow in the pipe cannot be stopped off completely.

   The American Concrete Institute (ACI) and the American Society of Testing and Materials (ASTM) define shotcrete as the method of placing concrete or mortar onto a surface at high velocity. ACI 506 adds "at sufficient high velocity to achieve full compaction." Shotcrete is a method of placing concrete upside down and vertically without the aid of forming. Table 10.2 presents a typical shotcrete mix design (Stier, 2000).

   In gunite, a trademark for more than 50 years, sand cement mix of specific aggregate gradation and cement content is used.

Material	Quantity
Cement (can be type I, II, or III)	700 lb (7.5 sacks)
Concrete sand (SSD)	2790 lb
Water (added at predampener and nozzle)	39 lb (38 gal)
Total weight	3810 lb
Unit weight	144 pcf
Volume	1 cubic yard

Source: Najafi and Gokhale, 2005.

TABLE 10.2 Typical Shotcrete Mix Design

For spraying operation, a pressure vessel machine is used that conveys and applies the material at high velocity. Aggregate contains 3 to 6 percent moisture providing for a short period of prehydration. Balance of water is added at the nozzle. If the water to cement ratio is approximately 0.35 to 0.40, the resulting concrete will have superior compaction and bonding properties. During the laboratory testing at Lehigh University, it was indicated that a high impact velocity, high fine content, and low water-cement ratio will achieve a greater density and durability, low permeability, and higher resistance to attack by carbonates and sulfates (Stier, 2000). These tests further indicated that higher velocity would provide a better bonding.

Since the 1980s, with development of concrete pumps and better understanding of concrete mix, the application of shotcrete and gunite has increased. These developments include longer stroke for pumping, better nozzles for injection of air and acceleration, and updated material technology for pumpability, high range of water reducers to decrease water-cement ratio and microsilica to increase density. The same admixtures used in structural concrete are used in shotcrete with the same results. For example, adding silica fumes improves adhesion, increases cohesive properties of mix, reduces rebound (which is substantial for dry-mix), and improves washout resistance. For reinforcement, wire mesh, reinforcing bars, or steel or synthetic fibers are used.

3. *In Situ Reinforced Concrete.* Continuous in situ reinforced concrete SIPP systems for worker-entry pipelines from 900 mm to over 5,500 mm diameter use reinforcing steel attached to the pipe wall. Lightweight formwork is then installed, usually in 2-m lengths, which are locked together with pins. Concrete is pumped under pressure behind the formwork to create a monolithic SIPP. The process accommodates bends and slight deformation, and produces a smooth finish with low hydraulic roughness. Connections can be formed through the use of adapted shutters, and flows can be maintained without over pumping.

    - Process for worker-entry pipelines:
        - The pipe section is cleaned to remove encrustation and protrusions.
        - No high levels of infiltration and a relatively clean surface for adhesion of the SIPP are ensured.
        - Formwork is installed.
        - Concrete is pumped behind the formwork.
        - Concrete is allowed to set.

- Advantages:
    - Lateral connections are easily handled.
    - No maximum limitation on sewer size.
    - Variations in cross section can be readily accommodated.
    - Low cost.
- Limitations:
    - Relatively slow installation.
    - Requires safe conditions for worker entry.
    - High-level operator skill is required.
    - Potential for lateral restriction.
    - Water pH sensitive.

10.3.7.2 Polymers

1. *Non-Worker-Entry Pipelines.* Polymer SIPP function is to provide corrosion protection and a smooth bore. Epoxy, polyurea, or polyurethane materials are centrifugally sprayed to the inside walls of non-worker-entry circular pipes. The objective is to make the resin bond with the prepared internal surface of the pipe, forming a monolithic SIPP that protects against corrosion, and enhances flow. Epoxy SIPPs are generally much thinner than cement mortar SIPPs, and therefore, do not cause significant bore reduction. In comparison with cement, epoxies also cure more quickly, do not affect the pH of the water, do not wear, and can significantly enhance flow. However, any defect in the epoxy SIPP may allow corrosion to start and, unlike cement mortar, there is then no alkalinity to inhibit deterioration chemically. Epoxy resins are also slightly more expensive.

   Epoxy resins that have been approved for the SIPP of water mains by relevant certifications (such as NSF/ANSI 61 by National Sanitation Foundation or NSF International, ASTM, AWWA, and others) do not impair the quality of the conveyed water, provided they are mixed, cured, and applied properly. AWWA Manual 28 requires only the use of approved SIPPs.

   The resin is applied by a spraying machine, which usually has a rotating nozzle. The typical 1 mm (40 mils) thickness of the SIPP is controlled by the flow rate and the forward speed of the machine. In most systems, the resin base and hardener are fed through separate hoses, and are combined by a static mixer just behind the spray nozzle. Ideally, the cure time should be as short as possible to minimize the period during which the main is out of service and to reduce the risk of contamination of the resin prior to cure. However, too rapid a cure carries the risk of blocking the static mixer or the nozzle. Unlike with cement mortar SIPP, the resin is not smoothed or trowelled after spraying, and the surface quality depends on the application technique and the properties of the material.

- Installation process:
  - The pipeline is cleaned to remove encrustation and protrusions.
  - Inspection of the main is carried out to ensure no infiltration and a relatively clean surface for adhesion of the SIPP.
  - A two-component resin is mixed and pumped though lines to a spinner head.
  - The spinner head is winched through the main at a required speed.
  - The resin is cured for a specified time.
  - A uniform, smooth SIPP, typically 40 mils thick is achieved.
- Advantages:
  - No pH effect.
  - Higher flow.
  - No wear for longer life.
  - Service connections do not need reconnection.
  - Variations in cross section can be readily accommodated.
  - Relatively quick installation or high production.
- Limitations:
  - Control of infiltration is required to prevent SIPP disbondment or collapse.
  - No structural integrity.
  - Slightly higher cost.

2. *Worker-Entry Pipelines.* SIPP is normally carried out using specialized spray equipment with a hand-spray gun in worker-entry pipelines, lift stations, junction boxes, clarifiers, tanks, and other underground structures. Most polymer systems use plural component spray equipment customized to the characteristics of the particular polymer to ensure properly mixed and metered output. Such equipment can typically pump material over 500 ft to access difficult jobsites, or the equipment can sometimes be dropped into larger structures to accommodate extended application needs.

3. *Composite Systems.* The use of high-strength cementitious materials top coated with polymer SIPPs is a cost-effective solution to renewal of severely deteriorated structures. The products must be compatible and installed in accordance with manufacturer recommendations.

10.4 Inspection and Testing

Inspection and testing methods during installation of SIPPs contained within this section are available through the product manufacturers and associations such as NACE, ASTM, and SSPC. In general, some common procedures include the following:

10.4.1 Visual Inspection

Visual inspection and project documentation should be carried out at a minimum by the project inspector. A trained inspector who is knowledgeable in the products used and their installation process should conduct a final visual inspection. Any deficiencies in the finished SIPP should be marked and repaired according to the SIPP manufacturer's recommendations.

10.4.2 SIPP Thickness

For cementitious and epoxy SIPPs, thickness of the applied SIPP should be randomly checked prior to the set of the SIPP. A wet film thickness gauge, such as those meeting ASTM D4414—Standard Practice for Measurement of Wet Film Thickness of Organic SIPPs by Notched Gages—can be used with epoxies to ensure uniform thickness during application. Cementitious products generally will use a depth gauge to measure the applied thickness. This is not practical for the rapid setting urethanes and ureas. The use of an ultrasonic thickness gauge can be used after polymer SIPPs have set or cured. This testing should only be performed by personnel trained in ultrasonically evaluating the thickness of thick-film SIPPs (>100 mils) on nonferrous substrates. Equipment such as a Positector 100D should be used. The equipment should have SSPC PA-2 capability to calculate the average of a set of readings to determine specification compliance.

10.4.3 Spark Testing

Spark testing or holiday detection can be used to detect holidays or pinholes in a polymer SIPP and inadequately welded seams in sheet liners. After the protective SIPP has set hard to the touch or after welding has been completed, the system can be inspected with high-voltage holiday detection equipment. Surfaces should first be dried, an induced holiday is then made on to the coated concrete surface and serves to determine the minimum or maximum voltage to be used to test the SIPP for holidays at that particular area. The spark tester is initially set at 100 Volts/1 mil (0.025 mm) of film thickness applied, but may be adjusted as necessary to detect the induced holiday (refer to NACE RPO188-99). All detected holidays are marked and repaired following SIPP manufacturer's recommendations.

10.4.4 Adhesion

Bond strength of polymer SIPPs can be measured in accordance with ASTM D-4541 and/or ASTM D-7234 depending upon the substrate. Any areas detected to have inadequate bond strength should be evaluated by the project engineer. Further, bond tests may be performed in that area to determine the extent of potentially deficient bonded area and repairs made in accordance with the manufacturer's recommendations. An understanding between the owner and installer should be made prior to testing taking place regarding testing frequency, test equipment, testing/evaluation methods, the required failure mode, minimum pull off values, and remediation requirements if/when "inadequate adhesion" is encountered. There are many variables that play a part in the testing process and the analyzing of results of an adhesion test. Thus testing and evaluation should be performed by trained personnel only. Adhesion testing is a destructive test and consideration should be made as to the extent of testing in a structure in which the objective is to provide a monolithic SIPP.

10.4.5 Vacuum and Exfiltration Testing

Manholes and pipelines can be tested for sealing integrity in accordance with ASTM standards set for vacuum and exfiltration testing.

10.5 Chapter Summary

SIPP systems are bonded to existing pipelines and structures; therefore, proper surface preparation is essential. Many methods are available and are used to prepare surfaces, including pressure water jetting, abrasive blasting, degreasing and others. Inspectors should pay particular attention to the surface preparation. All infiltration must be stopped, flows diverted, and preparation performed in a manner that provides a uniform, sound clean neutralized surface suitable for the specified product. Testing and inspection of SIPP during and after installation is another essential ingredient to ensure maximum performance of the installed system. Variables including weather, access, confined spaces, temperature, humidity, leakage or flow affect the performance of SIPP methods. Inspectors should document ambient and application environment variables and verify installation processes are being conducted in accordance with specifications and product manufacturer guidelines.

10.6 Review Questions

1. Describe the various types of coatings and linings available and their applications.
2. Under which project conditions would you recommend coating and linings?
3. Compare cementitious and polymer coatings and linings.
4. Describe the gunite and shortcrete process. In what project conditions are they applicable?
5. Describe the importance of pipe cleaning process for coating and lining applications.
6. What are the advantages and limitations of coating and lining processes?

CHAPTER 11

Close-Fit Pipe

11.1 Introduction

Since 1988, close-fit pipe (CFP) has been used extensively in the United States. This technique can be used for sewers systems, potable water and gas supply lines, and industrial applications. The CFP can be used for structural or nonstructural purposes and pipelines from 4-in. to 65-in. diameter [larger diameters are possible with the deformed and reformed (D&R) method] and lengths up to 5,000 ft.

11.1.1 Primary Characteristics

This type of trenchless pipeline renewal uses a new polyethylene (PE) or polyvinyl chloride (PVC) pipe that is modified in cross-section before it is installed. After placement or insertion into the existing pipe, it is reformed to its original size and shape to provide a close-fit with the existing pipe. There are two methods of CFP as shown in Fig. 11.1.

In the first method, called fold and formed (F&F), plastic pipes are folded (EX method) or deformed (U-liner process) in the factory during production and then wound onto large reels. Both methods can be used for gravity and/or pressure pipelines. Following delivery to the renewal site, the new pipe is heated until it becomes flexible, allowing it to be pulled from manhole-to-manhole through the existing pipe. Once in place, the new pipe is forced against the inside surface of the existing pipe using heat and pressure to form a new CFP. In case of a PVC pipe, a rounding device may also be used. After installation is completed, lateral connections are reopened using a remote-controlled robotic cutter.

The second CFP method is D&R technique that involves the use of long, butt-fused sections of the PE pipe (Fig. 11.2). This method is primarily used for pressure pipelines for structural (Swagelining or rolldown process) and nonstructural (subline process) renewal. After the preparation of butt-fused pipe and construction of entry and receiving pits, the diameter of the PE pipe is mechanically reduced, resulting in a smaller circular shape or folded shape,

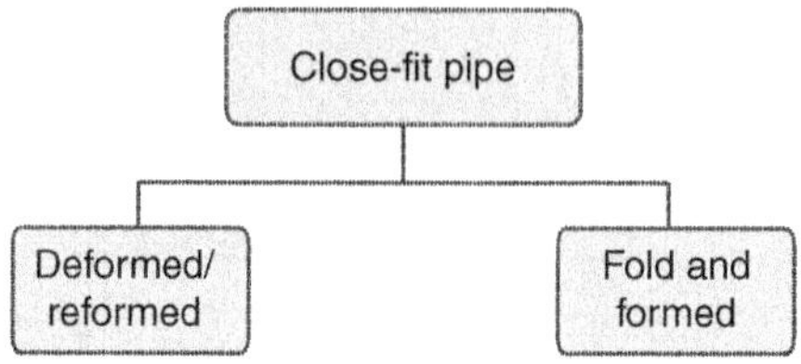

Figure 11.1 Two main variations of the CFP process.

Figure 11.2 The deformed & reformed process. (*Source:* Murphy Pipelines.)

depending on the process used. The diameter reduction is performed just before the new pipe is inserted into the existing pipe. After the new pipe is inserted into position, it expands to its original size and thus fits snugly into the existing pipe, making a close-fit with no gaps or annular space. Water pressure is required with the folded shape to burst the bands used to hold the folded shape in place during insertion. Figure 11.2 illustrates lateral connection and tapping for a water pipe. Figure 11.3 illustrates the diameter reduction, banding, and insertion of a D&R process.

11.2 CFP Applications

As for any other renewal method, it is important to choose the best renewal method for each individual project. The type of application, working space available, chemical composition of the fluid carried by

FIGURE 11.3 The diameter reduction and banding process for the D&R pipe. (*Source:* Plastic Pipe Institute.)

the pipeline, the pressure and temperature in the pipeline, number of laterals, depth, diameter, number of bends, any misalignments or joint settlements, the number of manholes, and so on must be assessed from project to project. As mentioned in Chap. 1, CFP can be used for structural or nonstructural purposes to renew both gravity and pressure pipelines. Table 11.1 provides an overview of the close-fit applications and possible limitations.

11.3 Construction Considerations

11.3.1 General

As with any construction project, it is the responsibility of individual contractors to comply with current work safety and environmental regulations. Product and technical specifications should be used to ensure that the chosen product is suitable for the renewal in question. For example, for waterlines, it is important to ensure that all product information and approvals, according to local and state laws, are available. Any surplus liner pipe must be safely disposed. Information on how the material used can be disposed of or recycled can be obtained from the material supplier or the manufacturer. Well-known products like PE, PVC, and modified PVC that

Pipeline Type	Suitable	Comments
Sewers	Yes	
Gas pipelines	Yes	
Potable water pipelines	Yes	See note A
Chemical/Industrial pipelines	Yes	See note B
Straight pipelines	Yes	
Pipelines with bends	No	See note C
Circular pipes	Yes	
Noncircular pipes	No	
Pipelines with varying cross-sections	No	
Pipelines with lateral connections	No	See note D
Pipelines with deformations and misalignments	No	See note E
Pressure pipelines	Yes	

A. Check all local and national codes for all materials that come into contact with potable water. Both F&F and D&R with butt-fused joint PE pipes are suitable for renewing potable water pipelines. Both methods usually require the excavation of entry and reception pits. In addition, valves, service connections, and sharp bends must be dug free before renewal work is begun.
B. Bends greater than 22 degrees cannot normally be negotiated with CFP methods.
C. The new pipe must fit the smallest diameter in the existing pipeline unless adapters are built into the line.
D. For sewers, it is not necessary to excavate lateral connections, as they can be reopened from inside the new pipe using robotic cutters. It is always necessary to excavate service-line connections on potable water pipelines.
E. The method is not directly suitable for pipelines with considerable deformations and misalignments.

Table 11.1 Overview of Close-Fit Applications and Possible Limitations

are currently used to manufacture CFPs can be delivered to approve waste collection points or returned to the manufacturer for reuse and/or recycling.

11.3.2 Preliminary Evaluation and Inspection

To achieve the best possible result of pipeline renewal using close-fit method, it is important to carry out a detailed preliminary survey to determine, at the very least, the following:

- The condition of the existing pipeline–using appropriate inspection method (see Chap. 3).
- The need for cleaning–based on inspection results.
- Existing pipeline material, length, and depth.
- Pipeline dimensions, deformations, and displaced joints.
- Water table depth and its fluctuations.

- Location of laterals.
- Traffic loads.
- Service-line data, including dimensions, unused service lines, service lines, and intruding service-line connections.
- Manhole condition.
- Location and size of access pits.

11.3.3 Preparatory Work

Once necessary preliminary surveying has been completed and renewal is about to commence, there are several routines for the way in which preparatory work should be carried out in order to achieve effective pipeline renewal. These include the following:

- Any planned over-pumping, its duration, and the number of consumers affected.
- Planned traffic diversions and their advertising.
- Pipeline shutoff (timing, duration, and number of laterals affected).
- Location and size of pits.
- Inspection of bends in the existing pipeline.

Usually, the preparatory work required would depend on what the pipeline is designed to carry, and whether it is a sewer or pressure line. For a sewer, for example, preparatory work would include

- inspection and possible grinding of intruding service-line connections, pipe sections, and displaced joints on the existing pipeline
- inspection and possible grinding or cutting away of scattered deposits
- water jetting/root cutting
- closing/blocking off unused service lines
- restoration of the existing pipelines or parts of it when badly deformed
- prevention of groundwater infiltration (temporary sealing to allow renewal)
- general grinding and cleaning of internal corrosion throughout the pipeline

11.3.4 Installation Phase

As for any renewal project, installation is an important phase of CFP process. It is therefore important for both the pipeline owner and the contractor to ensure that both the production of the new pipe and its installation are carried out under controlled conditions that can subsequently be documented in writing.

As swagelining is usually used for potable water and gas supply lines, it is necessary to excavate valves, service connections and sharp bends before installing the line. Folded plastic pipes for potable water pipelines are installed in more or less the same way as swagelining, while flooded pipes for sewers can usually be installed from manhole-to-manhole.

11.3.4.1 Insertion

After heating, the reel holding the folded pipe is positioned beside the manhole from which the new pipe is to be installed. Beforehand, a winch cable is pulled through the pipeline to be renewed. It is important to ensure that the new pipe has the correct insertion temperature, that is, the temperature at which the new pipe is most flexible. After fitting the towing head and cable, it must be ensured that the new pipe is not subjected to excessive bending during insertion. The new pipe must also be protected from damage caused by rubbing against the sides of the manhole or the entry and reception points. Use of rollers or assist wheels can help mitigate damage in a controlled fashion. The force used during insertion must never exceed the permissible pulling force for the new pipe.

11.3.4.2 The Heating Process

After insertion, a hot air plug or end cap is fitted over the end of the new pipe in the manhole so the new pipe can be heated by steam. The new pipe is then heated to the temperature recommended by the manufacturer. The temperature used will depend on the product, its dimensions, and the ambient temperature. When the recommended temperature has been reached, the new pipe is pressurized in accordance with manufacturer's recommendations. This pressure is usually between 0.6 and 3.0 bars depending on the material composition of the new pipe in question. The new pipe must then cool while still under pressure (e.g., using compressed air) to ensure that it remains tightly pressed against the existing pipe. After cooling, the hot air tool can be removed and necessary completion work is performed. Restraint of any exposed pipe, terminal ends, test caps, and associated testing equipment shall be utilized at all times during air pressurization process to protect property and workers in the event of an unanticipated catastrophic release of the stored energy when using compressed air.

11.3.5 Project Inspection and Delivery

After installation, it is often necessary to carry out completion work. This may, for example, take the following steps:

- Leak testing.
- Consumer information following service-line reinstatement.

- Manhole finishing ensuring continued smooth flow.
- Lateral reinstatement so that wastewater from individual consumers can again be led to the main sewer.
- For the case of water mains, installation of liner pipe end terminations and tees/branches must be carried out.
- Final inspection to document the completed renewal.

The main purpose of the above tasks is to document the renewal process in accordance with the contractor's own quality control and assurance standards and according to the technical specifications to contract documents.

11.4 Quality Assurance and Documentation

A final quality review compares the completed renewal work with the requirements and descriptions of technical specifications and contract documents. A final review would expose any variations and/or changes between the contract documents and the work of the contractor.

When a renewal project is completed, the contractor provides the pipeline owner with documentation of the quality of the work performed. In this respect, it is important that during the bidding phase, the pipeline owner ensures that such documentation can be prepared and submitted on completion of the renewal work. The documentation should include the following:

- Information on the products and installation method used.
- Pre- and post-inspection results.
- Test results.
- Any changes or deviations from contract documents and possible corrective work.
- Information on lateral reinstatements, pipe ends/manhole connection, and possible grouting process.
- Recommendations for future inspections, cleaning, and other renewal work to be completed as part of a comprehensive infiltration or inflow and/or structural plan.

11.4.1 Advantages

The CFP method has the following advantages over comparable renewal technologies:

- The product pipe is manufactured at the factory; therefore, the installation process in the field is faster.
- For the same reason as above, the quality assurance process is more efficient. Usually an original PVC or PE alloy with recognized physical and chemical properties is installed.

- Being a CFP, the reduction in the existing pipe cross-sectional area is minimal.
- The CFP method can solve pipeline internal corrosion problems, water quality problems associated with pipeline internal corrosion, leakage from corrosion holes, cracks, and failed pipeline joints, and flow capacity problems arising from pipe tuberculation and deposits.
- The new pipe can be installed via a manhole or insertion pit.
- The new pipe is capable of accommodating large radius bends.
- The new pipe can be installed up to 5,000 ft; therefore, it will have a smooth and jointless pipe between manholes with better flow characteristics than the excising pipe.
- Internal lateral connections are possible.

11.4.2 Limitations

CFP has the following limitations:

- Limited diameter range and installation length is available.
- For D&R pipe, a large working space is required to lay the butt-fused string of PE pipe before feeding into diameter reduction machine and insertion into the existing pipe.
- The method may not be suitable if there are variations in cross-sectional area of the existing pipe and for existing pipes with bends more than 22.5°.
- Usually an insertion pit is required.
- In most cases, bypassing the flow during the installation is required.
- For water mains, locations of valves and connections may have to be excavated.

11.5 Review Questions

1. What are primary characteristics of a CFP?
2. In what conditions would you recommend a CFP?
3. How CFP does differ from other renewal methods?
4. Describe the two main variations of the CFP method.
5. What type of preparatory work is usually performed for CFP?
6. What are construction considerations of CFP?
7. What are the environmental considerations of CFP?

8. Describe the CFP installation process in detail.
9. Check your local city phone directory to see what companies or contractors offer the CFP method. Interview one of these companies to see how many feet of pipe they have installed and which variation of CFP they have used. Check the prices they offer and compare it with other renewal methods.
10. Check your local city public works department of public works to see if they have used the CFP method. Obtain specific project conditions (location, diameter, existing pipe type, length, application, etc.) and their decision criteria to select this method over other methods.
11. List the main advantages and limitations of CFP.
12. An engineer is preparing bid documents for a small-diameter sewer project consisting of 20,000 ft of 6, 8, and 12 in. sewers. The sewers are typically 6 to 8 ft deep with minimal groundwater, possibly a few feet above the sewer during the wet season. There are house service connections approximately every 50 ft. The typical distance between manholes is 350 ft. The manhole rings have a clear opening of 22 in. The engineer is considering including F&F and D&R close-fit products in the bid documents. Is this a good application for these rehabilitation products?

CHAPTER 12

Sewer Manhole Renewal

12.1 Introduction

Manholes are called windows to the sewer system as they are the most visible point in identifying the condition of underground infrastructure. While manholes are easily accessible, they offer a variety of challenging problems that are frequently misunderstood and overlooked. Like pipelines, manholes come in a variety of materials and sizes. They are commonly made of brick mortar, precast concrete, concrete block, and cast-in-place concrete. At each major change of direction or grade in the sewer system, a manhole is usually constructed to facilitate flow, cleaning, and inspection. Open curved channels in the base of the manhole allow sewage to change direction with a minimum amount of friction from one pipe segment into another.

There are currently approximately 20 million manholes in the United States, and over half were installed prior to 1960. As we drive through communities, one can often determine the age of a system by looking at the manhole lid. In older systems, the lids may be 18 in. in diameter and have numerous vent holes. The manhole chamber below the lid of older systems is generally constructed of brick and mortar. Newer systems will generally have a lid of 24 in. or 30 in. with only one or two pick holes and be constructed of precast concrete. These early observations tell us a little about the potential problems we may encounter as old manholes suffer from a variety of serious problems. They are subject to erosion from surface runoff and groundwater intrusion, corrosion from liquids and gases, wearing from dynamic traffic loads, and general deterioration from age. Out of sight, their degradation is not easily monitored. When neglected, their complete collapse is likely to result as shown in Fig. 12.1.

Not considering the age, location of a manhole within a system often dictates the type of problems manholes will develop. These problems include erosion, corrosion, and wear from dynamic traffic loads, any of which will create structural fatigue and/or serious

FIGURE 12.1 Manhole collapse. (*Source:* Permaform.)

inflow or infiltration. Other influences which can contribute to the structural decline of a manhole include soil movement and water table fluctuation (hydrostatic loading). Soil movement can be due to natural occurring events, such as earthquakes, freeze/thaw cycles, or induced by nearby construction activities.

Once a rigid structure has experienced movement, there is a good possibility of foundation damage. A fracture in the wall of a manhole can also open up the possibility for erosion. Approximately half of the manholes that have been installed in the United States were never pressure tested to ensure water/air tightness. This constant water intrusion takes a wearing toll on a rigid structure over time. As shown in Fig. 12.2, for the case of a brick manhole, water movement can remove mortar between bricks, and aid in the loss of bricks and the potential for failure of the structure over time.

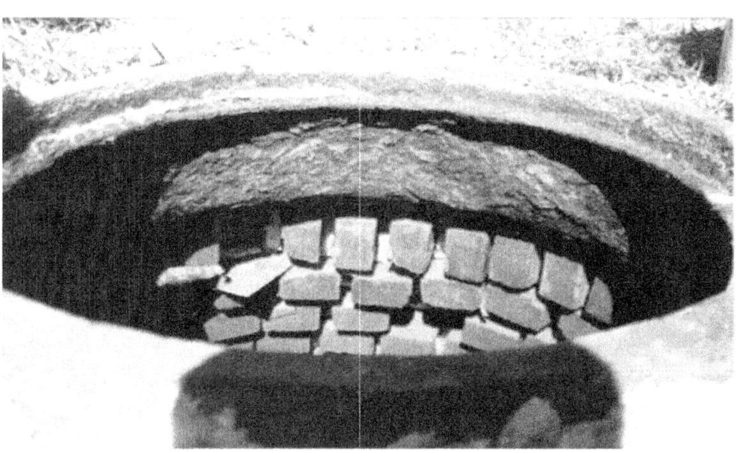

FIGURE 12.2 Structural failure of a brick manhole. (*Source:* Permaform.)

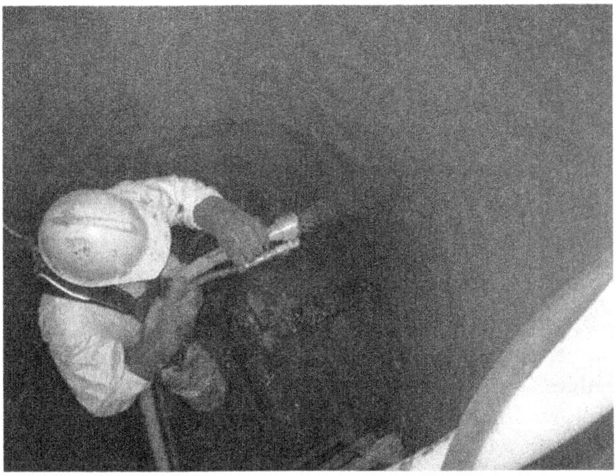

Figure 12.3 A manhole being renewed by cementitious lining. (*Source:* NASSCO.)

Manhole rehabilitation materials include Portland based cement mortar (with added chemicals to prevent hydrogen sulfide induced corrosion, see Fig. 12.3), calcium aluminates, geopolymers, polymeric (polyurethanes, epoxies, polyureas and their mixtures, see Fig. 12.4), fiberglass, epoxy-coated cement, coal tar epoxy, polymeric sheets (e.g., PVC), steel inserts, rubber inserts (used for chimney and joint rehabilitation), elastic polyurethane (for chimney restoration), cured-in-place composites (similar to CIPP used for pipeline rehabilitation), polymer concrete inserts, and fiber reinforced cement linings. Each of these materials and methods has its pros and cons, with a number of unknowns in terms of the structural support they can provide to a

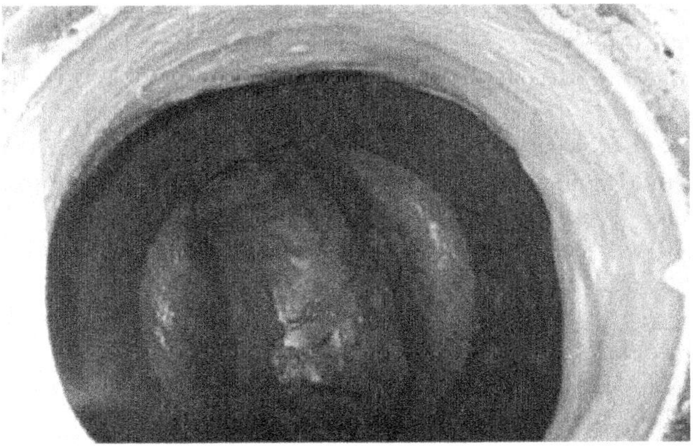

Figure 12.4 A manhole renewed with polymeric lining.

deteriorated manhole. A rehabilitated manhole is a system with the following parameters that play a role on the overall durability and life cycle:

- Residual strength of the manhole
- Mechanical properties of the lining material
- Adhesion between the lining and substrate (manhole component)
- Magnitude and type of loads exerted on the manhole
- Durability of manhole material against environmental effects (particularly to hydrogen sulfide induced corrosion). Manholes that have deteriorated and allowed I/I to breach also allows fines into the manhole can result in catastrophic consequences such as a sinkhole. Manholes that have deteriorated by corrosion may have material fall into the wastewater flow causing a backup that may result in a sanitary sewer overflow (SSO) and/or basement backups.

Accordingly, structural condition of a manhole is an important performance indicator and risk of failure should be minimized. However, as discussed above, mechanical strength of rehabilitation materials vary substantially, and to which extent manholes should be structurally supported is a matter of condition of the manhole to be rehabilitated and loads exerted on it. Little is known on the ability of rehabilitation materials in withstanding the dead and live loads that are exerted on them.

Manhole rehabilitation materials and methods can be divided into following:

Class A Rehabilitation Materials: Manhole relining methods that provide fully structural solutions in addition to stopping infiltration and inflow (I/I). Some of the rehabilitation solutions are apparently fully structural (e.g., polymer concrete manholes and some of the fiberglass inserts) based on the third party test data. Structural capacity of other types of linings will be investigated as a part of this study.

Class B Rehabilitation Materials: Class B manhole rehabilitation methods will be referred to as semi-structural; meaning that these methods can add to the residual strength of a deteriorated manhole, but they are not capable of withstanding the dead and live loads exerted on a manhole by themselves.

Class C Rehabilitation Materials: Class C manhole rehabilitation methods and materials will be deemed nonstructural. The purpose of using Class C rehabilitation materials is to provide preemptive protection for manholes that are in relatively good condition in addition to stop minor leaks into the manhole.

As durability and stiffness of a manhole lining material increases, so does the cost of it; therefore, selection of the optimum class of manhole renewal material should be based on a life cycle analysis that takes the residual strength of the deteriorated manhole into account, to achieve the most economical solution that will meet the expected service life (minimum 50 years) of the rehabilitated system.

12.2 Manhole Problems

Manholes consist of two sections: the floor and the shaft. The problems encountered in the two are often of different types. Subsequently renovation work can be prioritized on the basis of the extent of the problem and its location in the manhole.

A section of a defective manhole is shown in Fig. 12.5. Typical problems that may occur in such a manhole are as follows:

- Infiltration
- Corrosion
- Inflow

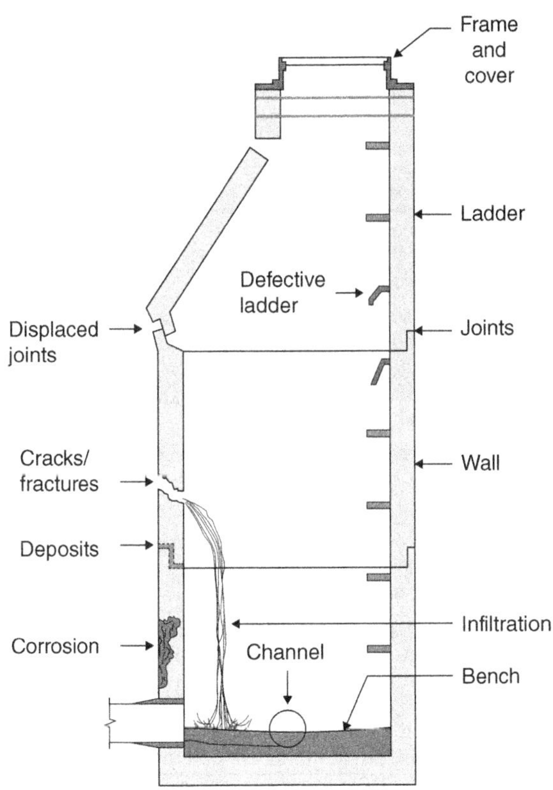

FIGURE 12.5 Section of a defective manhole.

- Displaced joints
- Cracks and fractures
- Defective or no floor
- No berms
- Poor flow
- Deposits
- Defective ladder
- Defective cover

Some manhole problems are found throughout the manhole, while others only affect part of it. Corrosion, for example, typically attacks the part of the manhole above the water level. Other problems are typically found in limited areas in the manhole. Leaks, for example, are most often found in joints. The most common manhole problems are described in Table 12.1.

The following parameters should be used to prioritize when and how individual manholes are renovated:

- Manhole strength (risk of collapse)
- Manhole capacity (flow)
- Floor tightness

Problem	Description
Corrosion	Hydrogen sulfide (H_2S) attacks above the water level. The floor may therefore remain intact while remainder of the manhole is subject to corrosion.
Cracks/ fractures	Cracks and fractures typically occur as the result of poor construction, inferior materials and/or outside influences. They result in leaks. Depending on their extent and location, cracks and fractures will reduce manhole strength and impair function.
Displaced joints	Displaced or open joints occur for the same reason as cracks and fractures. Manhole strength remains unaffected, but leaks result.
Deposits	Deposits often result from the action of aggressive groundwater. Lime is leached from the concrete, resulting in leaks and manhole weakening.
Infiltration	Infiltration presupposes the presence of leaks and groundwater. All the above-mentioned problems can cause infiltration. Typically, leaks result in greater quantities of water being delivered to wastewater treatment plants, thus reducing their efficiency. Exfiltration of wastewater may occur if the groundwater table is below the manhole.

Table 12.1 Common Manhole Problems

- Shaft tightness
- Corrosion protection
- Top rings/covers

This issue of corrosion chemical attack must be thoroughly understood by the evaluating entity. Chemical attack can be due to effluent discharges that eat away the channel and base of a manhole or due to corrosive gases. These discharges will have low pH levels of wastes discharging into the sewer system causing concrete materials to deteriorate. Most public agencies have been able to eliminate issues concerning chemical attack due to contributor discharges. Microbiologically induced corrosion (MIC) is a process where aerobic bacteria metabolize H_2S gas and oxygen to produce sulfuric acid (H_2SO_4). The MIC is generally most aggressive in manholes with a lot of turbulence.

Problems differ greatly from manhole to manhole and the renovation method used must be adapted to suit the manhole in question.

12.3 Classification of Manhole Problems

With the advancement of new pipe renewal technologies, manholes are receiving increased attention. Leaks, sealed out when the pipes are lined, are likely to travel to the next weakest part of the system—the manholes. Engineers have learned that neglecting such an integral part of the system only shifts the problem from the pipes to the manholes. When any part fails, the whole system fails. Thankfully, manholes are the easiest segment to investigate and the "most effective and least costly" to repair. Selecting the proper method is critical to a successful renewal program. Since each type of problem is different, each demands an engineered solution. Regardless of the renewal mechanism, bonding or adhering to the original substrata is critical for success. Understanding the environment, the steps needed in preparing the surface, and understanding the products being used are critical to a successful product.

Being aware of these various types of problems will aid in the investigation of manholes in a system. Once manhole defects have been identified, in accordance with NASSCO Manhole Assessment and Certification Program, they can be classified into different types. Each type of defect can then be related back to an effective repair method.

12.3.1 Type I: Frame and Cover

Type 1 defects are single repairable defects that can be easily and usually inexpensively repaired, such as frames and covers (see Fig. 12.6), which leak because of poor fit, multiple vent holes damaged adjusting rings or shifted frames. They can be replaced or fitted with under the cover inflow protectors.

FIGURE 12.6 Stainless steel insert. (*Source:* Permaform.)

Sewer dish inserts are installed under the cover creating a waterproof seal between the frame and underground structure. Inserts prevent the intrusion of rainwater through cover vent holes, damaged frames, or degraded covers. Effluent gases are vented through a nonmechanical device located in the insert.

The grade adjustment area is prone to develop inflow and infiltration problems due to traffic loading and weather conditions. Damaged adjusting rings can receive an internal or external chimney seal. Internal seals are available in mechanical or hand applied version made of a flexible membrane. Elastomeric urethane seals are available which seal the grade adjustment area with enough flexibility to withstand traffic loading and weather conditions. Mechanical seals (Figs. 12.7 and 12.8) are also available which create a watertight seal to the frame and another watertight seal on the top of the cone section. If the exterior of the structure is exposed during the renewal process, sealing the frame and grade adjustment area eliminates inflow and infiltration. Protective rubber jackets are on the market, which seal the frame and grade adjustment area to the cone section. Elastomeric urethane seals function on the exterior of the structure as well as the interior by sealing the frame and grade adjustment area with enough flexibility to withstand traffic loading and weather conditions (Fig. 12.9).

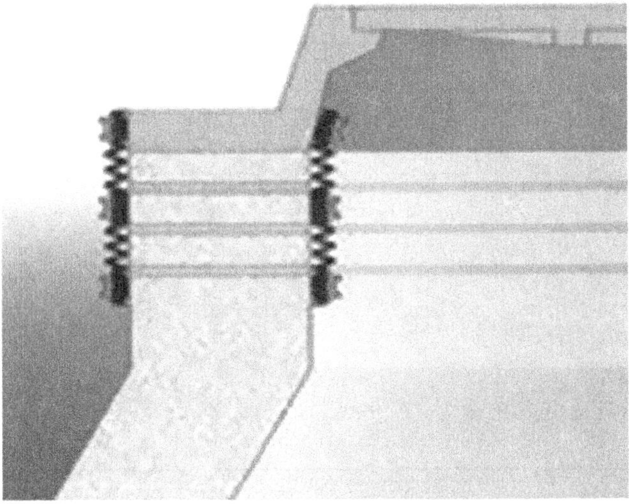

Figure 12.7 Mechanical seal. (*Source:* Permaform.)

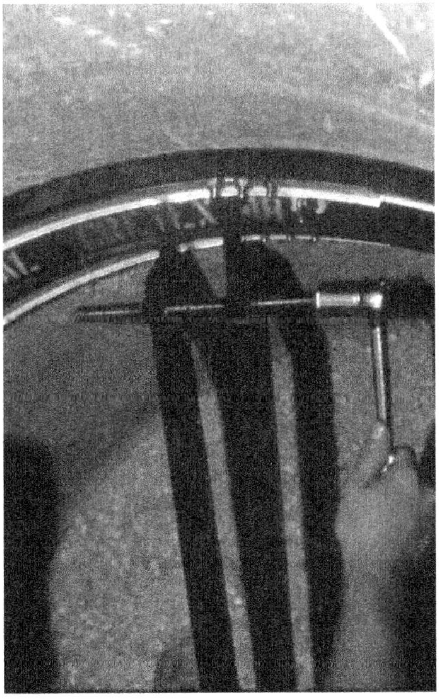

Figure 12.8 Installed mechanical seal. (*Source:* Permaform.)

FIGURE 12.9 External seal. (*Source:* NASSCO.)

Replacing degraded adjusting rings, frames, and covers reduce inflow and infiltration (Figs. 12.10 and 12.11). Standard products include precast concrete adjusting rings, cast iron frames, and covers. Newer adjusting ring and frame technologies include high-density polyethylene (HDPE) and rubber products, which are lightweight, impervious to structural erosion and effluent gases.

FIGURE 12.10 Flexible internal seal. (*Source:* Permaform.)

FIGURE 12.11 HDPE frame and risers. (*Source:* NASSCO.)

Individual leaks that have not yet resulted in structural erosion can be sealed with water plug or chemical grouts. Chemical grouts can permanently eliminate inflow and infiltration, which develop in underground structures. Gel type chemical grouts are injected from the interior of the structure to the exterior to form a gelatinous curtain around the structure. The gelatinous curtain prevents groundwater from entering into the structure and stabilizes the soil. In precast structures, urethane foam grouts are used to rebuild the joint gaskets (Fig. 12.12).

Eroded benches and leaking channels can likewise be patched and sealed. Bench and channel repair can be accomplished utilizing cementitious liners and epoxies that are hand sprayed or troweled on (Figs. 12.13 and 12.14). These processes structurally enhance the existing bench extending its life expectancy. Total replacement of the existing bench includes the complete removal of the old bench and channel material. Replace with new concrete while hand forming the new channel. New products on the market today include preformed fiberglass channels utilized when replacing the existing bench. The new preformed fiberglass channel is set in place and the concrete is poured around the new channel (Figs. 12.15 and 12.16).

Damaged steps create a serious hazard and should be removed. The trend within US cities is not to replace them in order to reduce liability for unauthorized entry and to force workers to use safety harnesses and tripods for all entry. These individual Type I repairs generally range from $50 to $500.

Figure 12.12 Rubber frame. (*Source:* Permaform.)

Figure 12.13 Chemical injection grout. (*Source:* Permaform.)

Figure 12.14 Grout injection. (*Source:* Permaform.)

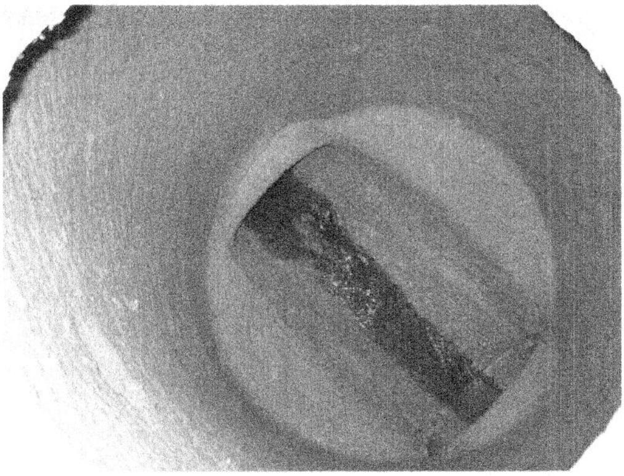

Figure 12.15 Bench repaired with mortar. (*Source:* Permaform.)

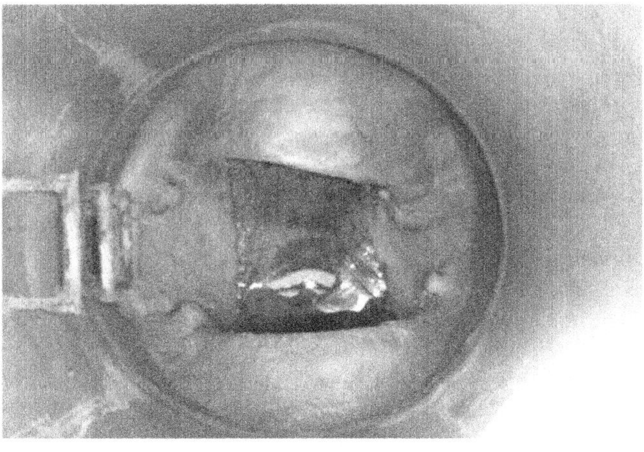

Figure 12.16 Preformed fiberglass channel. (*Source:* NASSCO.)

12.3.2 Type II: Structural Defects

Type II problems are more serious problems where some structural damage has occurred due to unplugged leaks resulting in voids, missing bricks, and mortar joints or when multiple Type I problems are evident. In these cases, a structural liner enforces and seals the damaged wall as warranted. Cementitious liners can be hand sprayed, troweled, or centrifugally cast onto the prepared interior of the old manhole to create a structural shell of sufficient strength to withstand the groundwater pressure and dynamic traffic loads. The thickness of the hardened, cementitious shell is determined by the diameter and condition of the manhole. Similar to DR ratings of pipe, large-diameter manholes require correspondingly greater thickness for the same strength value, that is, manholes that are 72 in. in diameter require the liner to have a thickness of 0.75 in. to attain a value equal to 1 in. thickness in a 48-in. diameter manhole. Likewise, manholes that are deeper and have greater groundwater pressures require a greater thickness than the shallower portion of the same manhole. Structural reinforcement and sealing with a cementitious lining cost about $8–12 per square foot of coverage.

12.3.3 Type III: Corrosion Problems

Type III problems relate to corrosion, the most common cause of deterioration in concrete sanitary sewer structures. Industry studies estimate that corrosion is the primary source of structural damage in three-fourths of all precast concrete manholes, and lift stations installed in the last 30 years. The damage and the corresponding repair and replacement costs are counted in billions of dollars.

The source of this corrosion can be either chemical or bacteriological. When corrosion occurs in the areas contacted by the flow, chemicals are present in sufficient concentrations and temperatures to corrode the lower portions of the manholes. This type of corrosion is common in industrial sewers. The most common corrosion in municipal sewerage is caused by acid generating bacteria, MIC. Thiobacillus bacteria thrive in the slime layer above the flows and secrete H_2SO_4. Wherever H_2S gas is produced, these bacteria will grow rapidly because the gas is their source of nourishment. Their rate of growth and subsequent damage is affected by the amount of nutrients (biochemical oxygen demand levels); long retainage times which allow the sewerage to go septic; very warm temperature; and, turbulence that releases the H_2S gas. This is why it is most severe in coastal plains in warm climates with lots of lift stations.

The corrective action for this type of manhole and lift station renewal is to place a protective barrier between the corrosive bacteria and the structural substrate. The most appropriate solution depends upon the severity of the deterioration (Figs. 12.17 through 12.20).

FIGURE 12.17 Brick manhole after leak plugging. (*Source:* Permaform.)

FIGURE 12.18 Centrifugal casting of cementitious lining. (*Source:* NASSCO.)

12.3.3.1 Type III-A: Light Corrosion

This is when the corrosion has done little damage, or before it has begun on new structures. In this case, a polymer coating may be applied to the properly prepared surface for protection. Thorough preparation and expert application are keys to the success of this method. High-pressure washing or sandblasting followed by acid neutralizing rinses is best. The industry offers a variety of coating materials, such as epoxies, polyurethanes, polyureas, and blended hybrid polymers, which can be brushed, roller applied, spray applied, or centrifugally cast. The thickness (commonly 65 to 100 mils) must

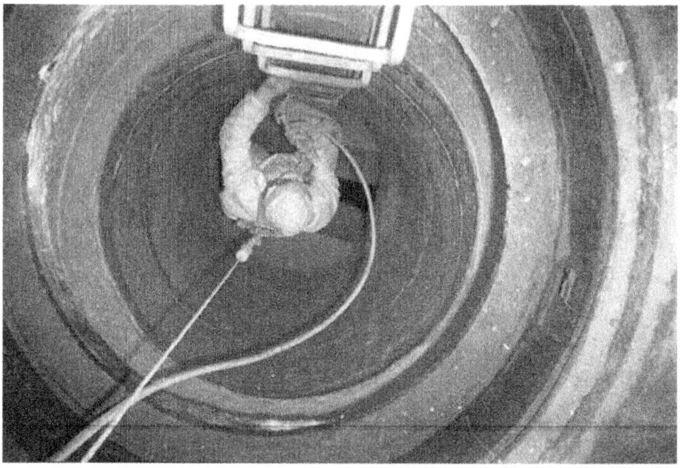

FIGURE 12.19 Spray applied epoxy. (*Source:* NASSCO.)

FIGURE 12.20 Finished spray applied urethane. (*Source:* SprayRoq, Inc.)

be sufficient to prevent vapor penetration and free from any holidays or pinholes that would permit bacteria to grow behind the coating. Prices for coatings used for Type III-A defects are commonly $10–15 per square foot of coverage.

12.3.3.2 Type III-B: Extensive Corrosion

These defects occur when the wall is structurally deficient. The coating in this case must be applied at a thickness that will compensate for the structural value of the lost wall. This can be done by using a polymer coating of sufficient strength and applied at a thickness equal to ¼ to 3/8 in. or by using a composite or laminated liner that combines structural reinforcement and corrosion protection at similar combined thickness and strength values. Common sources for the latter methods are cementitious liner/polymer composites and fiberglass/polymer composites (FGP). The FGP may be hand-laid or it may be a custom fitted bag that is expanded and cured-in-place with steam. Costs for these methods range from $20 to $30 per square foot of coverage (Figs. 12.21 through 12.25 illustrate different manhole renewal methods).

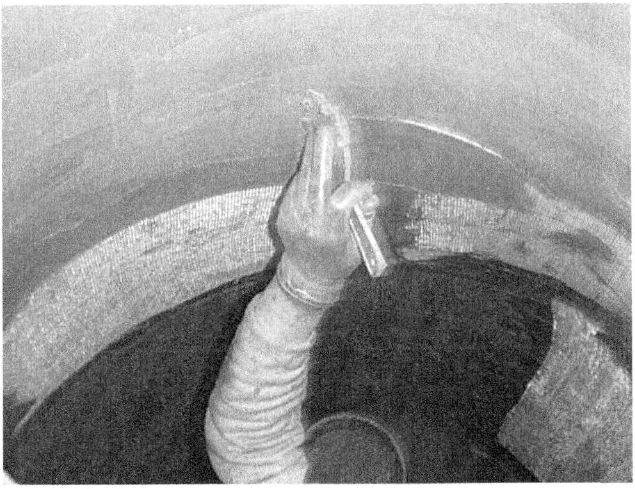

Figure 12.21 Hand applied FGP. (*Source:* Permaform.)

Figure 12.22 Cementitious liner/polymer composites. (*Source:* Permaform.)

A

B

Figure 12.23 Custom fabricated FGP. (*Source:* Permaform.)

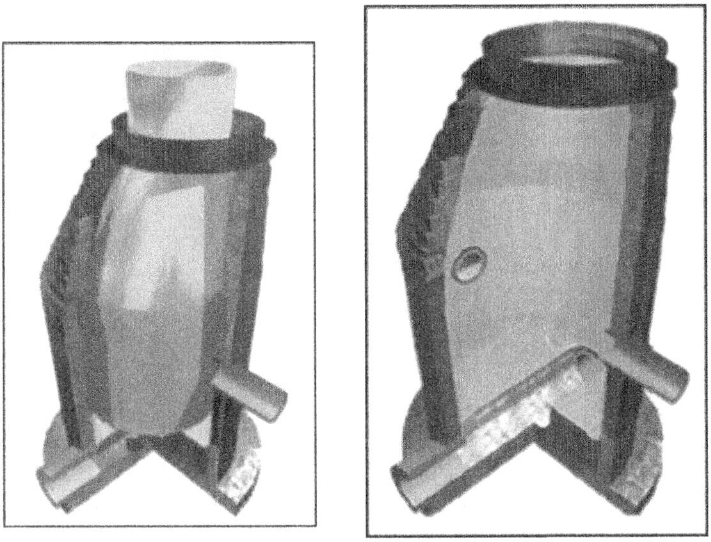

C D

FIGURE 12.23 Custom fabricated FGP. (*Continued*)

1. Liners are prefabricated to structure size. Onsite, structures are pressure-cleaned and grouted where necessary, while steps and obstructions are removed.
2. Epoxy-coated liner inserted into manhole interior.

FIGURE 12.24 Fiberglass sleeve.

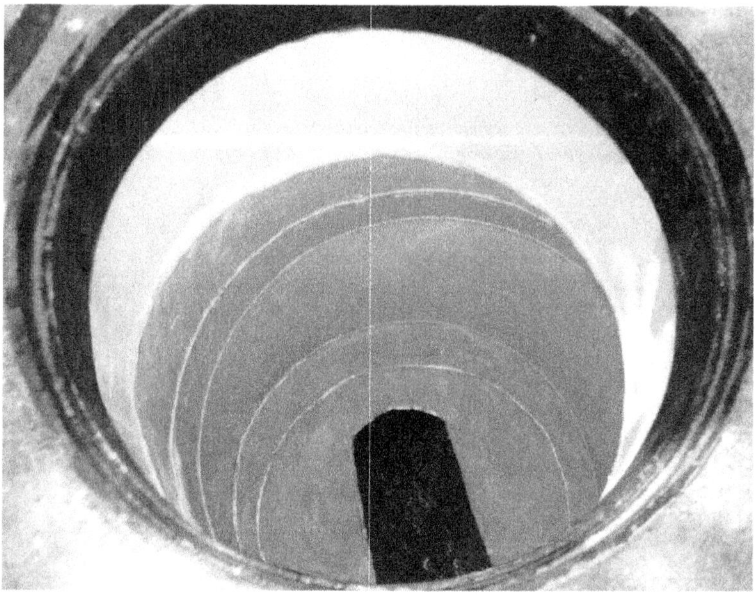

FIGURE 12.25 Poured-in-place manhole with embedded plastic lining. (*Source:* Permaform.)

3. Flexible diaphragm is inflated. Heat introduced within the system cures the epoxy, forming a protective barrier within the structure. The bond is enhanced due to inflation pressure during the cure.
4. Structures are virtually renewed, sealed from further corrosion and infiltration. Manholes and pump stations, for example, can be completely rehabilitated in 4 to 8 hours, with little or no downtime.

Separate and distinct alternatives to Type III-B defects are cementitious liners that retard or prevent production of acid-producing bacteria. Calcium aluminate cements generally retard the growth of bacteria by making surface conditions less favorable to their development than Portland cements. Over the past 6 years, a certain EPA registered antibacterial additive has proven to entirely prevent growth of the bacteria. In either case, the applied thickness is the same as for Type II repairs with correspondingly lower costs.

12.3.4 Type IV Severe Structural Damage

The Type IV category defines severe structural damage that requires replacement. These are easily determined because field staff can recognize structures in this category that are near collapse.

Replacement of some type is the only remaining option for this category. Excavation, removal, and resetting of a new structure is the most costly option, because it requires bypassing of flows, traffic interruption, and allowance for other buried utilities and general social disruption. Depending upon the location, size of the pipeline, and its depth and diameter, costs can range from $1,500 to $3,000/vertical ft.

No dig and partial dig solutions are available for replacement. No dig methods use forms, which can be passed through the existing opening and assembled entirely within the old manhole. An annulus space between the forms and the old wall is filled with standard DOT ready-mix concrete and vibrated into place creating a new, joint-free wall about 3-in. thick. In corrosive environments, a plastic liner with interlocking protrusions is fitted onto the form before placement of the concrete so that when the forms are removed, a fully plastic-lined concrete manhole remains in place. Partial excavation is required to sleeve the old structure. In this option, the entire cone is removed and a smaller-diameter fiberglass or plastic pipe is sleeved inside. The annulus between the sleeve and the old wall is similarly filled with a DOT ready mix concrete and then a new cone is installed and the pavement replaced. These options offer substantial savings with costs in the range of $500 to $1,000/vertical ft.

The most important part of a manhole renewal project is at the end of the project. When plastic linings, polymer coatings, or polymer composites are utilized, the project is not finished until the final repair product has been thoroughly tested to verify installation integrity. A visual inspection will give the owner's inspector the opportunity to verify that the manhole is leak free and coated as specified. Spark testing or the use of a holiday detector verifies that the new polymer coating or plastic lining is free from pinholes. The National Association of Corrosion Engineers' publication, *NACE RP0274-High Voltage Electrical Inspection of Pipeline Coating Prior to Installation,* is an excellent reference for inspectors. Verification of bond is another important test that can be conducted on polymer coatings after the installed products have cured. ASTM D-4541 "Pull-off Strength of Coatings Using a Portable Adhesion Tester" utilizes a portable adhesion testing device to verify the bond of the polymer coating to the prepared substrate. The bond may be tested to failure, a value recorded or predetermined bond strength may be selected, and the polymer coating tested up to that point.

The testing process gives data on two types of failure. A bond value is recorded when the polymer coating pulls cleanly away from the substrate, showing bond as a *psi* value. The other type of failure occurs when the polymer coating pulls free with some of the substrate attached. The value recorded at failure represents the tensile strength of the substrate surface. The inspecting engineer needs to plan how these two types of data will be used in the evaluation of bond. A drawback to this method is that it is destructive and will require that the test locations be repaired. However, it gives a good indication of how well the substrate surface was prepared and the polymer coating has bonded to the substrate. Any deficiencies can be rectified before the installation crews have left the jobsite.

12.4 Manhole Renewal Methods

This section gives an overview of the various principles for manhole renewal.

12.4.1 The Traditional Method

In this method, an ordinary craftsman or contractor is requested to renovate the manhole using traditional methods. Decaying areas are first chipped away and then replastered using quick-drying mortar. If floor channels and berms are to be repaired, the manhole must be drained.

Reasonable results can be achieved with this method if the work is performed carefully, the mortar is used, and the manhole is completely drained. It is difficult to achieve a satisfactory result when groundwater infiltration is the problem. However, in this case, it may help to lower the groundwater around the manhole using one or two suction points or by drilling a suitable number of drain holes to relieve water pressure until the manhole has been repaired. The drain holes are subsequently plugged.

This is seldom the method used when a manhole is to be completely renewed, but is used for partial renewal or repair. The method can produce the required result, but is very dependant on the tradesman who performs the work.

12.4.2 Lining

A fiberglass liner, desired especially for renewing manholes, is now on the market. Because of its special form, this liner adapts to the conical shape of the manhole and to a certain extent the floor channel and berms. The method allows the entire manhole to be renewed in one operation.

12.4.3 Panels

Several systems are available in which panels are fitted to the manhole walls, and in some cases, also to the cones and berms. When the panels have been securely attached, cavities are filled with a special grout, foam, or similar product. The panels are made of polyester, fiberglass-reinforced polyester (FRP), polyethylene (PE), polyvinyl chloride (PVC), and so on.

The manhole floor is similarly renewed using panels or pipe halves laid in a grout. Alternatively, a completely new concrete floor can be cast. The method is usually rather expensive, but for large custom-built manholes, it may be the only solution available apart from conventional repair using concrete or masonry.

12.4.4 Manhole Inserts

In this method, a smaller manhole is constructed within the existing manhole. Usually, the existing manhole floor is removed and a PVC or PE manhole insert is installed and connected to the pipeline using standard pipe joints.

The cavity between the insert and the original manhole is then filled with concrete or grout, and the opening is completed either by grouting beneath the existing manhole cover or by fitting a replacement cover that fits the new manhole size.

The advantage of this method is that it provides a new manhole that is completely watertight. The disadvantage is that the manhole size is reduced. The method can only be used in situations where the sewer traverses the manhole in a straight line, and cannot be recommended for deep manholes.

12.4.5 Grout Injection

This method is used for manholes that leak, and consists of injecting expanding grout at high pressure into joints and the cavities behind the walls of the manhole, through cracks or a number of specially drilled holes.

Following injection, the manhole sides and floor are renewed as required. It has sometimes proved beneficial to first inject polyurethane foam through the cracks in order to prevent water from infiltrating. Subsequently, an epoxy grout is injected through drilled holes. This method economizes on the use of expensive expanding grout, because the relatively cheap foam fills any cavities. The grout, subsequently injected, compresses the foam and permanently seals the leaks.

12.4.6 Spray lining

In these methods, a layer of grout is sprayed at high pressure on the inside surfaces of the manhole, thus providing the manhole with strength and tightness. Both cement grouts and epoxy grouts are used, as well as the more flexible polyurethane products.

The grout usually sets while the surface is being treated and forms a slightly undulating but very smooth surface. These methods make the manhole construction watertight, provide good hydraulic conditions, and allow easy cleaning of manhole surfaces. The methods are easy to use, and are suitable for complete manhole renewal.

12.5 Chapter Summary

This chapter presented the importance of manholes in the pipeline infrastructure system. The common manhole defects were described. While infiltration and inflow is the major problem with existing manholes, many times structural issues also must be addressed. The application, benefits, and limitations of manhole renewal and repair methods were discussed.

12.6 Review Questions

1. "While manholes are easily accessible, they offer a variety of challenging problems that are frequently misunderstood, oversimplified, and overlooked." Discuss.
2. Why is the understanding of chemical attack on manhole important?
3. List the main type of problems that occur in manhole renewal. Explain each of them in detail.
4. Describe the various methods used for manhole renewal.
5. A 24-in. vitrified clay sewer passes through a 4-ft diameter, 16-ft deep precast concrete manhole. An 8-in. PVC sewage force main discharges into the manhole 6 ft above the manhole flowline. Due to H_2S corrosion, several inches of the concrete manhole wall thickness have been lost and groundwater is infiltrating into the manhole. In fact, in two locations, the entire manhole wall has corroded away. The city engineer has received a recommendation to rehabilitate this manhole using a spray on polymer with a thickness of 100 mi (0.100 in.). Is this a suitable rehabilitation method for this manhole?

CHAPTER 13

Sanitary Sewer Lateral Renewal

13.1 Introduction

Service laterals are pipes that branch off from a sewer main and connects building sewers to the public sewer main lines (Figs. 13.1 and 13.2). Laterals may be as small as 4 in. in diameter, normally ranging from 15 to 100 ft in length. Laterals are built with any of the pipe materials listed in Chap. 5, but most of the lateral pipes being renewed today were usually built with concrete, clay, and orange burg (bituminous fiber) pipes. Today, most laterals and main lines are built using plastic (PVC) pipes. They are normally constructed at a minimum self-cleansing grade from the building to the immediate vicinity of the main sewer. At the main sewer, the grade may change abruptly for the line to descend to the main sewer[1]. Laterals normally enter sewer mains at angles ranging from 30 to 90 degrees from the axial flow direction and at vertical angles ranging from 0 to 90 degrees. In older housing developments (prior to potable water and sewer line separation laws), the same route was used to install potable water service lines and the lateral sewer. In this case, any leaks in the potable water line can enter the lateral sewer line if it is not watertight. It is common for older lateral pipes to have multiple bends, diameter changes, shifted joints, cracks, deposits, and roots which create considerable challenges to repair or renewal. The lateral pipe connection to the main sewer is also a concern for water tightness.

To ensure a water tight collection system, renewing the lateral pipe and reconstructing the lateral connection is the next step after renewing the main sewer lines. Infiltration is frequently found at the juncture where the service lateral pipe connects to the main sewer pipeline. Another significant source of infiltration is groundwater that continues to penetrate the original main pipe post CIPP renewal.

---

[1]See ASTM F3097-15, Standard Practice for Installation of an Outside Sewer Service Cleanout through a Minimally Invasive Small Bore Vacuum Excavation.

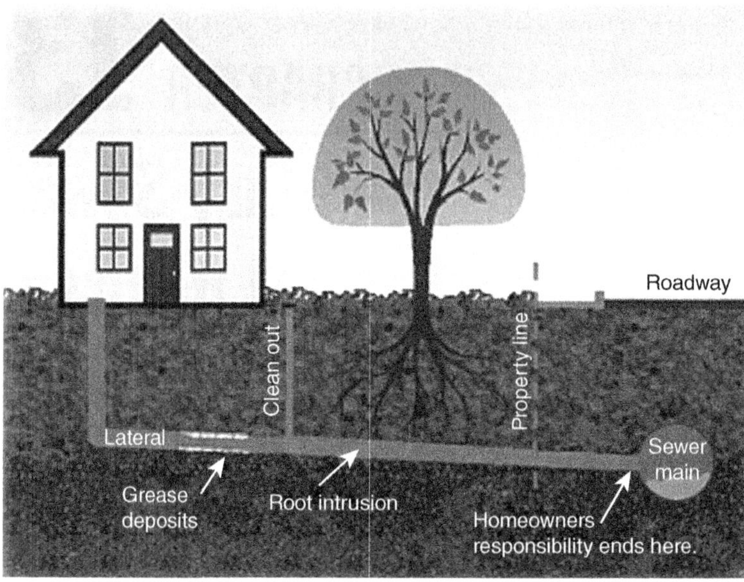

FIGURE 13.1 Service lateral pipe. (*Source:* LMK Technologies.)

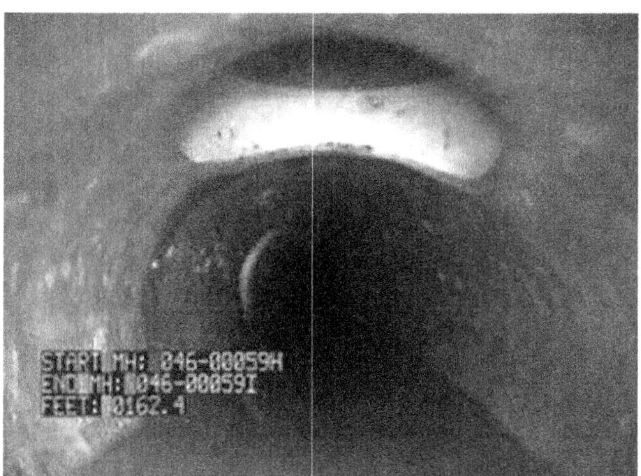

FIGURE 13.2 Service lateral connecting to main sewer line. (*Source:* NASSCO.)

Due to shrinkage, earth movements, the inability to provide long term chemical bonding for CIPP, an annulus is often the result. The infiltrating water migrats through the annulus, simply to reenter the collection system where a hole has been cut in the new CIPP mainline for the purpose of reestablishing service to the customer's service line (see Fig. 13.3).

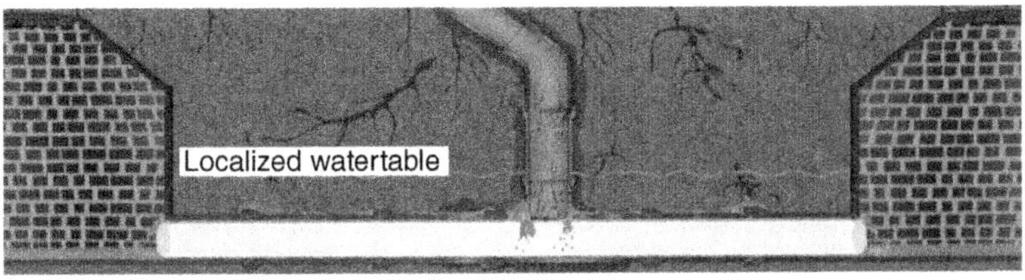

FIGURE 13.3 Groundwater tracking in annulus. (*Source:* LMK Technologies.)

For many years the impact of leaking lateral sewers on the collection system and the treatment facilities was considered insignificant. Research studies sponsored by the U.S. Environmental Protection Agency (USEPA) have indicated that in many cases a significant percentage of the extraneous flows is the result of the defects in lateral pipes and leaking lateral connections.

These defects include cracked, broken, or open-jointed pipes. Laterals historically have transported water from inflow sources such as roof drains, cellar and foundation drains, basement sump pumps, and clean water from commercial and industrial effluent lines. With new EPA regulations, such as CMOM (see Chap. 3) there will be tighter controls of sanitary sewer overflows (SSOs), which means an even greater emphasis on stopping I/I.

The construction and maintenance of service laterals is complicated because separate entities may have jurisdiction over different portions of the laterals. Usually the portion between the building and the property line is the building owner's responsibility and considered an extension of the municipal in-structure facilities; therefore, it is ordinarily installed under plumbing or building codes and tested and approved by the building inspectors. The section of the lateral between the property line and the main sewer line, including the sewer main connection, usually is installed under sewer use rules and is a property of the city or municipality. Typically, the inspection and maintenance of this section of laterals is the responsibility of the public works departments and engineering division of the city, township, or county. Figure 13.4 illustrates defective (a) and renewed (b) lateral connection with the main sewer.

13.2 Service Lateral Renewal Methods

Some of the renewal methods used for larger main sewers may be suitable to small-diameter laterals because of bends and diameter changes in the lateral, possible loss in hydraulic capacity, and the inability to prepare pipe surface for long-term bonding. These methods include chemical grouting, cured-in-place pipe (CIPP), close-fit pipe (CFP), pipe bursting (PB), and spray-on lining. Such trenchless renewal methods such as sliplining may not be applicable to small-diameter laterals

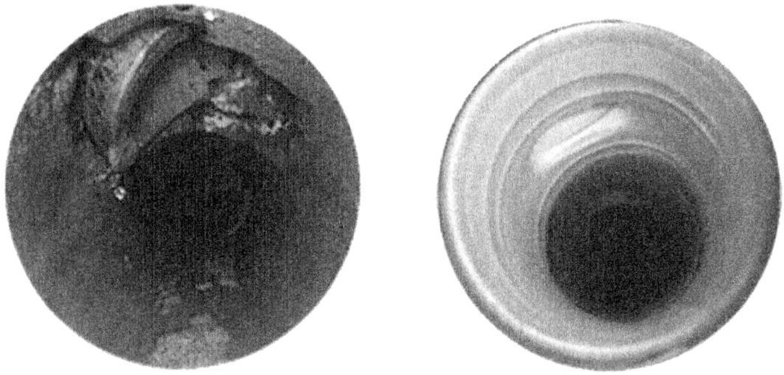

A B

FIGURE 13.4 Defective (left) and renewed (right) lateral connection with the main sewer. (*Source: LMK Technologies.*)

because of possible loss in hydraulic capacity. Unlike renewal of main sewers, access to laterals can often be limited, making renewal more difficult. This is combined with the additional problems of tree roots, existing surface and subsurface structures, and landscaping over laterals that also makes open-cut and digging pits difficult.

13.2.1 Preparatory Work

Before commencements of any lateral renewal project, the following preparatory work is usually conducted:

- Preliminary inspection of all laterals (usually a few months prior to renewal project).
- Decide on renewal method. Note that lateral conditions can vary widely, even in a local area.
- Mobilization.
- If necessary, pit excavation at the property line.
- If necessary, cleanout installation.
- Lateral cleaning and reinspection (preinstallation inspection).
- Renewal installation (see the following sections for specific method).
- Reconnection of lateral and reestablishment of sewer service to the building.
- Post inspection.
- Surface restoration and demobilization.

To measure effectiveness of I/I reduction, a continuous flow monitoring program before and after lateral renewal can be conducted, such as air testing, and ongoing visual inspection during wet weather flow.

13.2.2 Cured-in-Place Pipe

The technique of renewing laterals with cured-in-place pipe (CIPP), is similar to the methods used to renew main pipelines (see Chap. 6). Using the same materials (tube and resin) and some modifications to the original CIPP process, lateral CIPP provides a structural, jointless, new pipe (see Fig. 13.5). The new liner pipe can be manufactured in many ways, but in principle it comprises a resin impregnable tube typically constructed with a needle punched felt or a knitted tube that is saturated with a thermoset resin. A variety of resin systems are applicable for lateral renewal including polyester, vinyl-ester, epoxy, urethane, and silicate. These resin systems can be thermal-cured, light cured, or cured at ambient temperatures. The new lateral pipe can be installed from cleanout or the main sewer. To consider applicability of CIPP for lateral renewal, such factors as number of bends (usually more than three bends makes CIPP method impractical) and offset joints, collapsed pipe, and other obstructions should be evaluated.

13.2.2.1 Primary Characteristics

An important difference between installation from a cleanout and installation from a main sewer is that it is only possible to form a watertight seal between the new lateral CIPP and the main CIPP in one operation if the new lateral pipe is installed from the main sewer. This is achieved by fitting the lateral CIPP with a connection liner including a connection gasket that reconstructs the entire main/lateral fitting as described in ASTM F2561-11.

Dependent on the condition and the design of the lateral renewal, a new lateral liner can increase the hydraulic capacity of the lateral as this is determined by the cross-sectional area of the lateral pipe and the roughness of the renewed lateral. Reducing the diameter of laterals usually reduces their capacity, but this is counterbalanced by the fact that new liner pipes are considerably smoother than the existing laterals and a renewal method may improve the hydraulic capacity in the displaced joints, fractures, and the like. Cured-in-place pipe can usually be used for renewing most laterals where there have been no problems relating to self cleaning ability or insufficient capacity. Typically, the wall thickness of lateral CIPP is 4.5 mm.

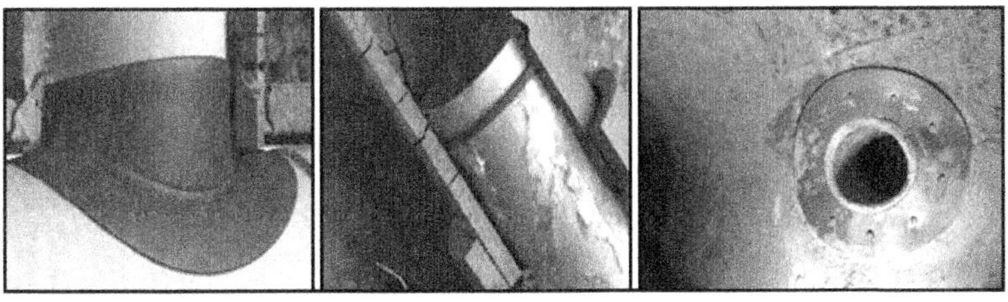

FIGURE 13.5 Lateral pipe renewed with CIPP. (*Source:* NASSCO.)

A lateral pipe can be renewed from a single access point, such as a cleanout located at the property line or in a basement. This process does not include any connection liner, the liner does not extend into the main, and the lateral CIPP is not connected or sealed to the main CIPP. However, gasket sealed connection liners can be added as a second step to renew and seal the main/lateral connection when a lateral pipe is renewed by inserting a liner from a cleanout.

There are predominantly two lateral CIPP installation processes: inversion and pull in-place. The inversion process for inserting CIPP into a lateral is similar to the technique used for renewing mainline sewers. Although, for lateral renewal it is necessary to invert two tubes: the resin saturated liner tube and a closed end inflatable bladder. The two tubes are combined by inserting the liner into the bladder to form a liner/bladder assembly. The liner is open ended so no cutting is required, while the bladder's end is closed so it can be inflated causing both the liner and the bladder to invert as an assembly. This liner/bladder assembly is used in the process described above to renew laterals from the main, and the liner bladder/assembly process is also used for repairing isolated defects found in a lateral pipe. Any section of a lateral pipe can be repaired by inverting a short liner from a cleanout (see Figs. 13.6 through 13.8).

A lateral pipe can also be renewed using CIPP pull-in-place (PIP) process. The PIP technique requires two access points such as a cleanout where the resin saturated liner is inserted, and a second point of access such as a manhole where a cable is pulled drawing the liner into the pipe. Installation of an outside cleanout can be performed using vacuum excavation for forming a small diameter borehole, a minimally invasive method for installing an outside cleanout as described in ASTM F3097-15. Traditional excavation is always an option for installing an outside cleanout or making an access pit at the property line, but most all lateral CIPP technologies today do not require an excavation.

13.2.2.2 Cleaning and Inspection

Before starting any renewal process, it is necessary to thoroughly clean the existing pipe prior to inspection. Typically a high-pressure water jet is used to free the service lateral of roots, mineral deposits, grease, sand, and sludge. The inspection process involves inserting a camera into the pipe through a cleanout or remotely propelling a small-diameter camera into the service lateral from a robot positioned in the main line. This inspection provides valuable information regarding the condition of the existing lateral pipe. The inspection is usually conducted by working through a cleanout or remotely from the main line, thereby avoiding any digging or disruption.

13.2.2.3 Installation Process

For small diameter mainline pipes, that is, 8-24 in., an installation assembly is winched through the main line pipe to the lateral service

FIGURE 13.6 Inversion lateral CIPP. (*Source:* LMK Technologies.)

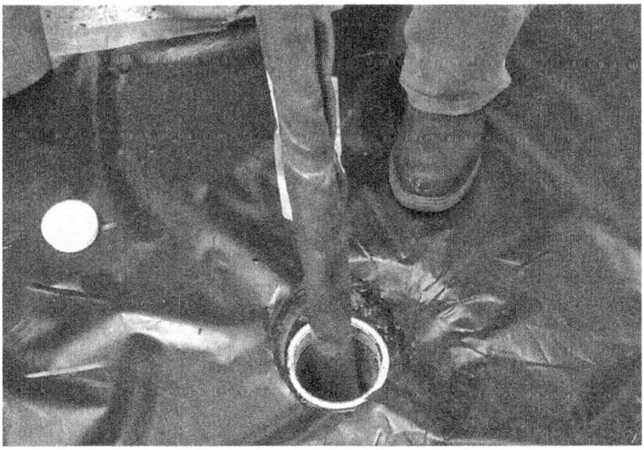

FIGURE 13.7 Pull-in-place lateral CIPP. (*Source:* Perma-Liner.)

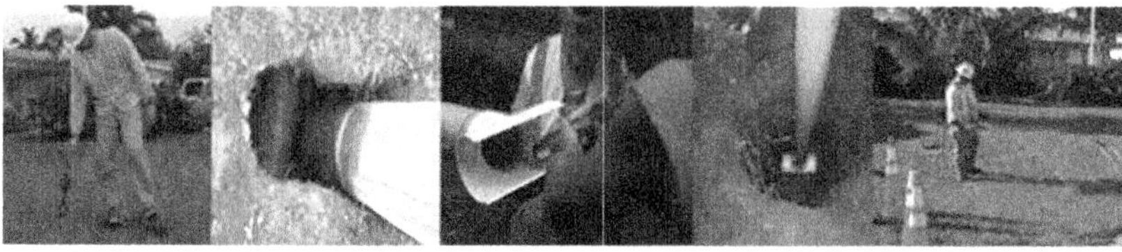

FIGURE 13.8 Process for a trenchless cleanout installation. (*Source:* LMK Technologies.)

connection. For large diameter mainline pipe applications, the lateral CIPP liner is manually positioned by worker entry and anchored to the mainline pipe by use of mechanical fasteners and a gasket seal as shown in Fig. 13.5. The assembly unit is then used to invert the liner pipe into the service lateral. Then, a curing process is initiated, during which the liner is held in place by the assembly unit's inversion hose. Once the new lateral pipe has cured, the assembly unit is removed, leaving the service lateral renewed and ready for use. Figure 13.9 shows the renewed lateral from the main sewer line.

13.2.3 Pipe Bursting

Pipe bursting (PB) method for renewing service laterals is same as PB for renewing main pipelines (see Chap. 8). PB is mainly used for laterals that are out of capacity or have structural problems. Lateral PB is the only trenchless renewal method that can significantly increase the hydraulic capacity of laterals by installing a new pipe. Lateral PB usually requires excavation at two points, one at a private property (next to the building) and another at the main line. Thus, this method requires more preparation and renewal time.

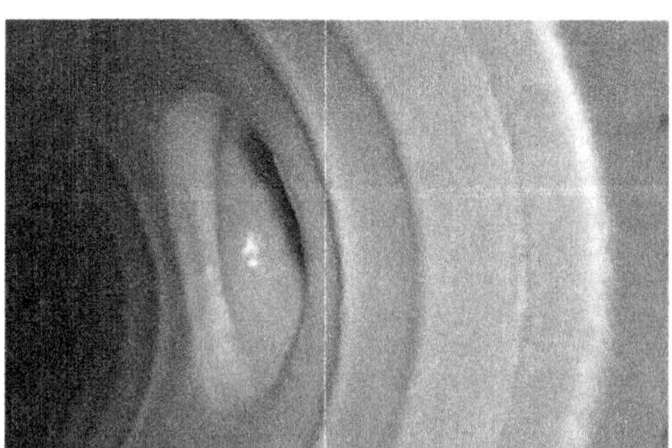

FIGURE 13.9 View of the renewed lateral from the main sewer line. (*Source:* LMK Technologies.)

13.2.3.1 Primary Characteristics

The pneumatic (air powered) PB process accounts for majority of lateral PB performed in the United States and worldwide. As stated in Chap. 8, common PB applications include water, sewer, and gas lines, as well as single and multiple duct telephone and power conduit. PB lateral lines has increased dramatically over the past several years with advances in guide head technology and the development of smaller, space efficient tools.

Pneumatic tools can be used to burst fracturable pipes (cast iron, clay, nonreinforced concrete, Acrylonitrile-Butadiene-Styrene (ABS), and other plastics) with diameters between 4 and 54 in. In the bursting procedure, a pneumatic tool is placed into the existing pipe at an entry pit or manhole. A winch line, based at the exit area pit and attached to the head of the tool, guides the head through the existing lateral pipe. Continuous percussion from the pneumatic tool fractures the existing pipe. The tool effectively hammers its way through, forcing the fragments into the surrounding soil, while pulling the new pipe into place behind it. Same size or larger pipe can be installed with the pneumatic PB system.

Several factors dictate whether PB is appropriate for service lateral replacement. These considerations include: existing lateral pipe material, diameter, condition, depth, and length; new lateral pipe diameter requirements; soil conditions; and location of nearby utilities. By using lateral PB process, trenching green areas, driveways and sidewalks can be avoided. As for main line pipe bursting, small entry and exit pits are much less disruptive and costly than long deep trenches. Additional PB considerations and risks include previous point repairs, pipe sags, turns and bends, collapsed lines, joint settlements and misalignments, pipes with various diameters, underground water, and difficult access.

13.2.4 Close-Fit Pipe (CFP)

Close-fit pipe can be used for lateral renewal. This method slightly reduces the cross-sectional area of the lateral in the same way as CIPP. However, slight reduction in the cross sectional area does not result in reduction in hydraulic capacity as the new pipe will improve hydraulics capacity by providing a smooth surface. The CFP method is often easier to install than CIPPs and are mainly used in laterals where project conditions, such as access points, bends, deformations, and so on, allow. For more information on CFP, refer to Chap. 11.

13.2.5 Chemical Grouting

Three chemical grouting methods (see also Chap. 14) are currently available for sealing lateral sewers for nonstructural purposes: pump full method, sewer sausage method (patented process), and camera-packer method (patented process).

Pump full method: This method involves injecting a chemical grout through a conventional sealing packer from the sewer main up the lateral sewer to an installed plug. As the grout is pumped under pressure, it is forced through the pipe faults into the surrounding soil where a seal is formed after the gel has set. After the sealing has been accomplished, excess grout is augured from the lateral sewer and it is returned to service.

Sewer sausage method: This method uses a camera-packer unit in the sewer main with the injection of grout from the sewer main up the lateral sewer to seal the lateral pipe. A tube is inserted into the lateral sewer before sealing to reduce the quantity of grout used and to minimize the amount of cleaning required after the sealing has been completed. The grout is pumped under pressure around the tube, up the lateral sewer, and through any pipe faults into the surrounding soil where the seals are formed after the gel sets.

Camera-packer method: Unlike the other methods described, this method only repairs faults seen through a television camera. First, a CCTV camera and a specialized sealing packer are inserted into the lateral sewer. Using a towline previously floated from the lateral sewer access to the downstream manhole of the sewer main, the camera-packer unit is pulled into the lateral sewer. The camera-packer is then slowly pulled back out, repairing faults that are seen through the television camera. First, the deepest leaking joints are sealed. Joints and cracks are sealed in a manner similar to the conventional methods used for sealing joints in sewer mains. Once the repairs have been completed, the equipment is removed and the lateral sewer is returned to service. The costs for this method vary, depending on the difficulties encountered while repairing lateral sewers. While estimating costs, allowances should be made for difficult site access and excavation dewatering.

13.2.6 Spray-in-Place Pipe

Recent developments in spray-in-place pipe (spray-on lining, see Chap. 10) technology show potential application for renewing small-diameter sewers such as service laterals for nonstructural purposes. The basic technology is the same as for larger sewers, although with smaller equipment. Difficulty may arise with bonding in the areas of high infiltration.

13.2.7 Other Renewal Considerations

The above service lateral renewal methods will eliminate or reduce infiltration/inflows (I/I) resulting both from high groundwater levels and from rainstorms. In addition to these renewal methods, efforts need to be made to remove sources of inflows that may be connected to the lateral sewer. Inflow sources connected are usually located on private property. In these situations, a public awareness or public relations program is often needed. Such programs are intended to persuade property owners (without threat of legal consequences) to make the needed changes to help correct a community problem.

13.3 Chapter Summary

Service laterals connections branch off the sewer mains and connect building sewers to the public sewer main lines. Laterals may be as small as 4-in. diameter, normally ranging from 15 to 100 ft in length. Existing laterals are built with any of the pipe materials listed in Chap. 5, but usually they are built with plastic (PVC or PE), concrete, clay, and ductile iron pipes. They are normally constructed at a minimum self-cleansing grade from the building to the immediate vicinity of the main sewer. At the main sewer, the grade may change abruptly for the line to descend to the main sewer. This chapter presented the renewal methods available for service laterals, which include CIPP, CFP, grouting (point repair), and possible SIPP. These methods usually can be used through a clean-out without digging a pit. For lateral replacement, PB method can be used which may require excavation of small pits.

13.4 Review Questions

1. Why is lateral renewal important? Explain the main objectives of lateral renewal projects.
2. Describe the characteristics of service laterals (common lengths, diameters, slope, existing pipe types, and so on).
3. Describe lateral renewal methods.
4. In what conditions would you recommend lateral CIPP?
5. In what conditions would you recommend PB?
6. In what conditions would you recommend chemical grouting?
7. Explain why even after lateral renewal there might be other sources of inflow to the sanitary sewer system and how can it be resolved?

CHAPTER 14
Localized Repairs

14.1 Introduction

The term *repair* is used when a pipe defect is fixed without necessarily extending the design life of the existing pipe. Local repair falls into this category under which a defect may be temporarily and/or locally fixed without renewing the whole section of the pipeline.

Local defects may be found in sound pipelines, as results of cracking or joint settlements or other types of failures. Localized repair (LOR) (also called point source repair methods) are used to address a number of *isolated* problems, such as cracks, broken pipe, hammer taps, root intrusion, infiltration, debris, soil erosion, exfiltration, and misaligned pipe sections. LOR methods cover a broad range of techniques, such as robotic repair, grouting, cured-in-place pipe (CIPP), internal seal, shotcrete, and so on. Several LOR systems have been developed that are aimed at sewer pipes, but some methods are also designed to seal joints in pressure pipelines. Many of these techniques are adaptations of full-length pipe lining systems presented in previous chapters; especially those based on cured-in-place resin or felt tubes.

As for any construction project, the economics of LOR techniques should be considered case by case only, but clearly in many situations repair of individual defects may be more cost-effective than renewing the entire section of the pipeline such as from one manhole or access point to another. This is especially true where damage is localized. As a general rule, it has been suggested that LOR can be economically viable if less than 25 percent of the pipe length contains defects, although this rule will vary according to the project and site conditions. Just as for any renewal method, preinspection and cleaning are essential before commencing LOR systems. For those LOR methods, which require a bond with the existing pipe, all traces of deposits, corrosion, grease and debris must be removed.

There are four broad categories of LOR system (Fig. 14.1): robotic repair, grouting, internal seal, and point CIPP.

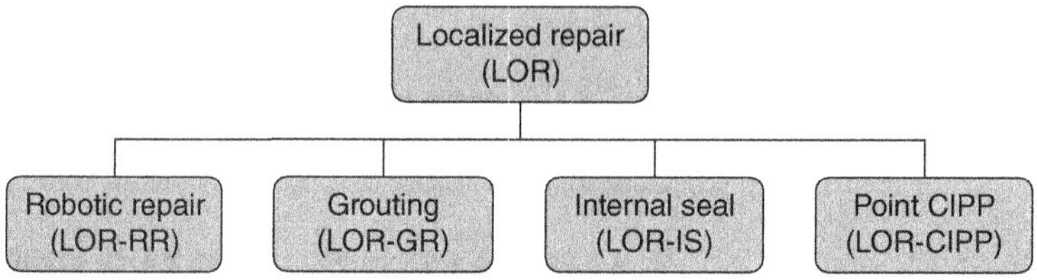

FIGURE 14.1 Categories of localized repair methods.

14.2 Primary Characteristics

As stated previously, when local defects are found in otherwise sound pipeline, localized repairs are considered. Linings are applied to the interior wall of an existing pipe by adhesion, or in the case of robotic repairs, the defect is filled with epoxy. Systems are available for remote-controlled resin injection to seal localized defects in the range of 4 to 30-in. diameter.

Localized repair methods may be used to provide an existing unreinforced pipe with additional load-bearing capacity. Another use is to provide a seal against infiltration and exfiltration and in the cases of missing pipe sections, internal seal methods can be used to fill in these gaps.

All forms of localized repairs provide some structural enhancement. Point CIPP systems, as well as internal-seal methods, are basically a new pipe inserted or cured within the existing one. Although in practice the restraint offered by the existing pipe will enhance the *stand-alone* performance of the new pipe section significantly, for a conservative design, the structural properties of new pipe can be designed in isolation from the existing pipe. For example, the ring stiffness of a close-fit pipe or CIPP may be increased by a factor of up to 7 by restraint from the existing pipe considering a good support, although for design purposes a much lower factor is used.

Both point CIPP and internal-seal methods are regarded as *structural* repairs. However, other methods such as resin or grout injection behave in a composite manner with the existing pipe and the surrounding ground support. In this case, because the strength of the repair is more complex and difficult to estimate, these systems are often termed *nonstructural*. These methods can potentially increase structural performance of the existing pipes by stabilizing and sealing the surrounding ground. The elimination of erosion resulting from the infiltration or exfiltration cycle is also important in extending the useful life of existing pipelines.

14.3 Robotic Repairs

14.3.1 Primary Characteristics

Robotic repair is relatively a new technology. Robotic repair systems, largely developed in Switzerland, are used mainly in the gravity pipelines and comprise a grinding robot and a filler robot. The former removes encrustation and intrusions, and also mills out cracks to provide a good surface key for the repair materials. The filler robot applies an epoxy mortar into the slot formed by the grinder, and trowels off the material to a smooth finish. Robots are available for use in pipes from 8 to 32-in. diameter. Typically, the smaller versions will operate in diameters up to 8 to 16 in., and the larger ones from 12 in. upward. Various axle and wheel sets are used to position the robot centrally within the pipe.

Robotic point repair is used either as stand-alone or as a precursor to renewal methods. As a stand-alone, robotics point repair is used to repair radial, longitudinal, and spider cracks. This method also can be applied to repair broken joints, slip joints, open joints, protruding service connection (Fig. 14.2), recessed service connections, roots, and other foreign objects that are usually found in collection pipeline systems and sewer interceptors. If chemical sealant was

FIGURE 14.2 Cutting or grinding robot removing an intruding lateral connection. (*Source:* KA-TE System AG.)

applied and all grease removed prior to the application of the applied epoxy, it will bond to the existing pipe substrate and permanently seal the wall from further infiltration of outside matter.

The grinding head is hydraulically driven, giving high torque at low speeds, and can be fitted with various shapes of diamond or carbide cutters suitable for clay, concrete, polymeric materials, and even steel pipes. Some of the more powerful grinding robots can cut through steel reinforcement as well. It is common for the cutters to be cooled by a water spray carried through a hose.

The tractor wheels are usually driven by electric motors, as are the head rotation and extension functions. The operator through a closed-circuit television (CCTV) camera, attached to the cutterhead, monitors the operation of the robot, and a second camera can be added for forward view. Some grinding robots have the capability to inject a sealing compound through a hollow drill, which prevent infiltration from affecting the grout applied by the filler robot. The equipment may be fitted with a high-pressure water jet to remove the debris created by the grinding operation.

Typically, cracks are milled out to a width and depth of between 1 and 1.5 in. It is essential to clean the area of the repair thoroughly after grinding, because any dust, mud, or debris will prevent the grout from adhering to substrate. Intruding laterals, grout deposits, and hard encrustation must also be removed.

The properties of the epoxy grout are important, because the material is normally applied to wet surfaces. The two-part mortar may either be mixed prior to filling the canister on board the filler robot, or, in some designs, the components are loaded into the robot separately and mixed at the outlet as they are used. As with the grinding robots, the filler robots are self-propelled with on-board cameras. The epoxy is applied by a system of remotely-controlled nozzles and spatulas, the material being forced out of the canister by a piston driven by compressed air. Alternatively, the grout can be injected through a flexible plate pressed against the pipe wall (Fig. 14.3).

In addition to the filling slots milled by the grinding robot, the filler robot can apply epoxy around poorly made service lateral connections, to seal these connections to the main pipe. Some systems allow the use of special shields, which act as a temporary shutter and allow defective lateral connections to be remolded in epoxy grout. They can also insert an inflatable stopper over a branch, to assist in reforming a connection and to temporarily stop the flow.

All robot functions are controlled from a central console housed in a truck, which also contains hose reels, a compressor, a hydraulic power pack, and other ancillary equipment. There is a hoist for lifting the robots into and out of access points or manholes. The main source of power is a generator, which is usually trailer-mounted. Robotic repair systems are versatile, but they need a steady program of work

FIGURE 14.3 Filler robot injecting epoxy resin through a flexible plate pressed against the pipe wall. (*Source:* KA-TE System AG.)

to be economically viable. They have found favor in regions, which offer a predictable workload for localized repairs, but have been less successful in commercial terms where the demand is only periodic.

14.3.2 Method Description

Robotic point repairs are carried out by an operator manipulating the robotic functions by a remote-controlled CCTV. As a first step, the robot is positioned at the defect area and is surveyed for the best starting position. Chemical grout is implemented if any infiltration of water is present. The operator then begins to grind out the crack(s). This accomplishes two goals: (1) the crack is cleared off all foreign material stopping further cracking because of the groove cut; and (2) the groove created, gives a larger surface area to inject the epoxy resin.

The second step is to fill the void area with epoxy. This is accomplished by making sure that the groove is fully filled and is flush with the pipe wall. Once the epoxy cures, (1 to 2 hours, complete 8 days) the pipeline is restored and returned to service with minimum reduction in flow.

14.3.3 Advantages

- The point repair is carried out at the defect location using a remote-controlled camera, with precision.
- Pipe structural integrity is restored and an infiltration barrier is created.

14.3.4 Limitations

- Because of robotic design, elliptical pipelines cannot be navigated.
- Pipelines that have extensive defects, that is, crushed or caved-in walls, or slip joints that have fallen greater than 2 in., usually are not possible to repair.

14.4 Grouting

14.4.1 Primary Characteristics

Grouting is one of the oldest methods of pipeline repair. In recent years, there have been new advances in products and equipment for a wide variety of grouting techniques.

14.4.2 Chemical Grouting

Chemical grouting is normally used for sewer lines and is applied to leaking joints and circumferential cracks. Chemical grouting may also seal small holes and radial cracks. Chemical grouting can be applied to pipeline joints and manhole walls using special tools and techniques. Some chemical grouting is used in concrete, brick, clay sewers, and other pipe materials to fill pipe embedment voids. Such embedment voids, if not corrected, can reduce pipe's lateral support and allow outward movement resulting in the rapid deterioration of the integrity of the pipe. Chemical grouting does not provide an effective seal where joints or circumferential cracking problems are present, because of the ongoing settlement or shifting of pipe sections. Therefore, it is not effective to use chemical grouting to seal longitudinal cracks or to seal joints where the pipe near the joint is longitudinally cracked. Chemical grouting is normally undertaken to control groundwater infiltration in nonpressure pipeline caused by leaking pipe joints or circumferential cracking of pipe walls. Grouting mainly should be considered for pipelines in good structural conditions. Table 14.1 presents a sample grouting payment schedule.

14.4.2.1 Chemical Grouting for External Repairs

External repair is performed by excavating from outside the pipe or applied from inside the pipe depending upon the pipe diameter and worker access. The common methods are variations of chemical and cement grouting. These methods are appropriate for solving problems of significant groundwater movement, washouts, soil settlement, and soil voids.

Chemical grouts for external repairs comprise a mixture of three or more water-soluble chemicals, which produce stiff gels from properly catalyzed solutions. The grouts produce a rapid repair as opposed

Item	Description	Quantity	Unit	Unit Price	Total Price
1	Pre CCTV and cleaning of main lines including root removal		Linear Foot		
2	Reaming of mineral deposits in mainlines		Linear Foot		
3	Cutting of protruding taps		Each		
4a	Testing of 8 in. pipe joints		Joint		
4b	Testing of 10 in. pipe joints		Joint		
4c	Testing of 12 in. pipe joints		Joint		
5a	Grouting 8 in. pipe joints		Joint		
5b	Grouting 10 in. pipe joints		Joint		
5c	Grouting 12 in. pipe joints		Joint		
6a	Testing lateral connections 8 in. pipe		Each		
6b	Testing lateral connections 10 in. pipe		Each		
6c	Testing lateral connections 12 in. pipe		Each		
7a	Grouting of lateral connections 8 in. pipe		Each		
7b	Grouting of lateral connections 10 in. pipe		Each		
7c	Grouting of lateral connections 12 in. pipe		Each		
8	Pre CCTV and cleaning of laterals connected to manholes		Each		
9	Testing and grouting of laterals connected to manholes		Each		
10	Post construction CCTV inspection		Linear Foot		
11	Post lateral connection residual grout cleaning		Each		
12a	Option A: Warranty testing of pipe joints		Joint		
12b	Option A: Warranty testing of lateral connections		Each		
12c	Option B: Warranty CCTV Inspection of mainlines grouted		Linear Foot		
13	CCTV of laterals connected to mainline		Each		
14	Cleaning of laterals connected to mainline		Each		
Supplementary Unit Prices					
15	Chemical grout (Price to be determined by Engineer)		Gallon		

(*Source:* NASSCO)

TABLE 14.1 Sample Grouting Payment Schedule

to cement; or cement or clay grouts that consist of suspensions of solid particles in a fluid. The reaction in the solution may be either chemical or physiochemical and may involve only the constituents of the solution or these constituents reacting with other substances encountered. In the latter type, the reaction causes a decrease in fluidity and to solidify and form a seal to fill voids in the material into which the grout has been injected.

14.4.2.2 Chemical Grouting for Internal Repairs

Internal repair methods are performed from within the pipeline, either remotely or through worker-entry. The various installation methods are discussed in this section.

Chemical grouting for internal repairs is most commonly used to reduce infiltration. It can be used to seal deteriorated and leaking pipe joints, service connections, open joints, maintenance holes, and so on. Grouting will not provide a structural repair, and is inappropriate for longitudinal cracks and broken or crushed pipes. Although chemical grouting is normally used in small-diameter pipes, intermediate and large-diameter pipes can also be successfully grouted at the joints with special equipment. Grouting of pipelines is generally accomplished using a sealing packer and CCTV camera. The packer is usually made of a hollow metal cylinder having an inflatable rubber sleeve on each end of a center band. The grout is pumped into the space created between the two inflated sleeves straddling the pipe joint. Depending upon the type of packer, the grout and the initiator solutions are mixed together, applied into this space and forced through the joint leak into the surrounding soil. The grout displaces the groundwater and fills the voids between the soil particles. Figure 14.4 presents a grouting process.

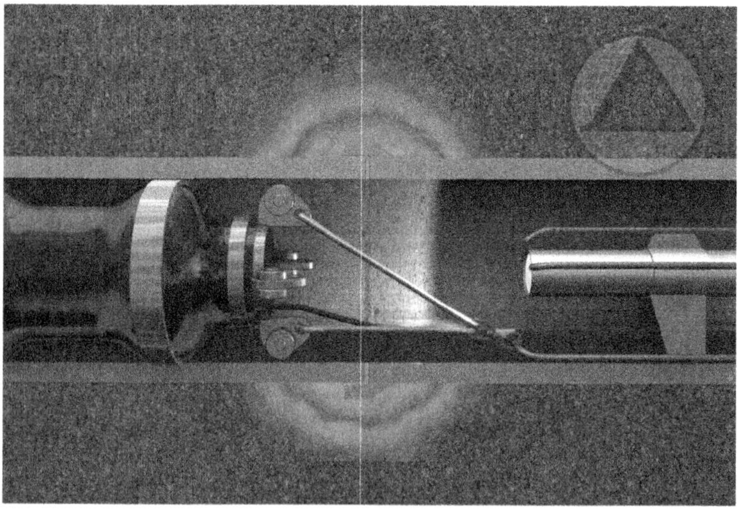

FIGURE 14.4 A grouting process. (*Source:* NASSCO.)

A remote CCTV is used to position the packer at the pipe joint and to inspect the joints before and after the sealing operation. A winch cable through a pipe section from an access point or manhole to another pulls the sealing packer and the CCTV. In addition, air or water pressure testing equipment is used to determine the effectiveness of the sealing. For worker-entry sewers, maintenance holes and structures, leaking joints and/or walls can be injected with chemical grouts using a nozzle-type applicator.

There are several different types of chemical grouts, which are categorized as either gel or foam. With each of the grouts there are a multitude of different types of additives, for example, initiators, activators, inhibitors, and various fillers. The general grout formulations are chemicals and water. When there is groundwater present, a higher concentration of chemicals are used because of the dilution potential. Because of soil and moisture variability, formulating the correct mixture is largely dependent on trial and error, rather than engineering and material principles.

The various parameters that affect the performance are

a. Viscosity control
b. Gel time variables
c. Influence of temperature
d. Influence of pH
e. Presence of entrained oxygen in the solution
f. Contact with particular metals
g. Influence of ultraviolet rays
h. Presence of mineral salts
i. Velocity of groundwater flows
j. Capabilities of placement equipment
k. Other soils and water conditions

The properties of the grout vary in

a. Appearance
b. Solubility
c. Swelling and shrinkage
d. Corrosiveness
e. Stability
f. Strength

The various additives for grouts affect

a. Viscosity
b. Density

c. Color
d. Strength
e. Shrinkage

Therefore, the environmental conditions must be taken into consideration for a proper formulation and must be done on a case-by-case basis. Another critical aspect of effective grouting is the proper operation of the equipment, that is, the grout packer, pumps, tanks, formulation, mixing, and application. The premixing of the final grout mixture is conducted using two separate tanks. One is the grout tank and the other is the catalyst tank. The grout tank contains water, grout, and buffer, whereas the catalyst tank contains water, oxidizer catalyst, and sometimes fillers.

14.4.2.3 Chemical Gel Grouts

The most commonly used gel grouts are of the acrylamide, acrylic, acrylate, and urethane-base types. All of these gels are resistant to most chemicals found in sewer pipelines but produce a gel-soil mixture that is susceptible to shrinkage cracking. Using chemical additives can reduce shrinkage cracks due to dehydration.

There are differences in composition of gel grouts. Acrylamide-base gel is more toxic than the others (grout toxicities are of concern only during handling and placement or installation). The nontoxic urethane-base gels are approved by U.S. Environmental Protection Agency (USEPA) for potable water pipelines. Urethane gels use water as the catalyst; whereas the other gels use chemicals. Therefore, the urethane gels require avoiding additional water contamination for proper curing.

14.4.2.4 Acrylamide-Base Gel

Acrylamide grout is mixed in proportions that produce stiff gels from dilute water solutions when properly reacted. Before a successful application can be made with acrylamide grout, the following criteria shall be evaluated:

a. The desired result
b. The nature of the grout zone
c. The application equipment
d. Alternative procedures
e. The plan of injection

The grout injection plan shall include gel times to be used, quantities of acrylamide, and the probable layout for grout injection points. After work starts, the work plan may be modified based on the additional information that is developed during application. No natural soil or rock formations has yet been found in which a gel will not form. However, the injected solution must remain in the grout zone until gelation occurs. In dry soils, there is often a tendency for the grout to disperse. Gravity and capillary forces in dry soils disperse the grout solution and may cause the gel to be ineffective. As in pipe joint grouting, dispersion can be avoided by saturating the soil prior to grouting and by the use of very short gel times. In soil void work, however, a dry soil mass cannot be as efficiently stabilized as a soil mass below the groundwater.

Dilution around the outer edges of the grout bulb may occur in wet soils when long gel times are used. The flowing groundwater distorts the normal shape of the stabilized mass and can displace it in the direction of flow. In turbulent flow conditions, dilution can be minimized by short gel times. In open formations or in fissures, solids such as bentonite or cement may be added to the grout solution to help produce a more complete block to flowing water. Acrylamide application is most successful in saturated or partially saturated soils.

Acrylamide is used primarily to reduce leakage rather than increase strength. It does, however, improve the structural integrity indirectly by stabilizing the surrounding soil mass. Acrylamide is a toxic chemical that can be absorbed into the body through broken skin, inhalation, and swallowing. Because of this toxicity, potential hazards in handling and usage can occur if not supervised by technically qualified personnel.

14.4.2.5 Acrylic- and Acrylate-Base Gel

The acrylic grouts are water solutions of acrylic resins with several types for different applications. After being mixed with catalysts a cohesive gel is formed. The set time can be closely controlled from a few seconds for flowing water condition to several hours for normal conditions.

These grouts are good for use in sewer pipe joints, maintenance holes, and structures. The acrylic grouts have a viscosity similar to water, prior to gelation. These grouts have a tendency to swell in water allowing a watertight sealing effect.

14.4.2.6 Urethane-Base Gel

Urethane grout is a solution of a prepolymer, which cures upon reaction with water. During the reaction the gel remains hydrophilic, that is, it absorbs water plus holds it within a cured gel mass. Upon being cured, the resultant gel is resistant to the passage of water. As the prepolymer is cured by water, premature contamination by other water must be avoided during application. The formulation obviously is water sensitive and various water-to-grout ratios provide various strengths. The ratio needed to provide a strong gel ranges from 5:1 to 15:1, by volume. Ratio less than this will produce a foam reaction, where greater ratios will produce a weak gel. An example formulation would involve an 8:1 ratio of compounded material for pipe joint grouting.

14.4.2.7 Urethane-Base Foam

Urethane foam is used primarily to stop infiltration into the pipe. These leaks occur through cracks in the foundation or base (or wall), through joints formed by the base, corbel or upper frame interfaces, or through pipe penetrations into the maintenance hole wall. Grout injections are placed into predrilled holes and the grout is injected under pressure. When cured it forms a flexible gasket or plug in the leakage path. When mixed with an equal amount of water, the grout expands and quickly cures to a tough, flexible closed-cell rubber. In some applications the material is used without premixing with water, however, it will eventually require an equal amount of water for total cure.

14.4.3 Resin Injection System

Another method of grouting is resin injection using epoxy resin. A tube is lowered into the pipe and moved by either pulling or pushing it to the location of damaged area. The tube is then inflated to pipe diameter to fit tightly against the walls of the pipe. The resin is released around the tube as well as around the damaged area. Depending on the size of the damaged area, the excess resin is forced out through the pipe into the soil, which creates a seal on the outside of the pipe. The repair head remains in place for approximately 90 to 120 minutes to allow the resin to harden enough to remove the head cleanly. The entire process is performed by remote control with the aid of a CCTV. The curing process takes about 24 to 36 hours and is not affected by lack of applied pressure or by the presence of water or air.

Resin injection systems fall into two categories: those whose principal function is to seal the pipeline against infiltration and exfiltration, and those, which aim to restore the structure of a damaged pipe. A common method of sealing leaking joints in gravity pipelines is to use a special packer, which combines the functions of leakage testing and grout injection. Joint testing and sealing may or may not be *localized*, depending on how many joints fail, but can be included under this heading as it aims to identify and cure a specific defect. A packer with inflatable end elements is positioned across a pipe joint and pressurized to isolate the joint. Air or water pressure is then applied to the center section of the packer, and the rate of pressure loss through the joint is measured. If the loss exceeds a specified limit, a sealing gel is injected into the joint through the packer, and the joint is retested.

There are numerous variations in packer design and sophistication, and most systems use either a two-part acrylic grout or a water-reactive polyurethane (PU) resin. In either case the grout has little intrinsic strength, but turns the ground around the leaking joint into an impermeable mass, which prevents leakage and also enhances structural stability.

It should be noted that acrylamide sealing grouts must be handled with care, because of the toxicity of the unreacted components. However, acrylate grouts, despite their similar name, have different chemical properties and they are safer to handle.

Polyurethane (PU) grouts are hydrophilic and react either with free water in the soil or with water injected through the packer at the same time as the grout. Although plain water can be used, a styrene-butadiene rubber (SBR) solution is often mixed with the water in the ratio 1:4 to add resilience to the cured grout and to reduce shrinkage. The proportion of grout to water affects the properties of the material—ratios stronger than 1:5 tend to produce foam—whereas slightly weaker mixes create a gel. A ratio of 1 part PU grout to 8 parts water (or water-SBR) is generally recommended for pipe sealing.

Many PU grouts contain acetone to reduce viscosity and increase miscibility. The flammability of acetone should be considered when storing the material. Certain manufacturers offer PU grouts, which are acetone free. A different form of resin injection system, normally using an epoxy resin, may be used to stabilize and rebond the existing pipe structure, in addition to sealing against exfiltration and infiltration. Although aimed originally at pipes where damage was not too severe or extensive, the technique has evolved to be suitable for more serious defects. An inflatable packer isolates the defect and injects a rapid-setting epoxy resin into the crack, fracture, or hole in the pipe wall. The packer is left in the position until the resin has cured, and a thin internal collar of resin usually remains after packer withdrawal.

14.4.4 Fill-and-Drain Systems

A different approach to leak sealing is taken by *fill and drain* technique, which repairs the main sewer, branches, and manholes in one operation. The section to be sealed is first isolated and then filled from a manhole with an environmentally safe chemical solution (usually sodium silicate). After a predetermined interval to allow the chemical to permeate through leaking joints and cracks, the solution is pumped out. The section is then filled with a second proprietary chemical solution, and this reacts with the residue of the first chemical to form a waterproof gel. The second chemical is then pumped out and the pipe is cleaned to remove any residue. As with packer-injected sealants, the effect of the system is to turn the manhole and pipe surround into an impermeable mass around the points of leakage. Because of plant requirements and the volumes of materials, fill-and-drain systems are aimed more at large-scale leakage control projects than at the treatment of isolated lengths. They have the advantage of treating leaks throughout the whole system in a single operation.

14.4.5 Cement Grouting

Cement grout consists of slurry (particulate suspension) of cement and water. Materials such as sand, bentonite, fly ash or accelerating admixtures may be added. Because of its particulate nature the use of cement grout is normally restricted to fractured rock and large-grained soils, where the voids are large enough to facilitate penetration and permeation.

14.4.5.1 Cement

Portland cement grout can be used to form impermeable subsurface barriers but is restricted in application to medium sands or coarser materials because of the larger size of cement particles (see microfine cement). For purposes of filling voids and washouts adjacent to sewer pipelines, various Portland cements have been used successfully. Type III cement is often selected because of its high and early strength properties. It usually produces design strengths at 7 days or less.

A range of water cement ratios can be used depending on subsurface conditions. Strength characteristics are generally not important where grouting is primarily intended to fill voids surrounding a buried sewer pipeline. Proper design of the water to cement ratio results in grout mixtures, which are easy to mix and inject, and have strengths of 500 to 1,000 psi. Clays can be added to cement for grouting, to form gels, and to prevent settlement of the cement from suspension. They do, however, have the problem of no well-defined setting time and they have a slow strength development. Because of this, they are normally not used in sewer renewal void grouting when groundwater is present.

Portland cement grouts may also be used as a filler and accelerator in silicate grouts and may be compatibly mixed with acrylamide to improve water shutoff capabilities and injection characteristics. For extremely large void filling applications, other cements such as pozzolan and fly ash mixes can be used and are more economical than straight Portland cement and soil-cement mixtures.

14.4.5.2 Micro-Fine Cement

Microfine cement consists of finely ground cement that allows it to penetrate fine sands that ordinary Portland cement cannot penetrate. Beside excellent permeability, it provides the needed strength and durability with a set time of 4 to 5 hours. When combined with sodium silicate, with no organics, a 1 to 3 minutes gel time can be attained for underground water control.

14.4.5.3 Compaction Grouting

Compaction grouting is the injection of very stiff, low-slump, mortar-type grout under relatively high pressure to displace and compact soils in place. Compaction grouting acts as a radial hydraulic jack, physically displacing the soil particles, and moving them closer together. The technique is used to strengthen loose, disturbed, or soft soils or for the control of settlement. Compaction grouting is used primarily on large pipelines applied through the pipe wall into the surrounding soil.

14.4.5.4 Reinforced Shotcrete Placement

Shotcrete is the application of concrete or mortar conveyed through a hose and pneumatically projected at high velocity onto a surface. Shotcrete includes both the wet and dry mix processes, but the term *shotcrete* usually refers to the wet process, whereas the dry mix process commonly is referred to as *gunite*. Shotcrete and gunite linings can provide structural strength and improved hydraulic performance, with some improvement in corrosion resistance.

Shotcrete and gunite linings are used in large-diameter (usually 42 in. and larger) sewers and pipelines. These linings can also be used in maintenance holes and other structures. Various segmental liners can be incorporated onto the finished cement for total corrosion protection. The shotcrete processes, are similar to reinforced concrete and generally use steel or mesh for reinforcing purposes. Also, various latex polymers can be added for improved bond strength, reduced absorption and permeability, and increased chemical resistance. See Chap. 10 for more information on SIPP including shotcrete and gunite.

14.5 Internal Seal

Internal seals can be used for structural repair of pipe joints and missing pipe sections. It can be used in both worker-entry and non-worker-entry pipes.

14.5.1 Primary Characteristics

Internal seals can be conducted on pipe diameters ranging from 6 to 110 in. For diameters between 6 and 24 in., a stainless steel sleeve, which is wrapped in polyethylene foam, is used. This sleeve and an inflatable sewer plug lock are placed in the defective pipe section. The plug is then deflated and a visual inspection takes place.

For pipe diameters between 24 and 110 in., a polyvinyl chloride (PVC) sleeve is used. It consists of six longitudinal segments. The edges are grooved to lock each segment together. The sleeve is wrapped in a polyethylene foam gasket. This gasket provides the radial force to secure the sleeve in place. If grouting is required, small portholes are drilled in the sleeve and filled with grout upon completion. This system can be used to repair any length. The sleeves come in lengths of 12, 24, and 35 in., and any number of them can be used to repair a pipe.

14.5.2 Mechanical Joint Sealing

Another method of joint sealing involves installing, across the joint, a metal band or clip faced with an elastomeric material, which forms a seal with the inner surface of the pipe.

Systems of this type are available for both gravity and pressure pipes. Mechanical sealing systems have the advantage of not relying on in-situ chemical reactions, and can also be installed quickly. The cost of the materials is, however, higher than for cured-in-place methods. Mechanical systems are available for sealing joints in man-entry pressure pipes rated at up to 300 psi. Only 30 to 45 psi pressure is needed to tighten the seal, so low-modulus pipe materials such as PE and PVC are not overstressed. Repair modules are available for pipes from 25 to 120 in. diameter, and the seal can be made of nitrile butadiene rubber (NBR) for gas applications or ethylene propylene diene monomer (EDPM) rubber for potable water. Tapered versions are manufactured for use between pipes of varying diameter, and to seal the annulus at the ends of sliplined pipes (Fig. 14.5).

Mechanical repair modules are also available for non-worker-entry gravity pipes from 8 to 25 in. diameter. The stainless steel inner sleeve is in the form of a scrolled clip, which can expand in diameter, and an integral ratchet mechanism prevents the unit from contracting again. The outer rubber sleeve also has bands of hydrophilic rubber to give a watertight seal with the existing pipe. The repair modules are installed by means of an inflatable packer, which expands the clip

Figure 14.5 Stainless steel and rubber internal repair sleeve for gravity pipes. (*Source:* LMK Technologies.)

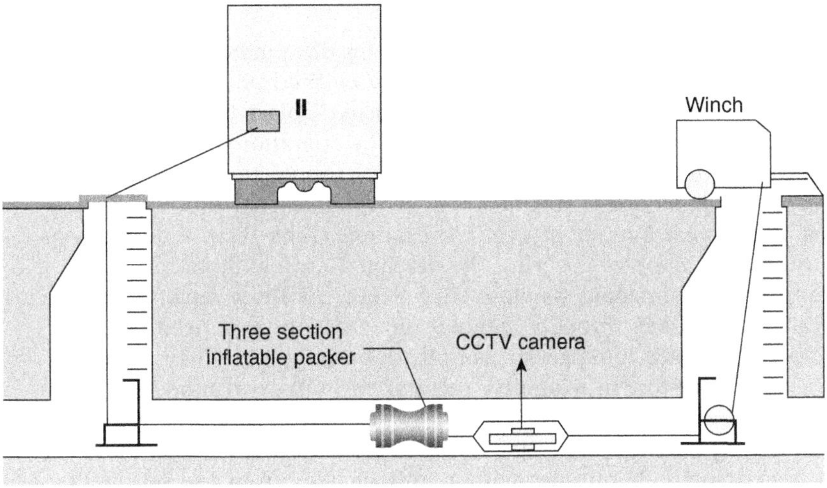

FIGURE 14.6 Joint sealing process using an inflatable packer. (*Source:* NASSCO.)

and presses the rubber against the pipe wall. The packer can then be deflated and withdrawn. Variations are available for sealing around connections, and for structural renovation in conjunction with a cured-in-place sleeve. Figure 14.6 illustrates a joint seal process using an inflatable packer.

14.5.2.1 Pipe Rerounding

Rerounding is not a standalone technique, but is intended to reshape a deformed pipe prior to patch repair or relining. An expander unit is used to reround the pipe and install a metal or plastic clip, which holds the pipe fragments in position until a patch or liner is installed. The expander can be made from an elastomeric material inflated with hydraulic pressure, or it may be a variant of a hydraulic mole with steel *petals* that are forced outward by hydraulic rams. The plastic or metal clip is scrolled around the expander prior to insertion, and held in place with bands or tape. The clip usually has some form of ratchet or locking arrangement to keep it at its expanded size. After positioning under CCTV control, the clip is expanded with sufficient pressure to reround the pipe.

Although a useful trenchless technology technique, pipe rerounding is not always successful. During expansion, the unit will take the path of least resistance. For example, if there is a void beneath the pipe invert, the clip may be forced down when expanded, rather than reshaping the soffit of the pipe. It is also important to follow with relining or patch repair as soon as possible after the rerounding operation, because the clip has limited structural strength and may soon become deformed if left without support.

14.5.3 Method Description

There are different propriety techniques for this method. Prior to the operation, the pipe must be thoroughly cleaned and inspected by CCTV to identify all the obstructions such as displaced joints, crushed pipes, and protruding service laterals. The operation then continues according to instruction from the system manufacturer and type of application. Figure 18.5 illustrates an internal seal process using stainless steel internal sleeve. For example, Link-Pipe, a propriety technique, involves covering the damaged area with an internally applied prefabricated stainless steel sleeve. As these repair sleeves can be installed through manhole no excavation is needed. The sleeves are used to repair structurally damaged pipes. They can also be used to restore missing pipe sections without excavation. Infiltration and exfiltration sealing is made recreating a sound and long-lasting pipe. Not counting the set-up time, an individual repair takes approximately 20 minutes with a crew of two. From one set-up 25 repairs have been made in a 5-hours work period. Link-Pipe process offers different products for sewers, water mains, culverts, gas mains, and industrial pipes. Figure 14.7 presents an application of link seal process where it is installed under water.

FIGURE 14.7 Under water culvert repair using link seal. (*Source:* Link-Pipe.)

14.5.4 Advantages

- This method can be used for structural localized repairs.
- Such problems as pipe breakage; joint settlement, and cracks can be fixed with this method.
- Link seal can be used for variety of applications, such as gravity sewers, pressure pipes, culverts, and gas mains.
- Usually no excavation is required.

14.5.5 Limitations

- An internal pipe repair will leave a ridge on the inside circumference of the pipe, decreasing the overall diameter as much as 1 in. at point of repair, which may decrease hydraulic capacity of the existing pipe.
- The existing pipe must be circular.

14.6 Point CIPP

The applications of CIPP are for pipelines that are sound but may contain isolated pipe lengths that have failed. The materials used in point CIPP repair are the same as the regular CIPP methods, which have more than four decades of proven performance.

14.6.1 Primary Characteristics

The point CIPP includes a liner reinforcement, which is made up of two layers of fiberglass woven roving with a layer of polyester sandwiched between them. The three layers are sewn together with polyester thread at 1 in. (25 mm) spacing using a zigzag pattern. The liner is cut and impregnated on site with epoxy resin to suit the particular requirements of the host pipe. The liner material holds extra resin, and when compressed gives up it to fill the cracks and surface roughness, thereby providing a better seal between the liner and the host pipe. The usual diameter range is 2 to 24 in. and the maximum length is up to 50 ft long with most liners being between the range of 3 and 15 ft in length. The minimum thickness is usually 1/8 in. and the maximum thickness is normally 3/8 in. with thicker liners available for specific conditions. The curing time is normally 1 to 4 hours depending upon the type of epoxy resin used and pipe diameter. Most CIPP contractors provide LOR systems using standard CIPP tube and resin materials to form short length of CIPP. After curing, the liner withstands external hydrostatic pressure of 30 psi or more according to design considerations. Figure 14.8 illustrates preparation and installation of a point CIPP.

A B

FIGURE 14.8 Preparation and installation of a point CIPP. (*Source:* LMK Technologies.)

14.6.2 Method Description

Most point CIPP techniques include impregnating a fabric with a suitable resin, pulling this into place within the sewer around an inflatable packer or mandrel, and then filling the packer with water, steam, or air under pressure to press the patch against the existing sewer wall while the resin cures. Both thermal-cure and ambient-cure systems are available. Although point CIPP methods are short versions of cured-in-place liners, often the fabrics and resins are stronger because material economy is less significant in the overall installation cost. The fabric is commonly polyester needle-felt (nonwoven), either on its own or in combination with glass-fiber. Some systems use a multilayer sandwich, the glass fiber providing strength while the felt acts as a resin carrier.

Although polyester resin may be used as in the full-length liners, epoxy resin is a common alternative for localized repairs. Epoxies can be formulated to be water-immiscible, whereas polyester resins are adversely affected by water prior to cure. This may be particularly relevant in techniques, which are designed for installation without diverting the flows in the pipeline. The drawback of epoxy resins, apart from their higher cost, is that the cure regime is more critical. Most epoxy-based systems require thermal cure, and ambient cure is generally restricted to polyester resins.

The patch may be preformed into a tube of the correct circumference, or it may simply be a rectangle of material, which is wrapped around the packer and scrolls out to the wall of the pipe when the packer is inflated. In the latter case an overlap will be visible in the finished tube, but this is of little consequence to the performance of the repair (Fig. 14.9).

Impregnation of the fabric is usually, but not always, carried out on site. With some systems the patch is preimpregnated and refrigerated for delivery to site. While on-site impregnation is generally acceptable, care is needed to avoid health risks and the spillage of

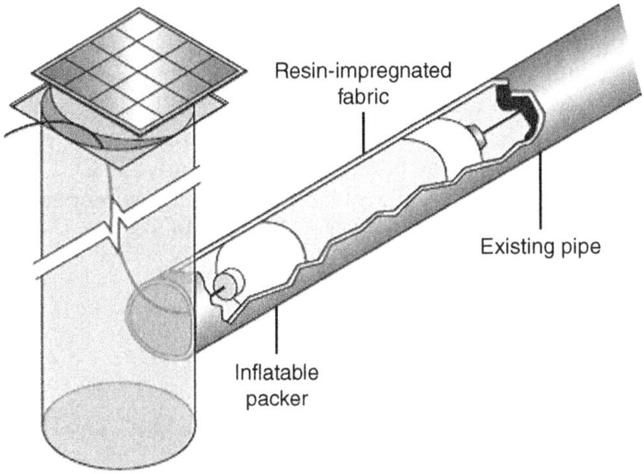

Figure 14.9 Typical arrangement for installing a point CIPP inflatable packer. (*Source:* Subterra.)

chemicals, some of which are toxic in their unreacted state. The styrene in most polyester resins dictates the need for adequate ventilation; the matter is discussed more fully in Chap. 6, "Cured-in-Place Pipe." During mixing and impregnation, it is important to exclude air from resin as far as possible. Entrained air weakens the material and, in severe cases, can result in porosity. It may be impossible to avoid air entrainment, especially with viscous resins, and some systems overcome this with vacuum impregnation.

With both ambient and thermal cured systems, it is essential to limit the rise in temperature of the materials until the patch is inflated within the pipe. One of the most common causes of failure is premature cure, where the patch has started to harden before it is in position. Exothermic cure starts as soon as the components of the resin are mixed, and the rate of temperature rise increases with the volume of resin. Mixed resin should therefore be applied to the fabric immediately and not left in the mixing vessel. Due regard should also be paid to the temperature of the surface on which the fabric is placed during impregnation and, once impregnated; the patch should be positioned and inflated without delay.

Packers are generally made from an elastomer such as rubber, and internal pressure first expands the packer and then presses the patch against the pipe wall. Compressed air is used for inflation with most ambient-cure systems. Thermal cure methods may use a mixture of compressed air and steam, or hot water, which is recirculated between the packer and a boiler unit on the surface. Care is needed to avoid overpressurization, especially with water-inflated systems where the packer is subjected to both a static head and an additional pumping head, which is related to fluid friction in the return hose.

The curing time depends on the resin formulation, the thickness of the patch, the temperature within the packer (in thermal-cure systems), and the temperature of the existing pipe wall. A high groundwater table will tend to act as a heat sink, cooling the outer surface of the patch, and additional curing time should be allowed. After curing, the packer is deflated and removed. CCTV should then inspect the patch, and any lateral connections reopened using the same techniques available for full-length liners.

14.6.3 Advantages

The point liner is installed according to ASTM 1216.

- Diameters up to 60 in. possible.
- The point CIPP repair is installed according to ASTM 1216.
- Lengths from 3 to 50 ft possible.
- Point CIPP bonds to the existing pipe structure providing a tight fit.
- Point CIPP eliminates infiltration and prevents root growth at the location of application.
- Usually bypass pumping is not required.
- Point CIPP provides a structural repair at the location of application.
- The point CIPP provides tapered smooth ends.

14.6.4 Limitations

- While point CIPP provides a structural solution at the location of application, it may not enhance the overall structural capability of a full section of the pipeline from one access point or manhole to another.
- The price of point CIPP against other localized repairs needs to be evaluated.
- Any possible reduction in hydraulics capacity of the existing pipe must be evaluated.

14.7 Chapter Summary

When a defect may be temporarily and/or locally be fixed without renewing the whole section of the pipeline, localized repairs methods can be considered. Local defects may be found in an otherwise sound pipeline, as a result of cracking or joint settlements or failure. As for any construction project, the economics of LOR techniques should be considered case by case only, but clearly in many situations repair of individual defects may be more cost-effective than renewing an entire section of a pipeline. As described in this chapter, there are four broad categories of localized repair system: (1) robotic repair, (2) grouting, (3) internal seal, and (4) point CIPP. The application for each of these methods depends on project and site conditions, but generally, localized repairs can stabilize the defects and stop infiltration/exfiltration, but will not extend the design life of the pipeline. Local repairs sometimes are required before other methods of pipeline renewals are used. In other applications, local repairs, such as may be considered as a temporary fix compared with a permanent solution.

14.8 Review Questions

1. In what project conditions localized repairs are applicable? Explain.
2. In what project conditions would you recommend grouting?
3. In what project conditions would you recommend internal seals?
4. In what project conditions would you recommend point CIPP?
5. In what project conditions would you recommend robotic repairs?
6. What are the advantages and limitations of localized repairs?
7. What is meant by *renewal* versus *repair*?

APPENDIX A
Acronyms

Abbreviation	Description
AASHTO	American Association of State Highway Transportation Officials
ABS	Acrylonitrile-butadiene-styrene
AC	Asbestos cement
ACI	American Concrete Institute
ACM	Asbestos-containing material
AM	Asset management
AMP	Asset management plan
ANSI	American National Standards Institute
API	The American Petroleum Industry
AREA	American Railway Engineering Association
ASCE	American Society of Civil Engineers
ASCII	American Standard Code for Information Interchange
ASTM	American Society of Testing and Materials
AWWA	American Water Works Association
BEM	Broadband electromagnetic
BPD	Barrels per day
BTU	British thermal unit
CAD	Computer-aided design
CAPEX	Capital expenditure
CCFRPM	Centrifugally cast fiberglass reinforced polymer mortar
CCTV	Close-circuittelevision
CFP	Close-fit pipe
CI	Cast iron
CIPP	Cured-in-place pipe
CML	Cement mortar lining
CMOM	Capacity, management, operation, and maintenance

Abbreviation	Description
CMP	Corrugated metal pipes
CoF	Cost and consequence of failure
CP	Closed profile
CSA	Canadian Standards Association
CUIRE	Center for Underground Infrastructure Research and Education
D&R	Deformed and reformed pipe
DD	Drawdown
DDM	Design Decision Model
DI	Ductile iron
DIA	Diameter
DIPRA	Ductile Iron Pipe Research Association
DOT	Department of Transportation
DR	Dimension ratio
DWCP	Dual-wall corrugated profile
EBITDA	Earnings before interest, taxes, depreciation, and amortization
ECAT	External condition assessment tool
ELL	Electrical leak location
EPA	Environmental Protection Agency
EPDM	Ethylene propylene diene monomer
ERW	Electric resistance welded
ESCR	Environmental Stress Crack Resistance
FAA	Federal Aviation Agency
F&F	Fold and formed pipe
FEC	Field eddy current
FGP	Fiberglass/polymer
FHWA	Federal Highway Administration
FPVCP	Fusible PVC pipe
FRP	Fiberglass reinforced polyester
FWC	Fiber wound collar
GASB	Government Accounting Standards Board Statement
GAAP	Generally accepted accounting principles
GIS	Geographical Information System
GPR	Ground-penetrating radar
GPS	Global positioning system

Abbreviation	Description
GRC	Glass reinforced cement
GRP	Glass reinforced pipe
GWT	Groundwater table
GWUT	Guided wave ultrasonic testing
HAB	Horizontal auger boring
HDB	Hydrostatic design basis
HDD	Horizontal directional drilling
HDPE	High-density polyethylene
HPPE	High-pressure polyethylene
I/I	Infiltration and inflow
ICRI	International Concrete Repair Institute
ID	Inside diameter
IEISS	Interdependent Energy Infrastructure Simulation System
IIMM	International Infrastructure Management Manual
ILI	In-line inspection
ILR	In-line replacement
ILR-PB	In-line replacement—pipe bursting
ILR-PR	In-line replacement—pipe reaming
iPVC	Improved polyvinyl chloride
IRR	Internal rate of return
LCCA	Life cycle cost analysis
LDPE	Low-density polyethylene
LOF	Likelihood of failure
LOR	Localized repair
LOR-CIPP	Localized repair—Cured-in-place pipe
LOR-GR	Localized repair—grouting
LOR-IS	Localized repair—internal seal
LOR-RR	Localized repair—robotic repair
LOS	Level of service
MCF	Million cubic feet
MDPE	Medium-density polyethylene
MFL	Magnetic flux leakage
MIC	Microbiologically induced corrosion
MIMS	Multimedia integrated modeling system
MJ	Mechanical joint
MRS	Minimum required strength

Abbreviation	Description
MSL	Modified sliplining
MT	Microtunneling
MTBM	Microtunnel boring machine
N/A	Not applicable, not available
NACE	National Association of Corrosion Engineers
NASSCO	National Association of Sewer Service Companies
NBR	Nitrile butadiene rubber
NDT	Non-destructive testing
NIOSH	National Institute of Occupational Safety and Health
NPV	Net present value
NRC	National Research Council of Canada
NSF	National Sanitation Foundation
O&M	Operations and maintenance program
OC	Open-cut
OD	Outside diameter
OP	Open profile
OPEX	Operating expenses
ORM	Operation, reliability and maintenance
OSHA	Occupational Safety & Health Administration
PACP	Pipeline Assessment Certification Program
PARMS	Pipeline asset and risk management system
PB	Pipe bursting
PC	Polymer concrete
PCCP	Prestressed concrete cylinder pipe
PE	Polyethylene
PHMSA	Pipeline and Hazardous Materials Safety Administration
PI	Inflection point
PIP	Pull in-place
PL	Panel lining
PLM	Panel-lining method
PPI	Plastics Pipe Institute
PPV	Peak particle velocity
PR	Pipe removal
PSR	Point source repair (same as localized repair)
PU	Polyurethane
PVC	Polyvinyl chloride

Abbreviation	Description
PVC-M	Modified poly-vinyl-chloride
PVC-O	Molecularly oriented poly-vinyl-chloride
PVC-U	Unplasticized poly-vinyl-chloride
QA	Quality assurance
QC	Quality control
RACM	Regulated asbestos-containing material
RCCP	Reinforced concrete cylinder pipe
RCP	Reinforced concrete pipe
RD	Rolldown
RDBMS	Relational database management system
RFEC	Remote field eddy current
RFT	Remote field technology
RTV	Room temperature vulcanization
SASW	Spectral analysis of surface waves
SBR	Styrene butadiene rubber
SCG	Slow crack growth
SDR	Standard dimension ratio
SDS	Safety data sheets
SIPP	Spray-in-place pipe
SIPP-CM	Spray-in-place pipe—cement mortar
SIPP-EL	Spray-in-place pipe—epoxy lining
SIPP-GU	Spray-in-place pipe—Gunite
SIPP-PL	Spray-in-place pipe—panel lining
SIPP-SH	Spray-in-place pipe—shotcrete
SL	Sliplining
SMR	Sewer manhole renewal
SSET	Sewer scanning and evaluation technology
SSO	Sanitary sewer overflow
SSPC	Society of Protective Coatings
SWM	Spiral wound method
SWMM	Storm water management model
TAMP	Total asset management program
TBL	Triple bottom line
TCM	Trenchless construction methods
TRB	Transportation Research Board
TRM	Trenchless renewal methods

Abbreviation	Description
TTC	Trenchless technology center
USCS	Unified Soil Classification System
USDOT	U.S. Department of Transportation
USEPA	United States Environmental Protection Agency
UT	Ultrasonic testing
UV	Ultraviolet
VCP	Vitrified clay pipe
VOC	Volatile organic compounds
WaterRF	Water Research Foundation
WISE	Water infrastructure simulation environment
WRc	Water Research Center, United Kingdom
WRF	Water Research Foundation

APPENDIX B

Glossary of Terms

ABS Acrylonitrile-butadiene-styrene; a form of thermoplastic.

AC Asbestos cement; a composite material used in pipe construction.

AM Asset Management; the combination of management, financial, economic, engineering, and other practices applied to physical assets with the objective of providing the required level of service in the most cost-efficient manner.

AMP Asset management plan; a structured approach for utilities to achieve long-term defined service standards or an external bearing used to isolate the final drive from the thrusting force of the machine.

ASCE American Society of Civil Engineers

ASTM American Society of Testing Materials.

Auger A flighted drive tube having hex couplings at each end to transmit torque to the cutting head and transfer the spoil back to the machine.

Auger boring Also horizontal auger boring, a technique for forming a bore from a drive pit to a reception pit, by means of a rotating cutting head. Spoil is removed back to the drive shaft by helically wound auger flights rotating in a steel casing. The equipment may have limited steering capability. See Guided auger boring.

Backstop Also thrust block, a reinforced area of the entrance pit wall directly behind the track or where the jacking loads will be resisted.

Bedding A prepared layer of material below a pipeline to ensure uniform support.

Bent sub An offset section of drill stem close behind the drill head that allows steering corrections to be made by rotation of the drill string to orientate the cutting head (used in horizontal directional drilling).

Bentonite Colloidal clay sold under various trade names that form a slick slurry or gel when water is added; also known as driller's mud. See Drilling fluid.

Bits Replaceable cutting tools on the cutting head or drill string.

Bore A generally horizontal hole produced underground, primarily for the purpose of installing services.

Boring (1) The dislodging or displacement of spoil by a TBM; a rotating auger or drill string to produce a hole called a bore. (2) An earth-drilling process used for installing conduits or pipelines. (3) Obtaining soil samples for evaluation and testing.

Boring machine An automated mechanism to drill earth.

Boring pit Also entrance pit, launch or drive pit; an excavation in the earth of specified length, depth, and width for placing the boring machine on required line and grade.

Bottom inversion The CIPP tube is inverted through a specially designed elbow located at the elevation of the pipe, typically in a manhole or excavated pit.

Breakout Controls the joint make and/or break mechanism (in HDD operations).

Burst strength The internal pressure required to cause a pipe or fitting to fail within a specified time period.

Butt-fusion A method of joining polyethylene and PVC pipe where two pipe ends are rapidly brought together under pressure to form a homogeneous bond.

Bypass An arrangement of pipes and valves whereby the flow may be passed around a hydraulic structure or appurtenance. Also, a temporary setup to route flows around a part of a sewer system.

Bypass pumping Taking all existing flow in a pipe and routing around the section of pipe to be renewed, replaced, or repaired.

Calibration hose A flexible hose inverted into a pulled-in liner and used to hold the resin saturated fabric tube up tight against the existing, old, or host pipe until final cure has been achieved.

Can A principal module, which is part of a shield machine as in microtunneling or tunnel boring machines (TBMs). Two or more may be used, depending on the installation dimensions required and the presence of an articulated joint to facilitate steering. It may also be referred to as a trailing tube.

Carbon fiber A reinforcing material used to strengthen cured-in-place pipe.

Carriage The mechanical part of a nonsplit boring machine that includes the engine or drives motor, the drive train, thrust block, and hydraulic cylinders.

Carrier pipe The tube, which carries the product being transported, and which may go through casings at highway and railroad crossings. It may be made of steel, concrete, clay, plastic, ductile iron, or other materials. On occasion, it may be bored direct under the highways and railroads.

Cased bore A bore in which a pipe, usually a steel sleeve, is inserted simultaneously with the boring operation. It is usually associated with horizontal auger boring or pipe ramming.

Casing See Casing pipe.

Casing adapter A circular mechanism to provide axial and lateral support of a smaller-diameter casing than that of the casing pusher.

Casing pipe A steel pipe usually installed by horizontal auger boring or pipe ramming methods to support boreholes under roadways or railroad tracks through which a carrier (or product) pipes or ducts are installed.

Casing pipe method Method in which a casing, generally steel, is pipe jacked into place, within which a product pipe is inserted later.

Casing pusher The front section of a boring machine that distributes the thrusting force of the hydraulic cylinders to the casing and forms the outside of the spoil ejector system.

Catalyst The component of a resin system that induces a reaction to form heat and subsequently cure the liner.

Cathodic protection Preventing corrosion of a pipeline by using special cathodes (and anodes) to circumvent corrosive damage by electric current. Also a function of zinc coatings on iron and steel drainage products is galvanic action.

Caulking General term which, in trenchless technology, refers to methods by which joints may be closed within a pipeline.

Cave-in The separation of a mass of soil or rock material from the side of an excavation, or the loss of soil from under a trench shield or support system, and its sudden movement into the excavation, either by falling or sliding, in sufficient quantity so that it could entrap, bury, or otherwise injure and immobilize a person.

CCTV See Closed-circuit television inspection

Cell classification Method of identifying plastic materials, such as polyethylene, as specified by ASTM D3350, where the cell classification is based on these six properties: (1) density of base resin, (2) melt index, (3) flexural modulus, (4) tensile strength at yield, (5) ESCR, and (6) hydrostatic design basis and color.

Cellar drain A pipe or a series of pipes that collect wastewater which leaks, seeps, or flows into subgrade parts of structures and discharges them into a building sewers; by other means, dispose of such wastewaters into sanitary, combined, or storm sewers. It also referred to as basement drain.

Chemical grouting Method for the treatment of the ground around a shaft or pipeline, using noncementitious compounds, to facilitate or make possible the installation of an underground structure.

Chemical stabilization A repair method in which a length of pipeline between two access points is sealed by the introduction of one or more compounds in solution into the pipe and the surrounding ground and, where appropriate, producing a chemical reaction. Such systems may perform a variety of functions such as the sealing of cracks and cavities, the provision of a new wall surface with improved hydraulic characteristics or ground stabilization.

Chimney The small vertical section between a manhole frame and cone, which is built from brick, masonry, or concrete adjusting rings.

Chippers See Bits.

Circumferential The perimeter around the inner surface of a circular pipe cross section.

Circumferential coefficient of expansion and contraction The fractional change in circumference of a material for a unit changes in temperature; expressed as inches of expansion or contraction per inch of original circumference.

Closed-circuit television inspection (CCTV) Inspection method using a closed-circuit television camera system with appropriate transport and lighting mechanisms to view the interior surface of sewer pipes and structures.

Close-fit Pipe (CFP) Description of a lining system in which the new pipe makes close contact with the existing defective pipe at normal or minimum diameter. An annulus may occur in sections where the diameter of the defective pipe is in excess of this.

Coefficient of thermal expansion and contraction The fractional change in length of a material for a unit change in temperature.

Collapse Critical failure of a pipeline when its structural fabric disintegrates.

Collection system A network of sewers that serves one or more catchment areas.

Collector sewer A sewer located in the public way collects the wastewaters discharged through building sewers and conducts such flows into larger interceptor sewers, pumping and treatment works (referred to also as street sewer).

Combined sewer system A single network of sewers designed to convey stormwater as well as sanitary flows.

Competent person One who is capable of identifying existing and predictable hazards in the surroundings, or working conditions that are unsanitary, hazardous, or dangerous to employees, and who has authorization to take prompt corrective measures to eliminate them.

Compressed air method In trenchless technology, refers to the use of compressed air within a tunnel or shaft to balance groundwater and prevent ingress into an open excavation.

Compression gasket A device that can be made of several materials in a variety of cross sections and that serves to secure a tight seal between two pipe sections (e.g., "O" rings).

Compression ring A ring fitted between the end bearing area of the bell and spigot to help distribute applied loads more uniformly. The compression ring is attached to the trailing end of each pipe and is compressed between the pipe sections during jacking. The compression rings compensate for slight misalignment, pipe ends that are not perfectly square, gradual steering corrections, and other pipe irregularities. Compression rings are also referred to as spacers.

Conduit A broad term that can include pipe, casing, tunnels, ducts, or channels. The term is so broad that it should not be used as a technical term in boring or tunneling.

Cone The section between the top of a manhole wall and chimney or the frame. The diameter of the manhole is reduced over the cone section to receive the frame. The cone section may be concentric or eccentric.

Continuous sliplining See Sliplining.

Conventional pipe jacking Jacking pipe sections simultaneously as tunnel excavation proceeds using various forms of TBMs or hand mining (not microtunneling or pilot tube microtunneling).

Conventional trenching See Open-cut.

Corrugated pipe Pipe with ridges (corrugations) going around it to make it stiffer and stronger. The corrugations are usually in the form of a sine wave and are usually made of galvanized steel or aluminum.

Creep The dimensional change, with time, of a material, such as plastic, under continuously applied stress after the initial elastic deformation.

Crossing Pipeline installation in which the primary purpose is to provide a passage beneath a surface obstruction, such as a road, railroad track, lake, or river.

Crown (1) Top of pipe segment, or (2) The highest elevation within a pipe.

Cured-in-place pipe (CIPP) A lining system in which a thin flexible tube of polymer or glass fiber fabric is impregnated with thermoset resin and expanded by means of fluid pressure into position on the inner wall of a defective pipeline before curing the resin to harden the material. The uncured material may be installed by winch or inverted by water or air pressure, with or without the aid of a turning belt.

Cut and cover See Open-cut.

Deformed and reshaped See Modified sliplining.

Diameter of reamer Largest diameter of reamer (in horizontal directional drilling).

Dimension ratio (DR) See Standard dimension ratio (SDR).

Dimple A term used in tight fitting pipeline renewal, where the new plastic pipe forms an external departure or a point of expansion slightly beyond the underlying pipe wall where unsupported at side connections. The dimples are used for location and reinstatement of laterals.

Directional drilling A steerable system for the installation of pipes, conduits, and cables in a shallow arc using a surface launched drilling rig. Traditionally the term applies to large-scale crossings in which a fluid-filled pilot bore is drilled using a fluid-driven motor at the end of a bend-sub, and is then enlarged by a washover pipe and back reamer to the size required for the product pipe. The positioning of a bent sub provides the required deviation during pilot boring. Tracking of the drill string is achieved by the use of a downhole survey tool.

Dog plate See Backstop.

Drill bit A tool that cuts the ground at the head of a drill string, usually by mechanical means, but may include fluid jet cutting.

Drill string (1) The total length of drill rods or pipe, bit, swivel joint, and so on in a drill borehole. (2) System of rods used with cutting bit or compaction bit attached to the drive chuck.

Drilling fluid or mud A mixture of water and usually bentonite and/or polymer continuously pumped to the cutting head to facilitate cutting, reduce required torque, facilitate the removal of cuttings, stabilize the borehole, cool the head, and lubricate the installation of the product pipe. In suitable soil conditions water alone may be used.

Drive or entry or launch or jacking shaft or pit Excavation from which trenchless technology equipment is launched for the installation of a pipeline. In pipe jacking, it incorporates a thrust wall to spread reaction loads to the soil.

Earth piercing (1) Term commonly used in North America as an alternative to impact moling. (2) The use of a tool, which comprises a percussive hammer within a suitable casing, generally of torpedo shape. The hammer may be pneumatic or hydraulic. The term is usually associated with nonsteered devices without rigid attachment to the launch pit, relying upon the resistance (friction) of the ground for forward movement. During operation, the soil is displaced not removed. An unsupported bore may be formed in suitable ground, or a pipe drawn in, or pushed in, behind the tool. Cables may also be drawn in.

Emergency repair An unscheduled repair that must be made during a pipe failure or collapse. This type of repairs may cost many times (usually 10 times more, not including social costs) of planned repair costs and may not be as effective and/or permanent.

Emergency stop A red, manually operated push button that, when activated, stops all functions of the machine.

Entrance pit See Boring pit or Drive shaft.

Entry/Exit angle Angle to horizontal (the ground surface) at which the drill string enters or exits in forming the pilot bore in a horizontal directional drilling operation.

EPDM (ethylene-propylene-diene monomer) Type of rubber that has excellent resistance to ozone, sunlight, and oxygen. It also has excellent resistance to acids, alkalis, and ketones. Plus, it has excellent heat resistance and aging. However, it has poor resistance to fuels and oils.

Epoxy Resin formed by the reaction of bisphenol and epichlorohydrin.

Epoxy lining A curable resin system based on epoxy resins.

Exfiltration The leakage or discharge of flows being carried by pipes or sewers out into the ground through leaks in pipes, joints, manholes, or other sewer system structures; the reverse of infiltrations.

Exit pit See Reception shaft.

Exit shaft See Reception shaft.

Existing pipe In trenchless renewal and replacement methods. Existing, old, or deteriorated pipe.

Expander A tool, which enlarges a bore during a pullback operation by compression of the surrounding ground rather than by excavation. Sometimes used during a pipe bursting process as well as during horizontal directional drilling.

Face stability Stability of the excavated face of a tunnel or pipe jack operation.

Felt A material specially designed to soak up, hold, and transport resins in place to produce a hard cured-in-place pipe.

Fiberglass A high strength material that is commonly layered with a felt material to add reinforcement for the resin cured-in-place pipe.

Fillers Materials used to enhance the capabilities of a resin system.

Film Either an outer or inner material to protect the resin impregnated tube from contaminants.

Flexural modulus The slope of the curve defined by flexural load versus resultant strain. A high flexural modulus indicates a stiffer material.

Flexural strength The strength of a material in bending expressed as the tensile stress of the outermost fibers at the instant of failure.

Flight The spiral plates surrounding the tube of an auger.

Fluid cutting (1) An old trenchless method where pressurized fluid jets are mainly used to provide the soil cutting action. (2) A process using high-pressure fluid to wash out the face of a utility crossing without any mechanical or hand excavation of the soils in the face. This method is no longer allowed.

Fluid-assisted boring or drilling A type of horizontal directional drilling technique using a combination of mechanical drilling and pressurized fluid jets to provide the soil cutting action.

Fold and form lining Method of pipeline renewal in which a liner is folded to reduce its size before insertion and reversion to its original shape by the application of pressure, or heat, or both.

Fold and form pipe A pipe renewal method where a plastic pipe manufactured in a folded shape of reduced cross-sectional area is pulled into an existing conduit and subsequently expanded with pressure and heat. The reformed plastic pipe fits snugly and takes the shape of the ID of the existing, old, or host pipe.

Force-main A pipeline that conveys sanitary, combined, or stormwater flow under pressure from a pumping (or lift) station to a discharge point (treatment plant).

Forward rotation The clockwise rotation of the auger as viewed from the machine end.

Gel time The time, at which a catalyzed resin system will begin to cure, creating internal heat and thus beginning to harden.

Geographical information system (GIS) A computer software system designed to store, manipulate, analyze, and print geographically referenced information.

Gravity sewer A sewer that is designed to operate under open channel conditions (below pipe full capacity) up to a maximum design flow at which point it will become surcharged.

Ground mat Usually used in horizontal directional drilling, metal mats rolled out on either side of drill rack for operators and crew to stand on during operation to give grounding protection in case of electrical strike.

Ground rod This is a copper or brass rod that is hand driven into the ground and is connected to the drill rack and mats to provide adequate grounding of an HDD rig.

Groundwater table (or level) Upper surface of the zone of saturation in permeable rock or soil (when the upper surface is confined by impermeable rock, the water table is absent).

Grout (1) Material used to seal pipeline and manhole cracks; also used to seal connections within pipe or sewer structures. (2) A material, usually cement or polymer based, used to fill the annulus between the existing pipe and the lining; and also to fill voids outside the existing pipeline. (3) A material such as cement slurry, sand, or pea gravel that is pumped into voids.

Grouting (1) Filling of the annular space between the existing, old, or host pipe and the carrier pipe. Grouting is also used to fill the space around laterals and between the new pipe and manholes. Other uses of grouting are for localized repairs of defective pipes and ground improvement prior to excavation during new installations. (2) The process of filling voids, or modifying or improving ground conditions. Grouting materials may be cementitious, chemical, or other mixtures. In trenchless technology, grouting may be used for filling voids around the pipe or shaft, or for improving ground conditions. (3) A method of filling voids with cementitious or polymer grout.

GRP Glass reinforced plastic, a family of renewal linings. Often generically known as reinforced plastic mortar (RPM) and reinforced thermosetting resin (RTR).

Guidance system The guidance system continuously confirms the position of the TBM.

Guide rail Device used to support or guide, first the shield and then the pipe within the drive shaft during a pipe jacking or utility tunneling operation.

Guided auger boring Pilot tube microtunneling. It is a modified version of horizontal auger boring method.

Guided boring This term is used in Europe for small diameter horizontal directional drilling method.

Guided drilling See Guided boring.

Gunite A renewal technique that employs steel reinforcement fixed to the inside surface of an existing sewer line, which is sprayed with dry concrete.

HDPE High density polyethylene, see Polyethylene.

Head (static) The height of water above any plane or point of references. (The energy possessed by each unit of weight of a liquid, expressed as the vertical height through which a unit of weight would have to fall to release the average energy posed.) Standard unit of measure shall be the foot. Head in feet for water is 1 psi = 2.310 ft.

Heat cure The application of either steam or hot water to cure a resin saturated tube.

Heaving A process in which the ground in front of a tunneling or pipe jacking operation may be displaced forward and upward, causing an uplifting of the ground surface.

Height of cover (HC) Distance from crown of a pipe or conduit to the finished road surface, or ground surface, or the base of the railway.

High density polyethylene A plastic resin made by the copolymerization of ethylene and a small amount of another hydrocarbon. The resulting base resin density, before additives or pigments, is greater than 0.941 g/cc.

Hoop stress The circumferential force per unit areas, psi, in the pipe wall owing to internal pressure.

Horizontal directional drilling (HDD) See Directional drilling.

Horizontal earth boring machine A machine used to bore horizontally through the earth by means of an MTBM (microtunnel-boring machine), cutterhead, rotating tool, or ramming tool.

Horizontal auger boring The use of auger boring machines to prepare holes by the installation of a casing whereby the spoil is removed by the use of augers.

Host pipe See existing pipe.

Hydraulic gradient line (HGL) An imaginary line through the points to which water would rise in a series of vertical tubes connected to the pipe. In an open channel, the water surface itself is the hydraulic grade line.

Hydrogen sulfide An odorous gas found in sewer systems with chemical formula of H_2S.

Impact moling Method of creating a bore using a pneumatic or hydraulic hammer within a casing, generally of torpedo shape. The term is usually associated with nonsteered or limited steering devices without rigid attachment to the launch pit, relying upon the resistance of the ground for forward movement. During the operation the soil is displaced, not removed. An unsupported bore may be formed in suitable ground, or a pipe drawn in, or pushed in, behind the impact moling tool.

Impervious coating The outer layer of a CIPP tube that will prevent the installation water or steam from mixing with the resin system in the tube.

Impregnated tube A felt tube fully saturated with a catalyzed resin system.

In-line replacement The process of breaking out of an existing, old, or host pipeline and the installation of a new pipeline at the same location. With this method, the existing pipeline will serve as a "pilot bore" for the new pipe installation, which might be a different pipe material with the same or larger diameter, and will be installed with the same alignment of the existing pipe.

Infiltration Penetration of groundwater into the sewer system through cracks and defective joints in the pipeline, or through lateral connections, or manholes.

Infiltration or inflow (I/I) The total quantity of water from both infiltration and inflow without distinguishing the source.

Inflow Storm water discharged into a sewer system and service connections from sources on the surface.

Interjack pipes Pipes specially designed for use with an intermediate jacking station used in pipe jacking and microtunneling operations.

Interjack station See Intermediate jacking stations.

Intermediate jacking method Pipe jacking or microtunneling method to redistribute the jacking force by the use of intermediate jacking stations.

Intermediate jacking stations A fabricated steel cylinder fitted with hydraulic jacks that are incorporated into jacking pipes strung between two pipe segments. Its function is to distribute the jacking load over the pipe string on long drives thereby decreasing the total jacking forces exerted on the thrust block and pipe sections near the shaft.

Internal seal Internal seals are used for structural repair pipe joints and missing pipe sections. It can be used in both worker entry and non-worker entry pipes.

Inversion The process of turning a fabric tube inside out with water or air pressure as is done at installation of a cured-in-place pipe (CIPP).

Invert (1) The lowest point on the pipe circumference; also the defined channel in the manhole platform that directs flow from inlet pipe to outlet pipe. (2) The inside bottom, lowest elevation, of a pipe.

Jacking The actual pushing of pipe or casing in an excavated hole. This is usually done with hydraulic cylinders (jacks), but has been done with mechanical jacks and air jacks.

Jacking force Force applied to pipes in a pipe jacking operation.

Jacking frame A structural component that houses the hydraulic cylinders used to propel the microtunneling machine and pipeline. The jacking frame serves to distribute the thrust load to the pipeline and the reaction load to the shaft wall or thrust wall.

Jacking pipes Pipes sections with smooth outside joints designed to be installed using pipe jacking techniques.

Jacking pit See Jacking shaft.

Jacking shaft (also launch or entry shaft) Excavation from which trenchless technology equipment is launched for the installation or renewal of a pipeline.

Jacking shield A steel cylinder from within which the excavation is carried out either by hand or machine. Incorporated within the shield are facilities to allow it to be adjusted to control line and grade.

Joint sealing Method in which an inflatable packer is inserted into a pipeline to span a leaking joint, resin, or grout being injected until the joint is sealed and the packer then removed.

Lateral A service line that transports wastewater from individual buildings to a main sewer line.

Lateral connection The point at which the downstream end of a building drain or sewer connects into a larger-diameter sewer.

Launch pit See Drive shaft or pit.

Lead pipe The leading pipe designed to fit the rear of a jacking shield and over which the trailing end of the shield is fitted.

Light cure The curing of a CIPP liner using UV light energy in lieu of either water or steam.

Liner A fabric tube that has been saturated with a liquid resin.

Liner plate A proprietary product, used to line tunnels instead of casing, and comes in formed steel segments. When these segments are bolted together they form a structural tube to protect the tunnel from collapsing. The segments are made so that they may be bolted together from inside the tunnel.

Lining A renewal process where a new pipe or coating material is inserted or cured in place to give an existing pipe a new design life.

Localized (spot) repair Repair work on an existing pipe, to an extent less than the run between two access points or manholes.

Locator An electronic instrument used to determine the position and strength of electromagnetic signals emitted from a transmitter (sonde) in a directional drilling operation, or from existing underground services, which have been energized, thereby identifying its location. In horizontal directional drilling, it is referred to as a walkover system.

Machine upset The inadvertent action of a horizontal auger boring machine that rotates the machine and track from its normal and upright position to another position.

Man-entry Also worker-entry, describes any inspection, construction, renewal, or repair process, which requires an operator to enter a pipe, duct, or bore. OSHA currently has no minimum size limit for worker-entry operations; however, they address a much broader concept of confined space in Title 29 Code of Federal Regulations Part 1910.146. Currently, the minimum size permissible in the United Kingdom is 900 mm (approximately, 36 in.). Many trenchless technologies do not require worker-entry inside the pipe.

Manhole A structure that allows access to the sewer system.

Manual inspection Method of sewer inspection that usually involves physical entry and hands-on examination.

Microtunnel-boring machine (MTBM) See Microtunneling.

Microtunneling A trenchless construction method for installing pipelines. Microtunneling uses all of the following features during construction: (1) Remote controlled—The microtunnel-boring machine (MTBM) is operated from a control panel, normally located on the surface. The system simultaneously installs pipe as spoil is excavated and removed. Personnel entry is not required for routine operation. (2) Guided—The guidance system usually references a laser beam projected onto a target in the MTBM, capable of installing gravity sewers or other types of pipelines to the required tolerance, for line and grade. (3) Pipe jacked—The process of constructing a pipeline by consecutively pushing pipes and MTBM through the ground using a jacking system for thrust. (4) Continuously supported—Continuous pressure is provided to the face of the excavation to balance groundwater and earth pressures.

Midi-HDD Steerable surface-launched horizontal directional drilling equipment for installation of pipes, conduits, and cables. Applied to intermediate-sized drilling rigs used as either a small directional drilling machine or a large-guided boring machine. Tracking of the drill string may be achieved by either a downhole survey tool or a walkover locator.

Mini-horizontal directional drilling (Mini-HDD) Small diameter horizontal directional drilling. In Europe, it is called guided boring.

Mixed face A soil condition that presents two or more different types of material in the path of the bore.

Modified sliplining A range of techniques in which the liner is reduced in cross-sectional diameter before insertion into the carrier pipe. It is subsequently restored close to its original diameter, generally forming a close-fit with the original pipe. There are different methods of cross-sectional area reduction.

Needled felt A highly absorbent felt material specially designed for the CIPP tube fabrication.

Non-worker entry Size of pipe, duct, or bore, less than that for worker-entry.

Occupational illness Any abnormal condition or disorder caused by exposure to environmental factors associated with employment. For excavations, this might include illnesses caused by the inhalation of toxic vapors.

Old pipe See Existing pipe.

On-site wet-out A non-factory tube wet-out that is performed over the hole in the field.

Open-cut The method by which access is gained to the required level underground for the installation, repair, or replacement of a pipe, conduit, or cable. The excavation is then backfilled and the surface restored.

Open-face shield Shield in which manual excavation is carried out from within a steel tube at the front of a pipe jacking operation.

On-site wet-out A non-factory tube wet-out that is performed over the hole in the field.

Ovality There are two definitions: (1) the difference between the maximum and mean diameter divided by the mean diameter, and (2) the difference of the mean and minimum divided by the mean, at any one cross section of a pipe, generally expressed as a percentage.

Overcut The annular space between the excavated borehole and the outside diameter of the pipe.

PACP Pipeline Assessment Certification Program by NASSCO.

Panel lining Panel lining is a modified sliplining method. The shape of the culvert is covered by preparing panels and fitting them to the culvert. It can be used to structurally renew large diameter pipes. This method can accommodate different shapes.

PE Polyethylene; a form of thermoplastic pipe.

pH A measure of the acidity or alkalinity of a solution. A value of seven is neutral; lower numbers indicate more acidity.

Physical pipe inspection The crawling or walking through manually accessible pipelines. The logs for physical pipe inspection record information of the kind detailed under television inspection. Manual inspection is only undertaken when field conditions permit this to be done safely. Precautions are necessary.

Piercing tool Similar to closed-face pipe ramming, but for small diameter (2–6 in.) boring; used for cable installations under roadways.

Pilot bore The action of creating the first (usually steerable) pass of any boring process, which later requires back-reaming or similar enlarging process to install the product pipe. Most commonly applied to horizontal directional drilling but also used in pilot tube microtunneling and guided auger boring systems.

Pinch rollers Used to control the thickness of the resin impregnated tube during wet-out.

Pipe bursting A pipe replacement method for breaking the existing pipe by brittle fracture, using force from within, applied mechanically, the remains being forced into the surrounding ground. At the same time, a new pipe of the same or larger diameter is drawn in behind the bursting tool. The pipe-bursting device may be based on an impact moling tool to exert diverted forward thrust to the radial bursting effect required, or by a hydraulic device inserted into the pipe and expanded to exert direct radial force or a static hammer. For new pipe, generally an HDPE pipe is used, but currently PVC, ductile iron, clay, and GRP are also used. Also known as pipe cracking and pipe splitting.

Pipe displacement See Pipe bursting.

Pipe eating See Pipe removal.

Pipe jacking A system of directly installing pipes behind a shield machine by hydraulic jacking from a drive shaft, such that the pipes form a continuous string in the ground.

Pipe ramming A nonsteerable system of forming a bore by driving an open-ended steel casing using a percussive hammer from a drive pit. The soil may be removed from the casing by augering, jetting, or compressed air.

Pipe reaming A variation of directional boring, pipe reaming can be used to replace existing clay, asbestos cement, non-reinforced concrete, and PVC pipe. A reamer is pulled through the existing pipe, which cuts the pipe into small pieces. The pipe pieces are flushed out the borehole with the drilling fluid.

Pipe removal A pipe replacement technique, usually conducted by use of a horizontal directional drilling rig in which a defective pipe is pulverized during the backreaming operation. Also microtunneling machines can be used where the existing pipe is excavated together with the surrounding soil as for a new installation. The microtunneling shield machine will usually need some crushing capability to perform effectively. The defective pipe may be filled with grout to improve steering performance.

Pipe segment A specific portion of the sewer or pipeline system; which usually runs between two structures (e.g., manhole, trap tanks, sumps); identified with unique sewer or pipe structure ID number.

Pipeline rehabilitation See Pipeline renewal.

Pipeline renewal The in-situ renewal of an existing pipeline, which has become deteriorated. The selection of appropriate renewal method is dependent on type of application and characteristics and types of defects of the existing pipe. See Chap. 3 for method selection criteria.

Pit See shaft.

Plastic Any of a variety of thermoplastic and thermoset material used in pipeline construction and renewal (e.g., polypropylene, PVC, fiberglass reinforced plastics, polyester felt reinforced pipe, epoxy and polyester mortars, and so on).

Point CIPP CIPP techniques entail impregnating fabric with a suitable resin, pulling this into place within the sewer around an inflatable packer or mandrel, and then filling the packer with water, steam, or air under pressure to press the patch against the existing sewer wall while the resin cures.

Point repairs Repair works on an existing pipe, to an extent less than the run between two access points or manholes.

Point source repair See Localized (spot) repair.

Polyester Resin formed by condensation of polybasic and monobasic acids with polyhydric alcohols.

Polyethylene A ductile, durable, virtually inert thermoplastic composed by polymers of ethylene. It is normally a translucent, tough solid. In pipe grade resins, ethylene-hexene copolymers are usually specified with carbon black pigment for weatherability.

Polymer coating It is a thermoset coating made up of inert plastic-like epoxies, urethanes and ureas, polyesters, which have a high resistance to corrosion; they are applied by trained professionals using special spraying equipments and as per the manufacturers' specifications.

Polyolefin A family of plastic material used to make pipes. Polypropylene (PP): A type of plastic pipe from the polyolefin family.

Potholing Digging a vertical hole to visually locate a utility.

Preparatory cleaning Internal cleaning of pipelines, particularly sewers, prior to inspection, usually with water jetting and removal of material where appropriate.

Preventative maintenance Routine maintenance designed to prevent pipeline system failures and resulting emergency repairs.

Product pipe Permanent pipeline for operational use. It is a pipe conveyance for water, gas, sewage, and other products.

Protruding To be projecting outward.

Pull-back force The tensile load applied to a drill string during the pull-back process. Horizontal directional drilling rigs are generally rated by their maximum pull-back force.

PVC Polyvinyl chloride; a form of thermoplastic pipe.

Quality assurance It includes developing inspection and testing methods to ensure products or services are designed and produced to meet or exceed owner's requirements.

Quality control It includes providing evidence needed (test results) to establish confidence among all concerned, that quality-related activities are performed.

Radian An arc of a circle equal in length to its radius; or the angle at the center measured by the arc.

Ramming A percussive hammer is attached to an open-end casing, which is driven through the ground. See Pipe ramming.

Receiving pit (1) See Exit pit. (2) An opening in the earth located at the expected exit of the cutting head or tunneling boring machine (TBM). (3) The pit that is dug at the end of the bore, opposite the jacking pit. See also Target shaft or pit.

Receiving shaft See Reception or exit shaft or pit.

Reception or exit shaft or pit Excavation into which trenchless technology equipment is driven and recovered following the installation of the product pipe, conduit, or cable. See Receiving pit.

Registered professional engineer It means a person who is registered as a professional engineer in the state where the work is to be performed. However, a professional engineer, registered in any state is deemed to be a "registered professional engineer" when approving designs for "manufactured protective systems" or "tabulated data" to be used in interstate commerce.

Rehabilitation See Renewal.

Reinstatement Method of backfilling, compaction, and resurfacing of any excavation to restore the original surface and underlying structures to enable it to perform its original function.

Remote-control system (microtunneling) The remote-control system monitors and controls the MTBM, the automated transport system, and the guidance system from a location not in the MTBM.

Renewal All aspects of rehabilitating, reconstructing, renovating, or upgrading with a new design life for the performance of existing pipeline systems.

Renovation See Renewal.

Repair Reconstruction of short pipe lengths, but not the reconstruction of a whole pipeline. Therefore, a new design life is not provided. In contrast, in pipeline renewal, a new design life is provided to existing pipeline system.

Replacement Replacing an old, existing, or deteriorated pipe with a new pipe by use of open-cut, in-line replacement (pipe bursting), and/or other new installation methods (HDD, microtunneling, etc.). The new pipe may have a larger diameter and different pipe material from the existing or old pipe.

Resin impregnation (wet-out) A process used in cured-in-place pipe installation where a plastic-coated fabric tube is uniformly saturated with a liquid thermosetting resin, while air is removed from the coated tube by means of vacuum suction.

Resins An organic polymer, solid or liquid; usually thermoplastic or thermosetting.

Reverse In horizontal auger boring, the counterclockwise rotation of the auger as viewed from the machine end.

Ring compression The principal stress in a confined thin circular ring subjected to external pressure.

Robot Remote-control device with closed-circuit television (CCTV) monitoring, used mainly in localized repair work, such as cutting away obstructions, reopening lateral connections, grinding and refilling defective areas, and injecting resin into cracks and cavities.

Robotic cutter A device used to reopen house connections after the installation of a liner without excavation.

Rod pushing Method of forming a pilot bore by driving a closed pipe head with rigid attachment from a launch pit into the soil, which is displaced.

Roller cone bit or reamer A bit or reamer in which the teeth rotate on separate, internal shafts that are usually aligned perpendicular to line. Used for boring rock.

Rollers The rollers that control the exact amount of resin material that is required to properly saturate a felt tube.

Rotary rod machine A machine used to drill earth horizontally by means of a cutting head attached to a rotating rod (not an auger). Such drilling may include fluid injected to the cutting head through a hollow rod.

Saddle In horizontal auger boring, a vertical support mechanism to hold the casing in position while starting (collaring) the bore.

Saturated tube See Impregnated tube.

SBR (Styrene butadiene) Type of rubber that has good abrasion resistance and excellent impact and cut-and-gouge resistance. Used as gasket material.

SDR See Standard dimension ratio.

Segmental lining See Sliplining.

Segmental sliplining See Sliplining.

Self-cleansing A consequence of good hydraulic design when the pipe invert is kept relatively free of sediments by ensuring adequate flow velocities.

Semi-structural liner A liner that in its own entity does not have the required strength to withstand internal, external, or both types of loading from soil column, traffic, and groundwater pressure for the design life of the product, but will offer some level of structural support against internal pressure.

Separate system A system that uses sanitary sewers to convey the wastewater and stormwater sewers to carry the stormwater.

Sewer An underground pipe or conduit for transporting stormwater, or wastewater, or both.

Sewer cleaning The use of mechanical or hydraulic equipment to dislodge, transport, and remove debris from sewer lines.

Sewer lateral A building sewer (sometimes referred to as a sewer lateral or house lateral) is the pipeline between the public sanitary sewer line, which is usually located in the street, and the indoor plumbing.

Sewer pipe A length of conduit, manufactured from various materials and in various lengths, that when joined together can be used to transport wastewaters from the points of origin to a treatment facility. Types of pipe are: acrylonitrile-butadiene-styrene (ABS); asbestos-cement (AC); brick pipe (BP); concrete pipe (CP); cast iron pipe (CIP); polyethylene (PE); polyvinyl chloride (PVC); and vitrified clay (VC).

Shaft An excavation from which trenchless technology equipment is launched for the installation or removal of a pipeline, conduit, or cable. May include a thrust wall to spread reaction loads to the ground.

Shield (shield system) It means a structure that is able to withstand the forces imposed on it by a cave-in and thereby protect employees within the structure. Shields can be permanent structures or can be designed to be portable and moved along as work progresses. Additionally, shields can be either pre-manufactured or job-built in accordance with 1926.652(c)(3) or (c)(4). Shields used in trenches are usually referred to as "trench boxes" or "trench shields."

Shoring (shoring system) A structure such as a metal hydraulic, mechanical or timber shoring system that supports the sides of an excavation and is designed to prevent cave-ins.

Shotcrete Spraying wet concrete (see also Gunite).

Skin friction Resistance to advancement caused by soil pressure around the pipe or casing.

Slipline A renewal technique covering the insertion of one pipe inside an existing pipe.

Sliplining (1) General term used to describe methods of lining with continuous pipes and lining with discrete pipes. (2) Insertion of a new pipe by pulling or pushing it into the existing pipe and grouting the annular space. The pipe used may be continuous or a string of discrete pipes. The latter is also referred to as segmental sliplining.

Slurry A fluid, mainly water mixed with bentonite and sometimes polymers, used in a closed loop system for the removal of spoil and for the balance of groundwater pressure during tunneling and microtunneling operations.

Slurry chamber Located behind the cutting head of a slurry microtunneling machine. Excavated material is mixed with slurry in the chamber for transport to the surface.

Slurry line A series of hoses or pipes that transport tunnel muck and slurry from the face of a slurry microtunneling machine to the ground surface for separation.

Slurry separation A process where excavated material is separated from the circulation slurry.

Slurry shield method Method using a mechanical tunneling shield with closed face, which employs hydraulic means for removing the excavated material and balances the groundwater pressure. See also earth pressure balance machine.

Social costs Costs incurred by society as a result of underground pipeline construction and renewal. These include but not limited to traffic disruptions, environmental damages, safety hazards, inconvenience to general public, and business losses owing to road closures.

Soft lining See Cured-in-place pipe (CIPP).

Spiral lining A technique in which a ribbed plastic strip is spirally wound by a winding machine to form a liner, which is inserted into a defective pipeline. The annular space may be grouted or the spiral liner expanded to reduce the annulus and form a close-fit liner. In larger diameters, the strips are sometimes formed into panels and installed by hand. Grouting the annular space after installation is recommended.

Spiral weld pipe (casing) Pipe made from coils of steel plate by wrapping around a mandrail in such a manner that the welds are a spiral helix.

Spiral wound In this process a new pipe is installed inside the existing pipe from the continuous strip of polyvinyl chloride (PVC). The strip has tongue and groove casting on its edges. It is fed to a special winding machine placed in a manhole, which creates a continuous helically wound liner that proceeds through the existing pipe. The continuous spiral joint is watertight. Upon completions of the annulus space between the lining and the existing pipe wall is usually required.

Spoil (muck) Earth, rock, and other materials displaced by a tunnel, pipe or casing, and removed as the tunnel, pipe, or casing is installed. In some cases, it is used to mean only the material that has no further use.

Spot repair See Localized (spot) repair.

Spray-in-place pipe A technique for applying a lining of cement mortar or resin by rotating a spray head, which is winched through the existing pipeline.

Spray-on lining See Spray-in-place pipe.

Springline (1) An imaginary horizontal line across the pipe that passes between the points where the pipe has its greatest cross-sectional width. (2) Midpoint of a pipe cross section (equal vertical distance between the crown and the invert of the pipe).

Standard dimension ratio (SDR) Standard dimension ratio is defined as the ratio of the outside pipe diameter to wall thickness. Same as Dimension ratio (DR).

Static mixer A computerized device that provides fast, uniform resin, and catalyst mixing.

Steering head In horizontal auger boring, a moveable lead section of casing that can be adjusted to steer the bore.

Styrene A component of polyester and many vinyl ester resin systems.

Subsidence The settlement of the ground, pipeline, or other structure. The effects may not be evenly distributed and/or immediately noticeable. Differential settlement may occur.

Sump A depression usually in the drive pit to allow the collection of water and the installation of a sump pump for water removal.

Swivel In horizontal directional drilling, it is used to attach product pipe (to be pulled into drilled hole) to drill pipe to prevent it from rotating.

Target shaft or pit See Reception or exit shaft or pit.

TBM See Tunnel-boring machine.

Televise Process by which a sewer, pipeline or lateral is inspected with a closed-circuit television camera.

The Global Positioning System (GPS) A space-based navigation system that provides location and time information in all weather conditions, anywhere on or near the Earth where there is an unobstructed line of sight to four or more GPS satellites.

Thermocouple A device used to measure the internal temperature of a resin-saturated felt tube during the installation process.

Thermoformed pipe A type of renewal method that uses polyvinyl chloride (PVC) or polyethylene (PE) pipe that is expanded by thermoforming to fit tightly to fit inside the existing, old, or host pipe.

Thermoplastic (TP) A polymer material, such as polyethylene, that will repeatedly soften when heated and harden and reformed when cooled. TPs are generally much easier to recycle than their thermoset (see below) counterparts.

Thermoset (TS) A polymer material, such as rubber, that does not melt when reheated. TS polymers can be formed initially into almost any desired shape, but they cannot be reformed at a later time.

Thermoset resin A material, such as epoxies, that will undergo or has undergone a chemical reaction by the action of heat, chemical catalyst, ultraviolet light, etc., leading to an infusible state.

Thrust Force applied to a pipeline or drill string to propel it through the ground.

Thrust block See Backstop.

Thrust pit See Drive pit.

Thrust ring A fabricated ring that is mounted on the face of the jacking frame. It is intended to transfer the jacking load from the jacking frame to the thrust bearing area of the pipe section being jacked.

Top inversion The CIPP tube is inverted from an inversion ring located at the top of the inversion platform.

Track A set of longitudinal rails mounted on cross members that support and guide a horizontal auger boring machine.

Trench (trench excavation) A narrow excavation (in relation to its length) made below the surface of the ground. In general, the depth is greater than the width, but the width of a trench (measured at the bottom) is not greater than 15 ft (4.6 m). If forms or other structures are installed or constructed in an excavation, so as to reduce the dimension measured from the forms or structure to the side of the excavation to 15 ft (4.6 m) or less (measured at the bottom of the excavation), the excavation is also considered to be a trench.

Trenching See Open-cut or conventional trenching.

Trenchless A technology that is used for renewal of existing pipelines, typically without any excavation.

Trenchless methods See Trenchless technology.

Trenchless rehabilitation See Renewal.

Trenchless technology Also NO-DIG, techniques for underground pipeline and utility construction and replacement, rehabilitation, renovation (collectively called renewal), repair, inspection, and leak detection, etc., with minimum or no excavation from the ground surface.

Tube The fabric material tube used to carry and hold the thermoset resin materials in place against the existing pipe prior to curing.

Tuberculation Localized corrosion at scattered locations resulting in knob-like mounds.

Tunnel An underground conduit, often deep and expensive to construct, which provides conveyance and/or storage volumes for wastewater, often involving minimal surface disruption.

Tunnel-boring machine (TBM) (1) A full-face circular mechanized shield machine, usually of worker-entry diameter, steerable, and with a rotary cutting head. For pipe jacking installation it leads a string of pipes. It may be controlled from within the shield or remotely such as in microtunneling. (2) A mechanical excavator used in a tunnel to excavate the front face of the tunnel (mole, tunneling head).

Ultraviolet light cure See Light cure.

Underground utility Active or inactive services or utilities below ground level.

Upset See Machine upset.

Upsizing Any method such as pipe replacement or pipe bursting that increases the cross-sectional area of an existing pipeline by replacing with a larger-diameter pipe.

VCP Vitrified clay pipe.

VCT Vitrified clay tile or vitrified clay tile pipe.

Velocity head For water moving at a given velocity, the equivalent head through which it would have to fall by gravity to acquire the same velocity.

Vinyl-ester Resin systems used for many industrial applications requiring a higher level of corrosion resistance.

Viscosity That property of a fluid which determines the amount of its resistance to a shearing stress.

Voids (1) Holes on the outside of the pipe in the surrounding soil or material. (2) A term generally applied to paints to describe holidays, holes, and skips in the film. Also used to describe shrinkage in castings or welds.

Walkover system See Locator.

Water jetting (1) Method for the internal cleansing of pipelines using high-pressure water jets. (2) An obsolete and unauthorized method for cutting earth with water jetting.

Water table (1) The depth of the groundwater. (2) The upper limit of the portion of ground wholly saturated with water.

Weatherability The properties of a plastic material that allows it to withstand natural weathering; hot and cold temperatures, wind, rain, and ultraviolet rays.

Wetout The process of injecting resin into, and distributing it throughout, a hose or tube, which will then be installed into the pipeline and cured in place.

Winch Mechanical device used to pull the CCTV cameras or cleaning tools through a pipe.

APPENDIX C

Abbreviations for Organizations Related to Trenchless Technology

Abbreviation	Description	Website
AASHTO	American Association of State Highway and Transportation Officials	www.transportation1.org/aashtonew
ACI	American Concrete Institute	www.concrete.org
ACIPCO	American Cast Iron Pipe Company	www.acipco.org
ACPA	American Concrete Pipe Association	www.concrete-pipe.org
ACPPA	American Concrete Pressure Pipe Association	www.acppa.org
AEM	Association of Equipment Manufacturers	www.aem.org
AGA	American Gas Association	www.aga.org
AISI	American Iron and Steel Institute	www.steel.org

Abbreviation	Description	Website
ANSI	American National Standards Institute	www.ansi.org
APC	American Plastics Council	www.americanplasticscouncil.org
APGA	American Public Gas Association	www.apga.org
API	American Petroleum Institute	www.api.org
APWA	American Public Works Association	www.apwa.net
ASA	American Shotcrete Association	www.shotcrete.org
ASCE	American Society of Civil Engineers	www.asce.org
ASDWA	Association of State Drinking Water Administrators	www.asdwa.org
ASME	American Society of Mechanical Engineers	www.asme.org
ASTM	American Society for Testing and Materials	www.astm.org
AWWA	American Water Works Association	www.awwa.org
WaterRF	Water Research Foundation	www.waterrf.org
BAMI-I	Buried Asset Management Institute-International	www.bami-i.com
BSI	British Standards	www.bsi-global.com
CSA	Canadian Standards Association	www.csagroup.org
CUIRE	Center for Underground Infrastructure Research and Education	www.cuire.org
DIPRA	Ductile Iron Pipe Research Association	www.dipra.org
EPA	Environmental Protection Agency	www.epa.gov
FHA	Federal Highway Administration	www.fhwa.dot.gov
ICRI	International Concrete Repair Institute	www.icri.org

Abbreviation	Description	Website
ISO	International Organization for Standardization	www.iso.org
ISTT	International Society for Trenchless Technology	www.istt.com
NACE	National Association of Corrosion Engineers	www.nace.org
NACWA	National Association of Clean Water Agencies	www.nacwa.org
NASSCO	National Association of Sewer Service Companies	www.nassco.org
NASTT	North American Society for Trenchless Technology	www.nastt.org
NCPI	National Clay Pipe Institute	www.ncpi.org
NCSPA	National Corrugated Steel Pipe Association	www.ncspa.org
NESHAP	National Emission Standards for Hazardous Air Pollutants	www.epa.gov/asbestos/asbestos-neshap
NFPA	National Fire Protection Association	www.nfpa.org
NIOSH	National Institute of Occupational Safety and Health	http://origin.cdc.gov/niosh/
NRC	National research council of Canada	www.nrc-cnrc.gc.ca
NSF	National Sanitation Foundation	www.nsf.org
NSF	National Science Foundation	www.nsf.gov
NTSB	National Transportation Safety Board	www.ntsb.gov
NUCA	National Utility Contractors Association	www.nuca.com
NWRA	National Water Resources Association	www.nwra.org
OSHA	Occupational Safety & Health Administration	www.osha.gov

Abbreviation	Description	Website
PHMSA	Pipeline and Hazardous Materials Safety Administration	www.phmsa.dot.gov
PCCA	Power & Communication Contractors Association	www.pccaweb.org
PPFA	Plastic Pipe and Fittings Association	www.ppfahome.org
PPI	Plastic Pipe Institute	www.plasticpipe.org
PRC	Pipeline Research Council	www.prci.com
SPE	Society of Plastic Engineers	www.4spe.org
SPI	Society of Plastics Industry, Inc.	www.plasticsindustry.org
SSPC	The Society for Protective Coatings	www.sspc.org
TRB	Transportation Research Board	www.trb.org
TSI	Transportation Safety Institute	www.tsi.dot.gov
TTC	Trenchless Technology Center, Louisiana Tech University	www.ttc.latech.edu
Uni-Bell	Uni-Bell PVC Pipe Association	www.uni-bell.org
USDOT	U.S. Department of Transportation	www.dot.gov
USEPA	United States Environmental Protection Agency	www.epa.gov
WEF	Water Environment Federation	www.wef.org
WERF	Water Environment Research Foundation	www.werf.org
WRc	Water Research Center	www.wrcplc.co.uk
WRF	Water Research Foundation	www.wrf.org

APPENDIX D

Related Documents

American Concrete Pipe Association. (2011). *Concrete Pipe Design Manual*, ACPA, Irving, Tex. Available at: http://www.concrete-pipe.org/pages/design-manual.html

American Iron and Steel Institute. (2007). *Handbook of Steel Drainage & Highway Construction Products*, 4th ed., AISI, Washington, D.C. Available at: http://www.ail.ca/site/media/ail/CSPI_Handbook%20of%20Steel%20Drainage%20&%20Highway%20Construction%20Products.pdf

American Iron and Steel Institute. (2016). *Modern Sewer Design*, 3rd ed., AISI, Washington, DC, Available at: http://www.ncspa.org/images/stories/technical/08blbk1.pdf, Accessed on April 28, 2016.

American Society of Civil Engineers. (1998). *Pipeline Route Selection for Rural and Cross-Country Pipelines*, ASCE Manuals and Reports on Engineering Practice No. 46, Reston, Va.

American Society of Civil Engineers. (2002). Standard Guidelines for the Collection and Depiction of Utility Data, Standard CI/ASCE 38-02, ASCE, Reston, Va.

American Society of Civil Engineers. (2004). *Horizontal Auger Boring Projects*, ASCE Manuals and Reports on Engineering Practice No. 106, Reston, Va.

American Society of Civil Engineers. (2005). *Pipeline Design for Installation by Horizontal Directional Drilling*, ASCE Manuals and Reports on Engineering Practice No. 108, Reston, Va.

American Society of Civil Engineers. (2007). *Geotechnical Baseline Reports for Underground Construction*, ASCE, Reston, Va.

American Water Works Association. (2014). *M45 Fiberglass Pipe Design, Third Edition*, AWWA, Denver, Colo.

American Water Works Association. (2004). *Manual of Water Supply Practices M11: A Guide for Design and Installation*, AWWA, Denver, Colo.

American Water Works Association. (2006). *Manual of Water Supply Practices M55: PE Pipe—Design and Installation*, AWWA, Denver, Colo.

American Water Works Association. (2014). *Manual of Water Supply Practices M28: Rehabilitation of Water Mains*, AWWA, Denver, Colo.

Association of Metropolitan Sewerage Agencies. (2002). *Managing Public Infrastructure Assets to Minimize Cost and Maximize Performance*, AMSA, Washington, D.C.

ASTM. (2006). "Standards Related to Trenchless Technology," 2nd ed. (Available in Print or CD-ROM), ASTM International, West Conshohocken, Pa.

Bennett, D. and Ariaratnam, S. T. (2008). *Horizontal Directional Drilling: Good Practices Guidelines*, 3rd ed., HDD Consortium, Arlington, Va.

Bizier, P. (ed.). (2007). *Gravity Sanitary Sewer Design and Construction*, American Society of Civil Engineers, Reston, Va.

Das, B. M. (1994). *Principles of Geotechnical Engineering*, PWS Publishing Company, Boston.

Deb, A. K., Snyder, J. K., Hammell, J. O., Tyler, E., Gray, L., and Warren, I. (2006). *Service Life Analysis of Water Main Epoxy Lining*, AWWA, Denver, Colo.

Federal Highway Administration. (2002). *Manual for Controlling and Reducing the Frequency of Pavement Utility Cuts*, Federal Highway Administration, Publication No. FHWA-IF-02-064, October.

Fisk, E. R. and Rapp, R. R. (2004). *Introduction to Engineering Construction Inspection*, Wiley, Hoboken, N.J.

GASB Statement 34. (1999). "Basic Financial Statements and Management's Discussion and Analysis—for State and Local Governments," Government Accounting Standard Series, No. 171-A, June, Norwalk, Conn.

Goodman, A. S. and Hastak, M. (2006). *Infrastructure Planning Handbook: Planning, Engineering, and Economics*, American Society of Civil Engineers, Reston, Va.

Heald, C. C. (1994). *Cameron Hydraulic Data*, 18th ed., Ingersoll-Dresser Pump Co., Liberty Corner, N.J.

Hovland, T. J. and Najafi, M. (eds.). (2009). *Inspecting Pipeline Installation*, American Society of Civil Engineers, Reston, Va.

Howard, A. (1996). *Pipeline Installation*, Relativity Publishing, Lakewood, Colo.

Howard, A. (2015). *Pipeline Installation*, Relativity Publishing, Lakewood, Colo.

Hughes, J. B. (ed.). (2009). *Manhole Inspection and Rehabilitation*, American Society of Civil Engineers, Reston, Va.

Hughes, D. (2002). *Assessing the Future: Water Utility Infrastructure Management*, American Water Works Association, Denver, Colo.

Iseley, D. T. and Gokhale, B. S. (1997). *Synthesis of Highway Practice 242: Trenchless Installations of Conduits beneath Roadways*, Transportation Research Board, National Research Council, Washington, D.C.

Iseley, D. T. and Najafi, M. (eds.). (1995). *Trenchless Pipeline Rehabilitation*, National Utility Contractors Association, Arlington, Va.

Iseley, D. T., Najafi, M., and Tanwani, R. (eds.). (1999). *Trenchless Construction Methods and Soil Compatibility Manual*, National Utility Contractors Association, Arlington, Va.

Janson, L. E. (2003). *Plastic Pipes for Water Supply and Sewage Disposal*, 4th ed., Stockholm, Sweden.

Kramer, R. S., McDonald, J. W., and Thomson, C. J. (1992). *An Introduction to Trenchless Technology*, Chapman & Hall, New York.

Liu, H. (2003). *Pipeline Engineering*, CRC Press, New York.

Meyer, E. G. (2005). *Polyethylene Pipes in Applied Engineering*, Brussels, Belgium.

Mohitpour, M. (2008). *Energy Supply and Pipeline Transportation: Challenges and Opportunities*, American Society of Mechanical Engineers, New York.

Morgan, K. C. (2012). "Managing Water Main Breaks Field Guide," American Water Works Association, Denver, Colo.

Moser, A. P. and Folkman, S. (2008). *Buried Pipe Design*, 3rd ed., McGraw-Hill, New York.

Najafi, M. (1994). "Trenchless Pipeline Rehabilitation: State-of-the-Art Review," Results of a two-year research conducted jointly by the U.S. Army Engineers–Waterways Experiment Station and the Trenchless Technology Center, Louisiana Tech University.

Najafi, M. (ed.). (2003). "New Pipeline Technologies, Security and Safety, Two-Volume," *Proceedings of International Pipeline Conference*, July 13–16, in Baltimore, Md., American Society of Civil Engineers, Reston, Va.

Najafi, M. (ed.). (2007). *Pipe Bursting Projects*, ASCE Manuals and Reports on Engineering Practice, No. 112, Reston, Va.

Najafi, M. (ed.). (2008). *Pipe Ramming Projects*, ASCE Manuals and Reports on Engineering Practice, No. 115, Reston, Va.

Najafi, M. and Perez, M. (Eds.) (2016). *Renewal of Potable Water Pipes*, ASCE Manuals and Reports on Engineering Practice, No. 132. Reston, Va.

Najafi, M. (2013). *Trenchless Technology: Planning, Equipment and Methods*, McGraw-Hill, New York.

Najafi, M. (2010). *Trenchless Technology Piping: Installation and Inspection*, McGraw-Hill, New York.

Najafi, M. and Gokhale, S. (2005). *Trenchless Technology: Pipeline and Utility Design, Construction and Renewal*, McGraw-Hill, New York.

Najafi, M. and Gokhale, S. (2005). *Trenchless Technology: Pipeline and Utility Design, Construction and Renewal*, McGraw-Hill, New York.

Najafi, M. and Rush, J. (eds.). (2008). *Guide to Water and Wastewater Asset Management, Benjamin Media Inc.*, Underground Infrastructure Management (UIM), Peninsula, Ohio.

National Association of Sewer Service Companies. (2001). *Pipeline Assessment and Certification Program*, NASSCO, Baltimore, Md.

National Association of Sewer Service Companies. (2006). *Jetter Code of Practice: A Guide for Selection and Operation of Sewer Jetter Equipment and the Selection of Nozzles for Different Applications*, NASSCO, Baltimore, Md.

National Association of Sewer Service Companies. (2008). *Inspector Training Certification Program for the Inspector of Cured-in-Place Pipe Installation, Version 2.0*, NASSCO, Baltimore, Md.

National Clay Pipe Institute. (2003). *Clay Pipe Engineering Manual*, NCPI, Lake Geneva, Wis.

National Utility Contractors Association. (2002). *NUCA Safety Manual, Fifth Revision*, NUCA, Arlington, Va.

Osborn, L. E. and Najafi, M. (eds.). (2010). *Trenchless Renewal of Culverts and Storm Sewers*, American Society of Civil Engineers, Reston, Va.

Pipe Jacking Association. (1995a). *Guide to Best Practice for the Installation of Pipe Jacks and Microtunnels*, Pipe Jacking Association, London.

Pipe Jacking Association. (1995b). *A Guide to Pipe Jacking and Microtunneling Design*, PJA, National Utility Contractors Association, London.

Plastics Pipe Institute. (2008). *Handbook of Polyethylene Pipe*, 2nd ed., Plastics Pipe Institute, Irving, Tex.

Public Works Standards, Inc. (2006). *Greenbook: Standard Specifications for Public Works Construction*, BNi Publications, Inc., Vista, Calif.

Rameil, M. (2007). *Handbook of Pipe Bursting Practice*, Vulkan-Verlag Gmbh, Germany.

Reinholtz, J. D. (ed.) (2006). *OSHA Standards for the Construction Industry*, CCH, Chicago, Ill.

Sarrgand, S. M., Mitchell, G. F., and Hurd, J. O. (eds.). (1993). *Structural Performance of Pipes 93*, A.A. Balkema, Brookfield, Vt.

Slavin, L. M. (ed.). (2009). Belowground Pipeline Networks for Utility Cables, Manual of Practice (MOP) 118, American Society of Civil Engineers, Reston, Va.

Stein, D. (2005). *Trenchless Technology for Installation of Cables and Pipelines*, Becker Druck, Arnsberg, Germany.

Stein, D. (2005). "Practical Guideline for the Application of Microtunneling Methods for the Ecological, Cost-Minimized Installation of Drains and Sewers." Stein und Partner, Bochum, Germany.

Sterling, R. L. (2000). "Utility Locating Technologies: A Summary of Responses to a Statement of Need." Report Prepared by the Trenchless Technology Center (TTC) for the Federal Laboratory Consortium for Technology Transfer, Washington, D.C.

Watkins, R. K. and Anderson, L. R. (2000). *Structural Mechanics of Buried Pipes*, CRC Press, Boca Raton, Fla.

Whidden, W. R. (ed.). (2009). *Buried Flexible Steel Pipe: Design and Structural* Analysis, Manual of Practice (MOP) No. 119, American Society of Civil Engineers (ASCE), Reston, Va.

WRc. (2013). Manual of Sewer Condition Classification, 5th ed., (MSCC), Blagrove, Swindon, Wiltshire, UK.

Publication Sources

American Concrete Pipe Association (www.concrete-pipe.org)

American Concrete Pressure Pipe Association (www.acppa.org)

American Gas Association (www.aga.org)

American Society of Civil Engineers (ASCE) Pipeline Division Annual Conference *Proceedings (www.asce.org)* and www.pipelinedivision.org

ASTM International (www.astm.org)

AWWA publications including AWWA Water Industry Database (WIDB) (www.awwa.org)
Ductile Iron Pipe Research Association (www.dipra.org)
Gas Technology Institute (www.gastechnology.org)
Louisiana Tech University Trenchless Technology Center (www.ttc.latech.edu)
National Association of Sewer Service Companies (NASSCO) (www.nassco.org)
National Utility Contractors Association (www.nuca.com)
National Corrugated Steel Pipe Association (www.ncspa.com)
North American Society for Trenchless Technology (www.nastt.org)
Occupational Safety & Health Administration (OSHA) (www.osha.gov)
Pipeline and hazardous Materials Safety Administration (PHMSA) (www.phmsa.dot.gov)
Plastic Pipe Institute (PPI) (www.plasticpipe.org)
Uni-Bell PVC Pipe Association (www.uni-bell.org)
Underground Construction Technology (UCT) and Exhibition (www.uctonline.com)
U.S. Environmental Protection Agency (USEPA) (www.epa.gov)
Water Environment Federation (www.wef.org)
Water Research Foundation (WaterRF) (www.waterrf.org)

Technical Journal

Journal of Pipeline Systems Engineering and Practice (*JPS*)
American Society of Civil Engineers (ASCE)
(www.edmgr.com/jrnpseng/)

Magazines and Newsletters

Civil Engineering
(www.asce.org/cemagazine)

Cost Engineering
The Journal of AACE International
(www.aacei.org)

Currently @ CUIRE
(www.cuire.org)

Governing
The States and Localities
(www.governing.com)

Journal
American Water Works Association
(www.awwa.org)

Municipal Sewer & Water
For Sanitary, Storm and Water Maintenance Professionals
(www.mswmag.com)

NASSCO Times
Setting the Industry Standards for the Rehabilitation of Underground Utilities
(www.nassco.org)

Opflow
(www.awwa.org/opflow)

Pipeline & Gas Journal
Pipeline News
(www.oildom.com)

Pollution Equipment News
(www.rimbach.com)

Shotcrete Magazine
(www.shotcrete.org)

Standardization News
(www.astm.org/standardization-news)

Trenchless International
The Official Magazine of the International Society for Trenchless Technology (ISTT)
(www.trenchlessinternational.com)
(www.istt.com)

Trenchless Today
North American Society for Trenchless Technology (NASTT)
(www.nastt.org)

TR News
Transportation Research Board
(www.trb.org)

Trenchless Technology
(www.trenchlessonline.com/)

Tube Products International
The World of Tube & Pipe Products, Materials & Ancillaries
(www.intras.co.uk)

Tunnels & Tunneling International
(www.tunnelsandtunneling.com)

UIM
Water Utility Infrastructure Management
(www.uimonline.com)

Underground Construction
(www.undergroundconstructionmagazine.com)

Underground Utilities
(www.uumagazine.com)

Utility Contractor
National Utility Contractors Association (NUCA)
(www.nuca.com)

WE&T
Water Environment and Technology
(www.wef.org)

World Tunneling/Trenchless World
(www.trenchless-world.com)

APPENDIX E

Conversion Table

From	To	Multiply by
Length		
mil	in.	0.001
in.	mm	25.4
ft	m	0.3048
yd	m	0.9144
ft	in.	12
mi	km	1.609
mi	ft	5280
Area		
in.^{2}	mm^2	645.16
ft^2	m^2	0.0929
yd^2	m^2	0.8361
mi^2	km^2	2.5889
acres	km^2	4.0469×10^{-3}
ha	km^2	0.01
acres	ft^2	43,560
m^2	ha	10^4
Pressure		
psi	kN/m^2	6.895
psi	atm	0.0680
psi	kg/m^2	9.80665
psi	Pa	6894.757
psi	lb/ft^2	144
N/m^2	Pa	1
atm	psi	14.696
atm	Pa	101.325
bar	psi	14.5

From	To	Multiply by
Specific Weight (Weight/Unit Volume)		
lb/ft^3	N/m^3	157.1
lb/in^3	lb/ft^3	1728
Forces		
lb	kN	0.004448
tons	kN	9.96401
kN	kgf	102.0
Velocity		
mi/h	km/h	1.609
ft/s	m/s	0.3048
ft/min	m/s	0.00508
ft/min	m/min	0.305
Volume Flow Rate		
ft^3/s	gal/min	449
m^3/s	ft^3/s	35.3
m^3/s	gal/min	15,850
gal/min	L/min	3785
m^3/s	L/min	60,000
m^3/s	ft^3/min	2120
m^3/h	L/min	16.67
ft^3/s	m^3/h	101.9
Temperature Conversion		
°F to °C: Subtract 32, multiply by 5, then divide by 9.		
°C to °F: Multiply by 9, divide by 5, add 32.		

APPENDIX F
Members of Center for Underground Infrastructure Research and Education (CUIRE)

Abu Abraham, P.E.
Technical Manager
Insituform Technologies17999 Edison Avenue
Chesterfield, MO 63005
Phone: 636-530-2872
Fax: 205-307-2799
Cell: 314-623-3273
Email: aabraham@insituform.com

Sam Arnaout, P.E.
Sr. Vice President of Engineering
Forterra Pressure Pipe
1003 N. MacArthur Blvd.
Grand Prairie, TX 75050-7943
Phone: 972-262-3600
Fax: 972-264-6236
Email: sam.arnaout@forterrabp.com

Robert A. Carpenter
Editor, Underground Construction Magazine
1160 Dairy Ashford, Suite 610
Houston, TX 77079
Phone: 281-558-6930
Fax: 281-558-7029
Email: rcarpenter@oildom.com

Walter Chiang, P.E.
Founder & Director
CP&Y Inc.
1820 Regal Row, #200
Dallas, TX 75235
Phone: 214-638-0500
Email: wchiang@cpyi.com

Frank Crumb, P.E.
Public Infrastructure Business Development
Halff Associates, Inc.
4000 Fossil Creek Blvd.
Fort Worth, TX 76137-2720
Phone: 817-764-7448
Cell: 817-480-4485
Email: fcrumb@halff.com

George H. Davis, R.G.
Geotechnical Liaison
Missouri Department of Transportation
P.O. Box 270
Jefferson City, MO 65102
Phone: 573-526-4344
Fax: 573-526-4345
Email: george.davis@modot.mo.gov

Hassan A. Elsaad, P.E.
Engineer Principal
Salt River Project
P.O. Box 52025
Phoenix, AZ 85072-2025
Phone: 602-236-4619
Email: hassan.elsaad@srpnet.com

Alan Goodman
Global Sales Manager
500 South C.P. Avenue
Lake Mills, Wisconsin 53551
Phone: 281-642-6854
Email: agoodman@hammerheadmole.com

Bruce Hollands
Uni-Bell PVC Pipe Association
2711 LBJ Freeway, Suite 1000
Dallas, TX 75234
Phone: 972-243-3902
Email: bhollands@uni-bell.org

Julia J. Hunt, P.E.
Assistant Regional Manager/Operations
Trinity River Authority of Texas
P.O. Box 60
Arlington, TX 76004
Phone: 817-493-5100
Email: huntj@trinityra.org

Larry Johnson
Vice President, Sales
Hobas
1413 E. Richey Road
Houston, TX 77073
Phone: 281-821-2200
Email: ljohnson@hobaspipe.com

Therese R. Kline, P.E.
Flexible Pipe Specialist
Special Structures, Bridge Design
Michigan Department of Transportation
P.O. Box 30050
Lansing, MI 48909
Phone: 517-241-0082
Email: KlineT@michigan.gov

Chad Kopecki, P.E.
Assistant Director
Public Works Department, City of Dallas
320 E. Jefferson Room 305
Dallas, TX 75203
Phone: 214-948-4259
Email: chad.kopecki@dallascityhall.com

David Marshall, P.E.
Engineering Services Director
Tarrant Regional Water District (TRWD)
808 East Northside Drive
Fort Worth, TX 76102
Phone: 817-720-4250
Email: david.marshall@trwd.com

Lynn Osborn, P.E.
Technical Advisor, NASSCO
LEO Consulting, LLC
2470 Longstone Lane, Suite M
Marriottsville, MD 21104
Phone: 410-442-7473
Cell: 314-378-1275
Email: lynn.osborn14@gmail.com

Anna Pridmore, Ph.D., P.E.
Vice President, Pipeline Solutions
Structural Technologies
1332 N Miller Street
Anaheim, CA 92806
Phone: 714-869-8824
Email: apridmore@structural.net

Camille George Rubeiz, P.E.
Director of Engineering
Plastic Pipe Institute
2839, Maple Lane
Fairfax, VA 22031
Phone: 469-499-1050
Cell: 703-472-4432
Email: crubeiz@plasticpipe.org

Jim Rush
Editor
Benjamin Media Inc. (Trenchless Technology)
1770 Main Street, P.O. Box 190
Peninsula, OH 44264
Phone: 330-467-7588
Fax: 330-468-2289
Email: jrush@benjaminmedia.com

Larry Slavin, Ph.D.
Principal
Outside Plant Consulting Services, Inc.
Rockaway, NJ 07866-1019
Phone: 973-983-0813
Fax: 973-983-0813
Email: lslavin@ieee.org

Harvey Svetlik
Staff Engineer—Water & Energy
Georg Fischer Central Plastics, LLC
GF Piping Systems
4949 Joseph Hardin Dr.
Dallas, TX 75236
Phone: 469-801-2225
Email: Harvey.Svetlik@georgfischer.com

Richard Williammee Jr., P.E.
District Materials Engineer
Fort Worth District
Texas Department of Transportation
P.O. Box 6868
Fort Worth, TX 76115
Phone: 817-370-6675
Fax: 817-370-6713
Email: Richard.Williammee@txdot.gov

References

Aggarwal, S. C. and Cooper, M. J. (1984). *External Pressure Testing of 'Insituform' Linings*, Internal Report, Coventry (Lanchester) Polytechnic.

Ambler, E., A, Thomas, B., and Matthews, J. C. (2014). "Casselberry Asbestos Cement Pipe Bursting Project: Moving Forward by Leaving it Behind." *North American Society for Trenchless Technology*, NASTT's 2014 No-Dig Show, April 13-17, 2014.

American Concrete Pipe Association. (2000). *Concrete Pipe Design Manual*, Irving, Tex.

American Society for Testing and Materials. (2006). "Standard Practice for Classification of Soils for Engineering Purposes (Unified Soil Classification System)." ASTM, D2487-06, West Conshohocken, Pa.

American Society for Testing and Materials. (2011). "Standard Practice for Underground Installation of Thermoplastic Pipe for Sewers and Other Gravity-Flow Applications." ASTM, D2321–89, West Conshohocken, Pa.

American Society for Testing and Materials. (2011). "Standard Practice for Underground Installation of Thermoplastic Pipe for Sewers and Other Gravity-Flow Applications." ASTM, D2321–11, West Conshohocken, Pa.

American Water Works Association. (2006). *Manual of Water Supply Practices M-55: PE Pipe-Design and Installation*, AWWA, Denver, Colo.

American Water Works Association. (2006). *Manual of Water Supply Practices C-950: Standard for Fiberglass Pressure Pipe*, AWWA, Denver, Colo.

ArcNews Magazine. (2005). "Edgecombe County, North Carolina, Maps and Inventories Water/Wastewater Infrastructure with GIS and GPS: Field Mapping Time Cut in Half." ArcNews Online, Summer. Available at: http://www.esri.com/news/arcnews/summer05articles/tofc-summer05.html, Accessed on February 15, 2016.

Atalah, A. (1998) "The Effect of Pipe Bursting on Nearby Utilities, Pavement, and Structures." *Technical Report TTC-98-01*, Trenchless Technology Center, Louisiana Tech University, Ruston, La.

Atalah, A. (2004). "The Ground Movement Associated with Large Diameter Pipe Bursting in Rock Conditions and Its Impact on Nearby Utilities and Structures." Bowling Green State University, Bowling Green, Ohio.

Clarke, N. W. (1968). *Buried Pipelines: A Manual of Structural Design and Installation*, Maclaren and Sons Ltd., London.

EPA. (1990). "40 CFR Part 61 Subpart M." Available at: http://www.epa.gov/asbestos/pubs/asbreg.html, Accessed on February 10, 2016.

ESRI. (2004). "ArcGIS 9, what is ArcGIS." www.esri.com, Available at: http://downloads.esri.com/support/documentation/ao_/698What_is_ArcGIS.pdf, Accessed on February 15, 2016.

ESRI. (2009). "Using GIS with GPS." www.esri.com, Available at: https://www.esri.com/library/bestpractices/using-gis-with-gps.pdf, Accessed on February 12, 2016.

Farshad, M. (2006). "Plastic Pipe Systems: Failure Investigation and Diagnosis." Elsevier, Maryland Heights, Mo.

Gabriel, L. H. (2006). *Corrugated Polyethylene Pipe Design Manual and Installation Guide*, Plastics Pipe Institute, Available at: http://drainage.plasticpipe.org/Resources.Design_Manual.asp, Accessed on April 10, 2016.

GAO. (2004). United States Government Accountability Office (GAO). Water Infrastructure: Comprehensive Asset Management Has Potential to Help Utilities Better Identify Needs and Plan Future Investments. United States General Accounting Office, Washington, D.C. Available at: http://www.gao.gov/products/GAO-04-461, Accessed on February 15, 2016.

GASB Statement 34. (1999). "Basic Financial Statements and Management's Discussion and Analysis for State and Local Governments." Government Accounting Standard Series, No. 171-A, June 1999, Norwalk, Conn.

Gumbel, J. E. (1998). "Structural design of pipe linings 1998-review of principles, practice and current developments worldwide." 4th ASTT Conference, Brisbane, Australia.

Hovland, T. J and Najafi, M. (Eds.). (2009). *Inspecting Pipeline Installation*, ASCE Manuals and Reports on Engineering Practice No. 117, Reston, Va.

Howard, A. (1996). "Pipeline Installation." Relativity Publishing, Lakewood, Colo.

Howard, A. (2015). "Pipeline Installation." Relativity Publishing, Lakewood, Colo.

Hu, Y., Wang, D., and Chowdhury, R. (2013). "Long Term Performance of Asbestos Cement Pipe." WRF Project No. 4093, Denver, Colo, 219 pp.

IIMM. (2006). *International Infrastructure Management Manual*, Association of Local Government Engineering NZ Inc. (INGENIUM), National Asset Management Steering (NAMS) Group, Thames, NZ.

International Pipe Bursting Association (IPBA). (2012). "Guideline for Pipe Bursting." National Association of Sewer Service Companies (NASSCO), Available at: http://www.ipbaonline.org/, Accessed April 28, 2016.

Investing Answers. (2013). "Earnings before Interest, Tax, Depreciation, and Amortization EBITDA." InvestingAnswers.com, Available at: http://www.investinganswers.com/financial-dictionary/financial-statement-analysis/earnings-interest-tax-depreciation-and-amortizatio, Accessed on February 15, 2016.

Investopedia. (2013). "Midstream" Definition of "Midstream." Investopedia.com/ask/answers, Available at: http://www.investopedia.com/terms/m/midstream.asp, Accessed on February 15, 2016.

Janson, L. E. (1999). "Plastics Pipes for Water Supply and Sewage Disposal." Borealis, Stockholm.

Khankarli, G. (2013). "CE 5389 Infrastructure Asset Management." Unpublished Lecture Materials, *The University of Texas at Arlington*, Arlington, Tex.

Kleweno, D. G. (1997). "Critical Buckling Behavior and Life Cycle Predictions of CIPP." *Conference Proceedings North American NO-DIG 1997*, NASTT, Seattle, Wash.

Lee, H., Najafi, M., and Matthys, J. (2007) "Cost Comparison of Pipeline Asset Replacement: Open-Cut and Pipe-Bursting." Pipelines 2007: pp. 1–11.

Lo, K. H. and Zhang, J. Q. (1994). "Collapse Resistance Modeling of Encased Pipes." *Buried Plastic Pipe Technology:* 2nd Volume, ASTM STP 1222, D Eckstein, Ed., 97- 110, American Society for Testing and Materials, Pa.

Makar, J. M. (2000). "A Preliminary Analysis of Failures in Grey Cast Iron Water Pipes." Engineering Failure Analysis 7, *Manual of Procedures for Pipe Construction (1998)*, Office of Highway Management, Ohio Department of Transportation, 43–53.

Makar, J. M., Desnoyers, R., and McDonald, S. E. (2001). "Failure Modes and Mechanisms in Gray Cast Iron Pipe," *International Conference on Underground Infrastructure Research: Municipal, Industrial and Environmental Applications*, Kitchener, Ontario, June 10–13, 2001, 1–10.

Massicotte, D. (2000). "Finite Element Calculations of Stresses and Deformations in Buried Flexible Pipes." *University of Ottawa*, Canada.

Matthews, J. C., Ryan S., and Jason L. (2015). Water Research Foundation, *Water Environment Research Foundation*, USEPA.

Moser, A. P. and Folkman, S. L. (2008). *Buried Pipe Design*, 3rd ed., McGraw-Hill, New York.

Moser, A. P. (2001). *Buried Pipe Design*, McGraw-Hill, New York.

Najafi, M. (2005). *Trenchless Technology: Pipeline and Utility Design, Construction and Renewal*, McGraw-Hill, New York.

Najafi, M. (2010). *Trenchless Technology Piping: Installation and Inspection*, McGraw-Hill, New York.

Najafi, M. (2013). *Trenchless Technology: Planning, Equipment, and Methods*, McGraw-Hill, New York.

Najafi, M. (Ed.). (2007). *Pipe Bursting Projects—ASCE Manuals and Reports on Engineering Practice No. 112*, American Society of Civil Engineers, Reston, Va.

Najafi, M. and Gokhale, S. (2005). *Trenchless Technology: Pipeline and Utility Design, Construction, and Renewal*, McGraw-Hill, New York.

National Corrugated Steel Pipe Association. (2000). *Corrugated Steel Pipe Design Manual*, NCSPA, Dallas, Tex.

NSF/ANSI 61-2013, Drinking Water System Components: Health Effects. (2013). *NSF International Standard*, Ann Arbor, Mich. Available at: https://www.nsf.org/newsroom_pdf/NSF_61-13_-_watermarked.pdf, Accessed on April 28, 2016.

O'Day, B. K., Weisss, R., Chiavari, S., and Blair, D. (1986). *Water Main Evaluation for Rehabilitation/Replacement*, American Water Works Association Research Foundation, Denver, Colo.

Petroff, L. J. (1995). "Installation Technique and Field Performance of HDPE Profile Pipe." Advances in Underground Pipeline Engineering II, 260–271.

PHMSA. (2016). "Regulations." Pipeline and Hazardous Materials Safety Administration, Available at: http://phmsa.dot.gov/regulations, Accessed on February 10, 2016.

PHMSA. (2013). Phmsa.dot.gov, Available at: http://phmsa.dot.gov/regulations, Accessed on February 10, 2016.

Plastics Pipe Institute (PPI), (2008). *Handbook of Polyethylene Pipe*, 2nd ed., Plastics Pipe Institute, Irving, Tex.

Project Management Institute Inc. (2004), *A Guide to the Project Management Body of Knowledge*, Project Management Institute Inc., Newtown Square, Pa.

Rajani, B. B. and Kleiner, Y. (2001). "Comprehensive Review of Structural Deterioration of Water Mains: Physically Based Models." Urban Water, Vol. 3, October, 151–164.

Rajani, B. B., Robertson, P. K., and Morgenstern, N. R. (1995). "Simplified Design Methods for Pipelines Subject to Transverse and Longitudinal Soil Movements." *Canadian Geotechnical Journal*, 32(2), 309-323.

Reddy, D. V. (2001). *Long-Term Performance of Buried High Density Polyethylene Plastic Piping*, Final Report, FDOT, Available at: http://www.dot.state.fl.us/structures/structuresresearchcenter/Final%20Reports/long%20term%20performance%20of%20buried%20high%20density%20polethylene%20plastic%20piping(inc).pdf, Accessed on April 28, 2016.

Rogers, C. (1995). *Ground Displacements Caused by Trenching and Pipe Bursting*, Loughborough University of Technology, UK.

Sever, V.F. and Barton D.B. (2009). "Adding Value to Sanitary Sewers." Utility Infrastructure Management (UIM), October 1, 2009, Available at: http://uimonline.com/index/archives?Issue=2009-10-01, Accessed on February 15, 2016.

Schrock, B. J. and Gumbel, J. (1997). "Pipeline Renewal 1997," Paper Presented at North American No-Dig 1997, Seattle, Wash.

Spangler, M. G. and Handy, R. L. (1982). *Soil Engineering*, 4th ed., Harper & Row, New York.

SPI Composites Institute. (1989). *Fiberglass Pipe Handbook*, Fiberglass Pipe Institute, New York.

Stier, B. (2000). *Shotcrete for Sewer Applications*, Presentation at the Midwest Society for Trenchless Technology (MSTT) Symposium, University of Missouri, St. louis, Miss. November 1-3.

Thomas, B. and Ambler, E. A. (2013). "Applicability of NESHAP to Rehabilitating Asbestos-Cement Pipelines." *North American Society for Trenchless Technology*, NASTT's 2013 No-Dig Show, March 3-7, 2013.

Timoshenko, S. P. and Gere, J. M. (1961). *Theory of Elastic Stability*, McGraw-Hill, New York.
Timoshenko, S. P. and Gere, J. M. (1988). *Theory of Elastic Stability*, 2nd ed., McGraw-Hill, New York.
Tiratsoo, J. N. H. (1999). "Cleaning, Inspection, Fitness-for-purpose." Pipeline Pigging Technology, 2nd ed., Woburn, Mass.
USEPA. (2009a). "Condition Assessment of Wastewater Collection Systems." United States Environmental Protection Agency. Available at: http://nepis.epa.gov/Exe/ZyPURL.cgi?Dockey=P1003ZQY.txt, Accessed on February 8, 2016.
USEPA. (2009b). "Condition Assessment of Ferrous Water Transmission and Distribution Systems State of Technology Review Report." United States Environmental Protection Agency. Available at: http://nepis.epa.gov/Exe/ZyPURL.cgi?Dockey=P1004Q38.txt, Accessed on February 8, 2016.
USEPA. (2010). "Report on Condition Assessment Technology of Wastewater Collection Systems." United States Environmental Protection Agency. Available at: http://nepis.epa.gov/Exe/ZyPURL.cgi?Dockey=P1008H44.txt, Accessed on February 8, 2016.
USEPA. (2012). "Condition Assessment Technologies for Water Transmission and Distribution Systems." United States Environmental Protection Agency. Available at: http://nepis.epa.gov/Exe/ZyPURL.cgi?Dockey=P100E3Y5.txt, Accessed on February 10, 2016.
Vaslestad, J., Johansen, T. H., and Holm, W. (1994). "Load Reduction on Buried Rigid Pipes; Load Reduction on Rigid Culverts beneath High Fills: Long-term Behavior of Flexible Large-span Culverts." *Directorate of Public Roads*, Norwegian Road Research Laboratory, Oslo, Norway.
Venkatesh, C., Paradkar, A., Hughes, D., and Najafi, M. (2016). "Preliminary Test Results of a New Structurally Improved PVC (iPVC) Water Pipe Material." *North American Society for Trenchless Technology (NASTT), 2016 No-Dig Show*, Dallas, Tex.
Von Aspern, K. (2009). "End of the Line." *Public Works Magazine*. Available at: http://www.pwmag.com/pipes/end-of-the-line.aspx, Accessed on April 28, 2016.
WERF. (2007). "Condition Assessment Strategies and Protocols for Water and Wastewater Utility Assets." Water Environment Research Foundation (WERF) 03-CTS-20CO, Available at: http://simple.werf.org/simple/media/documents/1_03-CTS-20CO_MWH_Final%20WERF%20Report_thru%20Appndx%20E.pdf, Accessed on April 28, 2016.
Young, O. C. and Trott, J. J. (1984). "Buried Rigid Pipes: Structural Design of Pipelines." Elsevier, London, UK.

Applicable Standards

ASME B-31.8-2014. Gas Transmission and Distribution Piping Systems, New York, NY.
ASTM D-3076 - 00(2010). Standard Test Methods for Effective Crimping on Outside Crimped Valves of Aerosol Containers. ASTM Vol. 15.10, West Conshohocken, Pa.
ASTM A-139-16 Standard Specification for Electric-Fusion (Arc)-Welded Steel Pipe (NPS 4 and Over). ASTM Vol. 01.01, West Conshohocken, Pa.
ASTM A-36-14 Standard Specification for Carbon Structural Steel. ASTM Vol. 01.04.
ASTM A-716 - 08(2014). Standard Specification for Ductile Iron Culvert Pipe. ASTM Vol. 01.02, West Conshohocken, Pa.
ASTM A-746 - 09(2014). Standard Specification for Ductile Iron Gravity Sewer Pipe. ASTM Vol. 01.02, West Conshohocken, Pa.
ASTM C-1208-11 Standard Specification for Vitrified Clay Pipe and Joints for Use in Microtunneling, Sliplining, Pipe Bursting, and Tunnels. ASTM Vol. 01.02, West Conshohocken, Pa.
ASTM C-14 - (2015). Standard Specification for Reinforced Concrete Sewer, Storm Drain, and Culvert Pipe. ASTM Vol. 04.05, West Conshohocken, Pa.

ASTM C-361 - 08(2008). Standard Specification for Reinforced Concrete Low-Head Pressure Pipe. ASTM Vol. 04.05, West Conshohocken, Pa.
ASTM C-425 - 04(2013). Standard Specification for Compression Joints for Vitrified Clay Pipe and Fittings. ASTM Vol. 04.05, West Conshohocken, Pa.
ASTM C-443-12 Standard Specification for Joints for Concrete Pipe and Manholes, Using Rubber Gaskets. ASTM Vol. 04.05, West Conshohocken, Pa.
ASTM C-655-15 Standard Specification for Reinforced Concrete D-Load Culvert, Storm Drain, and Sewer Pipe. ASTM Vol. 04.05, West Conshohocken, Pa.
ASTM C-76-15a Standard Specification for Reinforced Concrete Culvert, Storm Drain, and Sewer Pipe. ASTM Vol. 04.05, West Conshohocken, Pa.
ASTM C-990 - 09(2014). Standard Specification for Joints for Concrete Pipe, Manholes, and Precast Box Sections Using Preformed Flexible Joint Sealants. ASTM Vol. 04.05, West Conshohocken, Pa.
ASTM D-1598-15a Standard Test Method for Time-to-Failure of Plastic Pipe under Constant Internal Pressure. ASTM Vol. 08.04, West Conshohocken, Pa.
ASTM D-2412-11 Standard Test Method for Determination of External Loading Characteristics of Plastic Pipe by Parallel-Plate Loading. ASTM Vol. 08.04, West Conshohocken, Pa.
ASTM D-2837-13e1 (2013). Standard Test Method for Obtaining Hydrostatic Design Basis for Thermoplastic Pipe Materials or Pressure Design Basis for Thermoplastic Pipe Products. ASTM Vol. 08.04, West Conshohocken, Pa.
ASTM D-2990-09 Standard Test Methods for Tensile, Compressive, and Flexural Creep and Creep-Rupture of Plastics. ASTM Vol. 08.01, West Conshohocken, Pa.
ASTM D-2992-12 Standard Practice for Obtaining Hydrostatic or Pressure Design Basis for "Fiberglass" (Glass-Fiber-Reinforced Thermosetting-Resin) Pipe and Fittings. ASTM Vol. 08.03, West Conshohocken, Pa.
ASTM D-3262-16 Standard Specification for "Fiberglass" (Glass-Fiber-Reinforced Thermosetting-Resin) Sewer Pipe. ASTM Vol. 08.03, West Conshohocken, Pa.
ASTM D-3350-14 Standard Specification for Polyethylene Plastics Pipe and Fittings Materials. ASTM Vol. 08.02, West Conshohocken, Pa.
ASTM D-3517-14 Standard Specification for "Fiberglass" (Glass-Fiber-Reinforced Thermosetting-Resin) Pressure Pipe. ASTM Vol. 08.03, West Conshohocken, Pa.
ASTM D-4258 - 05(2012). Standard Practice for Surface Cleaning Concrete for Coating. ASTM Vol. 06.02
ASTM D-4259 - 88(2012). Standard Practice for Abrading Concrete. ASTM Vol. 06.02, West Conshohocken, Pa.
ASTM D-4414 - 95(2013). Standard Practice for Measurement of Wet Film Thickness by Notch Gages. ASTM Vol. 06.01, West Conshohocken, Pa.
ASTM D-4541-09e1 Standard Test Method for Pull-Off Strength of Coatings Using Portable Adhesion Testers. ASTM Vol. 06.02 West Conshohocken, Pa.
ASTM D-7234-12 Standard Test Method for Pull-Off Adhesion Strength of Coatings on Concrete Using Portable Pull-Off Adhesion Testers. ASTM Vol. 06.02 West Conshohocken, Pa.
ASTM F-1216-09 Standard Practice for Rehabilitation of Existing Pipelines and Conduits by the Inversion and Curing of a Resin-Impregnated Tube. ASTM Vol. 08.04, West Conshohocken, Pa.
ASTM F-1743-08 Standard Practice for Rehabilitation of Existing Pipelines and Conduits by Pulled-in-Place Installation of Cured-in-Place Thermosetting Resin Pipe (CIPP). ASTM Vol. 08.04, West Conshohocken, Pa.
ASTM F-2561-11 Standard Practice for Rehabilitation of a Sewer Service Lateral and Its Connection to the Main Using a One Piece Main and Lateral Cured-in-Place Liner. ASTM Vol. 08.04, West Conshohocken, Pa.
ASTM F-2620-13 Standard Practice for Heat Fusion Joining of Polyethylene Pipe and Fittings. ASTM Vol. 08.04, West Conshohocken, Pa.
ASTM F-3097-15 Standard Practice for Installation of an Outside Sewer Service Cleanout through a Minimally Invasive Small Bore Vacuum Excavation. ASTM Vol. 08.04. West Conshohocken, Pa.
AWWA C-104/A-21.4 ANSI Standard for Cement-Mortar Lining for Ductile-Iron Pipe and Fittings for Water, American Water Works Association, Denver, Col.

AWWA C-105-10 Polyethylene Encasement for Ductile-Iron Pipe Systems, American Water Works Association, Denver, Col.

AWWA C-116-09 Protective Fusion-Bonded Epoxy Coatings for the Interior and Exterior Surfaces of Ductile-Iron and Gray-Iron Fittings, American Water Works Association, Denver, Col.

AWWA C-150-08 Thickness Design of Ductile-Iron Pipe, American Water Works Association, Denver, Col.

AWWA C-151-09 Ductile-Iron Pipe, Centrifugally Cast, American Water Works Association, Denver, Col. American Water Works Association, Denver, Col.

AWWA C-200-12 Steel Water Pipe 6 in. (150 mm) and Larger, American Water Works Association, Denver, Col.

AWWA C-300-11 Reinforced Concrete Pressure Pipe, Steel-Cylinder Type, American Water Works Association, Denver, Col.

AWWA C-301-14 Prestressed Concrete Pressure Pipe, Steel-Cylinder Type, American Water Works Association, Denver, Col.

AWWA C-302-11 Design of Prestressed Concrete Cylinder Pipe, American Water Works Association, Denver, Col.

AWWA C-303-08 Concrete Pressure Pipe, Bar-Wrapped, Steel-Cylinder Type, American Water Works Association, Denver, Col.

AWWA C-304-07 Design of Prestressed Concrete Cylinder Pipe, American Water Works Association, Denver, Col.

AWWA C-900-07 Polyvinyl Chloride (PVC) Pressure Pipe and Fabricated Fittings, 4 in. Through 12 in. (100 mm Through 300 mm), for Water Transmission and Distribution, American Water Works Association, Denver, Col.

AWWA C-909-09 Molecularly Oriented Polyvinyl Chloride (PVCO) Pressure Pipe, 4 in. Through 24 in. (100 mm through 600 mm) for Water, Wastewater, and Reclaimed Water Service, American Water Works Association, Denver, Col.

AWWA C-950-13 Fiberglass Pressure Pipe., American Water Works Association, Denver, Col.

ICRI 03732 Technical Guideline for Selecting and Specifying Concrete Surface Preparation for Coatings, Sealers, and Polymer Overlays.

NACE No. 6/ SSPC SP-13 Standard Requirements for Surface Preparation of Concrete by Mechanical, Chemical, or Thermal Methods Prior to the Application of Bonded Protective Coating or Lining Systems.

Index

Note: Page numbers followed by *f* denote figures and *t* denote tables.

AASHTO. *See* American Association of State Highway and Transportation Officials
Abrasion, 74–75, 96, 222, 395
Abrasion resistance, 226, 234, 238, 250, 260, 308
AC. *See* Asbestos cement
Access shafts, 396, 423
ACI. *See* American Concrete Institute
ACM. *See* Asbestos-containing material
Acoustic correlation, for leak detection, 182–184
Acoustic monitoring systems, for PCCP, 157–158, 159*f*
Acoustic technologies. *See also* Sonar; Ultrasound
  for condition assessment, 156–159, 159*f*, 159*t*
  in-line leak detectors with, 184–185, 185*f*
  for leak detection, 182–184
Acrylamide grout, 474–475
Acrylate grout, 475
Acrylic grout, 475
Adhesion, 417, 466
  cathodic protection and, 74
  from epoxy, 399
  manholes and, 432, 449
  of SIPP, 401, 411, 417
Aggerwal, S. C., 111
Air inversion, for CIPP, 288
AM. *See* Asset management
American Association of State Highway and Transportation Officials (AASHTO), 35, 224
American Concrete Institute (ACI), 412
American Iron and Steel Institute, "Modern Sewer Design" of, 266
American Recovery and Reinvestment Act, 381
American Society for Testing of Materials (ASTM):
  A-139, 267
  A-716, 262
  A-746, 262
  C-14, 224, 226*t*
  C-76, 224, 226*t*
  C-361, 226*t*
  C-443, 227
  C-1208, 234
  C12, 235*t*
  C301, 235*t*
  C425, 234, 236–238
  C700, 235*t*
  C828, 235*t*
  C896, 235*t*
  C1091, 235*t*
  C1208, 235*t*, 236
  D-2837, 243
  D-4541, 417
  D-7234, 417
  D1598-15a, 30
  D2412, 110–111
  D2487-06, 205
  D2990-09, 30
  D2992-12, 30
  D4414, 416
  D2837013e1, 30
  F-1743, 117–118
  F-2620, 256
  F585, 318
  F1216-09, 25, 27, 29, 30, 96, 107–108, 117–118, 294

American Society for Testing of Materials (*Cont.*):
F2561-11, 457
on flexible pipe, 200–201, 203
on HDPE, 252
on PCP, 226
on PE, 252
on shotcrete, 412
American Society of Civil Engineers (ASCE), 29, 96, 344
American Standard Code for Information Interchange (ASCII), 193
American Water Works Association (AWWA), 64
on AC, 233
C-300, 225, 226*t*
C-301, 224, 226*t*
C-302, 224, 226*t*
C-303, 222, 225
C-900, 246–249
C-905, 246–249
C-950, 117–118, 259
C304, 224
C900-07, 245
C909-09, 245
on condition assessments, 139
M55, 68
AMP. *See* Asset management plan
Annular space, 29, 223, 305
manholes and, 446
MSL and, 15, 396
service laterals and, 453, 455*f*
API 1104, 74
Appendices, in AMP, 44–45
Aqua phones, 157
ArcGIS, 77, 147
Arching effect, of soils, 208–210, 209*f*
ArcPad, 146, 148
ArcWISE, 193, 193*f*
As-built drawings, for PB ILR, 372
Asbestos cement (AC), 201, 231–233
advantages and limitations of, 233
condition assessment and, 52
future handling of, 383
HDPE and, 383
joints for, 232, 233*f*
manufacturing of, 232
for PB ILR, 329, 331, 378–383
Portland cement in, 222
PR-PR for, 340
as semi-rigid pipe, 223*n*1
for water distribution systems, 222
water tightness of, 232
Asbestos-containing material (ACM), 381–382
ASCE. *See* American Society of Civil Engineers
ASCII. *See* American Standard Code for Information Interchange
ASME B31.8, 68
Asset management (AM), 35–91. *See also* Condition assessment
benefits of, 36–37
GASB-34 and, 45–52
GASB-34-modified approach and, 47
for oil and natural gas-related pipelines, 63–90
questions on, 37, 38*t*
risk management and, 52–63
service life extension and, 9–10, 9*f*
sustainability and, 56, 56*f*
Asset management plan (AMP), 37–45, 44*f*
appendices in, 44–45
asset management practices in, 43
background data in, 41
disposal plan in, 42
executive summary in, 39
financial summary in, 43
future demand in, 40–41
improvement monitoring in, 44
introduction to, 40
lifecycle management in, 41, 42*f*
LOS in, 40
new construction in, 42
PACP® and, 54
references in, 44–45
renewal and replacement in, 42
routine maintenance in, 41
TAMP, 194–195, 194*f*, 195*f*
Asset registry, for condition assessment, 142
ASTM C-655, 226*t*
ASTM. *See* American Society for Testing of Materials
AWWA. *See* American Water Works Association

Backfill, 2, 141*t*, 203*t*, 207*t*
arching effect and, 208–210
OC and, 107, 200
Background data, in AMP, 41
Backreaming tool, 340
Bacteria prevention, 65, 69, 271, 342, 446

Bar-wrapped steel-cylinder concrete pipe, 225
Battelle, 382
Beam failure, 4, 227, 231, 239
Bedding, 203*t*
- arching effects and, 209
- beam failure and, 239
- circumferential breaks from, 4
- concrete pipe and, 231
- in condition assessment, 141*t*
- flexible pipe deflection and, 111
- OC and, 200
- rigid pipe and, 211, 214

Bell-and-spigot joints, 227, 267, 310
- for GRP, 259–260, 259*f*
- for PVC, 239–240

BEM. *See* Broadband electromagnetic
Bench, of manholes, 440–441, 441*f*
Bending stress, 4, 124–125, 215
Bentonite, 325, 332, 338, 362–363, 475
Bidding form, for PB ILR, 359
Blisters, in flexible pipe pipeline failure, 214
Boulders, 205
Breakaway connector, for continuous sliplining, 313
Breaks, 2, 25
- in water mains, 188*t*

Brittle failure, 224, 226, 233, 367
Broadband electromagnetic (BEM), for water distribution system condition assessment, 180–181, 181*f*
Buckling, 29, 296*n*2
- CIPP and, 111
- of flexible pipe, 214, 218–219, 219*f*
- OC and, 200
- SL and, 316

Buckling equations, 30, 110, 111–112
Built-in pipeline defects, 2–4, 3*t*
Buoyancy correction factor, 122, 131*t*
Butt-fused joints:
- for continuous sliplining, 310
- for HDPE, 257, 326*f*
- for PB ILR, 360
- for PE, 360
- for PVC, 246–249
- for SL, 326*f*

Butt-welded joints, for steel pipe, 267, 267*f*
Bypass pumping:
- for CIPP, 290–293, 293*f*
- for PB ILR, 346–347, 363–364

C3A. *See* Tricalcium aluminate
CAD. *See* Computer-aided design
Calcium aluminate, 19, 398, 431, 446
Caliper pig, for oil and natural gas pipelines, 76, 87–88, 88*f*
Camera-packer method, for service lateral grouting, 462
Cameras. *See also* Closed-circuit television
- for condition assessment, 149–156, 157*t*
- Scout, 404

Canadian Standards Association (CSA), 246
Capacity, management, operation, and maintenance (CMOM), 49*t*, 455
Capital and Operations and Maintenance Program (O&M), 76, 88, 196
Capital assets inventory, AM for, 36
Casselberry, Florida, PB ILR in, 378–383
Cast-in-place:
- concrete, gamma-gamma logging for, 162
- for SMR, 21

Cast iron pipe, 262
- brittle failure of, 367
- circumferential failure in, 216*f*
- condition assessment and, 52
- corrosion protection for, 16
- longitudinal bending in, 215
- longitudinal breaks in, 216*f*
- longitudinal cracking in, 215
- for PB ILR, 329
- spiral cracking in, 215, 217*f*
- for water distribution systems, 263
- for water mains, 188*t*

Catastrophic failures, 4
- AM for, 36
- manholes and, 432

Catchment numbering, 144–145
Cathodic protection, 74–75, 76, 265*t*
Cavitation, 2
CCFRPM. *See* Centrifugally cast glassfiber reinforced polymer mortar
CCTV. *See* Closed-circuit television
Cement-based pipe. *See also* Asbestos cement; Concrete pipe
- construction of, 222–233

Cement mortar lining (CML), 16–19, 271

Cementitious materials:
  for grout, 477–479
  I/I problems and, 398
  for manholes, 431*f*, 443*f*
  for polymer linings, 18
  for sanitary sewer pipe, 398
  for SIPP, 398, 411–414
  for SMR, 21
  for water distribution system, 398
Centrifugal casting:
  of cementitious materials for manholes, 443, 443*f*
  for concrete pipe, 223
  of GRP, 258
Centrifugally cast glassfiber reinforced polymer mortar (CCRPM), 311–312
CFP. *See* Close-fit pipe
Channeline, 386, 387*f*
Chemical gel grouts, for LOR, 474
Chemical grouting, 439, 456, 461–462, 470–474
Chemical resistance, 395, 399
Chemical treatment, for oil and natural gas pipelines, 76
Chezy-Manning equation, 342, 343
CIPP. *See* Cured-in-place pipe
Circumferential breaks from, 4
Circumferential failure, 4, 216*f*
Clarke, N. W., 214
Clay, 205
  PB ILR and, 331
Clay pipe. *See also* Vitrified clay pipe
  ring flexure in, 215
Clean Air Act of 1970, 234, 381
Clean Water Act of 1977, I/I problems and, 398
Cleanout caps, broken, 25
Clogging:
  in flexible pipe pipeline failure, 214
  of natural gas pipelines, 86
Close-fit pipe (CFP), 20, 20*t*, 21*f*, 419–427
  advantages of, 425–426
  applications of, 420–421
  characteristics of, 419–420
  construction of, 421–425
  cost experience indicators for, 104*t*
  DD, 20, 20*t*
  D&R, 419, 420*f*, 421*f*, 426
  existing pipe and, 95–96, 97
  F&F, 20, 20*t*, 419, 420*f*
  heating process for, 424
Close-fit pipe (*Cont.*):
  inspection for, 422–425
  limitations of, 103*t*, 426
  preparatory work for, 423
  QA/QC for, 425–426
  RD for, 419
  for service laterals, 456, 461
  service life extension with, 24*t*
  usage statistics for, 108*f*
  for water mains, 425, 426
Closed-circuit television (CCTV):
  for CIPP, 294–295, 296
  for condition assessment, 149–152, 153*f*, 157*t*
    of water distribution system, 170–173, 174*t*
  with laser profiling, 162
  for link seals, 482
  for LOR, 468, 469, 472–473, 482
  for MSL, 389
  for PB ILR, 346, 349, 356, 359
  for QA/QC, 32, 33
  for renewal method selection, 23
  for sanitary sewer pipe, 164–165
  for service laterals, 106–107
  for SIPP, 402
  for TRM initial planning, 94
Closed profile (CP):
  honeycombing and, 244
  profile-wall PVC, 244
CML. *See* Cement mortar lining
CMOM. *See* Capacity, management, operation, and maintenance
CMP. *See* Corrugated metal pipe
Coatings, SIPP for, 397, 397*n*1
Cobbles, 205
Coefficient of elastic support, for CIPP, 119–120, 122, 124*t*
Coefficient of friction, 363, 395
Coefficient of soil reaction, 201
CoF. *See* Consequence of failure
Collapse failures, 1, 3, 74
  PB ILR and, 345, 353
  of sanitary sewer pipe, 7*f*
Commonwealth Scientific and Industrial Research Organization of Australia, 187
Compaction grouting, 478
Completion work, for PB ILR, 374–375
Composite materials, 272, 416, 431
  FGP, 445, 445*f*, 447*f*
Compression type rubber gasket joints, 228

Computer-aided design (CAD), for condition assessment, 145, 147
Concrete pipe. *See also* Prestressed concrete cylinder pipe; Reinforced concrete pipe
  advantages and limitations of, 227–231, 230*t*
  bedding and, 231
  bell-and-spigot joints for, 227
  centrifugal casting of, 223
  construction of, 223–231
  corrosion of, 272
  corrosion resistance of, 231
  dry casting of, 223
  as gravity pipe, 227–231, 230*t*
  impact echo for, 163–164
  joints for, 228*t*, 229*f*
  linings for, 231
  manufacturing of, 223–224
  non-reinforced, 224, 340
  OC and, 227
  packerhead process for, 223
  for PB ILR, 329, 331
  PCP, 225, 225*f*, 226, 314
  pipe-soil system and, 231
  as pressure pipe, 227–231, 230*t*
  RCCP, 225
  ring flexure in, 215
  SASW for, 163–164
  shear failure and, 227
  standards for, 226, 226*t*
  tamping for, 223
  types of, 224–226
  wet casting for, 224
Condition assessment, 51–52, 53*f*, 137–197. *See also* Sanitary sewer pipe; Wastewater systems; Water distribution system
  AC and, 52
  acoustic technologies for, 159*f*
  ArcGIS for, 147
  asset naming and numbering conventions for, 143–145
  asset registry for, 142
  CAD for, 145, 147
  cameras for, 149–156, 157*t*
  CCTV for, 149–152, 153*f*, 157*t*
  data collection and analysis for, 140, 187–194
  digital scanning for, 153–155, 154*f*, 155*f*, 157*t*
  in Edgecombe County, NC, 146–148, 149*f*
Condition assessment (*Cont.*):
  electrical and electromagnetic technologies for, 159–162, 162*t*
  ELL for, 159, 162*t*
  emergency repairs and, 138
  fieldwork for, 147–148, 147*f*
  gamma-gamma logging for, 162–163
  GIS for, 142, 145–147, 146*f*, 187
  GPR for, 163
  GPS for, 145–148, 146*f*, 147*f*
  for gravity pipe, 149–169
  GWUT for, 164
  impact echo for, 163–164
  infrared thermography for, 163
  input parameters for, 140, 141*t*
  inventory tools for, 145–146
  laser profiling for, 161–162, 161*f*, 162*t*
  for manholes, 167–169, 168*f*
  for Marco Island, Florida, 164–169, 166*f*
  MFL for, 160–161, 161*f*, 162*t*
  micro-deflection for, 163
  O&M and, 196
  pipeline failure and, 138
  for pressure pipe, 149–169
  push cameras for, 155–156, 156*f*, 157*t*
  RDBMS for, 145
  RFEC for, 160, 160*f*, 162*t*
  safety and, 138
  SASW for, 163–164
  software for, 146–147, 187
  sonar for, 158–159, 158*f*, 159*f*
  stages of, 141–142
  system integration and operation and, 194–196
  technologies for, 150*t*–151*t*
  ultrasonic pulse velocity for, 164
  for water mains, 188*t*
  zoom camera for, 152, 157*t*
Consequence of failure (CoF), 50
  calculation of, 62
  customers of significant importance and, 60–61, 61*t*
  economical costs and, 56
  environmental costs and, 57, 60, 60*t*
  environmentally sensitive features and, 60, 60*t*
  GIS for, 60, 61
  pipe accessibility and, 61–62, 61*t*
  pipe depth and, 58–59, 59*t*
  pipe diameter and, 58–59, 58*t*
  pipe location and, 59–60, 60*t*
  rating methodology for, 57–58

Consequence of failure (*Cont.*):
  relative network position and, 58–59, 58*t*, 59*t*
  risk management and, 55–63, 62*f*
  social costs and, 56–57
  traffic loads and, 59–60, 60*t*
Construction (installation), 199–265
  of cement-based pipe, 222–233
  of CFP, 421–425
  of CIPP, 287, 293
  of concrete pipe, 223–231
  of continuous sliplining, 312–314
  deflection and, 220, 221
  of OC, 200–201
  of PB ILR, 360–370
  pipe selection for, 222
  pipe-soil system and, 201–212
  of plastic pipe, 239–260
  of segmental sliplining, 314
  of service laterals, 455
  of SIPP, 401–416
  of SL, 312–315, 325, 326*f*
  terminology for, 203*t*
Construction defects:
  CIPP and, 283*f*
  hydraulic capacity and, 23
  of joints, 2
  in oil and natural gas pipelines, 85–86, 85*f*, 86*f*
Contingency plan, for PB ILR, 374
Continuous sliplining, 13, 13*t*, 14*f*, 307–310
  construction of, 312–314
  FPPVC for, 308, 309*f*, 310
  PE for, 308, 308*f*, 312–314
  PVC for, 310, 312–314
Contract documents, 30
Cooper, M. J., 111
Copolymerization, PE and, 253
Corrosion, 102, 272
  cathodic protection for, 74–75
  CFP and, 423, 426
  collapse failures from, 1, 3
  of DI, 265*t*
  in flexible pipe pipeline failure, 214
  hydraulic capacity and, 23
  long-term pipeline defects from, 2
  of manholes, 432, 434–435, 442–446
  NDT for, 170
  of oil and natural gas pipelines, 64, 65*f*, 81–83, 83*f*, 84*f*
  pipeline integrity assessment for, 100
Corrosion (*Cont.*):
  pit depth measurements for, 173
  radiographic testing for, 181
  SIPP and, 397, 398
  SMR for, 21
  sonar for, 159
  of steel pipe, 266
  TCM and, 4
Corrosion protection:
  for cast iron pipe, 16
  CML for, 16
  for DI, 261, 265*t*, 270–272
  GAB-34 and, 50
  MSL and, 395
  polymers for, 19
  for steel pipe, 272
Corrosion resistance, 33
  of AC, 233
  of concrete pipe, 231
  of GRP, 260
  of HDPE, 256
  of PCP, 225, 226
  of PE, 252
  of VCP, 238
Corrugated metal pipe (CMP), external loads on, 4
Corrugated steel pipe, 266
CP. *See* Closed profile
Cracks, 2, 25, 102
  in CML, 19
  ESCR and, 252, 254, 256–257
  HDPE and, 256
  impact echo and SASW for, 164
  in manholes, 434
  NDT for, 170
  OC and, 200
  pipeline integrity assessment for, 100
  quarter point, in rigid pipes, 215, 215*f*
  renewal methods for, 23
  in sanitary sewer pipe, 167
  SCG, 254
  sonar for, 159
  spiral, in cast iron pipe, 215, 217*f*
  in water mains, 188*t*
Creep, of viscoelastics, 241
Criticality, condition assessment and, 52, 53*f*
Crushing, 74, 210, 218, 219*f*
CSA. *See* Canadian Standards Association
Culverts, 209–210, 395, 482*f*, 483
Cumulative sum of services, CoF and, 59

Cured-in-place pipe (CIPP), 11, 11*t*, 275–304, 287*f*
advantages of, 298
air inversion for, 288
applications and limitations of, 277*t*
bending stress for, 124–125
buckling equations and, 111–112
bypass pumping for, 290–293, 293*f*
characteristics of, 276, 284
coefficient of elastic support for, 119–120, 122, 124*t*
construction of, 287, 293
construction defects and, 283*f*
cost experience indicators for, 104*t*
costs of, 300–304, 303*f*, 303*t*
curing schedule for, 32
design for, 107–134, 109*t*, 285*t*, 294, 295*t*
DR for, 111–112
enhancement factor for, 111–112
existing pipe and, 95–96, 97
external pressure on, 118–119
factor of safety for, 110
flexural modulus of elasticity for, 109
for full deterioration, 281*f*, 295*t*
of gravity pipe, 113–114, 117–118, 121*f*
of pressure pipe, 128–130, 129*f*
for gravity pipe, 26–27, 28
full deterioration of, 113–114
hydraulic design for, 131–133
partial deterioration of, 112–113
Hazen-Williams coefficients for, 133–134, 133*t*
hoop stresses and, 126
hydraulic design for, 131–134
hydrostatic pressure and, 27, 111, 113, 119–120, 126
initial setup for, 285–287
inspection for, 290, 291*t*–292*t*
inversion for, 458, 458*f*
lifts and, 293, 297*f*
limitations of, 103*t*, 298–299
linings for, 109, 289–290, 289*f*, 457–458
live loads and, 120–125, 121*t*
long-term flexural modulus of elasticity for, 109, 111
for LOR, 465, 466
Manning's coefficients for, 132, 132*t*
methods for, 285–290
Cured-in-place pipe (*Cont.*):
ovality reduction factor for, 112, 112*t*
for partial deterioration, 281*f*, 295*t*
of gravity pipe, 112–117, 116*t*, 117*f*
of pressure pipe, 125–127, 126*f*
PB ILR and, 343
PIP for, 459, 459*f*
pipe plugging for, 290–293
pipe stiffness for, 124
pipeline defects and, 276, 278*t*–280*t*, 281*f*, 282*f*, 294*t*
point, 483–486, 484*f*, 485*f*
preliners for, 288
for pressure pipe, 29
full deterioration of, 128–130, 129*f*
hydraulic design for, 133–134
partial deterioration of, 125–127, 126*f*
pull-in insertion for, 288
QA/QC for, 32, 291*t*, 293–298
scheduling, 299–300
for service laterals, 456–460, 457*f*, 458*f*, 459*f*
service life extension with, 24*t*
site compatibility and applications for, 275–276
for SMR, 21
soil load and, 119–120, 119*t*
soil modulus for, 109–110, 119–120, 129
steam curing for, 289–290
testing for, 295–296
tree roots and, 282*f*
tubes for, 109
usage statistics for, 108*f*
UV for, 11, 276, 290
water curing for, 289
water inversion for, 287–288
wet out process for, 285–287, 286*f*
winch-in-place, 11, 11*t*, 12*f*

D-load, for RCP, 224
Databases, 145, 187
DD. *See* Drawdown
DDM. *See* Design Decision Model
Dead load, 111, 120
Debeading tools, for continuous sliplining, 313
Deflection, 159, 188*t*, 245
creep and, 241
of flexible pipe, 111, 219–221, 220*f*
micro-deflection, 163

Deformations, 100, 214
CFP and, 422
CIPP and, 277*t*, 298
micro-deflection and, 163
MSL and, 388*t*
Deformed and reformed (D&R), 419, 420*f*, 421*f*, 426
Delaminations, 164, 214
Department of Health and Human Services (HHS), 284
Department of Transportation (DOT), PB ILR and, 359, 361
Department of Transportation 192 (DOT 192), 63–64, 68, 69, 74, 77, 78, 78*t*, 80
Department of Transportation 195 (DOT 195), 63–64, 69, 74, 77, 78, 78*t*, 80
Depreciation, GASB-34 and, 46–49, 48*t*–49*t*, 51*t*
Design, 31
for CIPP, 107–134, 109*t*, 285*t*, 294, 295*t*
hydraulic, for oil and natural gas pipelines, 72–74, 72*f*, 73*f*
for oil and natural gas pipelines, 68, 70–72, 70*f*, 71*f*
for PB ILR, 347–352, 378
for renewal methods, 25–29
for SL, 315–316, 324–325
for TRM, 93–135
Design Decision Model (DDM), of DIPRA, 270–271
Destructive testing, 170, 417
Dewatering, for PB ILR, 364
DIP. *See* Ductile iron pipe
Digital scanning, for condition assessment, 153–155, 154*f*, 155*f*, 157*t*
Dimension ratio (DR), 111–112, 245
DIPRA. *See* Ductile Iron Pipe Research Association
Directional drilling. *See also* Horizontal directional drilling
HDPE for, 254, 255*f*
segmental sliplining and, 310
Disposal plan, in AMP, 42
DOT. *See* Department of Transportation
DOT 192. *See* Department of Transportation 192
DOT 195. *See* Department of Transportation 195
DR. *See* Dimension ratio
D&R. *See* Deformed and reformed
DR 17, 352, 361
DR 21, 352
Drawdown (DD), 20, 20*t*
Dry casting, for concrete pipe, 223
Dual-wall corrugated profile (DWCP), profile-wall PVC, 244
Ductile iron pipe (DIP), 201
CML for, 271
corrosion protection for, 261, 265*t*, 270–272
as flexible pipe, 203, 262
for gravity pipe, 265*t*
for HDD, 262
joints for, 264, 264*f*, 265*t*
linings for, 262, 265*t*
manufacturing of, 262–263
for PB ILR, 262, 329, 342, 368, 368*f*
for PB-PB, 331
for PB-PS, 337
for pressure pipe, 265*t*
PSRs for, 345
for sanitary sewer pipe, 262, 272
standards for, 263, 263*t*
for water distribution system, 261
for water mains, 188*t*, 366
Ductile Iron Pipe Research Association (DIPRA), 261, 270–271
DWCP. *See* Dual-wall corrugated profile

Earnings before interest, taxes, depreciation, and amortization (EBITDA), 75
Earth loads, existing pipe and, 96, 97
EBITDA. *See* Earnings before interest, taxes, depreciation, and amortization
Edgecombe County, NC, condition assessment in, 146–148, 149*f*
Elastic zone, in PB ILR, 354, 354*f*
Electric resistance welded (ERW) pipe, 266–267
Electrical and electromagnetic technologies:
for condition assessment, 159–162, 162*t*
for water distribution system condition assessment, 177–181
Electrical leak location (ELL), 159, 162*t*
Electrofusion, saddle, 333–334, 369, 369*f*

ELL. *See* Electrical leak location
Embedment, 203*t*
  arching effect and, 208
  OC and, 200
  soils for, 206*t*–207*t*
Emergency repairs, 36, 59, 138
Engineering, of oil and natural gas pipelines, 68
Enhancement factor, for CIPP, 111–112
Environmental costs, CoF and, 57, 60, 60*t*
Environmental Protection Agency. *See* United States Environmental Protection Agency
Environmental stress crack resistance (ESCR), 252, 254, 256–257
Environmental testing, for water distribution system condition assessment, 182, 183*t*
Environmentally sensitive features, CoF and, 60, 60*t*
EPA. *See* United States Environmental Protection Agency
EPANET, 191, 191*n*7
EPDM. *See* Ethylene propylene diene monomer
Epoxy:
  chemical resistance of, 399
  for CIPP, 284
  for LOR, 467–468, 469*f*, 476–477
  for manholes, 431, 444, 444*f*
  for SIPP, 19, 399, 401
  for water distribution system, 399
Erosion, 2
  LOR and, 466
  manholes and, 430–431
  PB ILR and, 353
ERW pipe. *See* Electric resistance welded pipe
ESCR. *See* Environmental stress crack resistance
ESRI, 146
Ethylene propylene diene monomer (EPDM), 260
EX method, 419
Executive summary, in AMP, 39
Exfiltration testing, for SIPP, 417
Exfiltrations, in sanitary sewer pipe, 167
Existing materials:
  CFP and, 422
  LOR and, 466
Existing pipe:
  CIPP and, 111, 292*t*, 298
  hydraulic capacity of, 95–96
  PB ILR and, 341, 350, 378
  PB-PS and, 337
  PR-PX and, 340
  QA/QC and, 32
  renewal methods and, 95–97
  SIPP and, 411
External loads, 4, 141*t*
  for oil and natural gas pipelines, 74
  PCCP and, 224
  SL and, 316
External pressure, 25
  on CIPP, 118–119

Factor of safety, for CIPP, 110
Farshad, M., 214
Fatigue, 214, 429
Federal Emergency Management Agency, flood zones of, 57
Federal Highway Administration (FHWA), 35
F&F. *See* Fold and formed
FGP. *See* Fiberglass/polymer composites
FHWA. *See* Federal Highway Administration
Fiberglass. *See also* Glass-reinforced pipe
  for CIPP, 11
  for joints, 226
  for manholes, 431, 440–441, 441*f*, 448*f*, 450
  for MSL, linings of, 386, 387*f*
Fiberglass/polymer composites (FGP), for manholes, 445, 445*f*, 447*f*
Fiberglass reinforced polyester (FRP), 450
Field eddy current. *See* Remote field eddy current
Filament winding, for GRP, 258
Filament-wound collar (FWC), for pressure pipe jacking, 259–260
Fill-and-drain systems, for LOR, 477
Final backfill, 203*t*, 207*t*
Financial planning, AM for, 36
Financial statements and projections, in AMP, 43
Financial summary, in AMP, 43
Flexible pipe:
  arching effect for, 209, 209*f*
  behavior of, 212, 212*f*

Flexible pipe (*Cont.*):
- deflection of, 111, 219–221, 220*f*
- DI as, 262
- external loads on, 4
- for gravity pipe, 29
- horizontal passive resistance with, 204
- longitudinal tensile failure of, 221, 221*f*
- OC and, 200–201
- pipe-soil system and, 201–204, 202*f*, 204*t*
- pipeline failure with, 214, 218–221, 219*f*
- side soils and, 212
- side support of, 221*f*
- soil load for, 208, 208*f*, 212

Flexural failures, in rigid pipe, 215, 217*f*
Flexural modulus of elasticity, 30, 109
Flood zones, of Federal Emergency Management Agency, 57
Flow bypassing:
- for PB ILR, 356, 374
- for segmental sliplining, 312
- for TRM, 106

Flowtite, 259
Fold and formed (F&F), 20, 20*t*, 419, 420*f*
For oil and natural gas pipelines, 78, 87–88, 88*f*
Foundation, 203*t*
- longitudinal bending and, 217

FPVCP. *See* Fusible PVC
Fractures, 25
- in flexible pipe pipeline failure, 214
- in sanitary sewer pipe, 167
- of VCP, 167, 167*f*

Freshwater pipelines, oil and natural gas pipelines and, 72–74, 72*f*, 73*f*, 78–80, 79*t*
Frost loading, longitudinal bending and, 217
FRP. *See* Fiberglass-reinforced polyester
Full deterioration, 27–29
- CIPP and, 281*f*, 295*t*
- of gravity pipe, 28
  - CIPP for, 113–114, 117–118, 121*f*
- groundwater and, 113*n*2
- of pressure pipe, 29
  - CIPP for, 128–130, 129*f*

Funding strategy, in AMP, 43
Fusible PVC (FPVCP), for continuous sliplining, 308, 309*f*, 310
Future demand:
- in AMP, 40–41
- oil and natural gas pipelines and, 74

FWC. *See* Filament-wound collar

GAAP. *See* Generally accepted accounting principles
Gamma-gamma logging, for condition assessment, 162–163
GASB-34. *See* Government Accounting Standards Board Statement 34
GASB-34-modified approach, to depreciation, 46–50, 48*t*–49*t*, 51*t*
Gasketed joints, 228, 249 -250
- for HDPE, 257

Gauge plate, for oil and natural gas pipelines, 76, 87–88, 88*f*
Generally accepted accounting principles (GAAP), 45
GeoExplorer, 146
Geographic information system (GIS), 23, 60, 61
- for condition assessment, 142, 145–147, 146*f*, 187
- for oil and natural gas pipelines, 77
- WISE and, 191–194

Geophones, 157
Geopolymers, for manholes, 431
GeoXT, 146, 148
GIS. *See* Geographic information system
Glass-reinforced cement (GRC), for segmental sliplining, 314
Glass-reinforced pipe (GRP), 240
- advantages and limitations of, 260, 260*t*
- bell-and-spigot joints for, 259–260, 259*f*
- for jacking, 226
- joints for, 259–260
- for segmental sliplining, 310, 314
- for SL, 305
- standards for, 258–259, 259*t*

Global positioning system (GPS), 23
- for condition assessment, 145–148, 146*f*, 147*f*
- for manholes, 146

Gokhale, S., 107
Gouges, OC and, 200
Government Accounting Standards Board Statement 34 (GASB-34), 45–52, 48*t*–49*t*, 51*t*
GPR. *See* Ground-penetrating radar
GPS. *See* Global positioning system

GPS Analyst, for condition assessment, 147*f*, 148
GPScorrect, 146–147, 148
Graphical user interface (GUI), ArcWISE as, 193
Gravel, 205
Gravity pipe:
  CFP for, 421
  CIPP for, 26–27, 28, 112, 131–133
  collapse failures of, 1
  concrete pipe as, 227–231, 230*t*
  condition assessment for, 149–169
  DI for, 265*t*
  flexible pipe for, 29
  full deterioration of, 28
  GRP for, 240
  HDPE for, 256
  I/I problems in, 1
  link seal for, 480, 480*f*
  MSL for, 396
  partial deterioration of, CIPP for, 112–113
  PB ILR for, 363
  renewal methods for, 25, 27*f*
  SL for, 259, 318
  SWM for, 16
GRC. *See* Glass-reinforced cement
Grid reference numbering, 144
Ground movement:
  in condition assessment, 141*t*
  manholes and, 430
  from PB ILR, 332, 354–356, 354*f*, 355*f*, 364, 378
Ground-penetrating radar (GPR), 163, 185–187
Groundwater, 25, 29, 113*n*2, 141*t*
  CFP and, 422, 423
  CIPP and, 26–27, 110
  manholes and, 21
  MSL and, 389
  OC and, 200
  PB ILR and, 346, 351, 353, 364
  VCP and, 234
Grout:
  acrylamide, 474–475
  acrylate, 475
  acrylic, 475
  in annular space, 223, 305
  cementitious materials for, 477–479
  chemical, 439, 456, 461–462, 470–474
  for continuous sliplining, 308
  for I/I problems, 439
  for LOR, 470–479, 471*t*, 472*f*
  for manholes, 439, 440*f*, 450–451
Grout (*Cont.*):
  for MSL, 88, 391, 391*f*
  for PB ILR, 365
  for PE, 365
  PU, 477
  for PVC, 16
  for segmental sliplining, 310
  for service laterals, 461–462
  for SL, 312, 314–315, 327*f*
  urethane, 475–476
GRP. *See* Glass-reinforced pipe
GUI. *See* Graphical user interface
Guided wave ultrasonic (GWUT), 164, 176–177, 178*f*
Gunite, 412–413, 479
GWUT. *See* Guided wave ultrasonic
Gyro pig, for oil and natural gas pipelines, 76, 87–88, 88*f*

HAB. *See* Horizontal auger boring
Haunch zone, 202*f*, 203*t*, 220
Hazen-Williams coefficients, for CIPP, 133–134, 133*t*
HDD. *See* Horizontal directional drilling
HDPE. *See* High-density polyethylene
HDS. *See* Hydrostatic design stress
HHS. *See* Department of Health and Human Services
High criticality, condition assessment and, 52, 53*f*
High-density polyethylene (HDPE), 52
  AC and, 383
  advantages and limitations of, 256–257, 257*t*
  butt-fused joints for, 257, 326*f*
  for concrete pipe linings, 231
  for continuous sliplining, 308
  for directional drilling, 254, 255*f*
  for flexible pipe, 201, 203
  for freshwater pipelines, 72–74, 72*f*, 73*f*
  for manholes, 438, 439*f*
  manufacturing of, 253
  OD for, 255, 321*f*
  for oil and natural gas pipelines, 64
  for PB ILR, 333, 342, 361
  PR-PR for, 340
  for pressure pipe, 252, 256
  SL and, 318–327
  solid-wall, 254–255
  standards for, 254–255, 256*t*
  for water mains, 318–321
Hobas, 259

Holes, 102
  NDT for, 170
Holiday testing, 74, 75, 449
Honeycombing, 164, 244
Hoop stresses, 4, 30, 126, 215
Horizontal auger boring (HAB), 14, 261
Horizontal directional drilling (HDD):
  DI for, 262
  gyro pig for, 76
  HDPE for, 256
  PE for, 239–240, 252
  pipe removal ILR and, 14
  PR-PR and, 14–15, 339–340
  PVC for, 246
  spiral welded pipe for, 267
  steel pipe for, 261
Horizontal passive resistance, 204, 210
Hydrates, in oil and natural gas pipelines, 86, 87*f*
Hydraulic capacity:
  of CIPP, 298
  of existing pipe, 95–96
  I/I problems and, 1, 23
  inadequate, 102
  MSL for, 395
  pipeline integrity assessment for, 100
  renewal methods and, 23
  SL and, 316, 318
  TRM and, 95
Hydraulic design:
  for CIPP, 131–134
  for oil and natural gas pipelines, 72–74, 72*f*, 73*f*
Hydraulic loads, CIPP and, 26–27
Hydrogen sulfide, 2–3, 231
Hydrostatic design basis (HDB). *See* Unplasticized PVC
Hydrostatic design stress (HDS), 254
Hydrostatic pressure:
  CIPP and, 27, 111, 113, 119–120, 126
  from groundwater, 29
  service laterals and, 107

I/I problems. *See* Infiltration and inflow problems
ICRI. *See* International Concrete Repair Institute
ID. *See* Inside diameter
IEISS. *See* Interdependent Energy Infrastructure Simulation System
IIMM. *See* International Infrastructure Management Manual
ILI. *See* In-line inspection
ILR. *See* In-line replacement
Impact echo, for condition assessment, 163–164
Improvement monitoring, in AMP, 44
In-line inspection (ILI):
  BEM and, 181
  for oil and natural gas pipelines, 81–82, 87–88, 88*f*
    ultrasonic pigs for, 176
  of water distribution systems, ultrasonic pigs for, 181
In-line leak detectors, 156–157, 159*f*, 184–185, 185*f*, 186*t*
In-line replacement (ILR), 13–14, 15*t*. *See also* Pipe bursting; Pipe reaming
  applications for, 331
  categories of, 330*f*
  for existing pipe, 97
  limitations of, 103*t*
  method applicability for, 329–331, 330*t*
  pipe removal, 339–341
    HDD and, 14
  service life extension with, 24*t*
  usage statistics for, 108*f*
In situ reinforced concrete, for SIPP, 413–414
Infiltration:
  CIPP and, 298
  long-term pipeline defects from, 2
  in sanitary sewer pipe, 167
  SIPP and, 411
  in situ reinforced concrete and, 414
  VCP and, 233
Infiltration and inflow problems (I/I problems), 102
  cementitious materials and, 398
  CML and, 19
  condition assessment and, 52
  from corrosion, 272
  in gravity pipes, 1
  grout for, 439
  hydraulic capacity and, 1, 23
  for LOR, 466
  manholes and, 430, 432, 439
  pipeline integrity assessment for, 100
  polymers and, 398
  in sanitary sewer pipe, 165

Infiltration and inflow problems (*Cont.*):
  SIPP and, 406
  structural defect failures and, 25
  USEPA and, 234
  VCP and, 239
Inflection points (PIs), SL and, 316, 318
Infrared thermography, 163, 187
Initial backfill, 203*t*
Initial planning, for TRM, 94–95, 99, 99*t*
Inserta Tee, 369, 370*f*
Insertion pits:
  for CFP, 426
  for PB ILR, 333, 350, 360, 362, 362*f*
  for PB-PI, 338
  for PR-PX, 340
Inside diameter (ID):
  of HDPE, 253
  MSL and, 389
  for SL, 307, 324
Inspection, 31–32. *See also* Condition assessment; In-line inspection
  for CFP, 422–425
  for CIPP, 290, 291*t*–292*t*
  for MSL, 389
  QA/QC and, 32–33
  for service laterals, 459–460
  for SIPP, 402, 403*f*, 416–417
  TRM and, 95
Installation. *See* Construction
Interdependent Energy Infrastructure Simulation System (IEISS), 191
Intermediate pipe, 204
Internal loads:
  pipeline failure from, 213
  SL and, 316
Internal pressure:
  CIPP and, 130
  joints and, 226
  longitudinal tensile failure and, 221
  PCCP and, 224
  structural defect failures and, 25
Internal seals:
  for existing pipe, 97
  for LOR, 466, 479–483
  for MJ, 479–481, 481*f*
Internally restrained gasketed joint PVC pipe, 249 –250
International Concrete Repair Institute (ICRI), 406
International Infrastructure Management Manual (IIMM), 35, 39, 143
International Pipe Bursting Association (IPBA), 344
Introduction, to AMP, 40
Inventory, for oil and natural gas pipelines, 76–77
Inversion, for CIPP, 287–288, 458, 458*f*
Inverted-in place CIPP, 11, 11*t*, 12*f*
IPBA. *See* International Pipe Bursting Association
iPVC, 245, 245*f*, 246*t*

Jacking:
  bell-and-spigot joints for, 227
  FWC for, 259–260
  GRP for, 226
  PB-PI and, 337–338, 338*f*
  PCP for, 226
  PR-PX and, 340
  of pressure pipe, FWC for, 259–260
  segmental sliplining and, 310
  SL and, 317
  steel pipe for, 266
  VCP for, 233
Jeep testing, for oil and natural gas pipelines, 75, 84–85, 85*f*
Joints. *See also* Bell-and-spigot joints; Butt-fused joints
  for AC, 232, 233*f*
  butt-welded, for steel pipe, 267, 267*f*
  CFP and, 422
  CIPP and, 298
  for concrete pipe, 226–227, 228*t*, 229*f*
  construction defects of, 2
  for DI, 264, 264*f*, 265*t*
  fiberglass for, 226
  gasketed, for HDPE, 257
  for GRP, 259–260
  for HDPE, 257
  internal pressure and, 226
  manholes and, 434
  MJ, internal seal for, 479–481, 481*f*
  for PB ILR, 366
  for PE, 256, 366
  pipe selection and, 222
  pipeline integrity assessment for, 100
  for PVC, 246–250
  separation of, 102
  for SL, 312
  for steel pipe, 267–269, 268*f*, 269*f*
  structural defect failures and, 25
  for TCM, 227
  for VCP, 234–238, 235*f*, 236*f*, 237*f*
  water tightness of, 227, 227*n*2, 232

Klatte, Fritz, 241
Kleweno, D. G., 111

Laser profiling, 161–162, 161*f*, 162*t*, 173, 174*t*
LCCA. *See* Life cycle cost analysis
LDPE. *See* Low-density polyethylene
Leak detection, 156–157, 159*f*, 185*f*, 186*t*
  acoustic technologies for, 182–184
  for CFP, 424
  external, 182–184
  GPR for, 185–187
  infrared thermography for, 187
  radio-frequency interferometry for, 187
  for water distribution system condition assessments, 182–187
Leaks:
  from corrosion, 272
  infrared thermography for, 163
  long-term pipeline defects from, 2
  renewal methods for, 23
Level of service (LOS), 40, 52, 54, 193–194
Life cycle cost analysis (LCCA), 64
  for oil and natural gas pipelines, 68, 88–90
  pipe selection and, 222
Life expectancy, 187, 222
  of polymers, 19, 398
  of SIPP, 398
Lifecycle management, in AMP, 41, 42*f*
Lifts, CIPP and, 293, 297*f*
Likelihood of failure (LoF), risk management and, 54–55, 62–63, 62*f*
Linings. *See also* Sliplining
  adhesion of, 466
  for CIPP, 109, 275, 276, 289–290, 289*f*, 298, 457–458
  CML, 16–19, 271
  for concrete pipe, 231
  for DI, 262, 265*t*
  for existing pipe, 97
  hoop stresses on, 30
  for manholes, 431–433, 431*f*, 440–446, 450
  for MSL, 386–387, 387*f*, 392, 392*f*
  for natural gas pipelines, 317
  polymers for, cementitious materials for, 18
  of PU, for steel pipe, 272
  SIPP for, 397, 397*n*1
  spiral-wound, 388, 389*f*
Link seals, 480, 480*f*, 482*f*, 483
Live loads:
  CIPP and, 120–125, 121*t*
  existing pipe and, 96, 97
  RCP and, 224
Lo, K. H., 111
Localized repair (LOR), 102, 465–487
  categories of, 466*f*
  CCTV for, 468, 469, 472–473, 482
  chemical gel grouts for, 474
  chemical grouting for, 470–474
  CIPP for, 465, 466
  cost experience indicators for, 104*t*
  epoxy for, 467–468, 469*f*, 476–477
  fill-and-drain systems for, 477
  grout for, 470–479, 471*t*, 472*f*
  internal seals for, 479–483
  limitations of, 103*t*
  point CIPP for, 483–486, 484*f*, 485*f*
  robotic repairs for, 467–470, 468*f*, 469*f*
  shotcrete for, 479
LoF. *See* Likelihood of failure
Long-term flexural modulus of elasticity, for CIPP, 109, 111
Long term pipeline defects, 2–4, 3*t*
Long-term test methods, for renewal methods, 29–30
Longitudinal bending, 215, 217, 221
Longitudinal breaks, 4, 216*f*
Longitudinal cracking, in cast iron pipe, 215
Longitudinal flexure, bending stresses and, 215
Longitudinal stress, 4, 231
Longitudinal tensile failure, of flexible pipe, 221, 221*f*
LOR. *See* Localized repair
LOS. *See* Level of service
Low criticality, 52, 53*f*
Low-density polyethylene (LDPE), 252
Lubrication:
  for PB ILR, 332, 362–363
  for PB-PI, 338

Magnetic flux leakage (MFL), 160–161, 161*f*, 162*t*
  for oil and natural gas pipelines, 76, 87–88, 88*f*
  for water distribution system condition assessment, 177–178, 179*f*
Manhole Assessment and Certification Program, of NASSCO, 435

Manholes, 433–439, 434*t*
adhesion and, 432, 449
annular space and, 446
bench of, 440–441, 441*f*
calcium aluminate for, 431, 446
cementitious materials for, 431*f*, 443*f*
CFP and, 423, 426
collapse of, 429, 430*f*
condition assessment for, 167–169, 168*f*
corrosion of, 432, 434–435, 442–446
epoxy for, 444, 444*f*
erosion and, 430–431
fatigue of, 429
FGP for, 445, 445*f*, 447*f*
fiberglass for, 448*f*, 450
GPS for, 146
ground movement and, 430
groundwater and, 21
grout for, 439, 440*f*, 450–451
HDPE for, 438, 439*f*
holiday testing for, 449
I/I problems and, 430, 432, 439
inserts for, 450
linings for, 431–433, 431*f*, 440–446
longitudinal bending and, 217
mechanical seals for, 436, 437*f*
MSL and, 388, 392–393
PB ILR and, 346, 360, 364–366, 365*f*, 366*f*, 367*f*
PCP for, 226
PE and, 364–366
pipe accessibility with, 61
pipeline failure of, 429–431, 430*f*
PLM for, 450
polymers for, 431, 431*f*
polyurethane for, 444, 444*f*
Portland cement for, 431
renewal methods for, 449–451
sanitary sewer pipe and, 429–451
semi-structural techniques for, 432
SMR, 20–21, 21*t*, 22*f*
spark testing for, 449
structural defect failures of, 442
sulfuric acid and, 435, 443
testing of, 448–449
traffic loads and, 429
Manning's coefficients, 132, 132*t*, 342
Marco Island, Florida, 164–169, 166*f*, 168*f*
Marston, Anson, 205, 208, 210
Maximum allowable operating pressure, for PB ILR, 352
MCF. *See* Million cubic feet
MDPE. *See* Medium-density polyethylene
Mechanical joint (MJ), 368, 479–481, 481*f*
Mechanical seals, 106, 436, 437*f*
Medium-density polyethylene (MDPE), 252, 342
Metallic pipe, 261–270. *See also* Cast iron pipe; Ductile iron pipe; Steel pipe
Methanol injection, for oil and natural gas pipelines, 76
Meyer Pipe, 226
MFL. *See* Magnetic flux leakage
MIC. *See* Microbiologically induced corrosion
Micro-deflection, for condition assessment, 163
Micro-fine cement, 478
Microbiologically induced corrosion (MIC), 435
Microtunnel boring machine (MTBM), 340–341
Microtunneling (MT), 14, 227, 266
PCP for, 226
segmental sliplining and, 310
Million cubic feet (MCF), 68
MIMS. *See* Multimedia Integrated Modeling System
MJ. *See* Mechanical joint
"Modern Sewer Design," of American Iron and Steel Institute, 266
Modified PVC (PVC-M), 243, 422–423
Modified sliplining (MSL), 15–16, 18*t*, 104*t*, 385–396. *See also* Panel-lining method; Spiral wound method
access shafts for, 396
advantages of, 395–396
annular space and, 15, 396
applications for, 388, 388*t*
background on, 385–386
CCTV for, 389
characteristics of, 386–388
for existing pipe, 97
for gravity pipe, 396
grout for, 88, 391, 391*f*
inspection for, 389
limitations of, 396
linings for, 386–387, 387*f*, 392, 392*f*
manholes and, 392–393
preparatory work for, 389
safety for, 389–393

Modified sliplining (*Cont.*):
service life extension with, 24*t*
with steel pipe, 393, 393*f*, 394*f*
usage statistics for, 108*f*
Moisture tolerance, of SIPP, 401
Molecularly oriented PVC (PVC-O), 243
MSL. *See* Modified sliplining
MT. *See* Microtunneling
MTBM. *See* Microtunnel boring machine
Multimedia Integrated Modeling System (MIMS), 145
Muriatic acid, 407

NACE. *See* National Association of Corrosion Engineers
Najafi, M., 15, 107, 296
NASSCO. *See* National Association of Sewer Service Companies
National Association of Corrosion Engineers (NACE), 406, 449
National Association of Sewer Service Companies (NASSCO):
on condition assessments, 139
Manhole Assessment and Certification Program of, 435
PACP® of, 23, 32, 52–54, 165, 276
National Emissions Standards for Hazardous Air Pollutant (NESHAP), 381, 382–383
National Research Council of Canada (NRC), 189, 190, 191*t*
National Sanitation Foundation (NSF), 245
Natural gas pipelines. *See also* Oil and natural gas pipelines
CFP for, 422*t*
CIPP for, 277*t*
clogging of, 86
explosion from, 6*f*
linings for, 317
MSL for, 388*t*
PB ILR for, 333, 460
PE for, 251–252, 316–317
SL for, 316–317
NDT. *See* Nondestructive testing
NESHAP. *See* National Emissions Standards for Hazardous Air Pollutant
New construction, in AMP, 42
New York City, water mains in, 188
Non-friable ACM, 381–382
Non-reinforced concrete pipe, 224, 340
Nondestructive testing (NDT), 170, 175–177, 189
NRC. *See* National Research Council of Canada
NSF. *See* National Sanitation Foundation

OC. *See* Open-cut
Occupational Safety and Health Administration (OSHA), 284, 333, 362, 389
OD. *See* Outside diameter
O'Day, B. K., 4
Oil and natural gas pipelines, 78*t*
AM for, 63–90
cathodic protection for, 74–75, 76
chemical treatment for, 76
construction defects in, 85–86, 85*f*, 86*f*
construction oversight for, 77
corrosion of, 64, 65*f*, 81–83, 83*f*, 84*f*
design for, 68, 70–72, 70*f*, 71*f*
economic analysis for, 64, 66*t*–67*t*
engineering of, 68
external load design for, 74
financing for, 75
freshwater pipelines and, 72–74, 72*f*, 73*f*, 78–80, 79*t*
future demand and, 74
HDPE for, 64
holiday testing for, 74, 75
hydrates in, 86, 87*f*
hydraulic planning for, 72–74, 72*f*, 73*f*
ILI for, 81–82, 87–88, 88*f*
ultrasonic pigs for, 176
inventory for, 76–77
jeep testing for, 75, 84–85, 85*f*
LCCA for, 68, 88–90
material defects in, 84–85, 84*f*, 85*f*
MFL for, 76, 87–88, 88*f*
operations and maintenance procedures for, 76–80
pigging for, 65, 69–70, 69*f*, 76
pipeline failure of, 1
procurement oversight for, 77
record keeping for, 78
risk management for, 80–88, 89*f*
saltwater pipelines and, 78–80, 79*t*, 80*f*
steel pipe for, 266
system performance and tracking review for, 78, 80
wyes for, 70–72, 70*f*, 71*f*

Olefin gases, 251
O&M. *See* Capital and Operations and Maintenance Program
Open-cut (OC):
  backfill and, 200
  concrete pipe and, 227
  costs of, 107
  DI for, 263
  economic costs of, 9
  existing pipe and, 96, 97
  PB ILR and, 343–344, 344*f*, 348
Open profile (OP), profile-wall PVC, 244
Operation, reliability, and maintenance (ORM), 97
  for TRM, 95
OSHA. *See* Occupational Safety and Health Administration
Outside diameter (OD), 203*t*
  for HDPE, 253, 255, 321, 321*f*
  for PB ILR, 361
  for SL, 307, 321
Ovality reduction factor, for CIPP, 112, 112*t*
Over-pumping:
  CFP and, 423
  MSL and, 389
Overshoving, 2

Packerhead process, for concrete pipe, 223
PACP®. *See* Pipeline Assessment Certification Program®
Panel-lining method (PLM), 15–16, 17*f*–18*f*, 18*t*, 450
PANORAMO®, 155–156, 156*f*
PARMS. *See* Pipeline asset and risk management system
Partial deterioration:
  CIPP and, 281*f*, 295*t*
  of gravity pipe, CIPP for, 112–117, 116*t*, 117*f*
  groundwater and, 113*n*2
  of pressure pipe, 25–26, 125–127, 126*f*
PB. *See* Pipe bursting
PB-PI. *See* Pipe insertion
PB-PN. *See* Pneumatic pipe bursting
PB-PS. *See* Pipe splitting
PB-ST. *See* Static pipe bursting
PCCP. *See* Prestressed concrete cylinder pipe
PCP. *See* Polymer concrete pipe
PE. *See* Polyethylene
PE3408, for PB ILR, 342
PE3608, 252, 342
PE4710, 252, 342
Peak particle velocity (PPV), 354–355, 355*f*
Permalok™, 269
Permits, for PB ILR, 359
PHMSA. *See* Pipeline and Hazardous Materials Safety Administration
Pigging:
  for bacteria prevention, 65, 69
  for oil and natural gas pipelines, 65, 69–70, 69*f*, 76, 87–88, 88*f*
  for SIPP, 407, 411*f*
  ultrasonic:
    for oil and natural gas pipeline, 176
    for water distribution systems, 181
Pin and groove gasketed joint TerraBrute, for PVC, 249
PIP. *See* Pull in-place
Pipe accessibility, CoF and, 61–62, 61*t*
Pipe bursting (PB), ILR, 13–14, 15*t*, 329–384
  AC for, 329, 331, 378–383
  advantages and limitations of, 344–346
  as-built drawings for, 372
  bidding form for, 359
  butt-fused joints for, 360
  bypass pumping for, 346–347, 363–364
  case study for, 378–383
  CCTV for, 346, 349, 356, 359
  CIPP and, 343
  classifications of, 344, 345*t*
  collapse failures and, 345, 353
  completion work for, 374–375
  contingency plan for, 374
  cost estimating for, 376, 376*f*, 376*t*
  cost experience indicators for, 104*t*
  design for, 378
  dewatering for, 364
  DI for, 262, 368, 368*f*
  dispute resolution for, 359
  DOT and, 359, 361
  electrofusion saddle for, 333–334, 369, 369*f*
  erosion and, 353
  existing pipe and, 341, 350, 378
  flow bypassing for, 356, 374

Pipe bursting (*Cont.*):
- ground movement with, 354–356, 354*f*, 355*f*, 364, 378
- groundwater and, 346, 351, 353, 364
- grout for, 365
- HDPE for, 361
- insertion pits for, 333, 350, 360, 362, 362*f*
- joints for, 366
- limitations of, 103*t*
- manholes and, 346, 360, 364–365, 365*f*, 366*f*, 367*f*
- maximum allowable operating pressure for, 352
- methods of, 334–338
- MJ for, 368
- OC and, 343–344, 344*f*, 348
- OD for, 361
- operation layout for, 360
- PE for, 352, 361, 368–369, 368*f*
  - grooves on surface for, 333–334, 369, 369*f*, 373*f*
- permits for, 359
- plans and specifications for, 356–359
- potential problems with, 377–378, 379*t*–380*t*
- pre-design phase of, 347–348, 348*f*
- preparatory work for, 346–347
- process of, 333, 334*f*
- PSRs and, 345, 346
- pulling shaft for, 350, 360, 362
- PVC for, 246, 329, 331, 333, 342*n*1, 366–367, 368, 368*f*
- QA/QC for, 358, 375–376
- reception pits for, 333
- reconnection of service after, 333–334
- replacement pipe for, 342–343
- risk management for, 352–356
- safety for, 331, 361, 364, 374
- sags and, 346, 353
- with segmental pipes, 375
- for service laterals, 456, 460–461, 460*f*
- service laterals and, 332, 334, 345, 346, 347, 369–370, 370*f*
  - replacement of, 375
- soil and, 350–351
- submittals for, 357–358
- surveying for, 346
- thermal expansion and contraction and, 364–365
- traffic diversion for, 347, 359, 361
- unfavorable conditions for, 331–332

Pipe bursting (*Cont.*):
- for upsizing, 331, 335
- utility survey for, 349, 363
- for water mains, 366–369

Pipe depth, CoF and, 58–59, 59*t*
Pipe diameter, CoF and, 58–59, 58*t*
Pipe eating (PR-PE), 340–341
Pipe ejection (extraction) (PR-PX), 341
Pipe insertion (PB-PI), 337–338, 338*f*
Pipe jacking. *See* Jacking
Pipe plugging, for CIPP, 290–293
Pipe ramming (PR), steel pipe for, 261, 266
Pipe reaming (PR-PR), 15*t*, 16*f*, 331, 339–340, 339*f*
- cost experience indicators for, 104*t*
- HDD and, 14–15
- limitations of, 103*t*

Pipe removal (PR), ILR, 339–341
- HDD and, 14

Pipe rerounding, 481
Pipe-soil system, 201–212
- arching effect and, 208–210, 209*f*
- CIPP and, 111, 113
- concrete pipe and, 231
- flexible pipe and:
  - arching effect for, 209, 209*f*
  - behavior of, 212, 212*f*
  - horizontal passive resistance with, 204
  - OC and, 200–201
  - pipe-soil system and, 201–204, 202*f*, 204*t*
  - side soils and, 212
  - soil load for, 208, 208*f*, 212
- rigid pipe and:
  - arching effect for, 209, 209*f*
  - bedding factors for, 211
  - behavior of, 210–211, 211*f*
  - horizontal passive resistance with, 210
  - OC and, 200
  - pipe-soil system and, 201–204, 204*t*
  - soil load for, 208, 208*f*
  - three-edge bearing test for, 210–211, 211*f*

Pipe splitting (PB-PS), 337, 337*f*
Pipe stiffness, 110–111. *See also* Flexible pipe; Rigid pipe
- for CIPP, 124
- deflection and, 220

Pipeline and Hazardous Materials Safety Administration (PHMSA), 1, 63
Pipeline Assessment Certification Program® (PACP®), 52–54
- AMP and, 54
- CCTV and, 165
- CIPP and, 276
- QA/QC and, 32
- for renewal method selection, 23

Pipeline asset and risk management system (PARMS), 187, 188
Pipeline defects, 2–4, 102
- CIPP and, 276, 278*t*–280*t*, 281*f*, 282*f*, 294*t*
- in service laterals, 455
- soil and, 3–4
- sonar for, 159
- structural defect failures from, 25
- TRM and, 95

Pipeline failure, 1–2. *See also specific types*
- classification of, 4
- common modes of, 213–221, 213*t*
- condition assessments and, 138
- from corrosion, 272
- from creep, 241
- with flexible pipe, 214, 218–221
- from internal loads, 213
- of manholes, 429–431, 430*f*
- modes of, 4–7
- of oil and natural gas pipelines, 1
- with rigid pipe, 214–217
- traffic diversion from, 4
- of water distribution system, sinkholes from, 4

Pipeline integrity assessment, for TRM, 95, 98–99, 100, 100*t*
PIs. *See* Inflection points
Pit depth measurements, for water distribution system condition assessment, 173, 174*t*
Pits:
- NDT for, 170
- sonar for, 159

Plan numbering, 144
Plastic pipe. *See also* Glass-reinforced pipe; Polyethylene; Polyvinyl chloride
- buckling equations for, 30
- construction of, 239–260
- corrosion of, 272
- for PB ILR, 331

Plastic Pipe Institute (PPI), 64, 68, 318, 352
Plastic zone, in PB ILR, 354, 354*f*
PLM. *See* Panel-lining method
Pneumatic pipe bursting (PB-PN), 13–14, 15*f*, 335–336
Point CIPP, 483–486, 484*f*, 485*f*
Point source repairs (PSRs):
- cost savings from, 102
- for DI, 345
- existing pipe and, 96
- PB ILR and, 345, 346

Poisson's ratio, 110, 114, 188, 221
Polyesters. *See also* Glass-reinforced pipe
- for CIPP, 284
- for SIPP, 400–401

Polyethylene (PE), 251–257, 253*f*. *See also* High-density polyethylene
- abrasion resistance of, 308
- bacteria prevention and, 342
- butt-fused joints for, 360
- for CFP, 29, 422–423
- for continuous sliplining, 308, 308*f*, 312–314
- copolymerization and, 253
- for DI, 271
- engineering properties of, 254
- external loads on, 4
- grout for, 365
- for HDD, 239–240, 252
- joints for, 256, 366
- manholes and, 364–366, 450
- for natural gas pipelines, 251–252, 316–317
- for PB ILR, 329, 331, 342, 352, 361, 368–369, 368*f*
  - grooves on surface for, 333–334, 369, 369*f*, 373*f*
- PPI for, 64
- PR-PR for, 339
- for pressure pipe, 254*n*7
- profile-wall, 252
- PVC and, 365–366
- SCG of, 254
- for segmental sliplining, 310
- for SL, 252, 305
- thermal expansion and contraction of, 364–365
- for TRM, 252

Polymer concrete pipe (PCP), 225, 225*f*, 226, 314

Polymers. *See also* Epoxy; Polyurea; Polyurethane
- chemical resistance of, 399
- I/I problems and, 398
- for linings, cementitious materials for, 18
- for manholes, 431, 431*f*
- for PB ILR, 332, 362–363
- for PB-PI, 338
- for SIPP, 19, 398–399, 414–416

Polyolefins, 251
Polypropylene (PP), 284, 310
Polyurea, 19, 399–400, 401, 431
Polyurethane (PU):
- grout, 477
- linings of, for steel pipe, 272
- for manholes, 431, 444, 444*f*
- for SIPP, 19, 399–400, 400*f*, 401

Polyvinyl chloride (PVC), 241–251, 242*f*
- abrasion resistance of, 250
- advantages and limitations of, 250–251, 251*t*
- bell-and-spigot joints for, 239–240
- butt-fused joints for, 246–249
- for CFP, 419, 422–423
- for concrete pipe linings, 231
- condition assessment and, 52
- for continuous sliplining, 310, 312–314
- creep in, 241
- external loads on, 4
- as flexible pipe, 203
- for gamma-gamma logging, 163
- grout for, 16
- for HDD, 246
- internal seals for, 479
- internally restrained gasketed joint, 249 –250
- iPVC, 245, 245*f*, 246*t*
- joints for, 246–250
- for manholes, 450
- manufacturing of, 241–242
- for MLS, 387
- for PB ILR, 246, 329, 331, 333, 342*n*1, 366–367, 368, 368*f*
- PE and, 365–366
- pin and groove gasketed joint TerraBrute for, 249
- for PR-PR, 340
- profile-wall, 243–245, 244*f*
- sags in, 167
- for sanitary sewer pipe, 165, 167, 169

Polyvinyl chloride (*Cont.*):
- for segmental sliplining, 310
- for service laterals, 453
- for SL, 246, 305
- solid-wall, 243–245, 244*f*
- spline-lock gasketed joint Certa-Lok C-900/RJ for, 249 –250
- standards for, 246–250, 247*t*–248*t*
- SWM for, 16
- for TRM, 246–250, 247*f*–248*f*
- types of, 242–243
- for water distribution system, 245
- for water mains, 366

Portland cement:
- in AC, 222
- for CML, 19
- for concrete pipe, 223
- for grout, 478
- for manholes, 431

Positector 100D, 416
PP. *See* Polypropylene
PPI. *See* Plastic Pipe Institute
PPV. *See* Peak particle velocity
PR. *See* Pipe ramming; Pipe removal
PR-PE. *See* Pipe eating
PR-PR. *See* Pipe reaming
PR-PX. *See* Pipe ejection (extraction)
Pre-design, for PB ILR, 347–348, 348*f*
Preliners, for CIPP, 288
Pressure pipe:
- bar-wrapped steel-cylinder concrete pipe for, 225
- CFP for, 421, 422*t*
- CIPP for, 29, 112, 277*t*
  - hydraulic design for, 133–134
  - partial deterioration of, 125–127, 126*f*
- concrete pipe as, 227–231, 230*t*
- condition assessment for, 149–169
- DI for, 265*t*
- full deterioration of, 29
- GRP for, 240
- HDPE for, 252, 256
- jacking of, FWC for, 259–260
- link seals for, 483
- MSL and, 388*t*
- partial deterioration of, 25–26
  - CIPP for, 125–127, 126*f*
- PCCP for, 224
- PE for, 254*n*7
- RCCP for, 225
- renewal methods for, 25, 28*f*
- SL for, 318

Prestressed concrete cylinder pipe (PCCP), 215, 224
 acoustic monitoring systems for, 157–158, 159*f*
 condition assessment and, 52
 for MSL, 393
 SL for, 321–327
Profile-wall PE, 252
Profile-wall PVC, 21, 243–245, 244*f*
PSRs. *See* Point source repairs
PU. *See* Polyurethane
Pull-in insertion, for CIPP, 288
Pull in-place (PIP), for CIPP, 459, 459*f*
Pulling shaft, for PB ILR, 350, 360, 362
Pulmonary edema, from hydrogen sulfide, 3
Pump full method, for service lateral grouting, 461
Push cameras, for condition assessment, 155–156, 156*f*, 157*t*
PVC. *See* Polyvinyl chloride
PVC-M. *See* Modified PVC
PVC-O. *See* Molecularly oriented PVC
PVC-U. *See* Unplasticized PVC

Quality assurance/quality control (QA/QC), 32–33
 for CFP, 425–426
 for CIPP, 291*t*, 293–298
 for PB ILR, 358, 375–376
 for PVC, 242
Quarter point cracking, in rigid pipes, 215, 215*f*

RACM. *See* Regulated ACM
Radio-frequency interferometry, for water distribution system condition assessment, 187
Radiographic testing, for water distribution system condition assessment, 181
Rajani, B. B., 213
Rate shock, AM for, 36
RCCP. *See* Reinforced concrete cylinder pipe
RCP. *See* Reinforced concrete pipe
RD. *See* Rolldown
RDBMS. *See* Relational database management system
Reception pits, for PB ILR, 333
Record keeping, for oil and natural gas pipelines, 78
References, in AMP, 44–45
Regulated ACM (RACM), 381
Reinforced concrete cylinder pipe (RCCP), 225
Reinforced concrete pipe (RCP), 224
 condition assessment and, 52
 for PB ILR, 331, 342
 TRM of, 231
Relational database management system (RDBMS), 145
Relative network position, CoF and, 58–59, 58*t*, 59*t*
Remote field eddy current (RFEC), 160, 160*f*, 162*t*, 178–180, 180*f*
Renewal and replacement, in AMP, 42
Renewal methods. *See also* Trenchless renewal methods
 applicable uses for, 101*t*
 for buckling, 29
 cost experience indicators for, 104*t*
 defined, 9
 design for, 25–29
 durability, reliability, and strength of, 4–7, 8*t*
 emerging design concepts for, 29
 existing pipe and, 95–97
 hydraulic capacity and, 23
 limitations of, 103*t*
 long-term test methods for, 29–30
 for manholes, 449–451
 for pressure pipe, decision support system for, 28*f*
 project site and, 97
 for sanitary sewer system manholes, 429–451
 selection of, 23–25, 94–95, 97–106, 98*t*
 for service laterals, 453–463
 service life extension and, 24*t*
 usage statistics for, 108*f*
Repair, 9
Repair history, in condition assessment, 141
RFEC. *See* Remote field eddy current
Rigid pipe. *See also* Concrete pipe; Vitrified clay pipe
 arching effect for, 209, 209*f*
 bedding factors for, 211
 behavior of, 210–211, 211*f*
 flexural failures in, 215, 217*f*
 horizontal passive resistance with, 210
 longitudinal bending in, 217
 OC and, 200

Rigid pipe (*Cont.*):
PB ILR and, 342
pipe-soil system and, 201–204, 204*t*
pipeline failures with, 214–217
quarter point cracking in, 215, 215*f*
shear failure in, 217, 218*f*
side soils and, 210
soil load for, 208, 208*f*
three-edge bearing test for, 210–211, 211*f*
Ring flexure, 215
Risk management:
AM and, 52–63
CoF and, 55–63, 62*f*
LoF and, 54–55, 62–63, 62*f*
for oil and natural gas pipelines, 80–88, 89*f*
for PB ILR, 352–356
Robotic repairs, for LOR, 467–470, 468*f*, 469*f*
Rogers, C., 363
Rolldown (RD), 20, 20*t*
for CFP, 419
Rolled and welded pipe, 266
Room temperature vulcanization (RTV), 269
Routine maintenance, in AMP, 41
RTV. *See* Room temperature vulcanization

Safety:
condition assessments and, 138
for MSL, 389–393
for natural gas pipelines, 316
for PB ILR, 331, 361, 364, 374
Safety data sheets (SDS), QA/QC and, 32
Sags, 2
PB ILR and, 346, 353
in PVC, 167
in sanitary sewer pipe, 167
in water mains, 188*t*
Sahara®, 182, 185, 185*f*, 186*t*
Saltwater pipelines, oil and natural gas pipelines and, 78–80, 79*t*, 80*f*
Sand, 205
Sanitary sewer overflows (SSOs), 52, 57, 432
Sanitary sewer pipe:
beam failure of, 239
calcium aluminate for, 398
CCTV for, 164–165
cementitious materials for, 398
Sanitary sewer pipe (*Cont.*):
CFP for, 422*t*
CIPP for, 277*t*
CML for, 19
collapse failures of, 7*f*
condition assessment for, 164–169, 166*f*
ArcWISE for, 193, 193*f*
data content for, 189–190
data sharing and distribution for, 191–194
point-score protocols for, 190–191, 191*t*, 192*t*
continuous sliplining for, 308
DI for, 262, 272
exfiltrations in, 167
hydrogen sulfide in, 231
I/I problems in, 165
infiltrations in, 167
infrared thermography for, 163
manholes and, renewal methods for, 429–451
MSL for, 388, 388*t*
PB ILR for, 333, 363, 460
PVC for, 165, 167, 169
service laterals to, 454–455, 454*f*, 455*f*
shear failure of, 239
SIPP for, 398
smoke testing for, 165
VCP for, 165, 167, 169, 233–234, 239
SASW. *See* Spectral analysis of surface waves
SBR. *See* Styrene-butadiene rubber
SCG. *See* Slow crack growth
Scout camera, 404
SDS. *See* Safety data sheets
See Snake, 179–180, 180*f*
Segmental pipes, PB ILR with, 375
Segmental sliplining, 13, 13*f*, 13*t*, 310–312, 311*f*
bell-and-spigot joints for, 310
construction of, 314
flow bypassing and, 106
GRP for, 310
PE for, 310
PP for, 310
Semi-flexible pipe, 204
Semi-rigid pipe, 204
AC as, 223*n*1
Semi-structural techniques:
for manholes, 432
for SL, 316

Service and property identification, 143–144
Service laterals:
annular space and, 453, 455*f*
CCTV for, 106–107
CFP for, 423, 426, 456, 461
chemical grouting for, 456, 461–462
CIPP for, 456–460, 457*f*, 458*f*, 459*f*
existing pipe and, 96
grout for, 461–462
inspection for, 459–460
MSL and, 388, 388*t*
PB ILR and, 332, 334, 345, 346, 347, 369–370, 370*f*, 456, 460–461, 460*f*
replacement of, 375
pipeline defects in, 455
preparatory work for, 456
PVC for, 453
QA/QC for, 32
reinstatement of, 106–107
renewal methods for, 453–463
to sanitary sewer pipe, 454–455, 454*f*, 455*f*
SIPP for, 462
in situ reinforced concrete and, 414
SL for, 314, 318, 456
water tightness and, 453
Service life extension:
AM and, 9–10, 9*f*
renewal methods and, 24*t*
with TRM, 7–10
Sewer manhole renewal (SMR), 20–21, 21*t*, 22*f*
Sewer pipe. *See* Sanitary sewer pipe
Sewer sausage method, for service lateral grouting, 461–462
Sewer Scanning and Evaluation Technology (SSET), 155
Shear failure:
concrete pipe and, 227
in rigid pipe, 217, 218*f*
of sanitary sewer pipe, 239
Shearing stresses, 217
arching effect and, 208
Shotcrete:
for grout, 479
for LOR, 479
mix design for, 412*f*
for SIPP, 397, 412, 412*f*
Side soils:
flexible pipe and, 212
rigid pipe and, 210
Sidefill:
concrete pipe and, 231
PCCP and, 224
Siltation, 205
Sinkholes, 4, 5*f*, 6*f*, 432
SIPP. *See* Spray-in-place pipe
Sliplining (SL), 11–13, 13*t*, 305–328. *See also* Continuous sliplining; Modified sliplining; Segmental sliplining
advantages of, 317–318
annular space and, 305
bentonite for, 325
butt-fused joints for, 326*f*
capabilities and limitations of, 306*t*
case study for, 318–327
characteristics of, 307
construction of, 312–315, 325, 326*f*
cost experience indicators for, 104*t*
design for, 315–316, 324–325
for existing pipe, 96, 97
for gravity pipe, 259
grout for, 312, 314–315, 327*f*
GRP for, 305
HDPE and, 318–327
ID for, 307, 324
joints for, 312
limitations of, 103*t*, 318
methods for, 307–312
for natural gas pipelines, 316–317
OD for, 307
PCP for, 226
PE for, 252, 305
PVC for, 246, 305
for service laterals, 456
service laterals and, 314, 318
service life extension with, 24*t*
site compatibility and applications for, 305–306
steel pipe for, 266
usage statistics for, 108*f*
for water distribution system, 305–306
for water mains, 318–321, 320*f*
Slow crack growth (SCG), of PE, 254
Small System Asset Management Plan Best Practices Guide, of USEPA, 89
SmartBall®, 182, 184–185, 185*f*, 186*t*
Smoke testing, for sanitary sewer pipe, 165
SMR. *See* Sewer manhole renewal
Social costs, CoF and, 56–57

Society for Protective SIPPs (SSPC), 406
Software:
  for condition assessment, 146–147
  databases, 145, 187
Soil. *See also* Pipe-soil system
  arching effect of, 208–210, 209*f*
  CIPP and, 110
  classification of, 205
  in condition assessment, 141*t*
  deflection and, 221
  for embedment, 206*t*–207*t*
  longitudinal bending and, 217
  OC and, 201
  PB ILR and, 350–351
  pipeline defects and, 3–4
  sags from, 2
  strength of, 211–212
Soil load:
  arching effects and, 209
  CIPP and, 119–120, 119*t*
  for flexible pipe, 208, 208*f*, 212
  problems with, 102
Soil modulus:
  for CIPP, 109–110, 119–120, 129
  flexible pipe deflection and, 111
Soil movement. *See* Ground movement
Soil pressure, buoyancy correction factor and, 122
Soil prism pressure, 122, 123*t*
Soil stiffness, 201, 206*t*–207*t*, 212
Solid-wall HDPE, 254–255
Solid-wall PVC, 243–245, 244*f*
Sonar:
  for condition assessment, 158–159, 158*f*, 159*f*
  with laser profiling, 162
  for water distribution system condition assessment, 175
Spangler, M. G., 110–111, 208
Spark testing:
  for manholes, 449
  for SIPP, 416–417
Spectral analysis of surface waves (SASW), for condition assessment, 163–164
Spiral cracking, in cast iron pipe, 215, 217*f*
Spiral welded pipe, 267
Spiral-wound lining process, 387, 387*f*
Spiral wound method (SWM), 15–16, 18*f*, 18*t*
Spline-lock gasketed joint Certa-Lok C-900/RJ, for PVC, 249 –250
Split bell breaks, from transverse stresses, 4
Splits, 2
SPR, 387, 387*f*
Spray-in-place pipe (SIPP), 16–21, 19*t*, 20*f*, 397–418
  abrasion and, 96
  adhesion of, 401, 411, 417
  CCTV for, 402
  cementitious materials for, 398, 411–414
  cleaning methods for, 403–404, 404*f*, 405*f*, 405*t*
  CML for, 16–19
  for coatings, 397, 397*n*1
  composite materials for, 416
  construction of, 401–416
  corrosion and, 397, 398
  cost experience indicators for, 104*t*
  epoxy for, 399, 401
  equipment and process for, 410–416
  exfiltration testing for, 417
  I/I problems and, 406
  inspection for, 402, 403*f*, 416–417
  limitations of, 103*t*
  for linings, 397, 397*n*1
  moisture tolerance of, 401
  pigging for, 407, 411*f*
  polyesters for, 400–401
  polymers for, 19, 398–399, 414–416
  polyurea for, 399–400, 401
  PU for, 399–400, 400*f*, 401
  repair materials for, 408
  for sanitary sewer pipe, 398
  for service laterals, 462
  service life extension with, 24*t*
  shotcrete for, 397, 412, 412*f*
  site investigation and preparation for, 401–402
  in situ reinforced concrete for, 413–414
  spark testing for, 416–417
  surface preparation for, 404–407
  testing for, 416–417
  thermoset resins for, 398
  thickness of, 416
  vacuum testing for, 417
  VOCs in, 398–399
  voids and, 408
  for water distribution system, 398
  worker-entry pipelines for, 412–413, 415
Spray-on linings, for service laterals, 456
Springline, 203*t*

SSET. *See* Sewer Scanning and Evaluation Technology
SSOs. *See* Sanitary sewer overflows
SSPC. *See* Society for Protective SIPPs
Static pipe bursting (PB-ST), 13–14, 336–337, 336*f*
  utility survey for, 363
Steam curing, for CIPP, 289–290
Steel end ring joints, 228
Steel pipe, 266–270
  advantages and limitations of, 269, 270*t*
  cathodic protection for, 74–75
  condition assessment and, 52
  corrosion protection for, 272
  ERW, 266–267
  as flexible pipe, 212
  GWUT for, 177
  for HDD, 261
  joints for, 267–269, 268*f*, 269*f*
  manufacturing of, 266–267
  MSL with, 393, 393*f*, 394*f*
  for PB ILR, 329
  for PB-PB, 331
  for PB-PS, 337
  for PR, 261
  rolled and welded, 266
  spiral welded, 267
  standards for, 267, 267*t*
  for water distribution system, 266
Storm Water Management Model (SWMM), ArcWISE and, 193
Straight-line depreciation, 46, 48*t*–49*t*, 51*t*
Street sequential numbering, 143
Strength of soil, 211–212
Structural defect failures:
  of manholes, 442
  from pipeline defects, 25
Styrene, for CIPP, 284
Styrene-butadiene rubber (SBR), 477
Sublime process, for CFP, 419
Sulfate corrosion, long-term pipeline defects from, 2–3
Sulfuric acid, 2–3, 435, 443
Sustainability, AM and, 56, 56*f*
Swagelining, for CFP, 419, 424
SWM. *See* Spiral wound method
SWMM. *See* Storm Water Management Model
System performance and tracking review, for oil and natural gas pipelines, 78, 80
TAMP. *See* Total asset management program
Tamping, for concrete pipe, 223
TBL. *See* Triple Bottom Line
TCMs. *See* Trenchless construction methods
Technical specifications, 30–31
Testing:
  for CIPP, 295–296
  destructive, 170
    for adhesion, 417
  environmental, for water distribution system condition assessment, 182, 183*t*
  exfiltration, for SIPP, 417
  holiday, 74, 75, 449
  jeep, for oil and natural gas pipelines, 75, 84–85, 85*f*
  long-term methods for, for renewal methods, 29–30
  of manholes, 448–449
  NDT:
    data content from, 189
    for water distribution system condition assessment, 170, 175–177
  radiographic, for water distribution system condition assessment, 181
  for SIPP, 416–417
  smoke, for sanitary sewer pipe, 165
  spark:
    for manholes, 449
    for SIPP, 416–417
  three-edge bearing, for rigid pipe, 210–211, 211*f*
  vacuum, for SIPP, 417
Thermal expansion and contraction:
  circumferential breaks from, 4
  longitudinal tensile failure and, 221
  PB ILR and, 364–365
  of PE, 364–365
Thermoplastics. *See also* Polyethylene; Polyvinyl chloride
  for CIPP, 111
Thermoset resins, 239
  for CIPP, 111, 276
  in PCP, 225
  for SIPP, 398
  styrene and, 284
Third-party interference:
  circumferential breaks from, 4
  PB ILR and, 364

Three-edge bearing test, for rigid pipe, 210–211, 211*f*
Timoshenko, S. P., 30, 110, 111
Tomahawk System, 404
Tongue and groove joints, 227, 227*n*2, 228*t*
Total asset management program (TAMP), 194–195, 194*f*, 195*f*
Traffic diversion:
  CFP and, 423
  for MSL, 389
  OC and, 107
  for PB ILR, 347, 359, 361
  from pipeline failure, 4
Traffic loads:
  CIPP and, 110
  CoF and, 59–60, 60*t*
  condition assessment and, 52
  manholes and, 429
  problems with, 102
  structural defect failures and, 25
Transportation Research Board (TRB), 35
Transverse stresses, split bell breaks from, 4
TRB. *See* Transportation Research Board
Tree roots:
  CFP and, 423
  CIPP and, 282*f*
  hydraulic capacity and, 23
  longitudinal bending and, 217
  MSL and, 389
  in sanitary sewer pipe, 167
  service laterals and, 456
  structural defect failures and, 25
Trench:
  circumferential breaks from, 4
  deflection and, 220
  walls, 201, 203*t*
  width, 202*f*, 203*t*
Trenchless construction methods (TCMs):
  corrosion and, 4
  DI for, 263
  joints for, 227
  SL and, 317
Trenchless renewal methods (TRMs), 10–21, 10*f*. *See also* Close-fit pipe; Cured-in-place pipe; In-line replacement; Modified sliplining; Sewer manhole renewal; Sliplining; Spray-in-place pipe
  applicable uses for, 101*t*
Trenchless renewal methods (*Cont.*):
  cost experience indicators for, 104*t*
  decision support system for, 26*f*
  design for, 93–135
  DI for, 263
  economic costs savings from, 9
  for existing pipe, 97
  flow bypassing for, 106
  initial planning for, 94–95, 99, 99*t*
  limitations of, 103*t*
  ORM for, 95
  PE for, 252
  pipe rerounding for, 481
  pipeline integrity assessment for, 95, 98–99, 100, 100*t*
  PVC for, 246–250, 247*f*–248*f*
  of RCP, 231
  selection of, 94–95, 97–106, 98*t*
  service life extension with, 7–10
  usage statistics for, 108*f*
Tricalcium Aluminate (C3A), for concrete pipe, 223
Trimble GeoExplorer, 146
Trimble GPScorrect, 146–147, 148
Triple Bottom Line (TBL), 55, 57
TRMs. *See* Trenchless renewal methods
Tubes, for CIPP, 109, 275
Tunneling. *See also* Microtunneling
  PCP for, 226
  steel pipe for, 266

U-liner process, 419
Ultrasonic pigs, for ILI:
  of oil and natural gas pipeline, 176
  of water distribution systems, 181
Ultrasonic pulse velocity, for condition assessment, 164
Ultrasound:
  GWUT, 164, 176–177, 178*f*
  for oil and natural gas pipelines, 76
  for water distribution system condition assessment, 175–177, 175*f*, 177*f*
Ultraviolet (UV), for CIPP, 11, 276, 290
Uni-Bell, 244
Unified Soil Classification System (USCS), 205
United States Environmental Protection Agency (USEPA), 88
  on AC, 380, 382–383
  on bacteria prevention, 446
  on chemical gel grouts, 474
  on condition assessments, 139

United States Environmental Protection Agency (*Cont.*):
  I/I problems and, 234
  on service laterals, 454–455
  Small System Asset Management Plan Best Practices Guide of, 89
Unplasticized PVC (PVC-U), 242–243
Unsaturated polyester resins, for CIPP, 284
Upsizing, PB ILR for, 331, 334, 335
Urethane:
  foam, 476
  grout, 475–476
  seals, for manholes, 436
USCS. *See* Unified Soil Classification System
USEPA. *See* United States Environmental Protection Agency
Utility survey, for PB ILR, 349, 363
UV. *See* Ultraviolet

V-Bio, 271
Vacuum testing, for SIPP, 417
Valuation forecasts, in AMP, 43
VCP. *See* Vitrified clay pipe
Vertical earth pressure, arching effects and, 209–210
Vinyl ester, for CIPP, 284
Viscoelastics. *See also* Polyethylene; Polyvinyl chloride
  creep of, 241
  properties of, 240–241, 240*f*
Vitrified clay pipe (VCP), 233–239
  advantages and limitations of, 238–239, 238*t*
  fractures of, 167, 167*f*
  I/I problems and, 239
  joints for, 234–238, 235*f*, 236*f*, 237*f*
  manufacturing of, 234
  for PB ILR, 329, 331, 342
  for PR-PR, 340
  for sanitary sewer pipe, 165, 167, 169, 233–234, 239
  standards for, 234, 235*t*
VOCs. *See* Volatile organic compounds
Voids:
  in bedding, circumferential breaks from, 4
  in flexible pipe pipeline failure, 214
  impact echo and SASW for, 164
  infrared thermography for, 163
  SIPP and, 408
  sonar for, 159
Volatile organic compounds (VOCs):
  polymers and, 19
  in SIPP, 398–399
  water quality problems with, 25
von Pechmann, Hans, 251

w/c. *See* Water cement ratio
Wall buckling. *See* Buckling
Wall crushing. *See* Crushing
Wall thickness:
  buckling and, 296*n*2
  GWUT for, 177
  of HDPE, 255
  QA/QC for, 33
  radiographic testing for, 181
  SL and, 316
  in water mains, 188*t*
Wastewater systems:
  condition assessment for:
    GWUT for, 164
    raw data for, 186–201
    ultrasonic pulse velocity for, 164
  pipe selection for, 222
Water cement ratio (w/c), for concrete pipe, 223
Water curing, for CIPP, 289
Water distribution system. *See also* Freshwater pipelines; Water mains
  AC for, 222
  cast iron pipe for, 263
  cementitious materials for, 398
  CIPP for, 277*t*
  CML for, 19
  condition assessment for, 170–187, 171*t*–172*t*, 174*t*
    BEM for, 180–181, 181*f*
    CCTV for, 170–173, 174*t*
    data content for, 189–190
    data sharing and distribution for, 191–194
    electrical and electromagnetic technologies for, 177–181
    environmental testing for, 182, 183*t*
    GPR for, 185–187
    GWUT for, 176–177, 178*f*
    infrared thermography for, 187
    laser profiling for, 173, 174*t*
    leak detectors for, 182–187
    MFL for, 177–178, 179*f*
    NDT for, 170, 175–177
    pit depth measurements for, 173
    point score protocols for, 190–191, 191*t*, 192*t*

Water distribution system, condition assessment for (*Cont.*):
radio-frequency interferometry for, 187
radiographic testing for, 181
RFEC for, 178–180, 180*f*
sonar for, 175
ultrasound for, 175–177, 175*f*, 177*f*
DI for, 261
epoxy for, 399
ILI of, ultrasonic pigs for, 181
PB ILR for, 333
pipe selection for, 222
pipeline failure of, sinkholes from, 4, 6*f*
PVC for, 245
SIPP for, 398
SL for, 305–306
steel pipe for, 266
Water Environment Research Foundation (WERF), on condition assessments, 139, 170
Water Infrastructure Network, 37
Water Infrastructure Simulation Environment (WISE), 191–194
Water inversion, for CIPP, 287–288
Water mains:
cast iron pipe for, 188*t*
CFP for, 425, 426
condition assessment for, 188*t*
continuous sliplining for, 308
DI for, 188*t*, 366
HDPE for, 318–321
in New York City, 188
PB ILR for, 366–369
Water mains (*Cont.*):
PVC for, 366
radiographic testing for, 181
SL for, 318–321, 320*f*
Water Pollution Control Act of 1972, 234
Water quality problems, 25
Water Research Centre (WRc), 190, 191*t*, 385
Water Research Foundation (WRF), 139, 170, 189, 382
Water tightness:
of AC, 232
of joints, 227, 227*n*2, 232
QA/QC for, 33
Weathering, 204
WERF. *See* Water Environment Research Foundation
Wet casting, for concrete pipe, 224
Wet out process, for CIPP, 285–287, 286*f*
Winch-in-place CIPP, 11, 11*t*, 12*f*
WISE. *See* Water Infrastructure Simulation Environment
Worker-entry pipelines, for SIPP, 412–413, 415
WRc. *See* Water Research Centre
WRF. *See* Water Research Foundation
Wyes, for oil and natural gas pipelines, 70–72, 70*f*, 71*f*

Zhang, J. Q., 111
Zoning, in condition assessment, 141*t*
Zoom camera, for condition assessment, 152, 157*t*

CPSIA information can be obtained at www.ICGtesting.com
Printed in the USA
BVOW06*0210090616

451308BV00001B/1/P

9 780071 823340